COMPUTER-BASED
AUTOMATION

COMPUTER-BASED AUTOMATION

EDITED BY
JULIUS T. TOU
University of Florida
Gainesville, Florida

PLENUM PRESS • NEW YORK AND LONDON

Library of Congress Cataloging in Publication Data

Main entry under title:

Computer-based automation.

"Based on the proceedings of the International Conference on Advanced Automation, ICAA-83, held Dec. 19–25, 1983, in Taipei, Republic of China"–T.p. verso.
Includes bibliographical references and index.
1. CAD/CAM systems–Congresses. 2. Automation–Congresses. I. Tou, Julius T., 1940–   . II. International Conference on Advanced Automation (1983: Taipei, Taiwan). III. Title.
TS155.6.C66 1985   670'.28'54   84-26383
ISBN 978-1-4684-7561-6   ISBN 978-1-4684-7559-3 (eBook)
DOI 10.1007/978-1-4684-7559-3

Based on the proceedings of the International Conference on Advanced
Automation, ICAA-83, held December 19–25, 1983, in Taipei,
Republic of China

© 1985 Plenum Press, New York
Softcover reprint of the hardcover 1st edition 1985

A Division of Plenum Publishing Corporation
233 Spring Street, New York, N.Y. 10013

All rights reserved

No part of this book may be reproduced, stored in a retrieval system,
or transmitted in any form or by any means, electronic, mechanical,
photocopying, microfilming, recording, or otherwise,
without written permission from the Publisher

Preface

It has been recognized that productivity improvement is an important issue of the 80's. It is regarded as the most efficient way to improve national economy and to enrich the quality of life. The key to productivity improvement is advanced automation, especially computer-integrated automation for engineering design and office operations as well as manufacturing processes. This is the theme of 1983 International Conference on Advanced Automation, ICAA-83. This book contains the articles which are the revised and updated version of the papers presented at the ICAA-83 Conference.

Traditionally, automation is synonymous with mechanization; but this Conference has treated automation from a different point of view. We consider automation as a process to unify various automated information processing systems for performing business, administration, design, engineering and manufacturing functions, in addition to the traditional fixed automation in production. In other words, design automation and office automation form an integral part of factory automation to accomplish comprehensive computer-integrated manufacturing and production.

In engineering and manufacturing today, quality design and high productivity are synonymous with the use of computers, robots, expert systems, and other computer-based technologies. The greater the degree of computer-based automation exploited and implemented, the greater a nation's ability to survive in tomorrow's extremely competitive world market.

In addition to fostering information exchange among research engineers and scientists, the Conference aims at providing a forum for research engineers to learn what the real-world automation problems are and for practicing engineers from industrial firms to find out what has recently been accomplished at research institutions. The papers presented at the Conference were selected because of their contributions to a representative sample of the various specialized areas of computer-based automation. We realize that many other research workers have made or are making significant contributions to the understanding and applications of computer-based automation. Unfortunately, the omnipresent tyranny of time and space has prohibited the inclusion of their work in the Conference.

Credit for any success in this Conference must be shared with many people who contributed significantly of their time and talents. In organizing and conducting the Conference, which was attended by more than five hundred delegates from a dozen countries, the General Chairman received considerable help from Drs. H. C. Fang, Y. Kuo, K. S. Fu, C. C. Hsieh, K. Y. Cheng, J. S. Huang, K. J. Chen, C. S. Lin, Miss Joyce Chen, Astor Huang, and Jean Huang. It is the authors of individual papers whose contributions made possible the Conference and this book. The participation of Premier Y. S. Sun, Economic Affair Minister Y. T. Chao, Academia Sinica President T. Y. Wu, Dr. William Spurgeon of the National Science Foundation, and Dr. Seiuemon Inaba of FANUC LTD enhanced the stature of the Conference significantly. To all of them, the editor wishes to express his heartfelt appreciation.

Julius T. Tou

August 1984

Contents

Computer-Aided Design

Automatic Generation of Knowledge Base from Electronic
Diagrams for Computer-Aided Design ... 3
    Julius T. Tou and Jack M. Cheng

CAD Application on PC Board Design ... 37
    Chung-Jye Chang

SPICE-2P: A SPICE-2 Incorporated with a Partition Scheme ... 59
    Chung Len Lee, Wen Zen Shen, Ching An Liaw,
    and Chein Wei Jen

A CAD Program for VLSI Placement and Routing ... 73
    Jau-Yien Lee, J. M. Jou, T. T. Nian, C. Y. Chang,
    and H. C. Wu

BRUTUS: An Interactive Graphic Editor for IC Layout ... 95
    K. A. Hwang, D. C. Liu, and C. C. Lin

PLAMG: An Automatic PLA Minimizer and Generator for VLSI
System Design ... 119
    Cheng Chen, Wen Zen Shen, Young-Li Lee, Li-Wen Shih,
    and Yeh Jian Liang

CAD of Truss Structures by the Interactive Simplex Method ... 139
    V. Wuwongse and M. N. Nagraj

Low Cost CAD/CAM System for Machine Parts ... 153
    W. B. Ngai and Y. K. Chan

Computer-Assisted Methods for Defining 3D Geometric
Structure of Mechanical Parts ... 169
    Z. Chen and D. B. Perng

Computer-Aided Finite Element Analysis: Interfacing
Symbolic and Numerical Computational Techniques   203
    Paul S. Wang

A Theoretical Proposal for a CASD System Extending
Jackson's Method for Program Construction   213
    C. J. Lucena, R. C. B. Martins, P. A. S. Veloso,
    and D. D. Cowan

Robotics

Coordinating Multiple Robot Arms to Increase Productivity   245
    John Roach and P. Montague

Design of a Computer Aided Robot Design System   267
    Linfu Cheng

A CAD Tool for the Kinematic Design of Robot Manipulators   287
    Chi-haur Wu

Computer Vision and Image Processing

A Graph-Theoretic Approach to 3-D Object Recognition and
Estimation of Position and Orientation   305
    E. K. Wong and K. S. Fu

Industrial Computer Vision   345
    R. C. Gonzalez

Integrating Vision and Touch for Grasping of an Object   387
    Ruzena Bajcsy

Spherical Shading Correction of Eye Fundus Image by Parabola
Function   399
    Kozo Okazaki and Shinichi Tamura

Automatic Welding: Infrared Sensors for Process Control   411
    Bryan A. Chin and Nels H. Madsen

Computer-Aided Manufacturing

Present State and Future Trends in the Development of
Programming Languages for Manufacturing   433
    U. Rembold and W. K. Epple

On-Line Identification and Suppression of Time Varying
Machining Chatter in Turning Via Dynamic Data System (DDS)
Methodology   481
    Shing-Yuan Tsai and Shien-Ming Wu

CONTENTS

Microprocessor and Programmable Controller Based Industrial
Automation — 507
   M. F. Rahman

Production Scheduling Automation in an Aluminum Metal
Products Plant — 519
   Keh-Lon Thomas Lee and Arvind Jain

Machinability Data Base Systems for Automated Manufacturing — 535
   P. Balakrishnan and M. F. DeVries

An Integrated Approach to Design, Implementation, and
Testing of Digital System — 555
   Manzer Masud

V. Local Area Networks, Databases, and Graphics

Implementing Priority Functions in Local Area Networks — 569
   Lionel M. Ni

The Design and Implementation of a Distributed Database
System — 591
   Jyh-Sheng Ke, Ching-Liang Lin, Hsing-Lung Chen,
   Chiou-Feng Wang, Chia-Hsiang Chang, Yow-An Pan,
   Chien-Chun Lu, and Shyi-Ting Kang

A High Level Graphic Database for CAD — 607
   San-Cheng Chang

A Graph-Theoretic Approach to the Hidden Line Elimination
Problem — 617
   W. H. Chu and D. T. Lee

Contributors — 635

Index — 641

Computer-Aided Design

AUTOMATIC GENERATION OF KNOWLEDGE BASE FROM

ELECTRONIC DIAGRAMS FOR COMPUTER-AIDED DESIGN

Julius T. Tou and Jack M. Cheng

Center for Information Research
University of Florida
Gainesville, Florida, U.S.A.

ABSTRACT

An approach to automatic generation of knowledge base from electronic diagrams is presented. This approach consists of two major components: (1) Automatic interpretation of electronic diagrams and (2) organization of interpretation results into a knowledge base for CAD applications. The design technique for automatic interpretation is based upon the multiple-pass pattern extraction principles. The functional elements in a diagram and their interrelationships are extracted from the diagram. Picture manipulation languages are developed to read the functional elements and to describe the interrelationships. The results of interpretation are organized into a knowledge base from which input data to a CAD circuit analysis program are derived.

INTRODUCTION

The world of digital systems continues to advance at a rapid pace. The prime mover behind the advances is Computer-Aided Design of electronic circuits and systems [1]. It is now recognized that CAD will become an integrated part of production process, combining the design processes with automated manufacturing. A typical integrated CAD system for integrated circuits is shown in Figure 1. The CAD system performs a variety of functions: circuit analysis, logic simulation, test pattern generation, test tape editing, layout design, artwork data editing, artwork checking, which are accomplished by various CAD programs. The stiff competition in the electronics industries has sparked further advancement of extremely sophisticated CAD programs. The integrated CAD system automatically

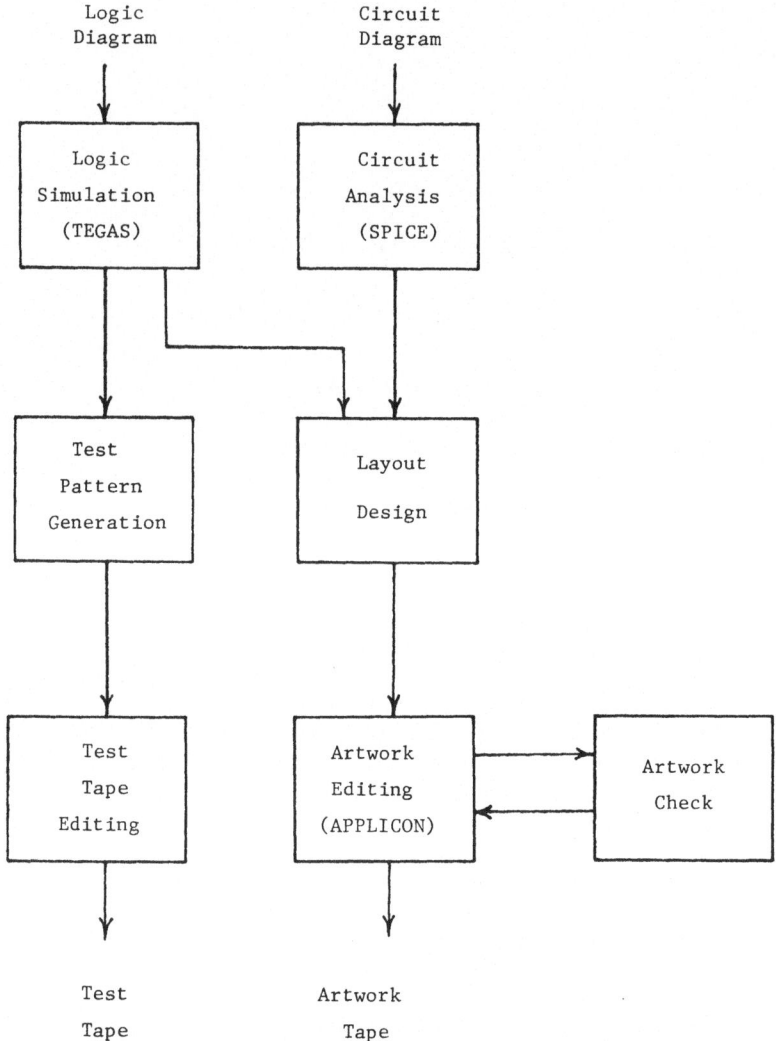

Figure 1. A CAD system for integrated circuits.

converts electronic circuit and logic diagrams into test tape and artwork tape. However, the initial design of the circuit and logic diagrams is creatively done by experienced engineers [2]. In current commercially available CAD/CAM systems, data entry of circuit and logic elements in an electronic diagram is limited to interactive editor mode through a special input device such as a light pen and tablet. We all realize that data entry by human operators is time-consuming and error-prone. The need for an automated machine for reading and interpretation of electronic diagrams is apparent if we wish to enhance the current CAD capabilities [3].

In recent years, several attempts have been made to develop computer techniques for automatic recognition of flow charts and logic diagrams. These proposed approaches, which are documented in references [4-12], are only applicable to circuit element symbols which are of simple geometrical shape such as a rectangular type. In this paper we present a computer system for automated interpretation of electronic circuit diagrams in printed documents or in neatly hand-printed forms. Our approach is not limited to symbols with a rectangular shape. In our design, electric and electronic symbols are taken from the standard drafting references [13-15].

By computer interpretation we mean the ability of the computer to read the electronic diagram, to identify the various functional elements in the diagram and to describe the interrelationships among these elements. The outcome of interpretation is organized into a knowledge base from which input data to CAD analysis or simulation programs can be derived [16,17].

REVIEW OF LITERATURE

During the past seven years, several attempts have been made to develop computer techniques for automatic recognition of flow charts and logic circuit diagrams. In 1976, Suen [4] proposed a method for recognition of hand-drawn flow charts, under the assumption that the flow chart symbols are limited to boxes with straight edges and the arrow heads must appear in the middle of horizontal line segments. His approach may be summarized in the following steps: (1) to obtain thinned binary image, (2) to trace and record all horizontal line segments from top to bottom, (3) using a sequential decision tree to identify a flow chart symbol after a line is traced in one of the eight possible directions starting from a vertex, (4) to perform connection tracing starting from a symbol which has been recognized or is to be recognized, and (5) to store the results in an output table. This method is not without shortcomings. Not only it is unable to handle curved symbols, but also it cannot distinguish between his Category R and Category D, nor can it recognize a diamong symbol. Furthermore, no computer printout results have been reported in the literature.

In 1978, Lin and Pun [5] proposed a method for machine recognition and plotting of hand-sketched line drawings. Their approach is based upon pattern recognition principles. They used a digitizing tablet as an input device. A set of features are derived from the flow chart and diagram symbols. The features they selected are the location of each symbol, the size, and a set of contour points.

A pattern vector is given by

$$\underline{X}(x_c, y_c;\ a, b, \theta;\ p_1, p_2, \ldots, p_n)$$

During the training phase, a set of training samples are used to generate a non-linear discriminate function. However, to obtain all necessary information for the pattern vector requires an experienced operator to perform curve tracing. Their objective is not computer-based interpretation of electronic diagrams.

A more advanced experimental system was developed by Kakumoto, et al. [6] at the Hitachi Central Laboratory in Japan. They employed a divide-synthesis method to recognize logical diagram drawn on a large sheet of paper at the size 420mm x 297mm. The logic symbols are drawn with special templates in rectangular form and described with the name on it. However, the logic symbols they used, such as a NAND gate or flip-flops, are limited in varieties and are described by rectangular boxes of predetermined sizes. The application of their work to read and interpret circuit diagrams seems to be quite limited.

In 1980, Zavidovique and Stamon [7] have proposed an automated process for electronic scheme analysis. They used a successive simplification method based on syntatic considerations to recognize and group a few symbols in electronic schematics. In their experiments, only rectangular logic elements can be recognized. Since size measurements are not taken, it is difficult to distinguish between the resistors in box forms and logic elements by their method.

In a recent approach by Bunke [8,9], the circuit components are decomposed into some primitive elements. The reconstruction sequence is represented by a decision tree. Recognition of a symbol is conducted as the traversal of the decision tree starting at the root and ending at a leaf. He introduced the concept of finite-state automata to deal with the denotation problem and used the relaxation method to correct errors introduced in his analysis. In this approach, the components of electronic schematics are limited to resistors, capacitors, diodes, and some logic elements in linear segment form. His input data consist of position values of the nodes read from superimposed coordinate paper, a list of line segments and classified characters; they are not in the form of a two-dimensional matrix of pixels generated by a scanner. For a more detailed review, the reader is referred to reference [18].

Anatomy of Electronic Circuit Diagrams

Electronic circuit diagrams may be considered as a graphic language to express and communicate ideas and designs among electrical, electronic, and computer engineers. Such diagrams are line drawings which consist of junction dots, line segments, symbols for circuit elements, and denotations for labeling. Standards for the drawing of electronic schematics are documented in several reference book [13-15].

An electronic circuit diagram is characterized by functional elements, connecting elements, and denotations. The functional elements are represented by symbols consisting of two or more terminals for connection to other elements. Resistors, capacitors, and diodes are two-terminal elements. Transistors and amplifiers are three-terminal elements. Flip-flops are four-terminal elements. Several basic symbols for electronic and logical diagrams are tabulated in Figure 2.

The connecting elements are represented by junction dots and horizontal and vertical line segments which are used to connect functional elements. Various configurations of connections are shown in Figure 3. The denotations are used to describe physical properties or names of the functional elements in a diagram. For instance, the denotation 10K attached to a functional element means that the elements is a resistor of 10,000 Ω resistance. The denotation C1 attached to a functional element indicates that the element is capacitor number one in the diagram. The use of denotations in a circuit diagram will simplify automatic recognition and interpretation tasks.

To expedite the recognition task, we classify the functional symbols into four categories according to their general shape. They are circular shape, rectangular shape, curved shape, and line-segment shape. The circular shape category is further divided into Transistor, Junction FET, and MOSFET types. Elements of these types can be distinguished by the orientations of the arrow markers as shown in Figure 4. The rectangular shape category is further divided into two major subcategories, the vertical bar shape and the horizontal bar shape. The former includes flip-flops, adders, multiplexers, and various VLSI circuits, and the latter includes symbol representations commonly used in Europe. The curved shape category is further divided into logic gates subcategory and inductor/transformer subcategory. The line-segment shape category consists of diodes, resistors, capacitors, and amplifiers. From the above classification scheme, we create a symbol tree structure which is built into the knowledge base to facilitate the recognition of functional elements as shown in Figure 5.

Design for Automated Interpretation

A typical electronic circuit diagram is shown in Figure 6. We want to design a computer to recognize and interpret the diagram so that the interpretation results may be used as the input to a CAD package such as SPICE or to a CAD database. Careful examination of

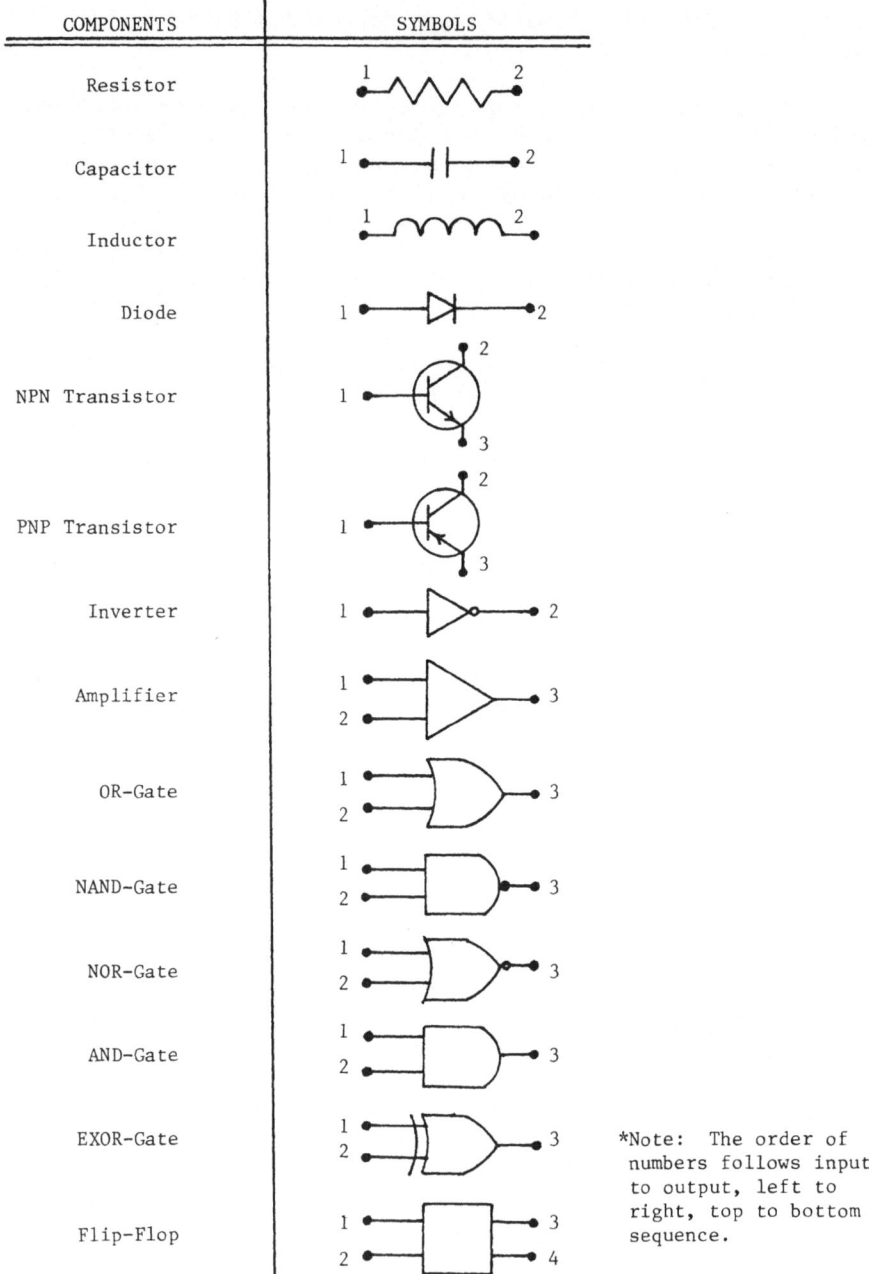

Figure 2. Standard position of symbols.

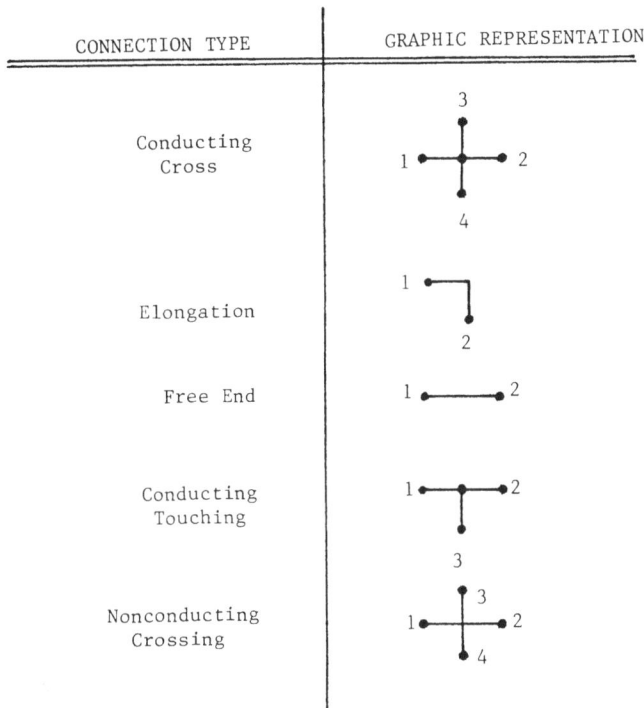

Figure 3. Standard configuration of connections.

the diagram reveals that the major difficulties of the recognition task are caused by the connecting elements. To automatically recognize an isolated functional element is not too difficult, which may be treated as a classical pattern recognition problem. If we liken an isolated functional symbol as a character, we may apply the character recognition techniques [19,20] to recognize the functional symbols. A major step in our design is to extract functional symbols from an electronic circuit diagram.

The design of the computer system AUTORED for automatic interpretation of electronic circuit diagrams is based upon the concept of multiple-pass pattern extraction, which we have developed [2]. Each pass of pattern extraction generates a page of information of one kind. Through multiple-pass pattern extraction, the electronic diagram is segmented according to the nature of the elements. The reading of isolated electronic symbols is then treated as a "character" recognition problem. The connecting elements provide the relational information for interpretation.

The system decomposes an electronic diagram into five sets of drawings, one on each page, via a multiple-pass pattern extraction

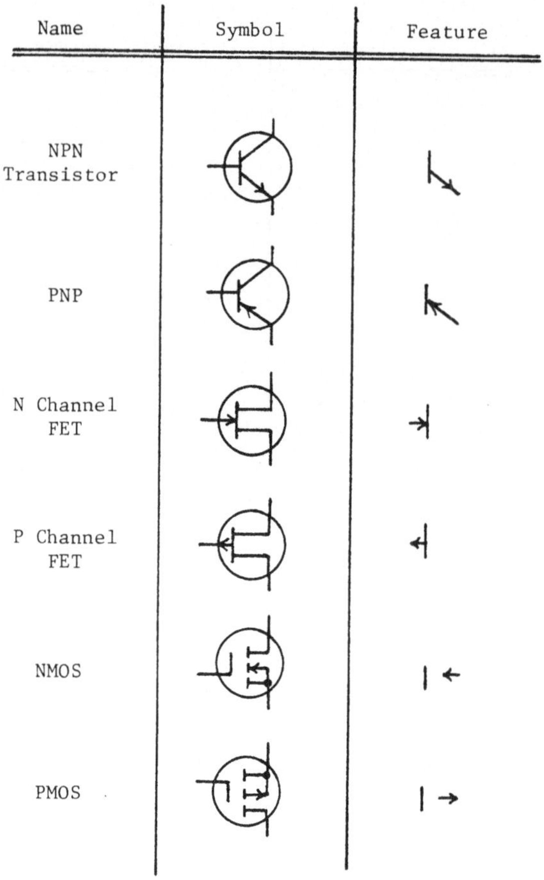

Name	Symbol	Feature
NPN Transistor		
PNP		
N Channel FET		
P Channel FET		
NMOS		
PMOS		

Figure 4. Key features of circular elements.

process. These drawings contain junction dots, connecting line-segments, functional symbols, denotations, and unrecognizable elements, respectively. The first page stores a drawing for the junction dots, the second page for connecting line segments, the third page for functional elements, the fourth page for denotations, and the fifth page for unrecognizable elements.

For the system to perform circuit diagram interpretation and pictorial database generation, we have developed two high-level picture manipulation languages to describe the symbol configuration and the interrelationships among them. The interrelationships are presented by an associative network structure.

The thinned symbols are decomposed into the primitives which will be stored in the knowledge base as feature vectors. The primitives which are currently used are shown in Figure 7. The new

AUTOMATIC GENERATION OF KNOWLEDGE BASE

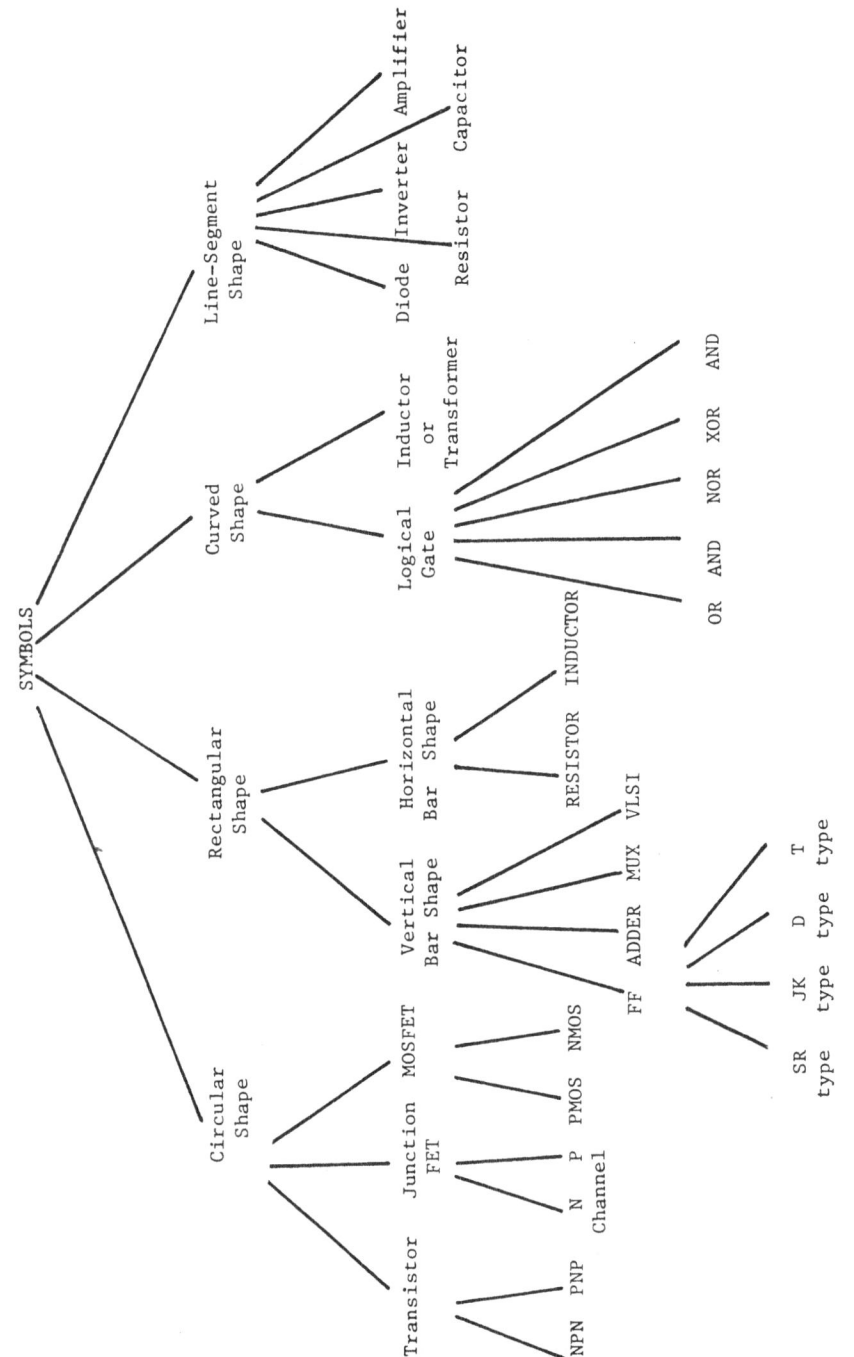

Figure 5. Symbol tree.

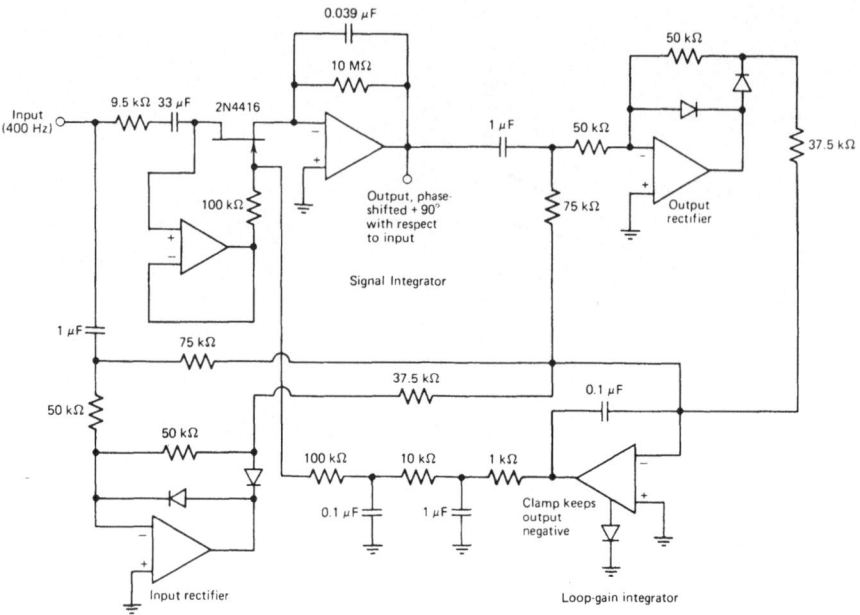

Figure 6. An electronic circuit diagram.

primitives of each symbol may be generated and updated in an interactive manner. The spatial relationships among the primitives of a symbol are extracted and then stored as feature vectors. The knowledge base is created during the training phase. In the following several sections, we will discuss the major tasks involved in multiple-pass pattern extraction which forms the basis for our approach to computer interpretation of diagrams and drawings.

Extraction of Junction Dots

The first step in the multiple-pass pattern extraction approach is to extract junction dots from an electronic circuit diagram. The junction dots consist of two types: crossing dots and touching dots which are illustrated in Figure 8, arranged according to the frequency of occurrence in a diagram. The black dots represent the skeleton pixels of a junction dot and its neighbors. These patterns form part of the primitives stored in the knowledge base. The sequence J1 to J5 is retrieved from the knowledge base when skeleton matching is performed during junction dots extraction. The white dots are the possible conducting pixels which are the results of imperfect preprocessing and poor quality of the paper.

The algorithm for junction dot extraction consists of the following steps:

AUTOMATIC GENERATION OF KNOWLEDGE BASE

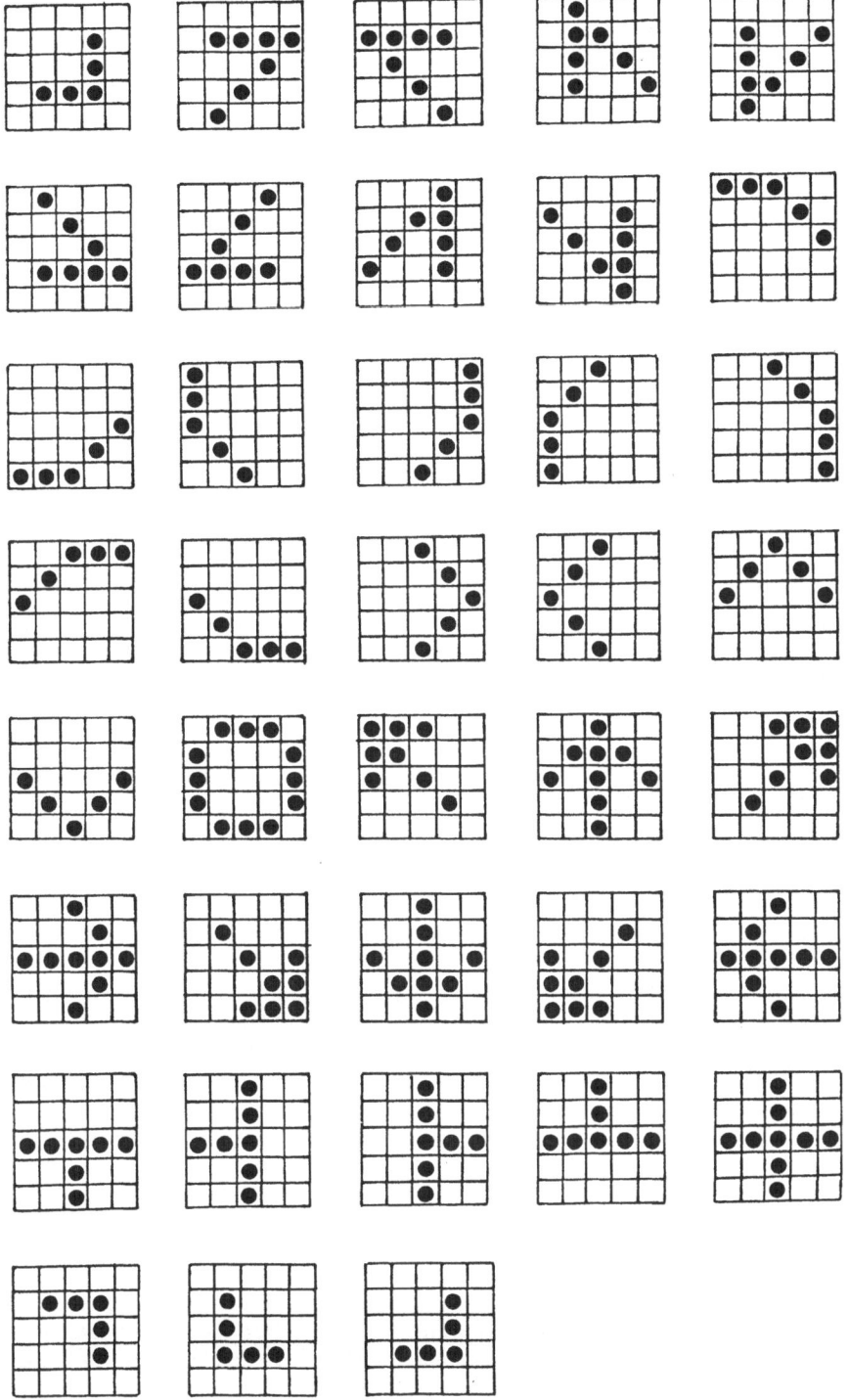

Figure 7. Primitives for electronic diagrams.

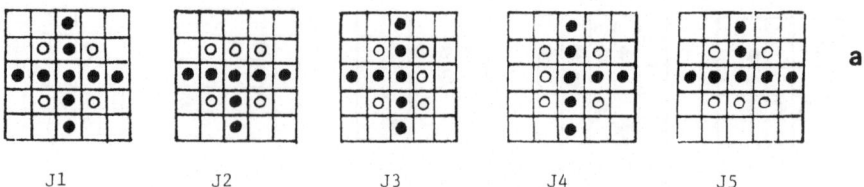

Figure 8a. Junction dot patterns.

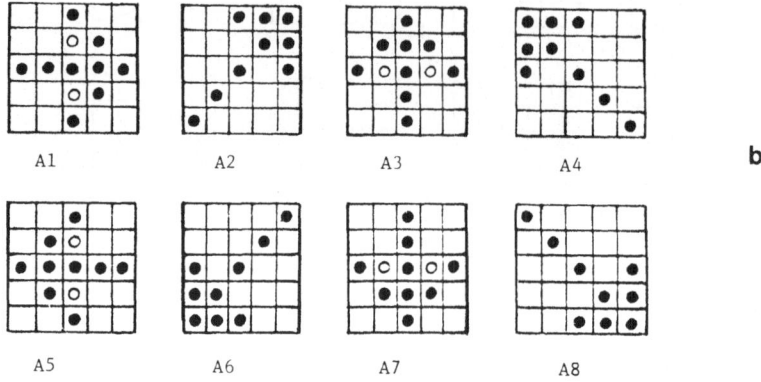

Figure 8b. Arrowhead patterns.

(1) Perform line-by-line scanning to determine the line segments with length exceeding a threshold θ_n. This threshold is used to eliminate line-segments of a capacitor or a diode and the associated junction dots. A threshold of 30 pixels has been chosen in our design, since the length of line-segments of a capacitor or a doide is about 12 pixels and the length of a shortest horizontal connecting line exceeds 30 pixels.

(2) Locate the junction dots in a horizontal line via template matching with the five primitives characterizing a junction dot. A 5x5 mask is centered on a horizontal line and moves from the left to the right. At each location four terminal points are checked. If all four terminal points are found, the junction dot at this location is a J1 function dot. If lower three terminal points are found, it is a J2 junction dot. If left three terminal points are found, it is a J3 junction dot. If right three terminals points are found, it is a J4 junction dot. If upper three terminal points are found, it is a J5 junction dot.

(3) Check the number of conducting pixels around the center pixel. Once we have located a junction dot, we want to make sure it is a junction dot. We count the number of conducting pixels

surrounding the central pixel. If this number exceeds a threshold θ_c, this central pixel is considered a junction dot. The x-y coordinates of this pixel are stored in page one with an ID code. Remove that junction dot from the binary image. In our design a threshold of 3 is chosen.

(4) Check the contents of the first page to resolve ambiguities and "flag" the junction dots falling within a 6x6 window. Preprocessing may make a junction dot blurred, enlarged, or slightly shifted. Ambiguities sometimes occur. If the differences of the coordinates of two junction dots are less than two pixels, these dots are merged. From ambiguous junction dots, we choose one and set a flag for the rest in page one.

(5) Recheck the contents of the first page to eliminate junction dots for arrowheads. The skeleton patterns for arrowheads in a diode, transitor-type element, or flow chart are used as templates for arrowhead identification.

A flow chart illustrating junction dot extraction is shown in Figure 9.

Extraction of Horizontal Connecting Lines

In printed documents, electronic diagrams contain only horizontal and vertical line segments connecting various circuit elements. Following the removal of the junction dots, the binary image contains disjoined connecting line segments and functional symbols. The second-pass pattern extraction is to extract the horizontal connecting line segments and to put them in the second page. We consider vertical connecting line segments as functional elements because image data are scanned horizontally before they are stored in random access disk. The lack of parallel processing facilities does not permit us to extract the vertical connecting line segments from the binary image files.

The second-pass pattern extraction is designed to trace and record the starting point and end point of each horizontal linesegment, if the length exceeds a threshold θ_L. A pair of parallel line-segments with almost equal length separated by a small spacing are not considered as connecting lines. Instead, they are candidates for capacitor, diode, or box-type elements which will be examined in the third-pass pattern extraction.

The algorithms for extracting horizontal connecting linesegments are given below:

(1) Trace horizontal line segments and record the starting coordinates and the length of each line-segment.

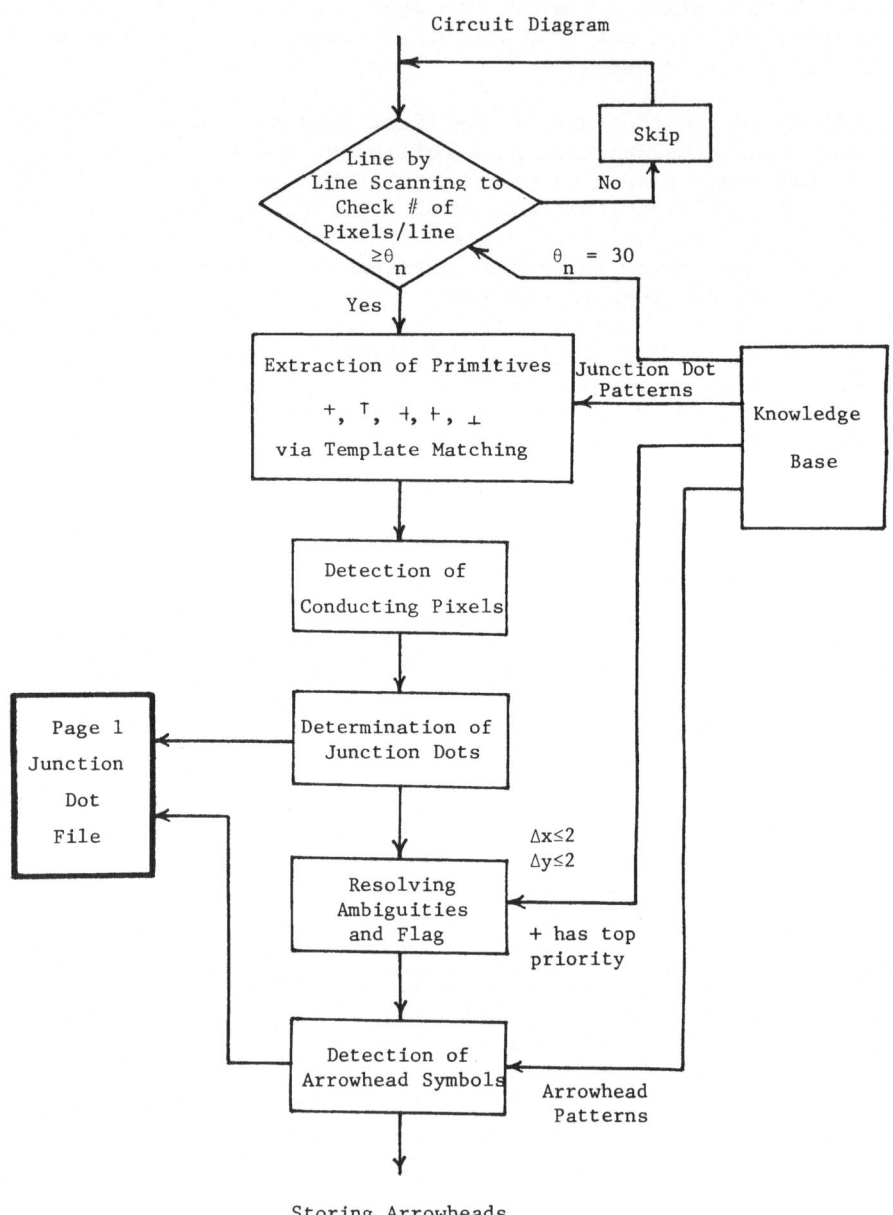

Figure 9. Junction dot extraction.

AUTOMATIC GENERATION OF KNOWLEDGE BASE 17

(2) Compare the length of each line-segment with a threshold θ_L. In our design, we choose $\theta_L=8$. Line-segments with length less than 8 pixels are considered noise segments. Capacitors and diodes contain line-segments with length equal to 9 to 12 pixels.

(3) Ambiguity resolving, flagging and linking. If two adjacent line-segments contain pixels with the same x-coordinates (i.e., $x_{i,j} = x_{k,j+1}$ and $|y_{i,j+1} - y_{k,j}| \leq 1$), these two line-segments will be merged and we choose the extreme points as the two end points of the combined line-segment. Let the coordinates of the two end points of a line-segment be (x_1^1, y_1^1) and (x_2^1, y_1^1), those for an adjacent line-segment be (x_1^2, y_2^2) and (x_2^2, y_2^2). These two adjacent line-segments satisfy the condition

$$|y_1^1 - y_2^2| \leq 1$$

Also, let the coordinates of the combined line-segment be (x_1, y) and (x_2, y). We have the following cases:

(a) If $x_1^1 \leq x_1^2 \leq x_2^1$ and $x_1^1 \leq x_2^2 \leq x_2^1$,

Then $x_1 = x_1^1$, $x_2 = x_2^1$ and $y = y_1^1$

(b) If $x_1^1 \leq x_1^2$ and $x_2^1 \leq x_2^2$,

Then $x_1 = x_1^1$, $x_2 = x_2^2$, and

$$y = \begin{cases} y_1^1 & \text{if } |x_2^1 - x_1^1| \geq |x_2^2 - x_1^2| \\ y_2^2 & \text{if } |x_2^1 - x_1^1| < |x_2^2 - x_1^2| \end{cases}$$

(c) If $x_1^2 \leq x_1^1$ and $x_2^2 \leq x_2^1$,

Then $x_1 = x_1^2$, $x_2 = x_2^1$, and

$$y = \begin{cases} y_1^1 & \text{if } |x_2^1 - x_1^1| \geq |x_2^2 - x_1^2| \\ y_2^2 & \text{if } |x_2^1 - x_1^1| < |x_2^2 - x_1^2| \end{cases}$$

(d) If $x_1^2 \leq x_1^1 \leq x_2^2$ and $x_1^2 \leq x_2^1 \leq x_2^2$,

Then $x_1 = x_1^2$, $x_2 = x_2^2$, and $y = y_2^2$

(4) Store the combined line-segment in page two and detect short parallel line pairs. Then determine candidate capacitors and diodes and transfer them to page three.

The tasks involved in horizontal connecting line extraction is summarized in a flow chart shown in Figure 10. Our program can also handle hand-drawn schematics as long as the line segment does not deviate from the medium line by 5° wave angle.

Extraction of Functional Elements

Following the removal of the junction dots and the horizontal connecting line-segments, the binary image of the electronic diagram contains disjoined functional elements, denotations, and vertical line-segments. The extraction of functional elements is conducted in three steps: blocking, grouping, and recognition. A block diagram illustrating functional element extraction is given in Figure 11. The blocking routine isolates a functional element, a denotation, or a vertical line segment by inscribing it with a rectangular or square block and assigning a code number to each isolated block in the order of the scanning sequence. The blocking routine for element isolation is the same as the blocking routine for character isolation in the automatic typewriter identification system which we have developed [20]. The basic principle consists of top-down and bottom-up pairwise comparison of two consecutive rows of binary image pixels. A rectangular block inscribing the element is then generated from these two data files.

The isolated blocks may be divided into two types, circular or rectangular, as shown in Figure 12. The circular symbols are extracted by using a circle denotation technique based upon a modified Hough transform [21]. The circular blocks represent such active elements as transistors, junction FET, and MOSFET, as shown in Figure 4. The rectangular blocks represent such functional elements as resistors, capacitors, inductors, diodes, amplifiers, logic gates, flip-flops, etc. The grouping routine is employed to categorize the isolated blocks by making use of a symbol tree. Since a vertical line-segment is characterized by an extremely small aspect ratio with width less than a threshold, it can readily be detected and transferred to the second page. Rectangular blocks are grouped according to their aspect ratios. Horizontal bar-shaped blocks represent resistors and inductors. Vertical bar-shaped blocks represent most of VLSI elements.

The recognition routine is to recognize the isolated functional elements in each group and to assign a label and rotation index to each element. Template matching techniques are used to identify the specific functional elements in a group of blocks. Size, strokes, symbol radicals, and number of terminals are the features which are used in the recognition process. As an example, we consider a resistor which may be described by its distinct features consisting of

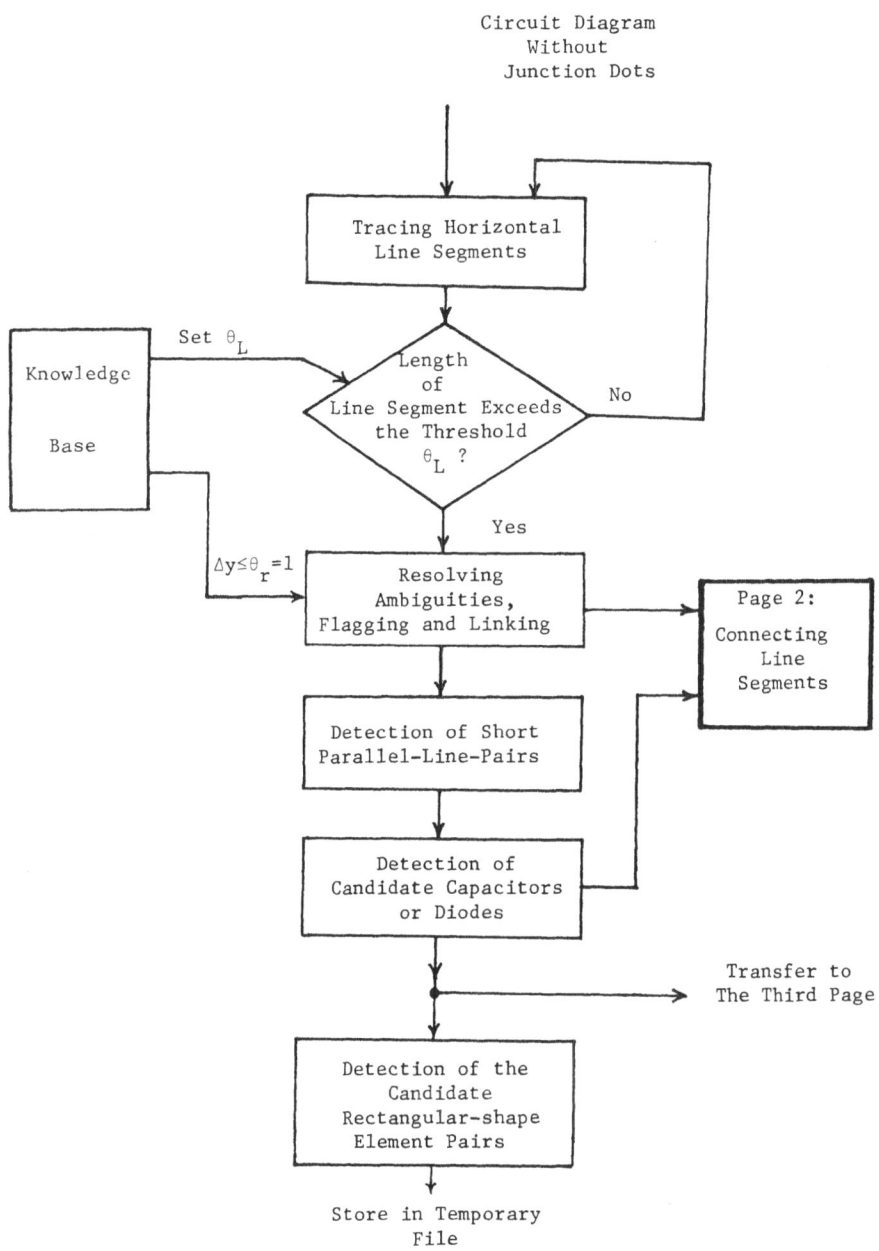

Figure 10. Horizontal connecting line extraction.

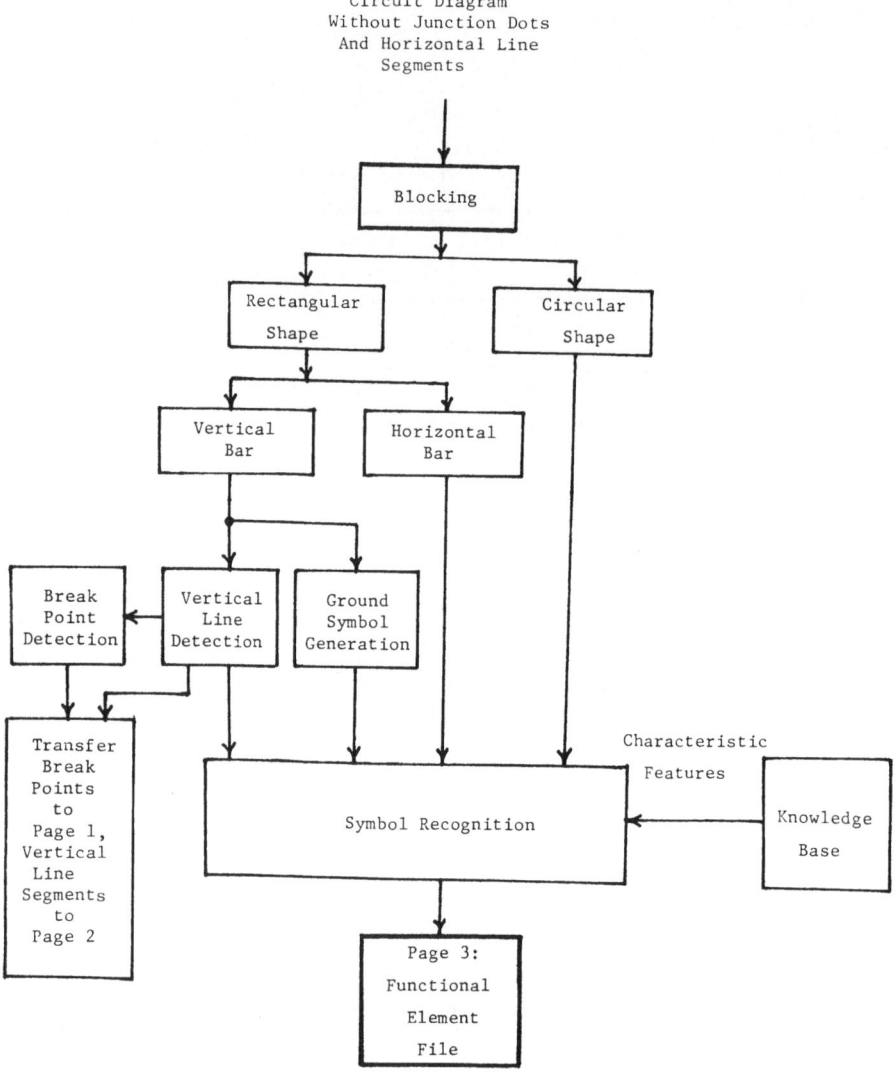

Figure 11. Functional element extraction.

```
# of primitive ∧ :   3
# of primitive ∨ :   3
# of terminals   :   2
repetition of  ∧ :   2
repetition of  ∨ :   2
```

AUTOMATIC GENERATION OF KNOWLEDGE BASE

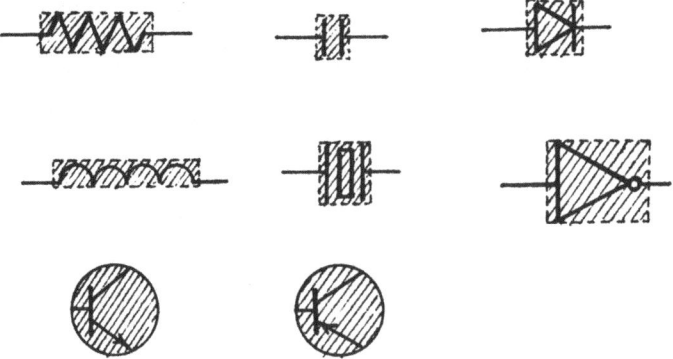

Figure 12. Symbol blocking.

For an operational amplifier, the distinct features are

```
         # of primitive ⊣:  2
         # of primitive >:  1
     # of input terminals:  2
    # of output terminals:  1
           repetition of ⊣: 1
```

Since some of the vertical bar-shaped elements are embedded in vertical line-segments after the blocking routine is performed, the vertical line detection must be done before the recognition routine is executed. The vertical connecting line-segments are stored in the second page. If a vertical bar-shaped element contains a corner feature, we detect that breaking point and transfer it to page one. To recognize a ground symbol, we make use of an inference mechanism. To recognize circular-shaped blocks, we examine the detailed structure inside the circle. In conducting partial feature matching on this group of blocks, we make use of information on the arrowhead. To avoid ambiguity, the terminal information is used for verification.

Recognition of Denotations

Following the removal of the function dots, the connecting line-segments, and the functional elements, we have a binary image of disjoined denotations and some unrecognizable symbols. The denotations are used to describe the physical properties or labels of the functional elements in a diagram. Since the denotations are expressed in terms of a character string, it appears that denotation recognition may be treated as character recognition in a text. However, the irregularities of denotations in a diagram complicates the recognition problem.

The block diagram of a denotation recognition scheme is shown in Figure 13. The first step is to recognize all the letters, numerals, and special characters such as Ω, µ if any. This step may be accomplished by character recognition techniques. The second step is to concatenate the character string to form a label for the functional element. The denotations are concatenated horizontally, thus resembling word recognition. The criteria for concatenation are (1) the x-coordinate of character string is increasing (that is, $x_n > x_{n-1} > \ldots > x_2 > x_1$, if n characters are concatenated), (2) the y-coordinate is almost the same (that is, $|\Delta y| \leq 2$), and (3) the spacing between two consecutive characters is limited to the width of a normal character.

The third step involves the labeling of a possible functional element. The labeling algorithm may be described by the following steps:

(1) Construct a denotation window.

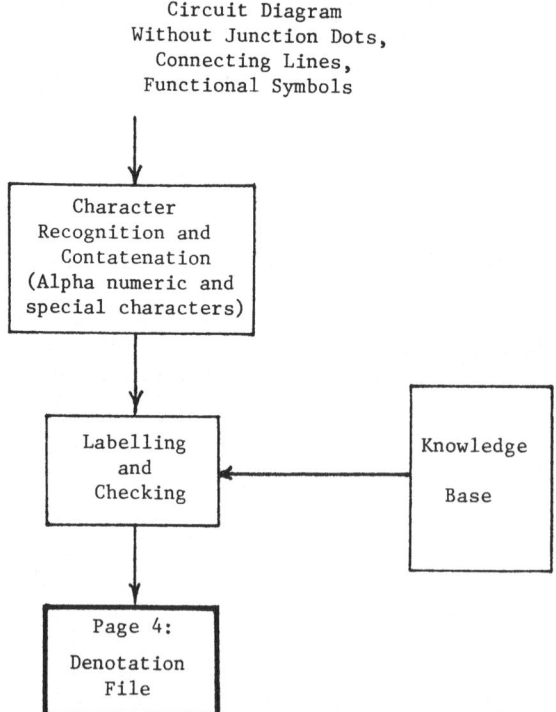

Figure 13. Denotation recognition.

(2) Shift the window horizontally and vertically. If the window touches a neighboring functional element, record the distance travelled, d_i, i = 1,2,3,4.

(3) Compare the recorded distances and choose the nearest functional element. Store the denotation in page 4 and mark the element as identified.

The above algorithm is illustrated in Figure 14, in which the window covers denotation 200KΩ, and the top resistor is the best candidate.

Following the above procedure, ambiguity may sometimes arise. For instance, the label for a capacitor may be assigned to a resistor on the basis of a minimum distance criterion. To avoid such ambiguity, we make use of a knowledge base which performs a semantic interpretation of the denotation. For instance, the character K or character Ω inside the denotation window implies that the label belongs to a resistor. The system will assign this denotation to a resistor which is in the nearest neighbor. Some denotations containing a single character, as exemplified by the denotations inside vertical bar-shaped elements. These denotations provide useful pin information which is needed in the identification and interpretation of certain functional elements.

Picture Manipulation Languages

An effective way for the computer to interpret electronic circuit diagrams and to generate a pictorial database for the diagrams is to develop high-level programming languages for the manipulation of pictures. Such languages are one-dimensional representation of two-dimensional pictures. The picture manipulation languages which we have developed consist of the symbol description language (SDL) and the picture generation language (PGL). The symbol description language is used to enable the

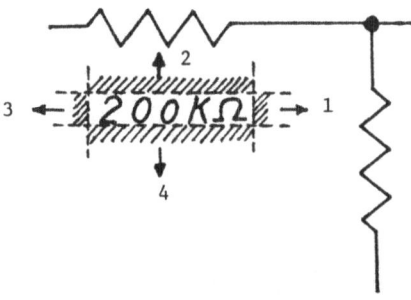

Figure 14. Labelling example.

computer to describe an isolated functional element in an electronic diagram. The picture generation language is employed to enable the computer to describe the interrelationships among various isolated functional elements in an electronic diagram. From the descriptions of the interrelationships, the electronic diagram may be generated by the computer.

The SDL is defined in terms of the syntax structure as follows:

$$
\begin{aligned}
&\langle SDL \rangle &&:= (ID\ code,\ label,\ rotation\ index) \\
&\langle ID\ Code \rangle &&:= \langle class\ name \rangle \\
&\langle class\ name \rangle &&:= \langle A \rangle | \langle C \rangle | \langle FF \rangle | \ldots | \langle R \rangle | \ldots \\
&\langle label \rangle &&:= 1|2|3|4|5|\ldots \\
&\langle rotation\ index \rangle &&:= \langle \theta=0° \rangle | \langle \theta=90° \rangle | \langle \theta=180° \rangle | \langle \theta=270° \rangle
\end{aligned}
$$

$$
\begin{aligned}
\langle \theta = 0° \rangle &:= 0 \\
\langle \theta = 90° \rangle &:= 1 \\
\langle \theta = 180° \rangle &:= 2 \\
\langle \theta = 270° \rangle &:= 3
\end{aligned}
$$

The ID code is a common denotation used in electronic diagrams. For instance, R is for resistor, C is for capacitor. The label code is assigned to each component for the purpose of linking and counting. The rotation index, RI, is used to indicate the orientation of the element. Examples of the four orientations are illustrated in Figure 15. In the normal position, $\theta = 0°$ and the input is at the left and the output is at the right. To facilitate the determination of the orientation of an element, we assign a number to each of the terminals of the element. The terminals are numbered from left to right and from top to bottom for an element in its normal position. Standard positions of symbols and standard configurations of connections are shown in Figure 2 and Figure 3, respectively.

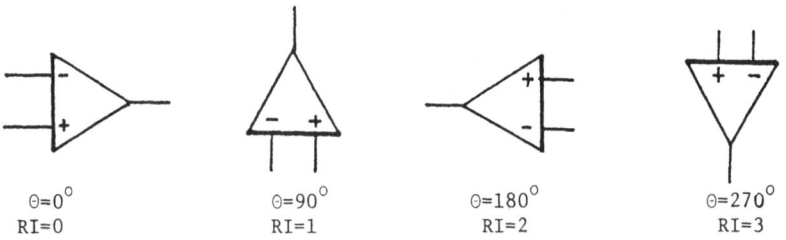

Figure 15. Four orientations of a symbol.

AUTOMATIC GENERATION OF KNOWLEDGE BASE   25

As an illustration, we consider the circuit diagram shown in Figure 16, which contains four functional elements and one connecting element. In terms of SDL, these elements are described by

```
R1:=(R,1,0)
R2:=(R,2,0)
R3:=(R,3,1)
C1:=(C,1,1)
J1:=(J,1,0)
```

The PGL is defined as

```
<PGL>:=(<Interrelation>; <symbol 1>; <symbol 2>
<symbol 1>:=<SDL 1>, <Term 1>|<Input symbol>, <Label>
<symbol 2>:=<SDL 2>, <Term 2>|<Output symbol>, <Label>|
     <Ground symbol>, <Ground label>
<Interrelation>:=<Input>|<Output>|<Between>
     <Input>    := I
     <Output>   := O
     <Between>  := B
<SDL 1>:=<Symbol description language for element 1>
<SDL 2>:<Symbol description language for element 2>
<Label>:<1|2|3|4|5|...
<Input symbol>:= $
<Output symbol>:= *
<Ground symbol>:= 0
<Term 1>:=<terminal code for element 1>
<Term 2>:=<terminal code for element 2>
```

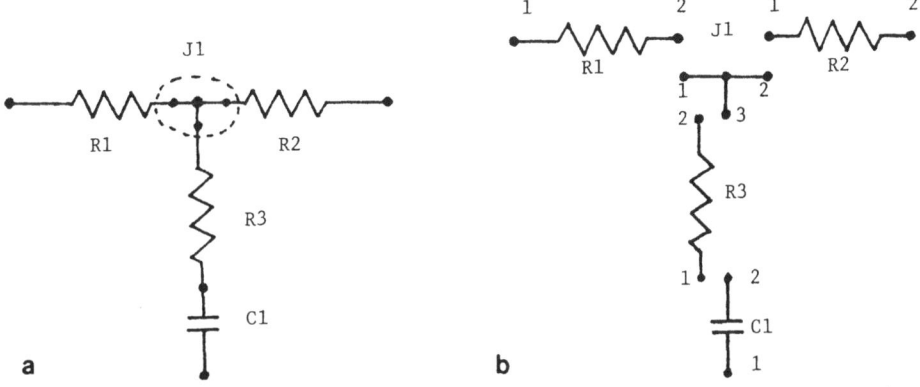

Figure 16a. Example of a circuit.   Figure 16b. Decomposition.

Note: The numbers assigned follow the
       left-right top-down sequence.

The interrelationships are categorized into three kinds: INPUT, BETWEEN, and OUTPUT. When there is a free-end segment connecting to the left or top side of an element, the interrelationship is INPUT. If the connection is to the right or bottom side of the element, the interrelationship is OUTPUT. The PGL of each branch will terminate at an output node, ground symbol, or an element which already appeared at the other branch.

As an illustration, we consider the electronic diagram in Figure 17. Based upon the above analysis, we obtain the associative network structure shown in Figure 18. The PGL for this diagram is

```
(I; $, 1; C, 1, 0, 1)
(I; $, 1; C, 1, 0, 1)
(B; C, 1, 0, 2; J, 1, 0, 1)
(B; J, 1, 0, 3; R, 1, 1, 2)
(B; J, 1, 0, 2; R, 2, 0, 1)
(B; R, 1, 1, 1; #, 0)
(B; R, 2, 0, 2; J, 2, 2, 1)
(B; J, 2, 2, 2; AR, 1, 0, 1)
(B; J, 2, 2, 3; J, 3, 1, 1)
(B; AR, 1, 0, 2; #, 0)
(B; AR, 1, 0, 3; J, 5, 2, 1)
(B; J, 5, 2, 2; R, 4, 0, 1)
(B; J, 5, 2, 3; J, 4, 3, 1)
(O; R, 4, 0, 2; *, 1)
(B; J, 4, 3, 2; CR, 2, 1, 1)
(B; J, 4, 3, 3; CR, 1, 0, 2)
(B; CR, 2, 1, 2; R, 3, 0, 2)
(B; R, 3, 0, 1; J, 3, 1, 2)
(B; CR, 1, 0, 1; J, 3, 1, 3)
```

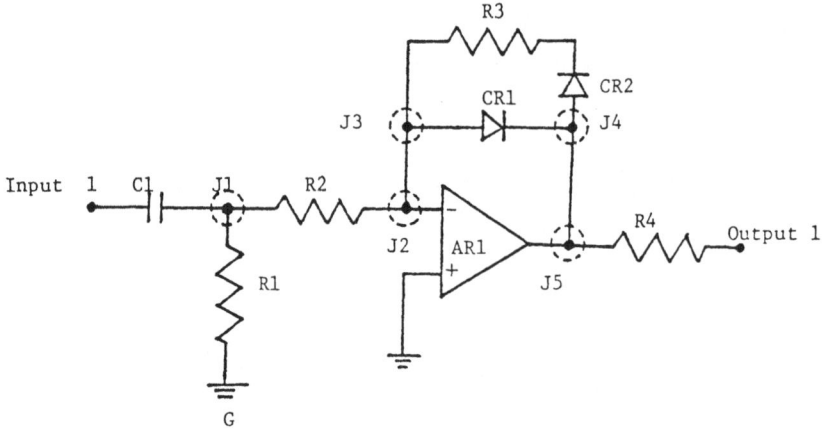

Figure 17. An electronic diagram.

AUTOMATIC GENERATION OF KNOWLEDGE BASE

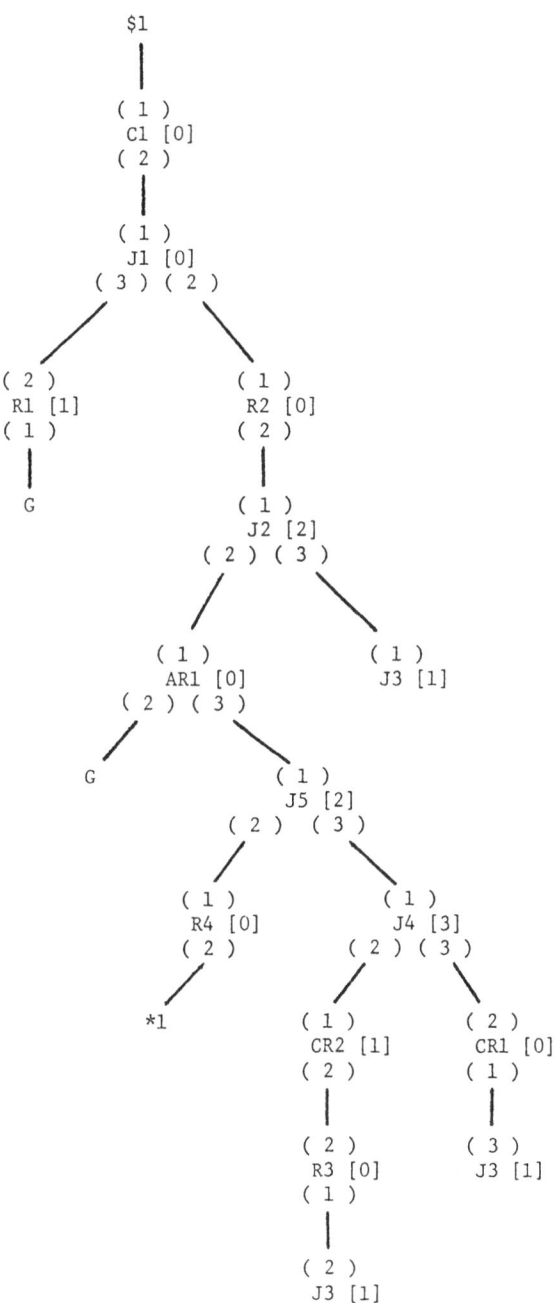

Figure 18. Associative network structure for the diagram in Figure 17.

The PGL offers the following advantages:

(1) It provides an easy way to describe electronic diagrams.

(2) The diagram can be systematically reconstructed from the descriptions by the PGL.

(3) It provides a ready mechanism to update or modify the circuit organization.

(4) It can also be applied to generate a symbol itself, thus serving as the symbol generation language. As an illustration of the symbol generation language, we consider an AND gate symbol which is shown in Figure 19. Using the proposed PGL, the system derives a simplified associative tree for the AND gate symbol, as shown in Figure 20.

(5) It can readily be translated automatically into topological data format to be used as input for some CAD programs such as SPICE.

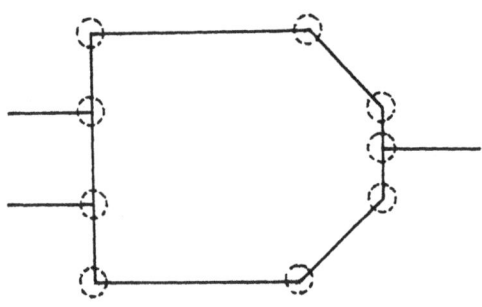

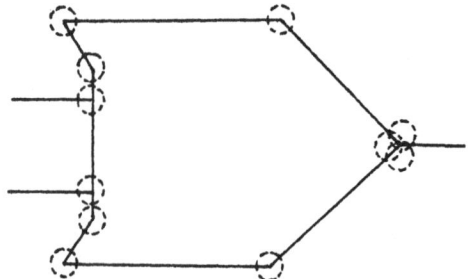

Figure 19. AND gate symbol.

AUTOMATIC GENERATION OF KNOWLEDGE BASE   29

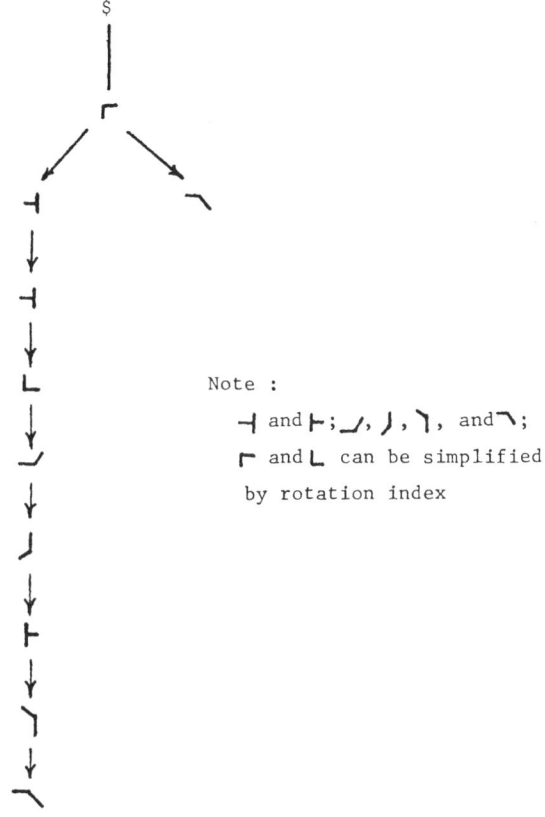

Figure 20. Simplified associative tree of the
AND gate symbol in Figure 19.

ILLUSTRATIVE EXAMPLE FOR AUTOMATIC INTERPRETATION

   The AUTORED system for automatic interpretation of electronic
diagrams has been programmed in FORTRAN and implemented on a PDP-
11/40 minicomputer under RSX-11M operating system. Experimental
results are summarized in Figure 21.

DATA FILES FOR CAD KNOWLEDGE BASE

   By making use of multiple-pass pattern extraction principles,
the AUTORED system identifies the junction dots, connecting lines,
functional elements and their interrelationships of an electronic
diagram. The results of interpretation are organized in data files
for junction dots, connecting line segments, functional elements,
and denotation, respectively. The data structures for these data
files are described as follows:

a THE COORDINATE OF THE CONNECTION COMPONENT
(AFTER THE EXTRACTION OF JUNCTION DOT)

REC #	X1	Y1	X2	Y2	TYPE
1	136	29	140	33	J-1
2	137	29	141	33	J-2
3	79	55	83	59	J-4
4	80	55	84	59	J-1
5	136	55	140	59	J-2
6	137	55	141	59	J-3
7	39	73	43	77	J-2
8	40	73	44	77	J-2
9	79	73	83	77	J-4
10	80	73	84	77	J-3
11	39	74	43	78	J-1
12	40	74	44	78	J-1
13	79	74	83	78	J-5
14	80	74	84	78	J-5

b THE COORDINATE OF THE CONNECTION COMPONENT :
(AFTER AMBIGUITY RESOLVING)

REC #	X1	Y1	X2	Y2	TYPE
1	137	29	141	33	J-2
2	79	55	83	59	J-4
3	137	55	141	59	J-3
4	39	73	43	77	J-2
5	79	74	83	78	J-5

c THE COORDINATE OF THE LINE SEGMENTS :
(BEFORE AMBIGUITY RESOLVING)

REC #	LINE #	POS1	POS2	TYPE
1	30	84	98	H
2	30	121	128	H
3	31	120	163	H
4	56	83	93	H
5	57	83	105	H
6	57	118	137	H
7	75	4	12	H
8	75	20	50	H
9	76	29	50	H
10	76	73	91	H
11	88	125	137	H
12	99	85	92	H

d THE COORDINATE OF THE LINE SEGMENTS :
(AFTER AMBIGUITY RESOLVING AND LINKING)

REC #	LINE #	POS1	POS2	TYPE
1	30	84	98	H
2	31	120	163	H
3	56	83	105	H
4	57	118	137	H
5	75	4	12	H
6	75	20	50	H
7	76	73	91	H
8	88	125	137	H
9	99	85	92	H

AUTOMATIC GENERATION OF KNOWLEDGE BASE 31

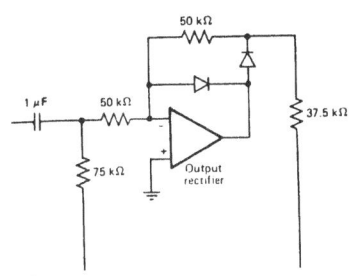

e THE COORDINATE OF THE CIRCUIT COMPONENTS :
(AFTER BLOCKING AND SCREENING)

REC #	X1	Y1	X2	Y2	TYPE
1	99	26	120	34	H -0
2	163	31	171	136	V -0
3	134	34	143	54	V -0
4	106	52	117	61	H -0
5	13	70	19	80	V -0
6	88	70	124	108	V -0
7	51	72	72	79	H -0
8	38	77	46	136	V -0
9	78	95	88	118	V -0

f THE COORDINATE OF THE CIRCUIT COMPONENTS :
(AFTER IDENTIFICATION)

REC #	X1	Y1	X2	Y2	TYPE
1	99	26	120	34	R -0
2	163	65	171	85	R -1
3	134	38	143	50	CR-1
4	106	52	117	61	CR-0
5	13	70	19	80	C -0
6	88	70	124	108	AR-0
7	51	72	72	79	R -0
8	38	95	46	116	R -1
9	78	116	88	122	G -0

g THE COORDINATE OF THE CONNECTION COMPONENT :
(AFTER THE EXTRACTION OF THE BENDED CORNER)

REC #	X1	Y1	X2	Y2	TYPE
1	137	29	141	33	J-2
2	79	55	83	59	J-4
3	137	55	141	59	J-3
4	39	73	43	77	J-2
5	79	74	83	78	J-5
6	81	30	83	32	J-6
7	164	31	166	33	J-7
8	138	85	140	88	J-9
9	81	99	83	101	J-6

h THE COORDINATE OF THE LINE SEGMENTS :
(AFTER THE EXTRACTION OF THE VERTICAL LINE SEGMENTS

REC #	LINE #	POS1	POS2	TYPE
1	30	84	98	H
2	31	120	163	H
3	56	83	105	H
4	57	118	137	H
5	75	4	12	H
6	75	20	50	H
7	76	73	91	H
8	88	125	137	H
9	99	85	92	H
10	81	33	54	V
11	166	34	65	V
12	167	86	136	V
13	81	60	72	V
14	139	60	85	V
15	42	79	94	V
16	42	117	136	V
17	31	102	115	V

Figure 21. Example of computer interpretation.

Junction Dot File (JD)

```
              1     2     3     4     5     6     7

              x₁    y₁    x₂    y₂    Type        Name
```

Words 1-4: coordinates of the junction dot
Words 5-6: type of the junction dot
Word   7: reserved for the name of the junction dot

Line Segment File (LN)

```
     1     2     3     4     5     6     7     8     9    10    11   12

   pos 1  pos 2  pos 3  Type  Line Seg 1  Line Seg 2   Node 1    Node 2
```

Words   1-3: coordinates of the line segment
Word   4: type of the line segment
Words   5-6: reserved for first line segment connected to
              the current line
Words   7-8: reserved for second line segment connected to
              the current line
Words 9-10: reserved for first node connected to the
              current line
Words 11-12: reserved for second node connected to the
              current line

Functional Element File (FE)

```
1   2   3   4    5   6    7      8   9    10  11  12   13    14    15

x₁  y₁  x₂  y₂   Type    Ornt    Name      Value       Nd 1  Nd 2  Nd 3
```

Words   1-4: coordinates of the functional element
Words   5-6: type of the functional element
Word   7: orientation of the functional element
Words   8-9: reserved for name of the functional element
Words 10-12: reserved for value of the functional element
Words 13-15: reserved for nodes connected to the functional
              element

Denotation File (DF)

```
              1     2     3     4     5     6     7

              x₁    y₁    x₂    y₂          Value
```

Words 1-4: coordinates of the characters
Words 5-7: value

AUTOMATIC GENERATION OF KNOWLEDGE BASE

The above data files are the ingredients for the knowledge base of electronic diagrams. From this knowledge base, the input data for CAD analysis or simulation programs may be automatically derived.

GENERATION OF INPUT FOR SPICE

The input data for CAD programs can be automatically generated from the above knowledge base by interpreting the data files for a specific application. For a SPICE program, the system determines the connection relations and fills in the reserved fields in the functional element file. The algorithms for generating the input data for SPICE program are summarized as follows:

(1) Read the values of a denotation file into the functional element file. This step creates a modified FEM file with the VALUE fields filled in.

(2) Determine line segments which are connected together and the junction dots which are connected to line segments. This step creates a modified LNM file which contains the connection relations.

(3) Merge junction dots. This step creates a modified LNM file which contains complete information on the connection relations.

(4) Label merged junction dots by filling in the NAME fields. This step creates a modified JD file which renames connected junction dots.

(5) Fill in the NAME fields of the FEM file with a unified name. This step creates a modified FEM file which contains unified names.

(6) Determine junction dots for functional elements and merge the modified FEM file and JD file. This step creates a modified FEM file which contains junction information for function elements.

(7) Generate input data for SPICE program from this modified FEM file.

ILLUSTRATIVE EXAMPLE FOR SPICE INPUT GENERATION

The above algorithms are applied to the electronic diagram in Figure 21. The computer printouts for each step and the generated input data for SPICE program are illustrated in Figure 22.

```
            1   137   29   141   33   j   2   0
            2    79   55    83   59   j   4   0
            3   137   55   141   59   j   3   0
            4    39   73    43   77   j   2   0
            5    79   74    83   78   j   5   0
```

(a) Junction Dot File, JD.DAT

```
 1    81    98    30   H   0   0   0   0   0   0   0   0
 2   121   136    30   H   0   0   0   0   0   0   0   0
 3   142   167    31   H   0   0   0   0   0   0   0   0
 4    84   105    57   H   0   0   0   0   0   0   0   0
 5   118   136    57   H   0   0   0   0   0   0   0   0
 6     1    12    75   H   0   0   0   0   0   0   0   0
 7    20    38    75   H   0   0   0   0   0   0   0   0
 8    44    50    75   H   0   0   0   0   0   0   0   0
 9    73    78    75   H   0   0   0   0   0   0   0   0
10    84    87    76   H   0   0   0   0   0   0   0   0
11   125   139    85   H   0   0   0   0   0   0   0   0
12    83    87   102   H   0   0   0   0   0   0   0   0
13   117   136    42   V   0   0   0   0   0   0   0   0
14    78    94    42   V   0   0   0   0   0   0   0   0
15   102   115    83   V   0   0   0   0   0   0   0   0
16    60    73    81   V   0   0   0   0   0   0   0   0
17    30    54    81   V   0   0   0   0   0   0   0   0
18    60    85   139   V   0   0   0   0   0   0   0   0
19    51    54   139   V   0   0   0   0   0   0   0   0
20    34    37   139   V   0   0   0   0   0   0   0   0
21    86   136   167   V   0   0   0   0   0   0   0   0
22    31    64   167   V   0   0   0   0   0   0   0   0
```

(b) Line Segment File, LN.DAT

```
1    99   26   120    34   R       0   0       0   0   0
2   163   65   171    85   R       1   0       0   0   0
3   134   38   143    50   C   R   1   0       0   0   0
4   106   52   117    61   C   R   0   0       0   0   0
5    13   70    19    80   C       0   0       0   0   0
6    88   70   124   108   A   R   0   0       0   0   0
7    51   72    72    79   R       0   0       0   0   0
8    38   95    46   116   R       1   0       0   0   0
9    78  116    88   122   G       0   0       0   0   0
```

(c) Functional Element File, FE.DAT

Figure 22. Example of computer generation of input for SPICE program.

CONCLUSION

This paper discusses the design of a computer interpretation system for automatic generation of knowledge base from electronic diagrams. The system reads the functional elements in an electronic diagram, determines their relative locations, and describes their interrelationships. The design technique for automatic interpretation is based upon the multiple-pass pattern extraction principles. The outcome of interpretation is represented in a knowledge base which may be converted to a data format to be used as input to a CAD analysis or simulation program or to a CAD database. The design approach is being applied to engineering drawings other than electronic diagrams.

AUTOMATIC GENERATION OF KNOWLEDGE BASE

ACKNOWLEDGMENT

The research presented in this paper was supported by the National Science Foundation under Grant No. IST-8305815.

```
1    99   26   120   34   5   0   K
2   163   65   171   85   3   7   K
3    13   70    19   80   1   U   F
4    51   72    72   79   5   0   K
5    38   95    46  116   7   5   K
```

(d) Denotation File, DF.DAT

```
1    99   26  120   34  R     0  R  11  5  0  K   12  11   0
2   163   65  171   85  R     1  R  12  3  7  K   22  11   0
3   134   38  143   50  C  R  1  D  13              13  11   0
4   106   52  117   61  C  R  0  D  14              12  13   0
5    13   70   19   80  C     0  C  15  1  u  F   32  14   0
6    88   70  124  108  A  R  0  X  16              10  12  13
7    51   72   72   79  R     0  R  17  5  0  K   14  12   0
8    38   95   46  116  R     1  R  18  7  5  K   21  14   0
9    78  116   88  122  G     0  G  19               0  10   0
```

(e) Complete Functional Element File, FEM3.DAT

```
GENERATED SPICE PROGRAM SEGMENT
R11      12    11    50K
R12      22    11    37K
D13      13    11
D14      12    13
C15      32    14    1uF
X16      10    12    13    OP16
.SUBCKT  OP16   1     2    3
R1        1     2
U1        3     0     1    2
.END
R17      14    12    50K
R18      21    14    75K
G19       0    10
```

(f) Computer Generated Input Data for SPICE Program

REFERENCES

[1] G. Musgrave (Ed.), Computer-Aided Design of Digital Electronic Circuits & Systems, North-Holland Publ. Co., The Netherlands, 1979.

[2] K. Kani, A. Yamada and M. Teramoto, "CAD in the Japanese Electronics Industry," Ibid, pp. 103-119, 1979.

[3] J. T. Tou and J. M. Cheng, "AUTORED: An Automated Electronic Diagram Reading Machine," Proc. of 1983 IEEE Int. Conf. on Computer-Aided Design, 1983.

[4] C. Y. Suen and T. Radhakrishnan, "Recognition of hand-drawn flowcharts," Proc. 3rd Int. Conf. Pattern Recognition, pp. 424-428, 1976.

[5] W. C. Lin and J. H. Pun, "Machine recognition and plotting of hand-sketched line figures," IEEE Trans. Syst. Man. Cybern., SMC-8(1), pp. 52-57, 1978.

[6] S. Kakumoto, Y. Fujimoto and J. Kawasaki, "Logic diagram recognition by divide and synthesize method," in Artificial Intelligence and Pattern Recognition in Computer Aided Design, J. C. Latombe (Ed.), pp. 457-476, North-Holland, Amsterdam, 1978.

[7] B. Zavidovique and G. Stamon, "An automated process for electronics scheme analysis," Proc. 5th Int. Conf. Pattern Recognition, pp. 248-250, 1980.

[8] H. Bunke, "Experience with several methods for the analysis of schematic diagrams," Proc. 6th Int. Conf. Pattern Recognition, pp. 710-712, 1982.

[9] H. Bunke, "Computer recognition of circuit diagrams," TR-EE 80-54, Purdue University, 1980.

[10] S. I. Shimizu, S. Nagata, A. Inoue and M. Yoshida, "Logic circuit diagram processing system," Proc. 6th Int. Conf. Pattern Recognition, pp. 717-719, 1982.

[11] Y. Fukada, "Primary algorithm for the understanding of logic circuit diagrams," Proc. 6th Int. Conf. Pattern Recognition, pp. 706-709, 1982.

[12] J. F. Jarvis, "The line drawing editor: Schematic diagram editing using pattern recognition techniques," Computer Graphics and Image Processing, Vol. 6, pp. 452-484, 1977.

[13] IEEE, Graphic Symbols for Electrical and Electronics Diagrams, ANSI Y32.2 - 1975.

[14] IEEE, Graphic Symbols for Logic Diagrams (Two-State Devices), ANSI Y32.4 - 1973.

[15] G. E. Rowbothan, Engineering and Industrial Graphics Handbook, McGraw-Hill, New York, 1982.

[16] E. Cohen, "Program reference for SPICE2," Report No. ERL-M592, Electronics Research Laboratory, University of California, Berkeley, 1976.

[17] Y. C. Yuan, D. Divekar and R. Dowell, "MODULAR-SPICE - A modular circuit simulation program," Proc. of 1983 IEEE Int. Conf. on Computer-Aided Design, 1983.

[18] J. M. Cheng, "Literature review of schematic diagram," Technical Report 82-11, Center for Information Research, University of Florida, 1982.

[19] J. T. Tou and R. C. Gonzalez, Pattern Recognition Principles, Addison Wesley Publ. Corp., Reading, MA, 1974.

[20] J. T. Tou, J. M. Cheng, D. Z. Lu and J. McCarthy, "Automatic type font identification for falsified documents," Proc. 6th Int. Conf. on Pattern Recognition, 1982.

[21] C. Kimme, D. Ballard and J. Sklansky, "Finding circles by an array of accumulators," Comm. ACM, 1975.

CAD APPLICATION ON PC BOARD DESIGN

Chung-Jye Chang

Telecommunication Laboratories
Ministry of Communications
Taiwan, R.O.C.

ABSTRACT

This paper presents a design process which enables a designer with limited computer background to accomplish PC board design interactively with CAD system. The primary process adopted in our system is the result of a close examination of what will be done by experienced artwork operator, who has received intensive training in layout and taping, with the very powerful computer aid tool in ratsnet checking and autoroute. This primary process includes physical creation, net list editing, initial placement, ratsnet checking, signal routing and editing line function. It has shortened the task-cycle time at Telecommunication Laboratories, even in early stage of application in PC board design. The performance of the system is satisfactory, and we are now in the process of upgrading from the primary process to an advanced process.

INTRODUCTION

There are two approaches used to fabricate the design of PC board: Design Automation (DA) and Computer-Aided Design (CAD). The CAD system relies on a human intelligence for the creative aspects of PC board layouts which concern many electronic constraints, such as balance of common impedance coupling, suppression of radiation, noise or crosstalk. PC board design is the first CAD application field involving graphics, and its approach is initially favored in the digital circuit design only. The existing literature currently appears to be gradually regarding what analog-oriented programs are available [1]. CAD systems have offered the combinations of data tablet, electronic pen and graphical display as a design station to allow the designers to enter schematic and physical symbols, drafting commands and X-Y positional data which are all strongly powerful assistance to PC board design.

A branch group of CAD members in the Telecommunication Laboratories is responsible for PC board design and its integration tasks with subsequent fabrications to meet research and development of communication-related fields in transmission, switching and computer applications. It is well-known that a properly functioning, suitable PC board is fundamental to the successful operations of electronic products. Our particular problems and challenges at this moment are not only the increasingly required quantities, which is about 200 runs of PC board designs per year, but the less lead time for design-to-manufacturing cycle. Besides, there are usually many

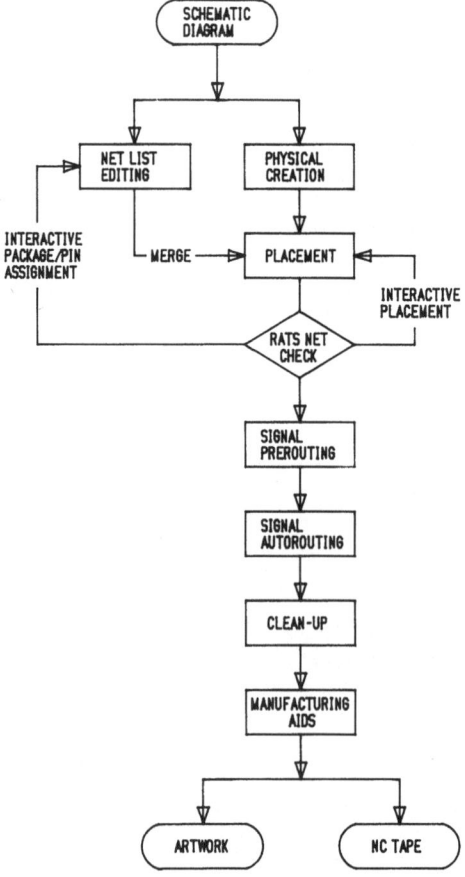

Figure 1. Primary process of PC board design.

CAD APPLICATION

times of engineering changes during the process of developing an acceptable PC board for complicated communication circuitry.

The PC board design process illustrated in Figure 1 has adopted since May, 1982 when the CAD system was set up at TL. The initial layout of PC boards design which contains IC component placement, with all mechanical and electronic engineerings constraints, is finished by circuit design engineers in free-hand graphical forms [2]. The initial placement guideline is shown as Figure 2. The detail layout drawings on grid paper are prepared and digitized into system as digital form database which has been merged with net list extracted from schematic diagram. The database is subsequently manipulated in many ways including comparing, debugging, and path routing. Both the final geometrical Design Rule Check (DRC) and electronic interconnection check are also implemented by CAD system. The manufacturing-aids output of photoplotted artworks

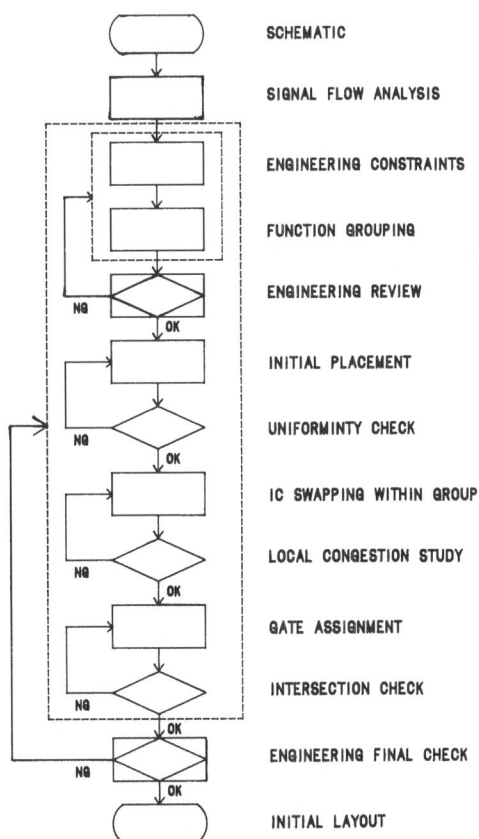

Figure 2. Initial placement guideline.

masks, NC drill tapes and other necessary engineering documentations could be released in PC board fabrication and assembly application. This process is called as "primary process" because it is apparently suitable for early stage of CAD system application on PC board design.

The PC board design has quickly linked into CAD/CAM system with postprocessor of autoinsertion, autowiring and autotesting in the electronic industry manufacturing fields. On the other hand, the packages of circuit analysis, logic simulation and automated interpretation of electronic diagram have enhanced the current CAD system capabilities, and provided Computer-Aided Engineering (CAE) functions. However, the advent of powerful CAD system has allowed the development of execute program which performs a sophisticated design more efficiently [3]. To upgrade our CAD system capability, we have developed advanced process [4] for converting graphical forms data into digital forms data from the very beginning input of schematic diagram, refer to Figure 3 [5], and will be adopted soon. There are other engineering studies and surveys in preprocessor and postprocessor proceeded to strive toward a smooth integrated development involving CAE, CAD, CAT, and CAM.

Factors Affecting PC Board Design

The main factor affecting PC board design in the application of CAD system are: productivity increasing and quality improvement.

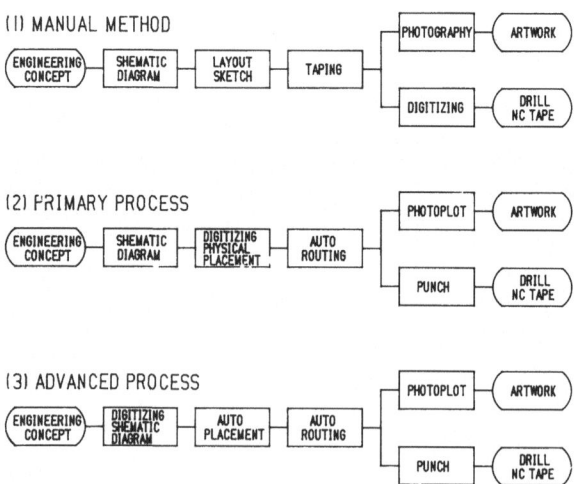

Figure 3. Different methods of PC board design (digitizing is the crucial step of converting graphical into digital form.

CAD APPLICATION

Let's discuss the productivity increasing first: It is well known that the manufacturer who issues new product to market first, generally captures the large market share. The pressure faced is to shorten lead time, while the complicated technological developments have tended to slow down the creation-to-manufacturing cycle. The designers have to shorten the engineering turnaround and implement the PC board design successfully in very first time, and deliver the master artworks and NC drill tapes which are all already manufacturable. The intense competition in the electronics industries has recognized that interactive CAD systems have become an integration part of production process. During this half decade, CAD systems are fast becoming an economic necessity for companies which design more than 100 kinds of PC board per year in U.S., Europe and Japan [6]. In designing a PC board, there are lots of repetitive work such as placements of the same physical component symbols in many locations, text annotation, interconnection routing and path geometrical checking. The engineers are faced with the challenge to optimize creativity and maximize productivity. The CAD system provides computer assistance to the designer for those work steps where repetition dominates and creativity is not required. And it has broken the biggest bottleneck of manual PC board design involving repetitive, time-consuming tasks of layout, placement, hand-drawn lines routing, checking and updating.

The current trend for design complex circuitry of communications and computer systems is toward large boards. For example, DTLS used in Los Angeles Olympic broadcasts is packaged in 19"x13" rackmounted PC board [7]. The bigger the board, the more time should be taken for complicated considerations of components placement and interconnections routing which all belong to error-proven operations. The interactive design station of CAD systems enable the engineers to try new ideas, evaluate alternative on visual feedback screen, and finally get an optimized result. Figure 4 shows the cycletime involved in design of PC boards by hand and by means of a CAD system as a function of PC board area. It has been shown that the reduction of design time required is of the order of 70%-80%.

The other subject is to describe quality improvement: Another current trend of PC board design is the packing of increasingly dense circuit in limited areas, typically from 1.2 to 0.75 square inches per IC, causing the interconnection path to get narrower, and the registration of layer to layer is more difficult. Unfortunately, the manual method in PC board design has its limitation in precise artwork of 3 mil accuracy [5]. There are some special needs for lightweight, complex, and very densely packed electronic equipment for aircraft and the space program creating a widespread trend toward the usage of multilayer PC boards [8]. Both the multiple layers and the packing of increasingly dense circuitry are required greatly for precise artworks. It takes more and more time

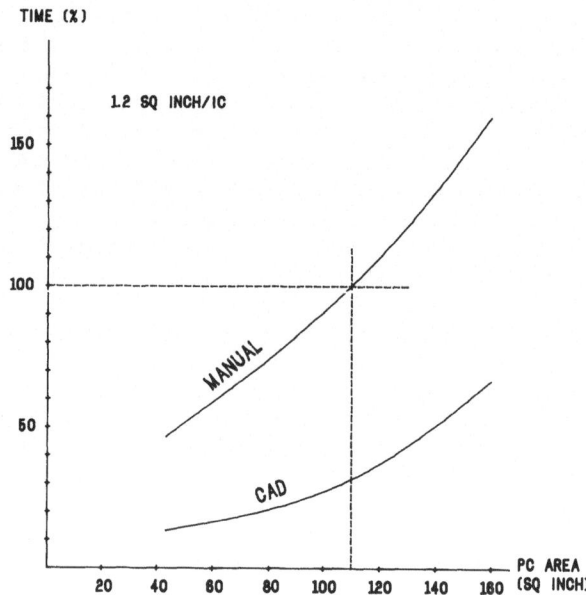

Figure 4. PC board design cycletime as a function of board size.

and high-paid talent to design these boards in a short turnaround time without sacrificing accuracy. Figure 5 shows how the average time required varies with IC density in PC board. Commonly, accuracy and resolution of 0.001" are required. To be off by a mil or two is to be out of specifications. In CAD system the designer created a digital form database, the X-Y positional data of physical components symbols, and retrieved it whenever the designer needed. The final output of photoplotted artworks masks, NC driller tape and other manufacturing aids are all derived from the same digital database. The corresponding accuracy and resolution of finished PC board are consequently guaranteed within 1 mil [9].

The manual design involves hand-drawn interconnection which is subject to high human error. It takes higher and higher probability of incurring a random defect as the current trend of larger, greater circuit density PC board. The spiraling difficulty of detecting and debugging the errors is another headache problem. CAD system has offered available packages to detect interconnections errors, path width or spacing problems. The designer could correct violations interactively in CRT screen which has indicated the positions of errors, if any. The results of free from human caused error are immediately finished. Figure 6 shows the errors caused by manual and by CAD method in the first implementation. In addition, the conventional method has practical difficulty in implementation of making even a small engineering change in a very

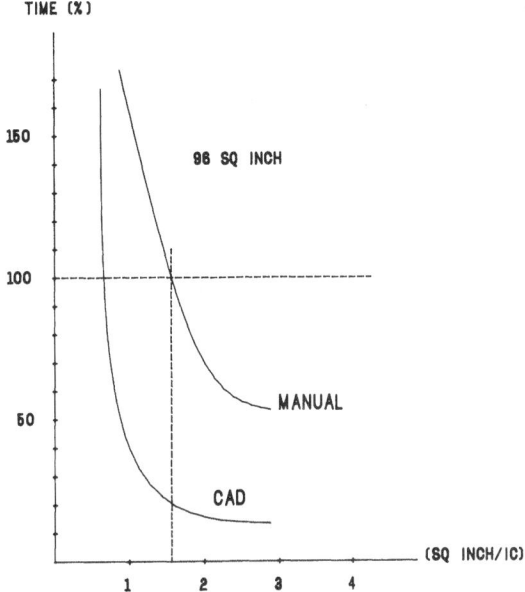

Figure 5. Time required varied with IC density.

complicated PC board and fully update all related documentations. It is obvious that the circuit engineers could not develop a mature electronic product directly without even an engineering change. The documentation updating followed is quite time-consuming and subject to very high human error. And it is a very serious problem for those R&D oriented institutes which may have many frequent

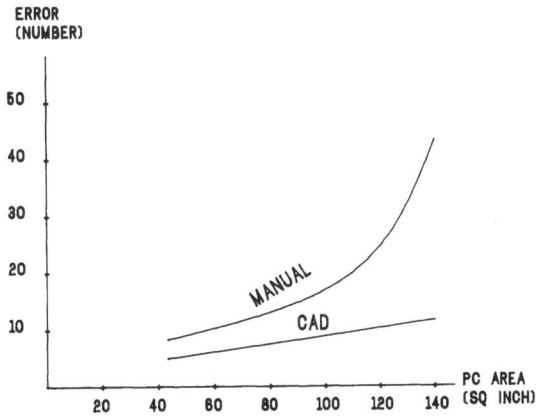

Figure 6. First implementation error.

engineering changes, especially in design the complex communications instruments. CAD systems assist the engineers to interrogate database to acquire the necessary information for engineering reviews, by a highly efficient means of correcting digital form database related to engineering change, if necessary and subsequently generate the updating documentations. The average modification portion of manual and CAD method during development is shown in Figure 7.

PHYSICAL CREATION

The creation of physical symbol is a small step in CAD approach, but one of important keys to cost-effectiveness. The frequently used symbols are stored as library parts and several basic stored symbols are shown in Figure 8. They can greatly simplify the preparation of symbol in other PC board design tasks by directly calling from libraries. The combination or modification of existing stored symbol is another advantage which allows the designer to get the desired symbol in a minimum time. The menu in digitizing tablet is a powerful facility for entering into symbol creation and retrieval. The text annotation can be assigned with symbol in preparation by layering technique. It is a very time-saving way that the designer can call up the annotations associated with the library parts without inserting text every time. Figure 9 shows a typical procedures in creating physical component symbol.

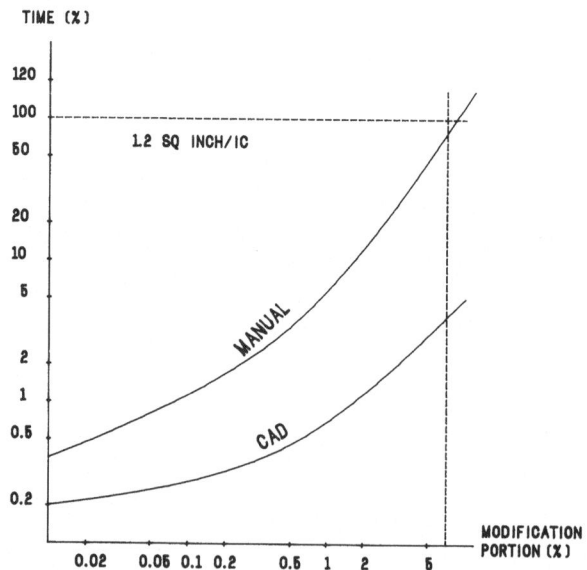

Figure 7. Modifications portion during development.

CAD APPLICATION

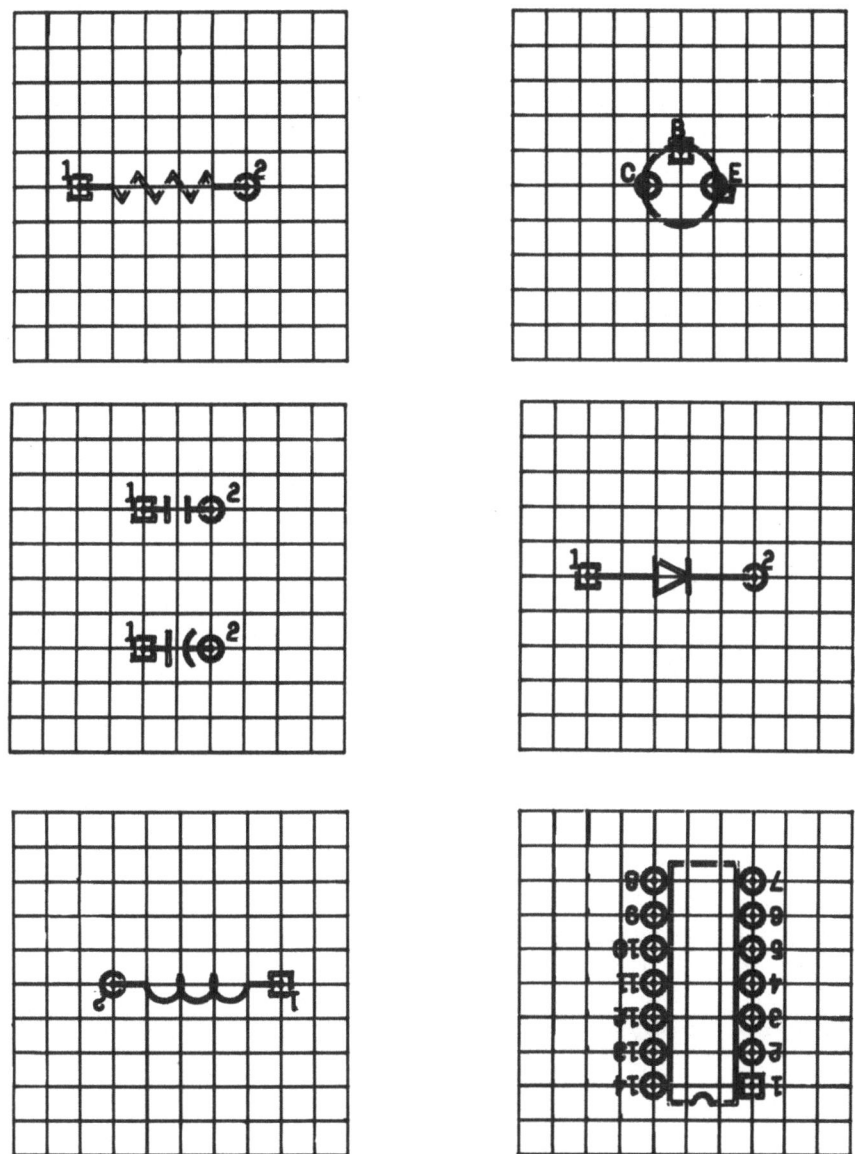

Figure 8. Basic stored symbol.

Generally, the reference codes in sequential order, such as R1, R2, R3 for resistors which are particular to different PC boards will be placed individually after placement.

PC board geometry is another physical creation that should be created by tablet with immediate visual feedback. The designer draws the PC board outline and other mechanical details, including

DISCRIPTION	COMMAND	GRAPHIC SYMBOL	REMARKS
INSERT CONNECTION NODE (CNOD) OF COMPONENT	INS CNOD		CNOD IS A PROPERTY FOR ROUTING CONNECTIONS BETWEEN COMPONENTS
INSERT FIGURES OF PADS	INS FIG	 ⊙ ⊙ ⊙ ⊙ ⊙ ⊙	FIG IS PHYSICAL PADS OF COMPONENT PIN
INSERT TEXT NODE OF EVERY PADS	INS TNOD	 .⊙ ⊙. .⊙ ⊙. .⊙ ⊙.	TEXT NODE SPECIFY THE LOCATION FOR PLACING TEXT
INSERT NODE TEXT TO SPECIFY THE PIN NUMBER OF COMPONENT	INS NTXT	 1.⊙ ⊙6 2.⊙ ⊙5 3.⊙ ⊙4	
INSERT THE STRING OF COMPONENT OUTLINE	INS STG	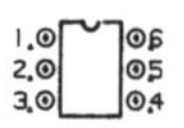 1.⊙ ⊙6 2.⊙ ⊙5 3.⊙ ⊙4	

Figure 9. Typical symbol creation procedure.

mounting hole, keep-out area, and other constraints. Similarly, a family of boards using the same board outlines, such as in a rack assembly, which has standardized size recommended by DIN 41494, can be stored as library parts.

NET LIST EDITING

The net list can be extracted logically or edited manually. In Primary Process, editing net list is quite an easy way. By following the schematic diagram, we can get the required textual data to build a computer database. The net list indicates the signal interconnection between related elements. Both the device code and its pin number should be clearly specified. Figure 10 shows an example of net list. For the sake of reducing complexity of subsequent signal routing task, the designer should very carefully take the reasonable pin assignments within any IC packages and the package assignment of logic family.

CAD APPLICATION

```
045 A0      U2-18 U4-3 U10-10 U17-10 U21-9 U22-18 U23-34 U24-6
046 A1      U2-16 U4-6 U16-9 U17-9 U21-5 U22-19 U23-33 U24-5
047 A2      U2-14 U4-10 U16-8 U17-8 U6-1
048 A3      U2-12 U4-13 U16-7 U17-7 U6-2
049 A4      U2-9 U5-3 U16-6 U17-6 U6-3
050 A5      U2-7 U5-6 U16-5 U17-5 U35-10
051 A6      U2-5 U5-10 U16-2 U17-4 U35-9
052 A7      U2-3 U5-13 U35-13 U16-3 U17-3
053 A8      U1-38 U4-2 U16-25 U17-25
054 A9      U1-39 U4-5 U16-24 U17-24
055 A10     U1-10 U4-11 U16-21 U17-21
056 A11     U1-1 U4-14 U16-23 U17-13
057 A12     U1-2 U5-2 U16-2 U17-2
058 A13     U1-3 U5-5 U30-1 U34-10
059 A14     U1-4 U5-11 U34-13
060 A15     U1-5 U5-14 U34-12
061 D0      U1-14 U7-2 U15-18 U16-11 U17-11 U21-19
062 D1      U1-15 U8-2 U15-16 U16-12 U17-12 U21-12 U22-26 U23-1
            U24-20 U25-4
063 D2      U1-12 U9-2 U15-14 U16-13 U17-13 U21-1 U22-27 U23-39
            U24-1 U25-7
064 D3      U1-8 U10-2 U15-2 U16-15 U17-15 U21-40 U22-28 U23-2
            U24-40 U25-8
065 D4      U1-7 U11-2 U15-9 U16-16 U17-16 U21-39 U22-1 U23-38
            U24-39 U25-14
066 D5      U1-9 U12-2 U15-7 U16-17 U17-17 U21-38 U22-2 U23-3
            U24-38 U25-14
067 D6      U1-10 U13-2 U15-5 U16-18 U17-18 U21-3 U22-3 U23-37
            U24-3 U25-17
068 D7      U1-13 U14-2 U15-3 U16-19 U17-19 U21-2 U22-4 U23-4
            U24-2 U25-18
069 A0-     U4-4 U7-5 U8-5 U9-5 U10-5 U11-5 U12-5 U13-5 U14-5
            U15-5 U16-5
070 A1-     U4-7 U7-7 U8-7 U9-7 U10-7 U11-7 U12-7 U13-7 U14-7
            U15-7 U16-7
071 A2-     U4-9 U7-6 U8-6 U9-6 U10-6 U11-6 U12-6 U13-6 U14-6
            U15-6 U16-6
072 A3-     U4-12 U7-13 U8-12 U9-12 U10-12 U11-12 U12-12 U13-12
            U14-12 U15-12
073 A4-     U5-4 U7-11 U8-11 U9-11 U10-11 U11-11 U12-11 U13-11
            U14-11 U15-11
074 A5-     U5-7 U7-10 U8-10 U9-10 U11-10 U10-10 U12-10 U13-10
            U14-10 U15-10
075 A6-     U5-9 U7-13 U8-13 U9-13 U10-13 U11-13 U12-13 U13-13
            U14-13 U15-13
077 A7-     U5-12 U7-9 U8-9 U9-9 U10-9 U11-9 U12-9 U13-9 U14-9
            U15-9 U16-9
081 +5V     R51-1 R52-1 R53-1 R54-1 R55-1 R56-1 R57-1 R58-1 R59-1
            R60-1 R61-1
082 +65     P-77 P-78 R63-1 R64-1
083 -12V    P-79 P-80 P-81 R68-1 R69-1
084 +12V    P-83 C11-1 C12-1 C13-1 C14-1
085 GND     P-85 P-86 A1-4 A2-4 A3-4 A4-4 A5-4 A6-4 A7-4 A8-4
            A9-4 A10-4 A11-4
```

Figure 10. Example of net list.

Generally, it is the responsibility of circuit design engineers, who have the idea of functional requirements of schematic circuit, to release a sketch of layout to meet the engineering constraints, with a matching package assignment, to acquire the optimum routing subsequently with no or little interaction by the designer.

The net list, having annotated and generated, should be approved by circuit design engineer who makes final confirmation on circuit, detects human error, and corrects immediately, if any. The textual database of net list will be used to check the relations between schematic diagram and PCB physical connections after PCB design is completed.

INITIAL PLACEMENT

The strategy of handling component placement on a PC board is to minimize total wire length. The designer should consider the alternative package, pin assignment in previous task of net list editing to get symmetry and even flow of signal paths to optimize subsequent routing results.

The physical symbols should be placed according to the function of circuit and the dependency of neighboring components. Generally, the components having close relation with connector are always treated prior to other electrical constraints. This task almost decides whether the PC board design is acceptable or not, it is a matter of trial and error, always requiring many interaction before a satisfactory placement is obtained.

The initial placement always needs interactively to operate on CRT. A successful configuration design should cooperately be accomplished by PC board designer and schematic circuit design engineer, with a depiction of component placement on a finer inch grid paper under the considerations of conceptual data, mechanical requirements, thermal problems, and sensitive circuit limitation. This is the basis to get through the task quickly. In primary process, this task is the crucial step for converting graphical data into digital form data (refer to Figure 3). Any error induced here will cause unrecovery failure in subsequent tasks.

RATSNET CHECKING

In the primary process by DIS NET, a ratsnet checking on CRT is a very useful tool to aid the designer to display the complexity of net list on PC board and to prove the initial placement which is acceptable. There are many cross-hair lines to indicate the interconnection relations of net list between IC chips. Figure 11 shows a pattern of ratsnet display on CRT screen. An even flow with no abnormal congestion in small area could assure the signal routing results. The designer could improve the component placement or assignment of package interactively with the system, especially with the help of circuit engineer who has more whole conceptual data with the circuit and more idea of constraints.

Ratsnet displays a very rough cross-hair pattern which simply shows the number of interconnection lines between IC chips, but it is sufficient configuration data to allow the designer to decide whether the currently initial placement is acceptable or not. Its

CAD APPLICATION

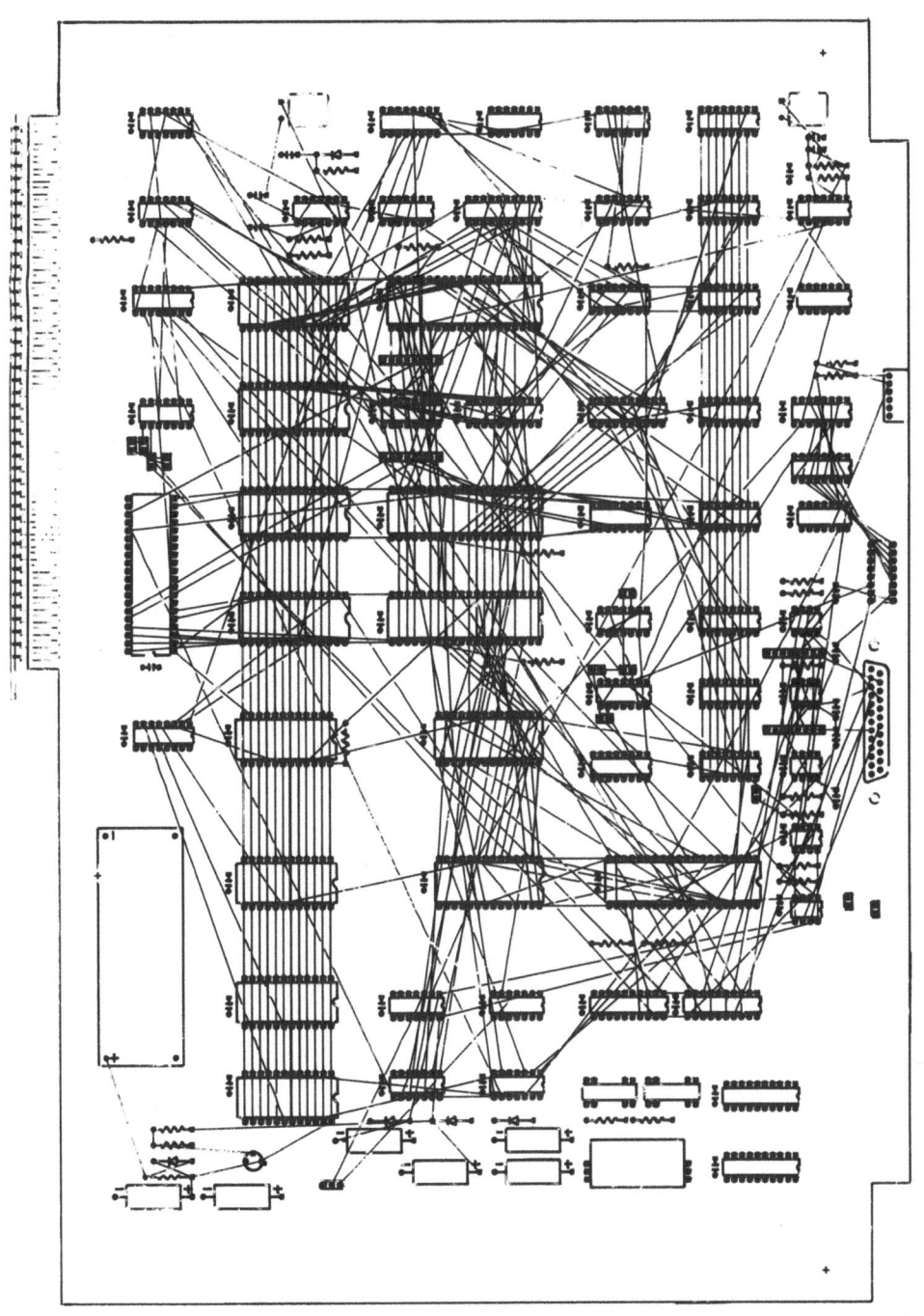

Figure 11. Ratsnet display.

cross-hair should be even, uniform with litter interlacing and no congestion.

The PC board designer should study the close relations between ratsnet and automatic signal routing. If designer can arrange the number of allowable paths to meet required routes and to get optimum placement by the help of computer-aid ratsnet display, the subsequent signal routing will be always succesful directly.

SIGNAL PREROUTING

The signal routing is the most time-consuming operation and lower-level repetitive task in manual PC board design. CAD system offers many features to assist the designer to accomplish signal routing. The most powerful computer aid is automatic router program. But there is no computer algorithm that can route the all interconnection to meet the engineering requirements. The power and ground path, address bus path, data bus path, and other paths of crucial circuit, such as clock, should be routing manually before autorouting.

All the dual-in-line IC chips on digital board have to be connected to the power and ground as a distribution system in grid form to reduce the coupling to signal path. The prerouting should primarily consider the engineering constraints which were released by engineer as a rough layout sketch.

The designers should take advantage of autorouting the partial signal as a prerouting to meet the engineering requirements and to reduce the changes due to improper routing. A suitable prerouting actually could make a higher completed ratio of autorouting and a direct-good routing.

AUTOROUTING

Computer-aided system offers automatic router program to accomplish signal interconnection more quickly and easily than manual design process. The strategy of autorouting includes maximum-percent of completion, minimization of through-holes, minimization of total conductor length, limitization of trace congestion. Before the operation, the engineering constraints, such as tool holes, keep-out areas, and other prerouting interconnection, are necessary to assign for meeting the original engineering requirements.

The designers should properly select the parameters under the consideration of electrical current, voltage, and the density of conductor to optimize the complexity of PC board. Generally, a complete ratio above 95% is optimally suitable, and the designer

CAD APPLICATION

always needs interactively to rearrange the placements to achieve the desired results, which having even flows of completed paths and leaving unconnection paths with available unused area for easy manual clean-up.

There is no router which has the algorithms that will route to the desired level all the time. A well-experienced designer, after reviewing the engineering concepts data, can get through the final autorouting task with little or no interaction. The designers should pay more attention to follow-up the relations between initial placement, ratsnet checking and autorouting to acquire a satisfactory signal route directly.

CLEAN-UP

There is editing function on CAD system to facilitate the additional signal routing not automatically completed by autoroute package. The editing command should interactively operate on raster CRT, which has partially erasing function, with the help of layer discrimination and color selection to allow the designer to handle the signal interconnection uncompleted. Even in inserting one uncompleted line, the designer always needs to change, i.e., move, other routes to assure the design requirements of PC board. The editing task is very time-consuming and human-dependent. It almost takes half of the full design cycle time, and many human errors are inevitably caused by eye fatigue.

The designer should be protected from fatigue caused by glare or reflection during interactively editing line on CRT. Since the color display used in PC board design always causes human eyes to be more uncomfortable, the reasonable workstation layout with suitable room lights will minimize the problem.

Before going to edit task on CRT screen, a sketch of additional routing, which is not completed by autoroute, should be prepared manually to make the clean-up task more quickly and more correctly. With the computer-aid editing function, line editing is essentially more accurate than manual taping method. The comparing net procedure within clean-up task is another computer-aid function, which can check out all the error cause by human operations during editing and assures the final PC board design is fully correct.

MANUFACTURING AIDS

In practical PC board, the dual-in-package IC should be texted in the sequential order by column and row. Before manufacturing-aids release, the designer should annotate the components reference code to corresponding matrix order.

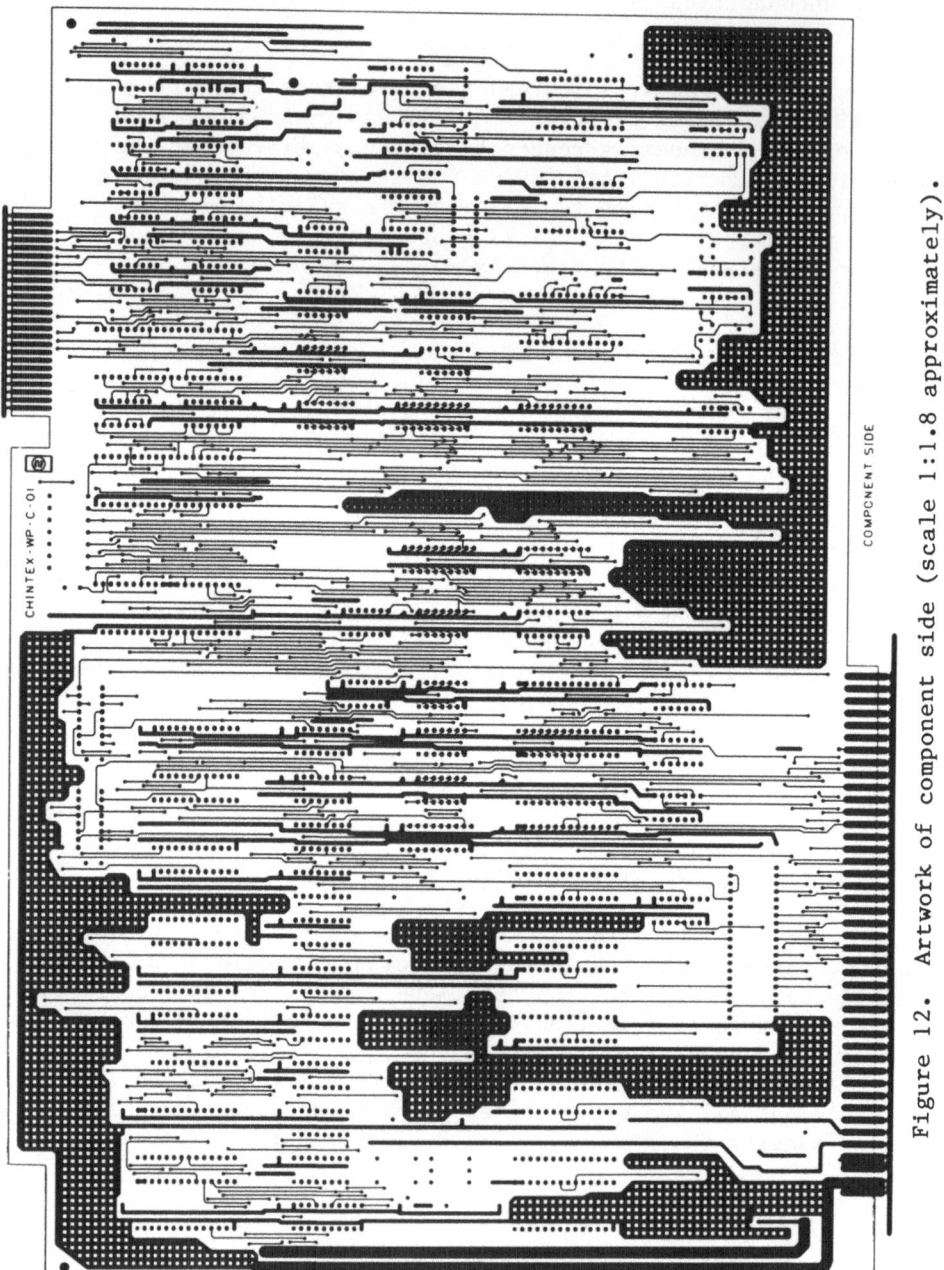

Figure 12. Artwork of component side (scale 1:1.8 approximately).

Manufacturing aids of artworks in Figures 12, 13, 14, 15 and driller tape is the most significant output of the CAD system at TL at this moment. The digital database, created previously in initial placement task, is the basis of PC board design and should be compatible with photoplotter and NC driller by the help of the full CAD system, including photoplotter and NC driller that are CAM facilities. The accuracy of the completed PC board, integrated with PC board fabrication which is one of the Model Shop selections at TL, has promoted to the range of one to two mil. The photoplotter being heat-sensitive equipment should be installed at air-conditioned room, and air filtration will protect flash head from dust damage.

There are many more useful and advanced manufacturing aids related to PC board design field, such as autotest and autoinsertion. The compatibility of these facilitlies with the CAD system installed will be surveyed and studied before setting up all of them as an integration system of CAD and CAM.

CONCLUSION

The primary process has shown a reasonable process that operator experience and intelligence in matching with the CAD system, can design PC board as perfect as technician or engineer in conventional method. We will continually develop the primary design process to integrate all the available functions that the CAD systems provided to PC board design and fabrication applications.

Our goal in CAD application on PC board design is that the operator eventually need not digitize anything, but extract the digital form data from the database which was initially edited by the engineer who designs electronic circuit interactively on work station of CAD system, which can perform a variety of functions: schematic recognition, circuit analysis and logic simulation. The final output of manufacturing aids should be all compatible with the CAM facilities in PC board fabrication and telecommunication instruments manufacturing which should accomplish other tasks of autowiring, autoinsertion and autotesting.

ACKNOWLEDGMENT

I would like to thank Dr. Shyue-Ching Lu, Mr. Wau-Kao Hu, Miss Ke and the CAD group members for their encouragement, help and support in writing this paper.

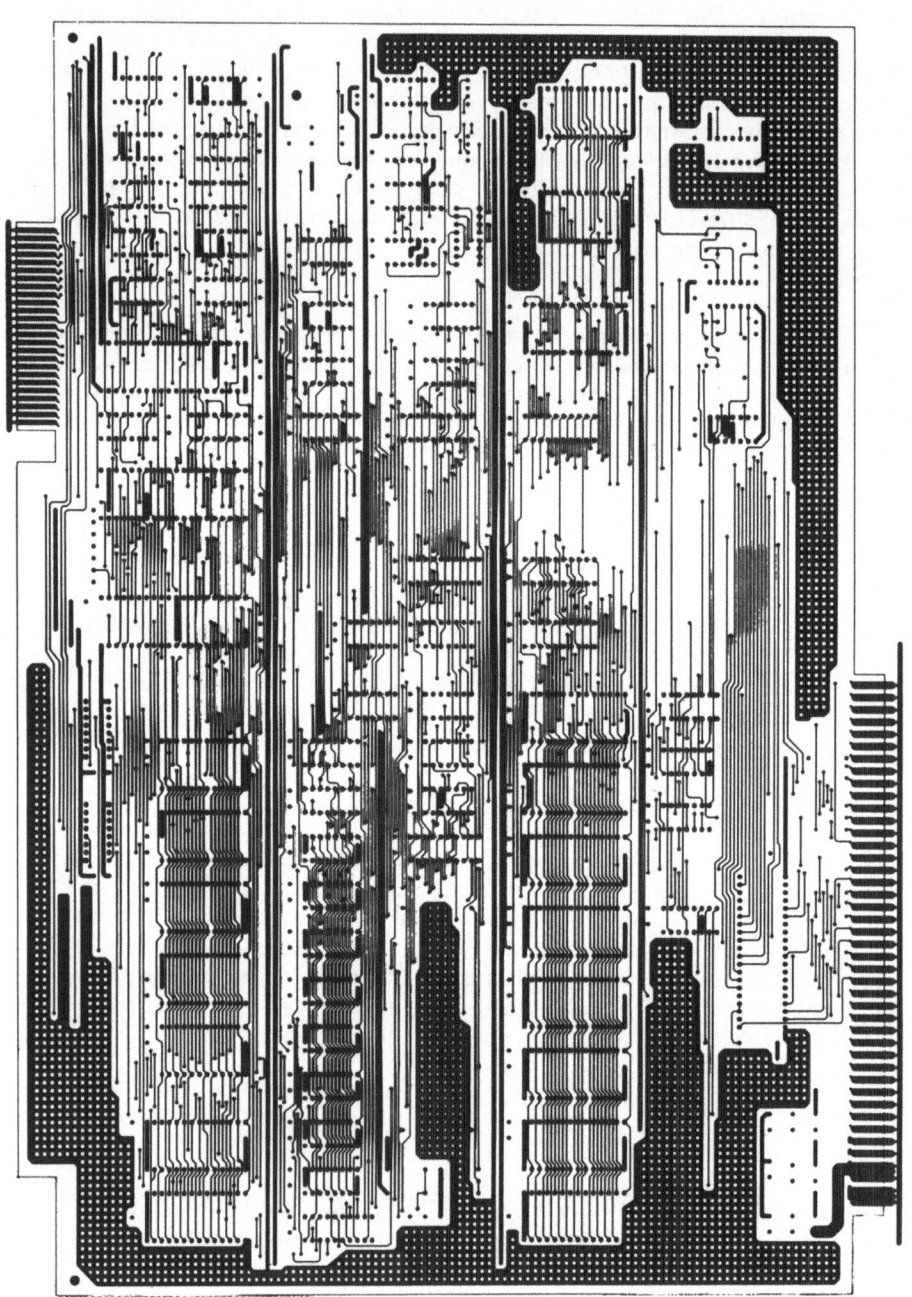

Figure 13. Artwork of solder side (scale 1:1.8 approximately).

CAD APPLICATION

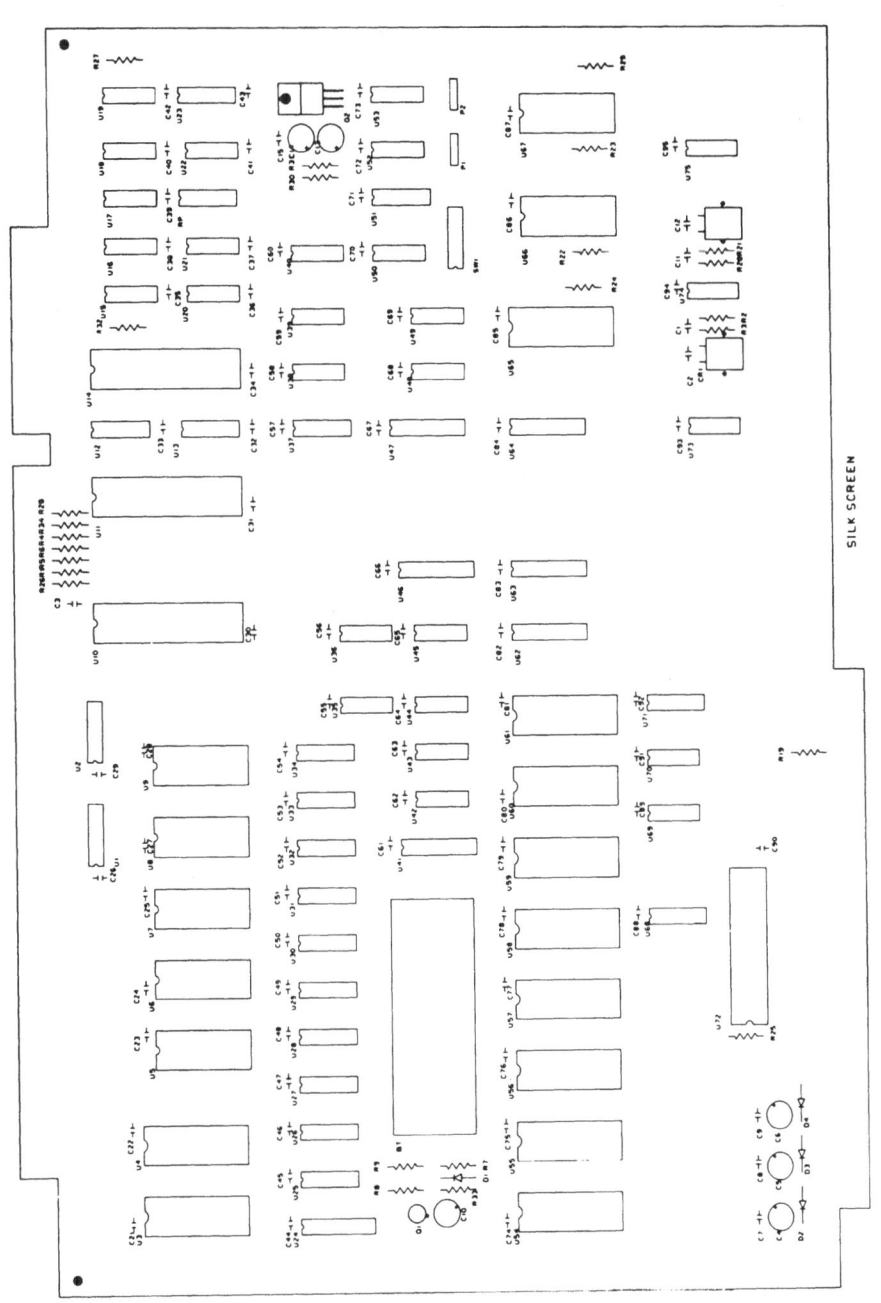

Figure 14. Artwork of silk screen (scale 1:1.8 approximately).

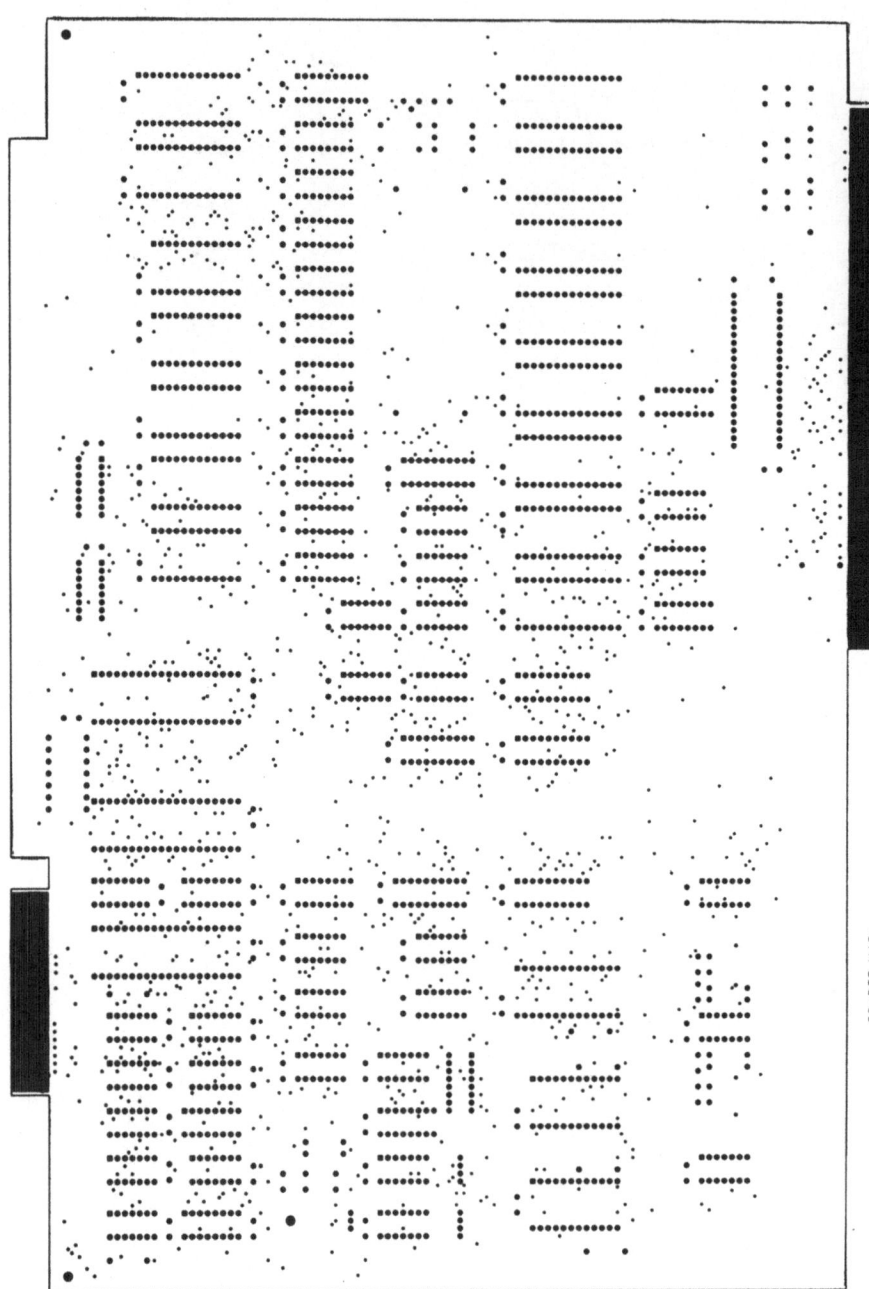

Figure 15. Artwork of solder mask (scale 1:1.8 approximately).

CAD APPLICATION

REFERENCES

[1] D. R. J. White, EMI Control in the Design of Printed Circuit Boards and Backplanes, Don White Consultants, Inc., pp. 3.6-3.17, 1981.

[2] Ing. Francesco Palazzi, PC/ES Applications in a Factor of the Ansaldo Group, European Computervision User Conference, Computervision Design System (H.K.) Ltd., Hong Kong, 1982.

[3] W. Macleod, CV/Redac Printed Circuit Board Interfacing, Fifth Annual International Computervision User Conference, Computervision Corp., Bedford, MA, 1984.

[4] Wilson, Autoplacement at GTE, PC/ES International Computervision User Conference, Computervision Corp., Bedford, MA, 1983.

[5] P. Lund, Generation of Precision Artwork for Printed Circuit Board, John Wiley & Sons, New York, p. 290, pp. 305-315, 1978.

[6] T. I. Ristine, PC Board Design and Documentation, Computervision Corp., Bedford, MA, 1980.

[7] News, Digital Television Lightware System Will Bring Olympic Broadcasts to the World, Bell Laboratories Record, Sept. 1983.

[8] M. Servit, Multilayer Interconnection Problem Complexity, CAD 82 Conference, Brighton, UK, 1982.

[9] K. Dables, Manager's Seminar, Computervision Corp., Bedford, MA, 1980.

SPICE-2P: A SPICE-2 INCORPORATED WITH A PARTITION SCHEME

Chung Len Lee, Wen Zen Shen, Ching An Liaw, and
Chein Wei Jen

Institute of Electronics
National Chiao Tung University
Hsin Chu, Taiwan, R.O.C.

ABSTRACT

SPICE-2, the circuit simulation program, is modified by incorporating a partition scheme with which a large circuit can be decomposed into several smaller but cascaded subcircuits, and each subcircuit is simulated separately in sequence with the outputs of the preceding subcircuit to be the inputs of the next succeeding subcircuit. This effectively changes the SPICE-2 which is a time incremental circuit simulator into a time waveform circuit simulator. Examples of circuits simulated by this SPICE-2P show significant savings on the simulation time (up to 50%) and the memory utilized (up to 80%). The loading effect introduced due to the circuit partitioning is also studied. A simple model for the loading effect is used and implemented in the program. Less than 5% of error on the simulated waveforms are usually obtained.

INTRODUCTION

The CPU time for circuit simulation usually increases more than linearly with the size of the circuit. It has been shown that the increase of the simulation time is proportional to N^m, where $1.2 \leq m \leq 2$ and N is the number of nodes [1,2]. As the circuit size increases, the time and memory requirements for a conventional circuit simulation become prohibitive. As a result, many talented professionals have been striving to find solution for "economic" and realistic simulation. Recent years, many techniques have been invented: Among them, there are 1) special purpose program - special technique is developed for specific type of circuit to efficiently perform the analysis [3]; 2) temporal sparsity technique - taking full advantage of digital circuit characteristics to identify the temporal and structure sparsity in constructing the program [4], the CPU time and memory storage may be saved significantly; and 3) time waveform technique - instead of using the conventional time incremental technique, the circuit is partitioned into subcircuits and every subcircuit is simulated at its optimum time step [2,5].

In this work, the conventional SPICE-2 [6], which is a time incremental circuit simulation program, is modified by incorporating with a partition scheme to be time waveform program. In the program, an interconnection subroutine is added so that the output waveforms of the preceding subcircuit can be automatically fed into the succeeding subcircuit as its input. In this way, the time waveform for each subcircuit is integrated independently, and the total time waveform is obtained by applying the analysis to each subcircuit in sequence. In considering the accuracy of the simulation of this program, the loading effect of each subcircuit to the preceding subcircuit is modelled by an effective capacitance which takes into account the gate oxide capacitance, the overlap capacitance and the stray capacitance. The inclusion of this effective capacitance makes the simulation results of this program differ from those obtained by the conventional SPICE-2 by less than 2%.

Many circuit examples simulated by this program have shown that one third to one half of simulation time and up to eighty percent of memory storage, depending on the partition scale of the subcircuits and complexity of circuits, can be saved.

PARTITION OF A GENERAL CIRCUIT

A Partitioning Model

Figure 1 is a general model for a circuit simulation, where U(t) and Y(t) are the inputs and outputs of the circuit, respectively. To describe the "connection" and "state" of the circuit, a circuit variable (node voltage or branch current) matrix is required in the simulator to be solved during the process of simulation. As the size of the circuit is large, the size of matrix becomes correspondingly large, and the solving of it

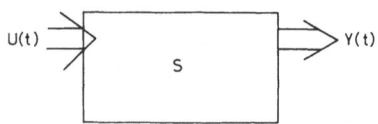

Figure 1. A general model for circuit simulation.

becomes extremely time-consuming. Hence, it is natural to consider
to break the circuit into several smaller subcircuits and to simulate each subcircuit separately. This makes the time for computing
and solving the matrix decrease drastically. However, all the
elements of the circuit are "coupled", namely, the "state" of one
element influences the "state" of the other element. When a circuit is broken into several subcircuits, the effect of this "coupling" should be considered. Figure 2 shows a "Decoupled Model" of
the circuit of Figure 1. In this partitioning scheme, the large
circuit S is decoupled into n small subcircuits S_1, S_2, ..., S_n.
In the figure, $W_i(t)$, $U_i(t)$, $O_i(t)$ and C_{ei} are defined as follows:

- $W_i(t)$: the state and output variables of the S_i subcircuit, which will be fed into the next succeeding subcircuit S_{i+1};

- $U_i(t)$: the input of the ith subcircuit S_i except the $W_{i-1}(t)$;

- $O_i(t)$: the output of the S_i subcircuit;

- C_{ei}: the effective loading capacitance of the succeeding subcircuits for the ith subcircuit.

The "coupling" effect, due to the succeeding subcircuits, for a
subcircuit S_i is lumped into an effective capacitance C_{ei}. With
the inclusion of this C_{ei}, each subcircuit can be simulated independently, that is, each subcircuit can be simulated with its own
incremental time step and time period. The output of one subcircuit is the input of the next subcircuit and the waveform of overall circuit is the output the subcircuit of the last stage. From
the above description, this is the time waveform simulation.

Loading Effect Consideration - The Effective Capacitance

The loading effect of each subcircuit can be demonstrated as
in Figure 3, where stage 1 could be one of the output stages of any
subcircuit and stage 2 be one of the input stages of any other

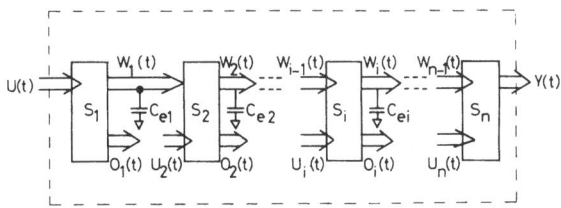

Figure 2. A decoupled model after partitioning the circuit of
Fig. 1 with the loading effect taken into consideration.

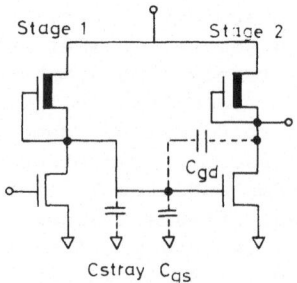

Figure 3. The loading capacitances of an inverter.

subcircuit. The loading effect to the stage 1 is the input capacitance of stage 2, namely, $C_{GS}+2C_{GD}+C_{stray}$, [7] where C_{stray} is the stray capacitance and C_{GS}, C_{GD} are defined in the figure. In the above, a factor 2 is added because of the Miller effect of the MOS transistor amplifier stage. Since an MOS transistor is an excellent buffer, only one stage needs to be considered. For an MOS transistor, the associated capacitances can be shown in Figure 4, where C_1 and C_3 are overlap capacitance and C_2 is the gate oxide capacitance. They can be calculated as following, respectively:

$$C_1 = C_3 = C_{ox} \times W \times L_D \quad \text{and}$$

$$C_2 = C_{ox} \times W \times (L-2L_D)$$

where C_{ox} is the unit area oxide capacitance, W is the channel width, L is the channel length and L_D is the overlap length. From ref. [8], the gate-to-substrate capacitance C_{GB}, the gate-to-drain capacitance C_{GD}, the gate-to-source capacitance C_{GS} for an MOS transistor are functions of the gate-to-source voltage V_{GS}. Besides, for most of the cases, an MOS transistor usually operates in the saturation region, so we can assume $C_{GS}=C_1+\frac{2}{3}C_2$, $C_{GD}=C_1$,

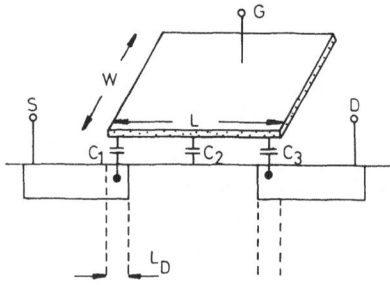

Figure 4. Capacitances in an MOS transistor.

respectively. Thus, the total effective loading capacitance can be expressed as

$$C_{eff} = C_{stray} + C_{ox} \times W \times (\frac{5}{3} L_D + \frac{2}{3} L)$$

Due to including this effective loading capacitance, which accounts for the interaction between linked subcircuits, the complete decoupling among subcircuits can be done.

General Rules for Partitioning

The following are the general rules for partitioning a circuit:

1) When a circuit is partitioned into several subcircuits, each subcircuit should form a separated connected-graph.

2) When there is a repetitive substructure in a large circuit, pick up the substructure circuit as a subcircuit.

3) Identify tightly coupled feedback loops in a large circuit. Every feedback loop should be included into one separate subcircuit.

4) When a subset of a large circuit can behave itself as an independent function, pick up that subset circuit as a subcircuit.

5) After the circuit is partitioned and levels are assigned to each subcircuit, some outputs (at least one, may be more than one) of the preceding level subcircuit must be the inputs of the succeeding level subcircuit.

IMPLEMENTATION OF SPICE-2P

There are two major modifications on the original SPICE-2 program. The first one is on the OUTPUT overlay. It is modified to be able to store the output variables of the subcircuit of the preceding level as user specified. The specified output variable data are saved in two arrays, VALUEX and VALUEY, which are added in common block /BLANK/. The second one is adding a subroutine, MSORUPD, into DCTRAN overlay so that it may feed the internal output variable vector W_i of the subcircuit of the preceding level into the subcircuit of the succeeding level as its input. As an user specifies, SORUPD subroutine can call another subroutine MSORUPD to update the internal input variable vector of the next stage subcircuit. An interpolation technique is used to obtain the input waveforms for each respective time point when unequal time steps are required for each subcircuit. MSORUPD is the subroutine that interacts with the OUTPUT overlay through a common block /BLANK/ to transfer the data.

The flowchart in Figure 5 illustrates how these two subroutines (the dashed line box) are added.

RESULTS AND DISCUSSIONS

This section presents the results of a number of circuits simulated by SPICE-2P program. These results are compared with those which are obtained on the same circuit but simulated with the conventional SPICE-2. All the simulations are run on a CDC CYBER 170/720 computer with NOS-Network Operating System.

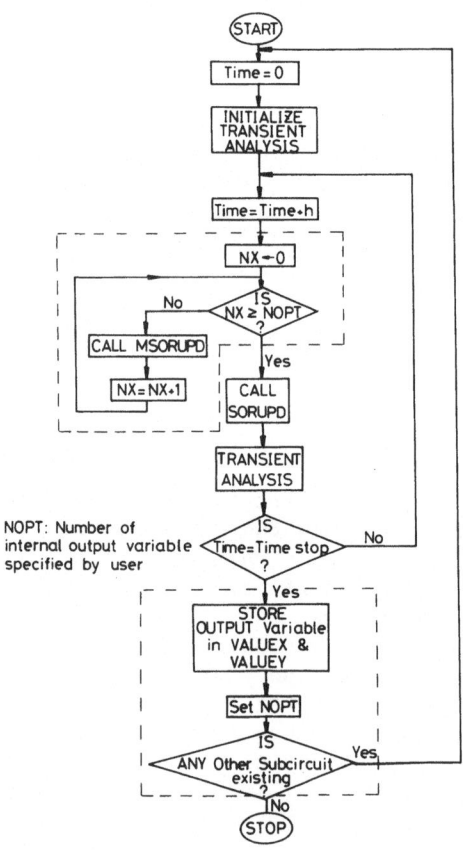

Figure 5. The flowchart to demonstrate where the SPICE-2 is modified.

The first circuit example is a 12-stage NMOS inverter cascaded in series (Figure 6). This is an ideal circuit to demonstrate how simulation time can be saved when different ways of partitioning are applied to the circuit. The circuit can be partitioned to be INV1×12, i.e., one inverter is a subcircuit but simulated in 12 stages, or INV2×6, or INV3×4, OR INV4×3, OR INV6×2, OR INV12×1. For the last case, the circuit is not partitioned but simulated as a whole, i.e., the conventional case. Table 1 summarizes the results on simulation time and memory used for each case. It is seen that the smaller the subcircuit is, the more saving on simulation time and on memory used is obtained. For the INV1×12 case, up to 48% saving in simulation time and 84% saving in memory size are obtained. This is a significant saving.

To illustrate the simulation accuracy for this circuit by SPICE-2P, the output pulse waveforms of the last stage for each case are shown in Figure 7. All the output waveforms for five cases are almost identical. This indicates that SPICE-2P gives simulation results good enough to be comparable with those obtained by the conventional SPICE-2.

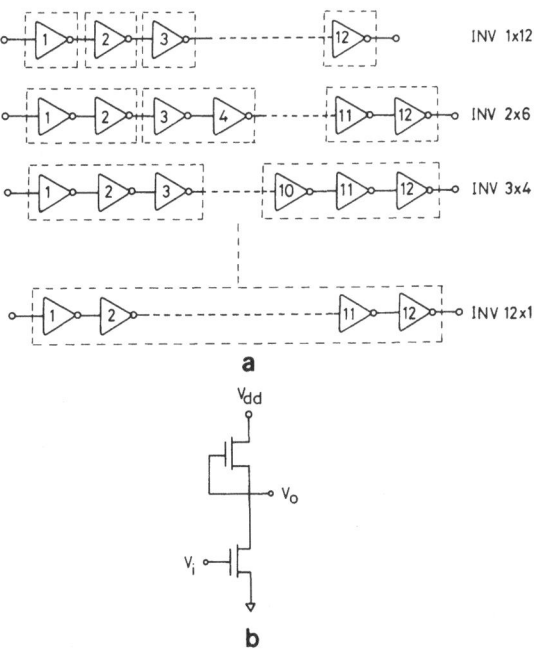

Figure 6(a) An NMOS inverter cascaded in 12-stage simulated with various ways of partitioning.

(b) The detailed circuit for a single stage.

Table 1. The simulation time and memory used for a 12-stage NMOS inverter train by using SPICE-2P with various partitioning (A~E) and by using a conventional SPICE-2 (F)

Partitioning type	Simulation time (CPU time in sec)	Memory used	Time saved %	Memory saved %
INV1 × 12 A	94.92	1573	(F-A)/F 47.69	(F-A)/F 83.60
INV2 × 6 B	97.29	2080	(F-B)/F 46.39	(F-B)/F 78.39
INV3 × 4 C	107.91	2653	(F-C)/F 40.54	(F-C)/F 72.43
INV4 × 3 D	112.94	3270	(F-D)/F 37.76	(F-D)/F 66.02
INV6 × 2 E	139.97	4620	(F-E)/F 22.89	(F-E)/F 52.09
INV12 × 1 F	181.47	9624	(F-F)/F 0	(F-F)/F 0

To illustrate how the loading effect of inverters with various driver/load ratios affects the simulation accuracy, Figure 8(a) to Figure 8(e) show the output pulse waveforms of the circuit of Figure 7(a), partitioned into 12 subcircuits, with loading capacitance taken into account (the "o" dot curves) and without the loading capacitance taken into account (the "w" dot curves). In figures, the results of the circuit simulated with the conventional SPICE-2 program are also included (the "*" dot curves). It is seen that for all the cases, the pulse waveforms of those with loading capacitances taken into account agree with the pulse waveforms of those simulated with SPICE-2. The pulse waveforms for those without the loading capacitances taken into account deviate significantly from the SPICE-2 waveforms. From these results, it can be concluded that the loading capacitance is very important in determining the simulated accuracy in waveforms.

In the figures, especially for Figure 8(b) and Figure 8(e), larger discrepancies are observed in SPICE-2 waveforms with respect to SPICE-2P waveforms. This indicates that the model in calculating the effective loading capacitance is rather simple. A more accurate approach should be that the loading capacitance is computed in terms of its variation with respect to the output voltage.

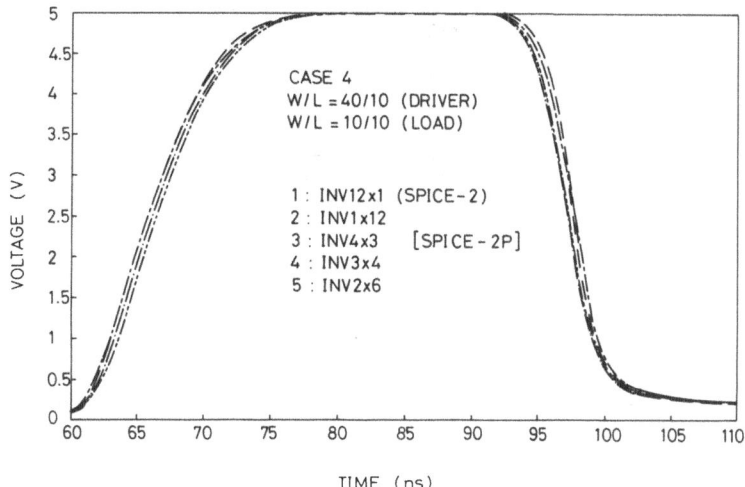

Figure 7. Output pulse waveforms for various partitioning of the circuit of Fig. 6.

Other examples run by the SPICE-2P are some circuits typically used in integrated circuit. They are (1) 1 of 4 decoder, (2) 2-bit shift register, (3) 3-stage flip-flop, (4) 4-bit full adder, (5) random logic, and (6) output pad-driver circuit. They have been simulated by SPICE-2P and SPICE-2, respectively, and the results are summarized in Table 2. In average, 43% of simulation time and 62% of memory size are saved by using SPICE-2P compared to by using SPICE-2. The simulated output responses by SPICE-2P and SPICE-2 for each circuit have also been compared. It is found that for all the cases, two groups of curves differ by less than 5%.

CONCLUSION

In this work, we have modified the conventional SPICE-2 into a new version: SPICE-2P, which employs a scheme to partition a large circuit into several small size subcircuits and each subcircuit is simulated in sequence to obtain the overall response. The partitioning of the circuit reduces the size of subcircuit which is to be simulated at one time and allows integration for each decomposed subcircuit to be performed independently at its own optimum time steps. The simulation time and memory needed for one simulation are reduced significantly. Hence, SPICE-2P is a time waveform circuit simulator which has the advantage of precision as a conventional time incremental SPICE-2. Circuit examples run by SPICE-2P have readily demonstrated this program can save simulation time up to 50% and memory size up to 80%. One potential merit in this program is that since relative smaller memory size is needed for running this program, it can be implemented on desktop computer like HP9845.

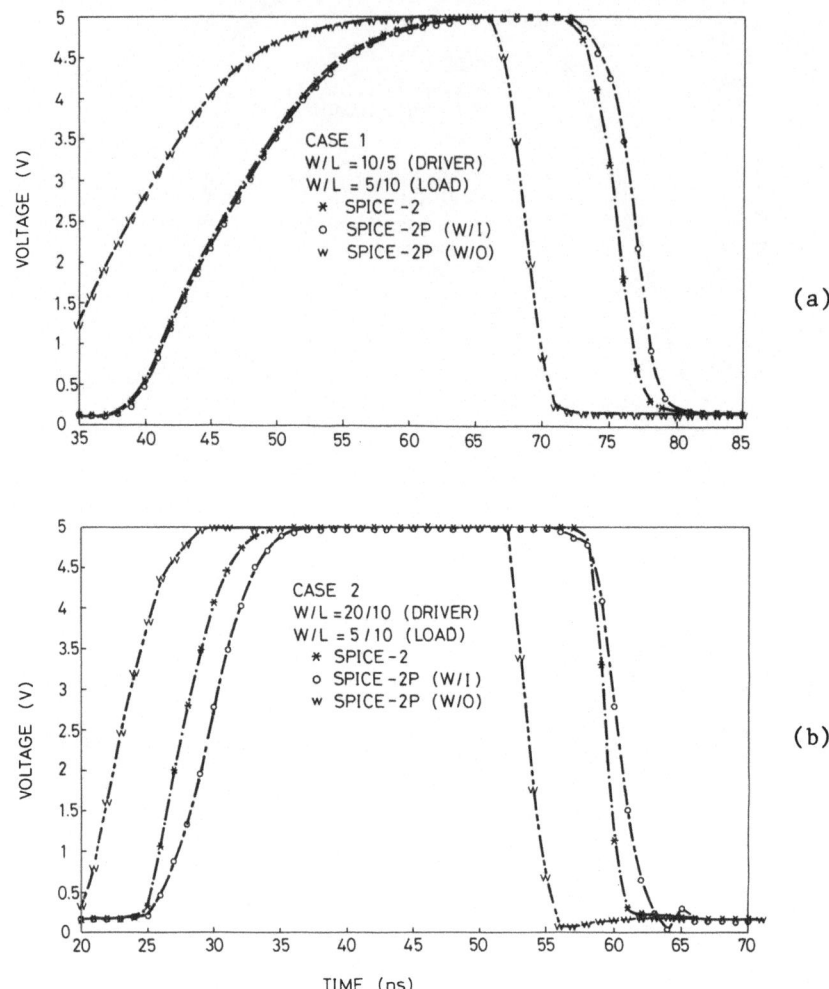

Figure 8(a) Output pulse waveforms for case 1: W/L = 10/5 (driver), W/L = 5/10 (load).

(b) Output pulse waveforms for case 2: W/L = 20/10 (driver), W/L = 5/10 (load).

REFERENCES

[1] D. J. Blattner, "Choosing the right programs for computer-aided design," Electronics, pp. 102-105, April 1976.
[2] A. E. Ruehli and A. L. Sangiovanni-Vincentelli, "The waveform relaxation method for the time domain analysis of large scale integrated circuits," ERL Memo No. ERL-M81/75, June 1981.

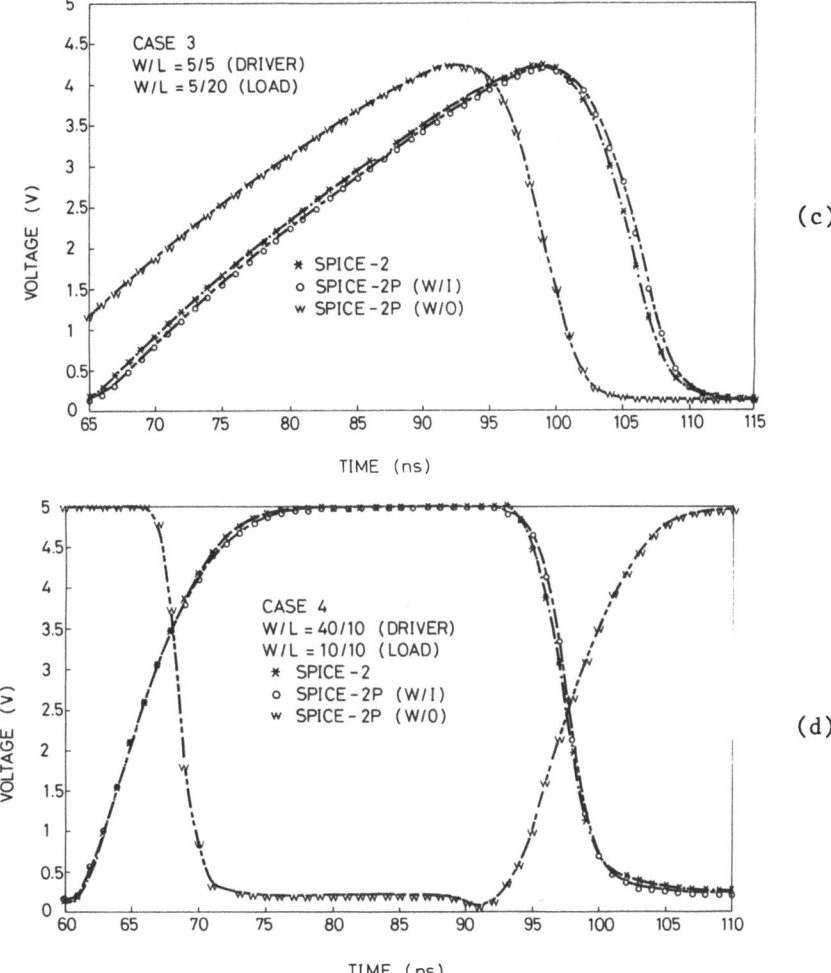

Figure 8(c) Output pulse waveform for case 3: W/L = 5/5 (driver), W/L = 5/20 (load).

(d) Output pulse waveform for case 4: W/L = 40/10 (driver), W/L = 10/10 (load).

[3] H. De Man, J. Rabaey, G. Arnout, and J. Vandervalle, "DIANA as a mixed-mode simulator for MOS LSI sampled-data circuits," in Proc. IEEE Intl. Symp. on Circuits and Systems, Houston, TX, pp. 435-438, April 1980.

[4] N. Rabbat and H. Y. Hsieh, "A latent macro-modular approach to large scale sparse networks," IEEE Trans. Circuit Syst., Vol. CAS-22, pp. 745-752, Dec. 1976.

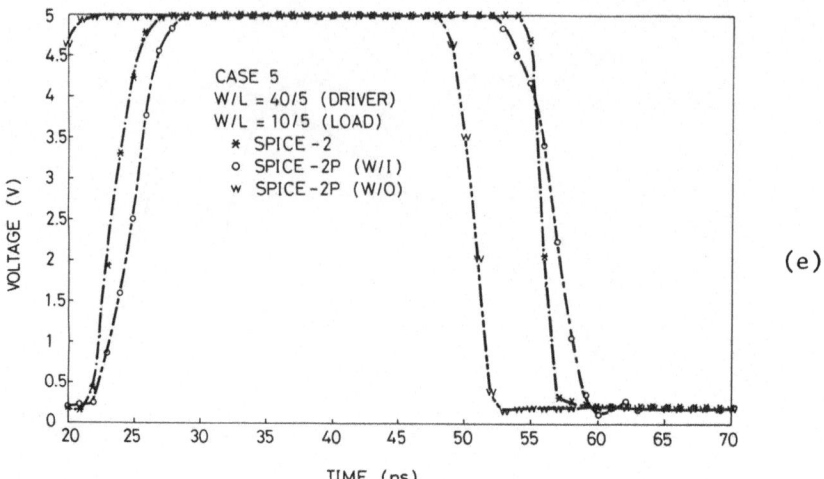

Figure 8(e) Output pulse waveform for case 5: W/L = 40/5 (driver), W/L = 10/5 (load).

[5] A. E. Ruehli, A. L. Sangiovanni-Vincentelli, and N. B. G. Rabbat, "Time analysis of large scale circuits containing one-way macromodels," IEEE Trans. Circuits Syst., Vol. CAS-29, pp. 185-189, March 1982.

[6] L. W. Nagel, "SPICE-2: A computer program to simulate semiconductor circuits," ERL Memo No. ERL-M520, University of California, Berkeley, May 1975.

[7] C. Mead and L. Conway, Introduction to VLSI Systems, Reading, MA: Addison-Wesley, 1980.

[8] R. S. C. Cobbold, Theory and Applications of FET's, Wiley-Interscience, pp. 272-304, 1970.

Table 2. A comparison of the simulation times and memories required for several circuits simulated by SPICE-2 and SPICE-2P, respectively

Circuit of	1 of 4 decoder	2-bit shift register	3-stage Flip-Flop	4-bit Full-adder	random logic	pad output driver
# of MOS devices	20	40	36	84	46	12
CPU time (in sec) of SPICE-2(A)	77240	876.19	625.97	1416.06	482.76	121.66
CPU time (in sec) of SPICE-2P(B)	325.97	569.32	448.91	589.90	237.59	86.18
memory used by SPICE-2(A)	12348	15336	14668	31892	15136	5896
memory used by SPICE-2P(B)	6772	5154	8253	8431	3509	1991
Time saved (A-B)/A %	57.80	35.02	28.28	58.34	50.97	29.16
memory saved (A-B)/A %	45.16	66.39	43.73	73.56	76.82	66.23

A CAD PROGRAM FOR VLSI PLACEMENT AND ROUTING

J. Y. Lee, J. M. Jou, H. T. Nian, C. Y. Chang, and
H. C. Wu

Institute of Electrical and Computer Engineering
National Cheng Kung University
Tainan, Taiwan, R.O.C.

ABSTRACT

In this paper, a set of algorithms for circuit layout is developed. Algorithms for automatic/interactive placement of standard cell blocks in parallel rows use the concept of blending and repartition. The grouping and linear placement techniques are also used. Constructive-initial placement is interlaced with iterative-improved placement for better results.

Routing is divided into two subproblems; global (loose) routing and channel (detailed) routing. Global routing is used to find a layout topology for the interconnection nets. Graph theory is used to model the layout. Channel routing is used to interconnect nets or subnets within a channel using some novel and constructive methods. Algorithms for both global and channel routing are proposed.

INTRODUCTION

With great advances in the technology of semiconductor fabrication and increasing demands for the sophisticated electronics system, the integration scale of an integrated circuit tend to ascend rapidly. The development of a VLSI circuit is expensive due to engineering design, simulation, layout, and extensive testing [1]. If development costs are to be kept low, efficiently automated or computer-aided IC design method must be used to reduce both the time and cost for custom circuit design [2].

A large, often dominant, part of the cost and time required to design an IC is consumed in the layout stage [1]. The standard cell layout scheme is widely used for MOS LSI's to simplify layout design by means of standard cell library [8,13]. It assumes a layout model as shown in Figure 1 in which a block may contain a logic gate or a set of gates. Various types of blocks can be defined, each characterized by its dimensions and the types of circuit elements belonging to it. Figure 1 shows two types of blocks, 'cell blocks' which can contain only standard cells, and "pad block" which can contain only chip I/O pads. Blocks are of almost the same height, but of distinct width.

Channel is a region between two adjacent rows in which 'cell blocks' are arranged and is used as working area for routing. There are a number of tracks in a channel on which the interconnection paths may lie.

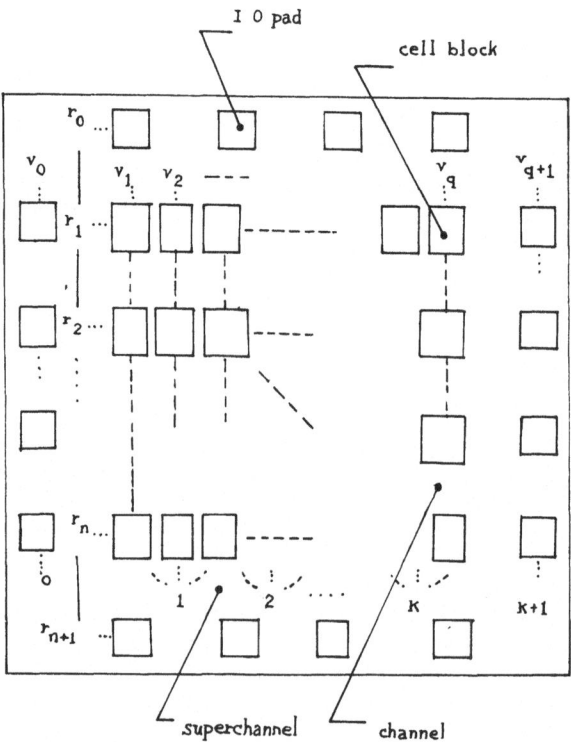

Figure 1. Standard cell layout model and two linear placement problems.

A CAD PROGRAM

The regions between cells and I/O pads are called superchannels on which some interchannel routing traffic are carried. These superchannels are treated similarly to other channels. The chip layout basically consists of the following three stages [3]:

- Topological Placement — to determine relative positions of blocks so that wiring nets do not waste much chip area.

- Global Routing — to determine a sequence of streets through which wiring nets run.

- Channel Routing — to assign each wiring net on tracks between two adjacent sequences of blocks.

PLACEMENT CONSIDERATION AND DEFINITIONS

Figure of Merit

In design algorithms, it generally depends on various "figures of merit" (objective function) to compare alternative arrangements or select the most promising candidate for the next operation. The selection of a figure-of-merit is a delicate choice between realism and computer time.

We will make this notion explicit after we develop some terminologies and notations. Let us define for a given logic circuit diagram D a bipartite graph $G = [\%, \$; H]$ representing D, where $\%$ and $\$$ are disjoint sets of vertices and H is a set of edges, such that each $e_i \in \%$ corresponds to a logic gate e_i in D and $s_j \in \$$ to a signal line s_j in D. There is an edge $(e_i, s_j) \in H$ connecting e_i and s_j if and only if signal line s_j is an input to or output from gate e_i in D. For example, for a logic diagram D of Figure 2, the corresponding graph G is shown in Figure 3. Thus, we have the following:

W_{s_i} is defined as the weight associated with the s_i signal. We assume it as one for all signals.

S_{e_i} is defined as the set of signals of which block (i.e., logic gate) e_i is a member.

Selection rule. The measure of connectivity between two distinct blocks, e_x and e_y is called the first conjunction which is the sum of weighted number of signal sets common to e_x and e_y. We denote this measure by

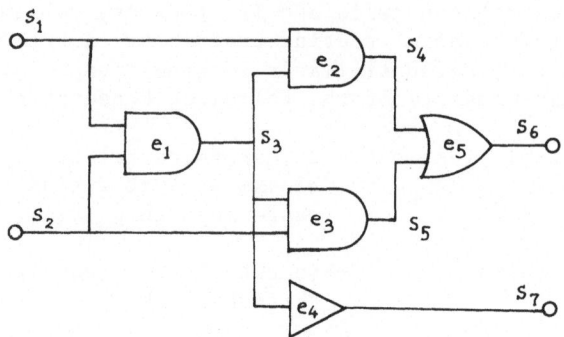

Figure 2. A logic circuit diagram D.

$$\text{FCON}(e_x, e_y) = \begin{cases} \sum_{s_i \in S_{e_x} \cap S_{e_y}} W_{s_i}, & \text{for all } e_x \neq e_y \\ 0, & \text{for all } e_x = e_y \end{cases} \quad (1)$$

Let $T(e_i)$ denote a set of vertices $e_j \in \%$, such that FCON $(e_i, e_j) \neq 0$. Then the second conjunction for an ordered pair (e_i, e_k) of e_i and e_k, denoted by SCON (e_i, e_k), is defined as

$$\text{SCON}(e_i, e_k) = \text{FCON}(e_j, e_k), \quad e_j \in T(e_i), e_j \neq e_k \quad (2)$$

With the use of FCON and SCON, we introduce an index AF (e_i, e_j) measured a kind of 'affinity' between vertices e_i and e_j of $\%$, which is defined by

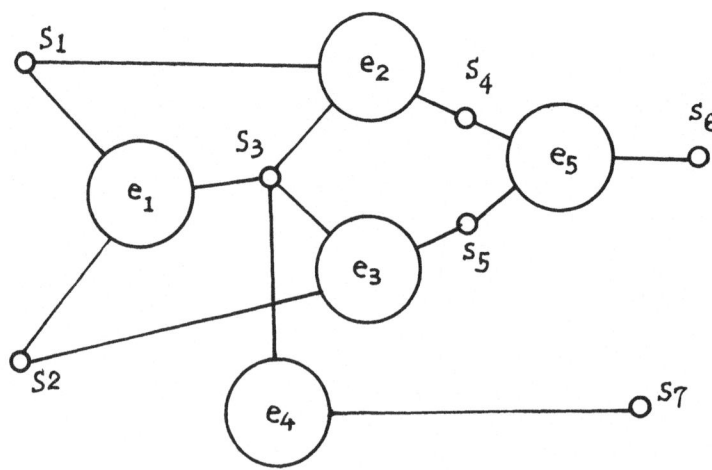

Figure 3. Graph G representing D.

A CAD PROGRAM

$$AF(e_i, e_j) = W * FCON(e_i, e_j) + SCON(e_i, e_j) \quad (3)$$

where W is a weight and is empirically fixed to be 10 in our algorithm for desired result. Thus, we construct our selection rule.

<u>Objective Function [11,12,13]</u>. Let % be a set of cells to be laid out on a chip. For any $\%_i \subset \%$, let

$$B(\%_i) = \bigcup_{e_i \in \%_i} S_{e_i} \quad (4)$$

Suppose that blocks are placed in the row form, and let $R = \{r_0, r_1, r_2, \ldots r_n, r_{n+1}\}$ be a set of rows, each r_k indicating the kth row, except that r_0 and r_{n+1} are virtual rows corresponding to sets $\%_0$ and $\%_{n+1}$ of I/O pads which are located at the bottom and the top edges, respectively, of a chip. Decompose all blocks into each row slot, such that the total sum of widths of a set $\%_k$ of blocks assigned to each row r_k ($1 \leq k \leq n$) is between two boundary parameters L1 and L2 with L1 < L2 and define CH_k as follows:

$$CH_k = \left| \bigcup_{1 \leq i \leq k \leq j \leq n} \{B(\%_i) \cap B(\%_j) - B(\%_k)\} \right| \quad (5)$$

where $|S|$ denotes element number of a set S. That is, CH_k indicates the number of signals which cross over row r_k, but have no terminals at any blocks in $\%_k$. Then $\sum_{k=1}^{n} CH_k$ is one of objective functions for the algorithm.

In each row cluster, we then decompose blocks in the same row into column clusters. A set of such column clusters is denoted by $V = \{V_0, V_1, V_2, \ldots, V_1, V_{q+1}\}$, V_0 and V_{q+1} are virtual columns for I/O terminals at the left and the right hand sides of the chip, respectively. Let $\%_h$ be a set of blocks in the cluster V_h, the number of signals CV_h which cross over V_h is defined by

$$CV_h = \left| \bigcup_{1 \leq i \leq h \leq j \leq q} \{B(\%_i) \cap B(\%_j) - B(\%_h)\} \right| \quad (6)$$

this is another objective function for the algorithm.

Problem Definition

Our approach differs from those described in [7]. Throughout the process we bear in mind and depend on what can be done in the much similar one dimensional placement problem. Then the placement problem with standard cell in VLSI is divided into two linear placement problems as illustrated in Figure 1 [9]. One is the decomposition of blocks into horizontal clusters (rows) to determine their Y-coordinates, and the other is the optimum ordering of blocks at each horizontal cluster (row) to fix their X-coordinates. Thus, the following two placement problems are formulated:

<u>Problem I</u> Given a set of blocks % and two numbers L1 and L2 (L1 < L2), decompose blocks into each row, such that

(i) the total sum of width of blocks assigned to row r_k is between L1 and L2, and

(ii) $\sum_{k=1}^{n} CH_k$ is minimum.

<u>Problem II</u> Given a set of blocks %, assign a collection of blocks into each vertical cluster, such that

(i) $\sum_{h=1}^{q} CV_h$ is minimum.

PLACEMENT ALGORITHM

Algorithm Overview

Now we are in the position to describe our placement progress, an outline is briefly shown in Figure 4. The entire process is divided into five phases each one is to achieve a specific goal. (1) Logic diagram processing and parameters determination that read the input circuit description to form signal nets and assign I/O pads, then determine required parameters, i.e., the chip's row number, ..., etc. (2) Blocks decomposition [14] that decomposes blocks into each row to determine Y-coordinates of blocks as stated in Problem I. (3) Blocks interchange and ordering that determine X-coordinates of blocks as stated in Problem II and locate the final positions of all blocks in the chip. (4) Interactive processing that lets user have some interactive commands to process blocks directly by user's experiments to get a best placement result. (5) Generating routing data and graphic diagram that output matrices to indicate the position of each block and signal set for each block, plot a graphic diagram for later routing.

Determination of Parameters

Before initially decompose blocks into rows, we should consider some parameters, such as the chip row number and the chip column number.

The number of rows is determined to force the chip to meet the desired shape as closely as possible. In general, a square-like shape is preferred for saving chip area. Let TA = total sum of area of blocks (exclude I/O pad blocks), AH = average height of

A CAD PROGRAM					79

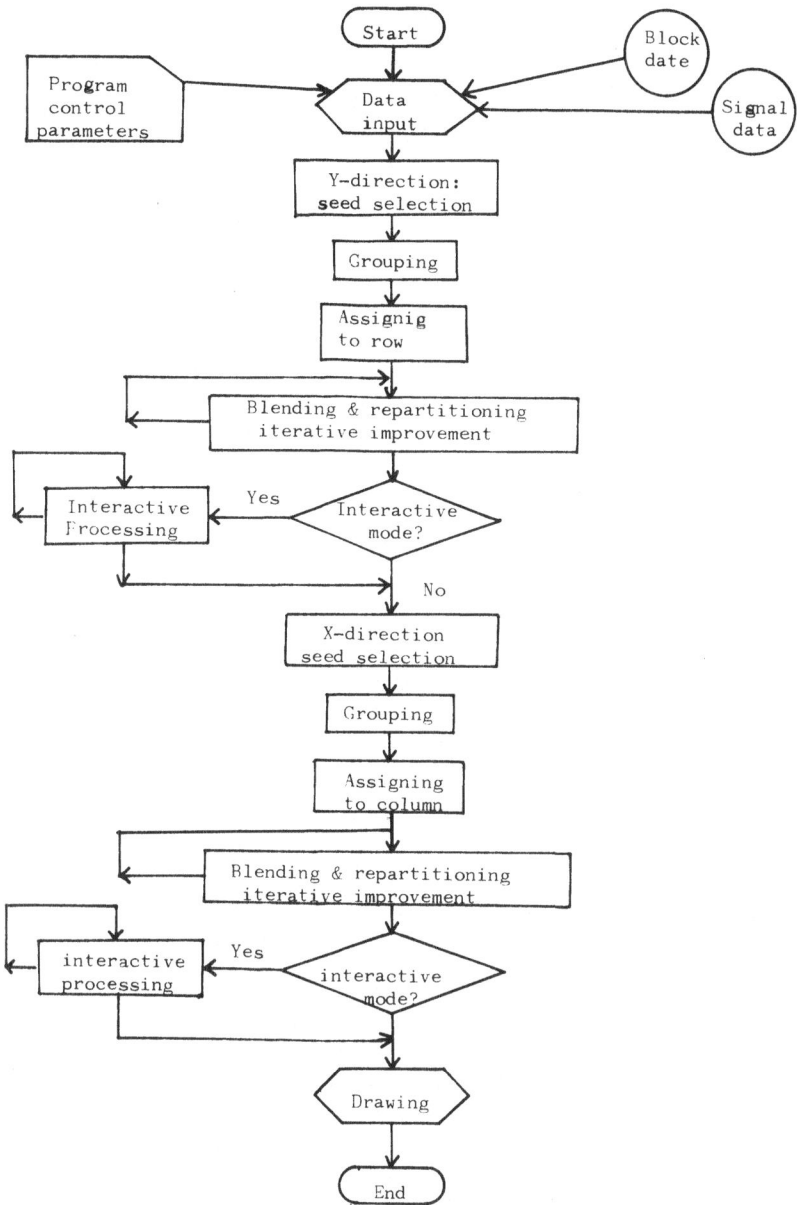

Figure 4. Placement program overview.

blocks, and AR = aspect ratio of a chip. And let L be a constant
for taking account of routing area, then basic equation to determine
the row number n for a square-like chip can be evaluated by

$$n = [L * \sqrt{TA} / AR * AH] \qquad (7)$$

where [x] denotes the largest integer number not greater than x.

Using the same concept, we determine the column number q by the following equation

$$q = [TN / n+1.0] \qquad (8)$$

where TN is the total block number.

However, generally speaking, it is hard to determine the optimal numbers of row and column for various types of input data, therefore, it is sometimes necessary to adjust n and q from the layout result obtained after the routing process.

Selection of Seeds

The first algorithm which selects seed blocks for each row cluster is conducted and described as follows:

[Selection of Seeds]

STEP 1: Sort all blocks in descending order according to the number of pins of each block.

STEP 2: Select the first block in the sorted list as an initial seed to form the initial seed set.

STEP 3: Again, select a block from the sorted list to minimize the affinity between the selected block and the seed set, then add the selected block to the seed set and form a new set.

STEP 4: If the seed number is equal to the number of rows, then halt, otherwise, return to STEP 3.

While in step 3, sometimes it may happen that two or more blocks have the same value of affinity, this tie can be broken by first selected block having higher priority. In addition, while in the search of seeds, we can get some search depth constraints to spare the processing time.

Grouping of Blocks and Assigning of Clusters to Rows

After the seed set has been generated, the development of each row cluster is then proceeded. The algorithm for generating row clusters is described in the following.

A CAD PROGRAM

[Generating Row Clusters]

STEP 1: Assign one seed block to each row as row cluster; initially every cluster has a seed block.

STEP 2: If width of a row cluster exceeds the average row width, then reject this row cluster and go to STEP 4.

STEP 3: Add an unplaced block which has the largest first conjunction related with the target row cluster for each row cluster.

STEP 4: If all blocks have been selected, then halt, otherwise, return to STEP 2.

Step 2 is for adjustment of the width of each row within a specified range. Right after grouping of blocks, the assignment of the grown clusters to rows is developed as follows.

[Assigning Grown Clusters to Rows]

STEP 1: Generate the signal sets of top and bottom I/O pads.

STEP 2: With the sequence of top to bottom assign one cluster to each row slot.

STEP 3: Have all clusters been positioned? If not, then return to STEP 2, otherwise, halt.

In step 2, the principle of assigning clusters to row positions is to maintain a minimum CH_k with positioned clusters. The order of which row slot to be occupied can be changed in order to get better results.

Iterative Improvement of Blending and Repartition [4]

The purpose of this stage is to improve the placement by applying some kinds of changes, such as the pair-wise interchange of blocks. Typically, there is a large number of trials and it is essential that the objective function is calculated on an incremental basis. Some of those have been described in [6], but they easily get stuck in a local minimum. In our algorithm, a better method is applied exchange blocks within a sliding window (see Figure 5(a)), i.e., blending and repartitioning scheme. This concept is illustrated briefly in Figure 5(b), where cluster $%_1$ indicates a set of blocks assigned to row r_1. In the beginning, clusters $%_1$ and $%_2$ are merged into $%_{1,2}$ and then $%_{1,2}$ is repartitioned into two clusters $%_1$' and $%_2$" so as to minimize ΣCH_k, subject to condition that other clusters are fixed. The same operation is repeated for a pair of clusters $%_2$" and $%_3$, and so on until for $%_1$', $%_2$', ..., $%_{n-1}$", then

STEP 3: If all clusters are processed, then go to STEP 4, otherwise, return to STEP 1.

STEP 4: Repeat STEP 1 to STEP 2 in the reverse order of the last time until all row clusters have been processed.

STEP 5: Is chip's total signal cuts decreased? If not, then reject the new cluster set and save the old, otherwise, accept the new one and delete the old one.

STEP 6: If the upper bound of the iterative number is reached, then halt, otherwise, return to STEP 1.

According to conventional placement methods [10] in which one block or a pair of blocks is selected one at a time to be moved or exchanged. However, for some cases such as a flip-flop may consist a set of strongly connected blocks which, using the conventional methods, cannot possibly be moved from row to row, as illustrated in Figure 6. Hence, the final placement will depend on an initial placement.

However, according to our method a pair of clusters are merged temporarily and then repartitioned into two clusters, and hence a set of strongly connected blocks can be moved from row to row to a common cluster.

Step 2 in the algorithm is for adjustment of the total sum of block widths in each row cluster within a certain range specified by L1 and L2.

Interactive Processing [15]

In the interactive mode the user has a set of available commands allowing him/her to do operations such as the following:

- To view the blocks having connections to a particular block in order to decide where to best place that block.

- To remove/add a block from/to current placement configuration.

- To exchange four or less blocks once by user's designation.

- To show the figure of merit (the value of ΣCH_k) based on the current placement configuration.

Thus, by using the interactive commands, the user has the capability to improve the placement configuration resulting from the automatic placement.

A CAD PROGRAM

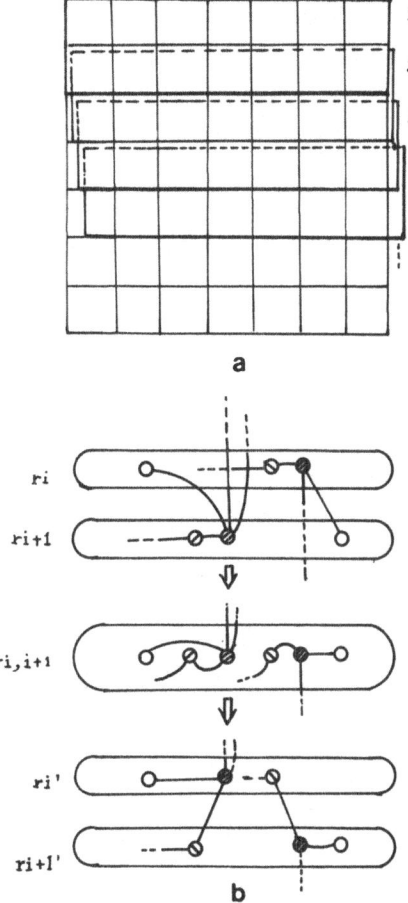

Figure 5. Placement improvement, using more complex exchanges within a sliding window. (a) Moving the window with an overlay; (b) A concept of blending and repartition.

starting off with a pair of clusters $\%_n$ and $\%_{n-1}''$, the same blending-repartitioning operation is iteratively conducted in reverse order to improve the assignment. When no improvement is attained or the upperbound for iteration number of this cycle is exceeded, the procedure is terminated. This algorithm is described in the following.

[Blending and Repartition]

STEP 1: Blend two adjacent row clusters into a new group.

STEP 2: Repartition the group to form two new row clusters to minimize CH_k; the width of row clusters must be between L1 and L2.

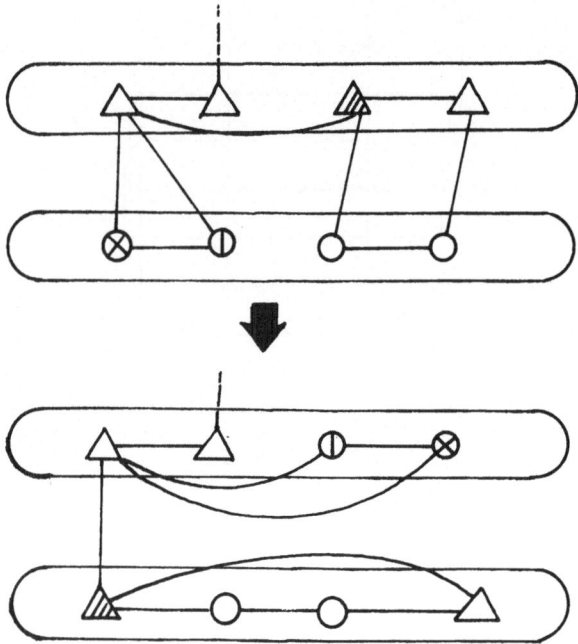

Figure 6. Strongly connected blocks.

Vertical Clusters Placement

In this phase, the final positions of blocks assigned to each row are determined. The algorithm applied in this stage is almost the same as that of block assignment stated in section on Interactive Improvement of Blending and Repartition.

Of primary importance is to determine the number K of vertical clusters. Suppose that each vertical cluster contains at most one block taken from a row, i.e., the vertical cluster number equals the column number. Then the blending-repartitioning results in only an exchange of two blocks, and hence, a set of strongly connected blocks may not possibly be moved into a common cluster, as illustrated in Figure 6. Thus, K is taken as q/Q and Q is taken as following:

$$Q = \begin{cases} 2, & \text{if column number} < 9, \\ 3, & 9 \leq \text{column number} < 18, \\ 6, & 18 \leq \text{column number}. \end{cases} \quad (9)$$

GLOBAL ROUTING

Global routing, as stated before, is used to interconnect all the terminals for each net loosely. If we take a set of terminals of

A CAD PROGRAM 85

the same net to be the vertices of an undirected graph, then the interconnection net is a tree defined on these vertices. The minimum Steiner tree is introduced, which is a tree of minimum length whose vertices include the terminals of the net. Because of the constraint of the layout model, this tree permits wires in horizontal and vertical directions only.

An algorithm is proposed by Kazuhiro [16], and an example is shown in Figure 7. According to this tree, feed-throughs for this net can be determined. Because a feed-through must not go through a cell, but between two cells, a comparison must be made to decide the position of the feed-through. This is illustrated in Figure 8. However, if a cell row contains any electrical equivalence for this net, no feed-through is needed for this cell-row. This is also illustrated in Figure 8.

The global router also determines the floating terminals [3] for each net with I/O pad around the chip. The ordinate of the floating terminal is not determined until the channel routing procedure is completed.

CHANNEL ROUTING

After all the net has been assigned to channels by the global routing procedure, nets within each channel must be interconnected and assigned to tracks. This is done by the channel routing procedure, which is also called the track assignment procedure [17].

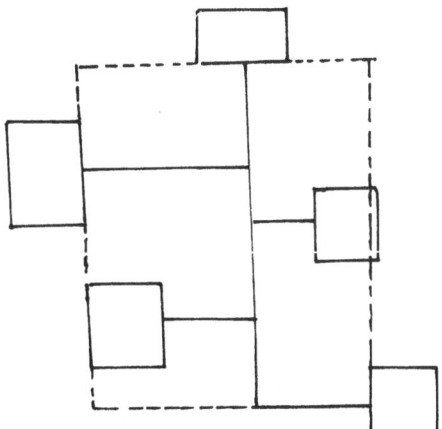

Figure 7. Minimum Steiner tree.

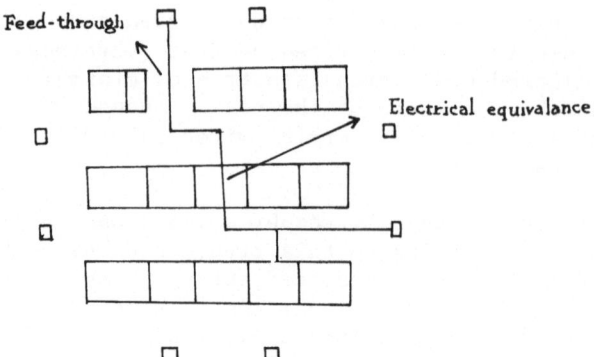

Figure 8. Global routing.

The purpose of channel routing is to assign nets within a channel to proper tracks without overlapping for different nets and with minimum number of tracks. The vertical constraint graph (VCG) [19] is needed for this purpose. In this graph, an arrow from node 1 to node 3 indicates that net 1 must be placed above net 3, otherwise overlapping may occur (see Figure 9).

Using the VCG and applying the left-edge algorithm [18] (i.e. from the left side to the right side), we select the node without any arrow incident on it. Place this net on the track and remove it from the VCG. If the current track is full, then proceed to the next track. Repeat until the VCG becomes a null graph (see Figure 10).

However, if there is any cycle in the VCG, as shown in Figure 11, this cycle must be broken, so that the process may continue. We may partition the net into two subnets at a proper position, so that the cycle may be broken. This is called "net partitioning."

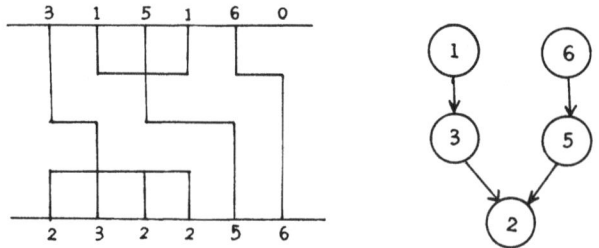

Figure 9. (a) A routed channel
          (b) Vertical constraint graph (VCG).

A CAD PROGRAM

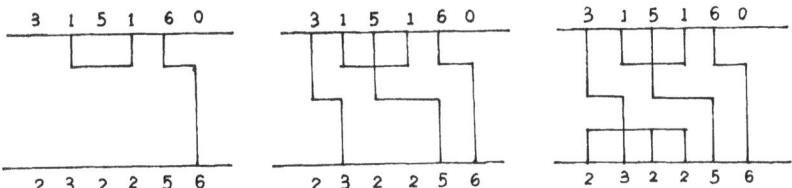

Figure 10. Procedure of channel routing using VCG.

In our system, we establish a table [15] to implement this graph. This is shown in Figure 12. In this table, suppose net 1 occupies column 2 and column 4, then the forward pointer (FP) for net 1 at column 2 is 4, and the backward point (BP) for net 1 at column 4 is 2. A negative number for FP or BP indicates the bottom edge of the column.

To route a channel according to the table is similar to that according to the VCG. Whenever a subnet is assigned to a track, subnet numbers, FP and GP are reset, which is similar to removing a node from the VCG. When all the net numbers are reset, routing procedure for this channel is completed (see Figure 13).

Based on the dogleg routing algorithm [3], the channel routing algorithm using this table is shown as follows:

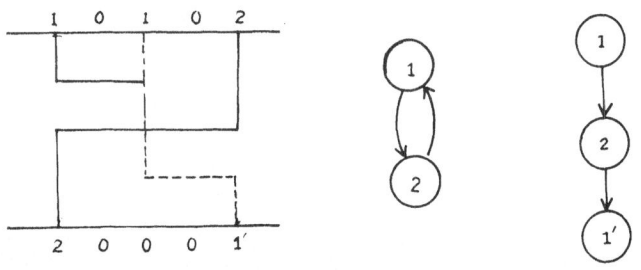

Figure 11. Constraint loop and net partitioning.

column	1	2	3	4	5
net	1	0	1	0	2
	2	0	0	0	1
FP	3	0	-5	0	0
	5	0	0	0	0
RP	0	0	1	0	-1
	0	0	0	0	3

Figure 12. Table representing Figure 11.

[Channel Routing]

```
Track = 1; Column = 1; Succeed = TRUE;
WHILE (succeed=TRUE) DO BEGIN
  WHILE (|column|<channel_length) DO BEGIN
  IF (net (column)=0) AND (column>0) THEN column = - column;
  IF net (column)=0   THEN column = |column| + 1
  ELSE
    IF (FP(column)=0) THEN BEGIN
      IF (BP(column)=0) THEN net (column) = 0;
      column = |column| + 1
    END ELSE BEGIN
      PLOT (column, FP(column)) on this track;
      IF (BP(column)=0) THEN net (column) = 0;
      BP (column) = 0;
```

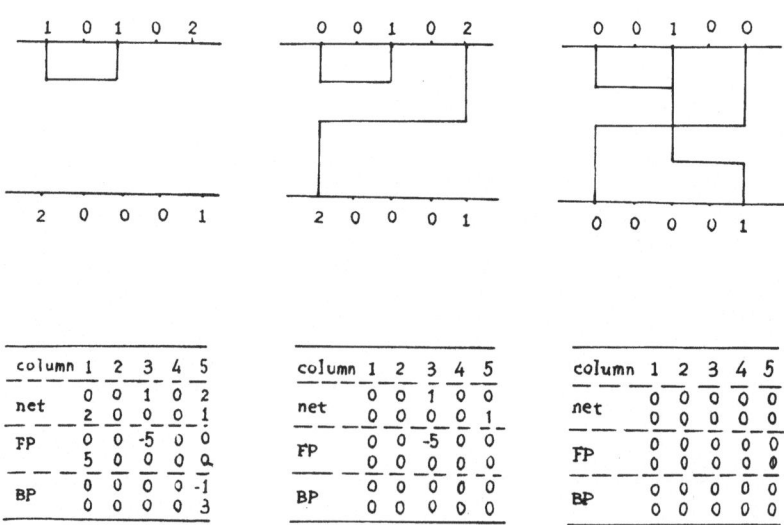

Figure 13. Procedure of channel routing using table.

A CAD PROGRAM

```
            column = FP (column);
            FP (column) = 0
         END
      END;
      IF (no net plotted in this track) THEN succeed = FALSE
      ELSE BEGIN track = track + 1; column = 1 END;
   END.
```

There are two more features for the channel router. First, rather than route from top to bottom, a more symmetric choice was made. We alternate the routing sequence between top and bottom tracks such that the total wire length may be reduced. Second, rather than route the tracks from the left edge, we may change the diretion for some tracks such that less track number is needed, as shown in Figure 14.

After all horizontal channels are routed, ordinates of the floating terminals are determined, which may be treated as terminals of the vertical channel, they are also routed by the channel router.

IMPLEMENTATION AND RESULTS

We have programmed placement scheme in PASCAL and FORTRAN 77 with 17 main procedures on VAX-11/780 and displayed the output

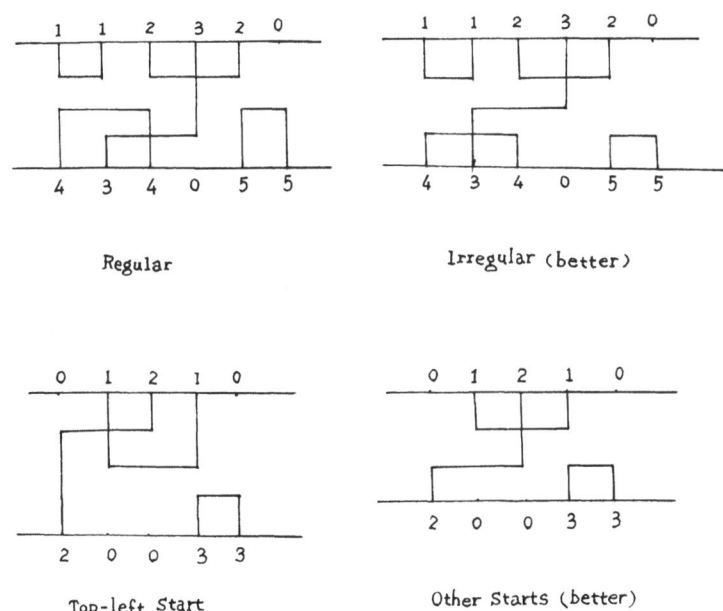

Figure 14. Two additional features to improve the result.

diagrams on Tektronix 4052 graphic terminal after some data are applied. A sample input logic diagram is shown in Figure 15. Its output data and results are shown in Figure 16.

By means of the blending and repartitioning scheme, a final solution obtained from an initial placement by applying the row grouping procedure is almost the same as the one with random clustering as shown in Figure 17. But the speed of convergence to a final value by the former procedure is rather faster than the latter.

Figure 18 shows the routing result drawn by the graphic plotter. The global router is applied to each net to determine channels occupied by this net; feed-throughs are determined if this net occupies more than one channel; floating terminals are also determined if this net must be connected to the I/O pad. The channel router is applied to each channel to assign tracks for net within this channel. The ordinates of the floating terminals are determined during this procedure. Finally, the regions between cell blocks and I/O pads on the left and right sides of the chip are treated as two vertical channels, which are also routed by the channel routing procedure.

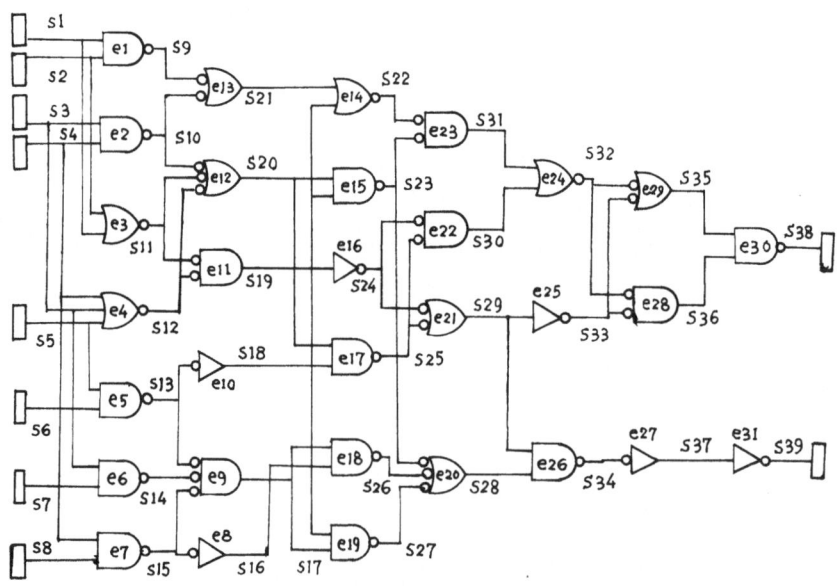

Figure 15. (a) A sample logic circuit.

A CAD PROGRAM

```
DATABEGIN
 4 31
 1NAD 2 1.29   9   1  2
 2NAD 2 1.29  10   3  4
 3NOR 2 1.29  11   1  2
 4NOR 3 1.72  12   3  4  5
 5NAD 2 1.29  13   5  6
 6NAD 2 1.29  14   3  7
 7NAD 2 1.29  15   4  8
 8INV 1 0.86  16  15
 9NOR 3 1.72  17  13 14 15
10INV 1 0.86  18  13
11NOR 2 1.29  19  11 12
12NAD 3 1.72  20  10 11 12
13NAD 2 1.29  21   9 10
14NOR 2 1.29  22  21 19
15NAD 2 1.29  23  20 19
16INV 1 0.86  24  19
17NAD 2 1.29  25  20 18
18NAD 2 1.29  26  17 16
19NAD 2 1.29  27  17 19
20NAD 3 1.72  28  23 26 27
21NAD 2 1.29  29  24 25
22NOR 2 1.29  30  24 25
23NOR 2 1.29  31  22 23
24NOR 2 1.29  32  30 31
25INV 1 0.86  33  29
26NAD 2 1.29  34  28 29
27INV 1 0.86  37  34
28NOR 2 1.29  36  33 32
29NAD 2 1.29  35  32 33
30NAD 2 1.29  38  35 36
31INV 1 0.86  39  37

 3 10
 1IOP$   1U
 2IOP$   2U
 3IOP$   3L
 4IOP$   4L
 5IOP$   5L
 6IOP$   6L
 7IOP$   7D
 8IOP$   8D
 9IOP$  38R 0
10IOP$  39R 0
DATAEND
```

Figure 15. (b) Input data format.

CONCLUSIONS

In this report, a set of new algorithms based on heuristic approach was proposed for the standard cell two-dimensional layout problem. The result fulfills the important layout requirements. In particular a tight and uniform packing is generated.

Some features of the layout algorithms are summarized in the following.

1. It takes advantages of what one can do in linear placement in tackling the two-dimensional placement problem.

2. It is capable of operating in a mixed manner of automatic mode and interactive mode to get the best placement result.

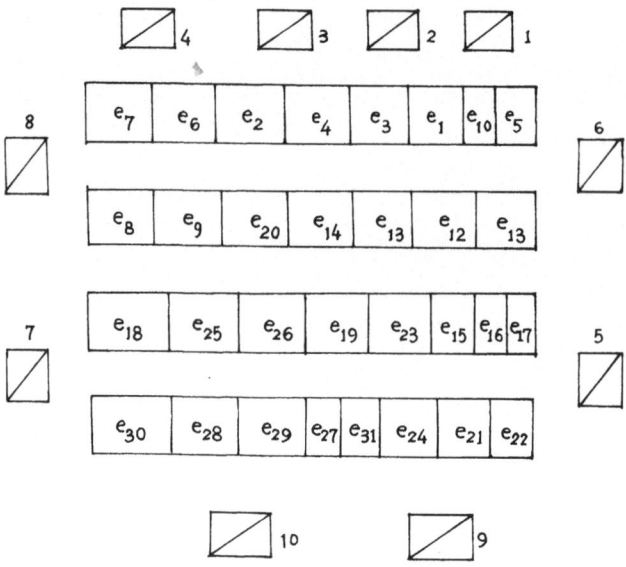

Figure 16. Implemented result for Figure 15.

3. It is able to produce a good final placement regardless of the initial placement (due to blending and repartitioning operation).

4. The tree graph used in the global routing is applied for the signal net whose length is shorter than that of other trees.

5. The table used in channel routing to implement the constraint graph has the function of net partitioning. To establish and access this table is very convenient.

REFERENCES

[1] S. Muroga, "VLSI System Design," John Wiley Company, 1982.
[2] A. R. Newton, "Computer-aided design of VLSI circuits," Proc. of the IEEE, Vol. 69, No. 10, Oct. 1981, pp. 1189-1199.
[3] G. Persky, et al., "LTX-A Minicomputer-Based System for Automated LSI Layout," Design Automation & Fault-Tolerant Computing, Vol. 1, No. 3, May 1977, pp. 217-256.
[4] J. Soukup, "Circuit layout," Proc. of the IEEE, Vol. 69, No. 10, Oct. 1981, pp. 1281-1304.
[5] I. Nishioka, et al., "An approach to gate assignment and model placement for printed wiring boards," Proc. 15th Annual Design Automation Conf., 1978, pp. 60-69.

A CAD PROGRAM 93

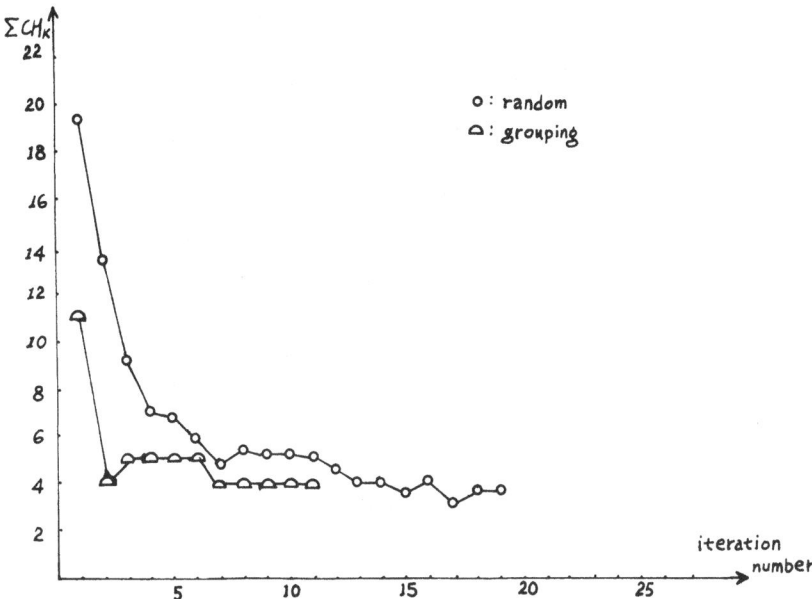

Figure 17. The comparison of grouping placement with random ones.

[6] M. Hanan and J. M. Kurtzberg, "Placement Techniques," Chap. 5 in Design Automation of Digital Systems: Theory and Techniques, Vol. 1 (M. A. Breuer, Ed.), Prentice-Hall, 1972, pp. 213-282.

[7] M. Hanan, et al., "A Study of Placement Techniques," Design Automation & Fault-Tolerant Computing," Vol. 2, No. 2, May 1978, pp. 145-164.

[8] D. G. Schweikert, "A 2-dimensional placement algorithm for the layout of electrical circuits," Proc. 13th Annual Design Automation Conf., 1976, pp. 408-416.

[9] S. Goto and E. S. Kuh, "An approach to the two-dimensional placement problem in circuit layout," IEEE Trans. on Circuits and Systems, Vol. CAS-25, No. 4, April 1978, pp. 208-214.

[10] M. Breuer, "Min-Cut Placement," Design Automation & Fault-Tolerant Computing," Vol. 1, No. 4, Oct. 1977, pp. 343-362.

[11] B. W. Kernighan and S. Lin, "An efficient procedure for partitioning graphs," Ball System Technical Journal, Feb. 1970, pp. 291-307.

[12] D. C. Schmidt and L. E. Duffel, "An iterative algorithm for placement and assignment of integrated circuits," Proc. 12th Annual Design Automation Conf., 1975, pp. 361-368.

[13] T. Kambe, et al., "A placement algorithm for polycell LSI and its evaluation," Proc. 19th Annual Design Automation Conf., 1982, pp. 655-662.

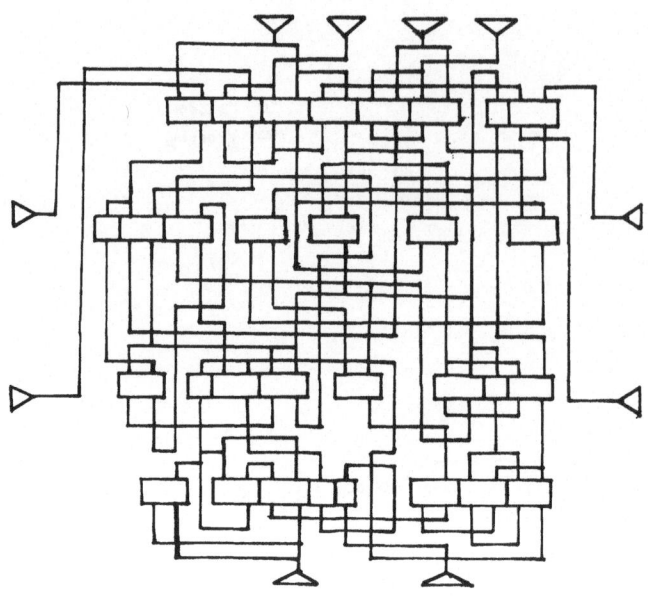

Figure 18. Complete layout for Figure 15.

[14] D. M. Schuler and E. G. Ulrich, "Clustering and linear placement," Proc. 9th Annual Design Automation Workshop, 1972, pp. 50-56.
[15] H. Beke and W. Sansen, "CALMOS: A portable software system for the automatic and interactive layout of MOS/LSI," Proc. 16th Annual Design Automation Conf., 1979, pp. 102-108.
[16] K. Ueda, Y. Sugiyama, and K. Wada, "An automatic layout system for masterslice LSI: MARC," IEEE J. Solid State Cir., Vol. SC-13, No. 5, Oct. 1978, pp. 716-721.
[17] H. Kanada, K. Okazaki, M. Tachibana, B. Kato, and S. Murai, "Channel order router," J. Dig. Sys., Vol. 4, Issue 4, 1981, pp. 427-441.
[18] A. Hashimoto and J. Stevens, "Wire routing by optimizing channel assignment within large apertures," Proc. 8th Annual Design Automation Conf., 1971, pp. 155-169.
[19] T. Yoshimura and E. S. Kuh, "Efficient algorithms for channel routing," IEEE Trans. on CAD & ICAS, Vol. CAD-1, No. 1, Jan. 1982, pp. 25-35.

BRUTUS: AN INTERACTIVE GRAPHIC EDITOR FOR IC LAYOUT

K. A. Hwang, D. C. Liu, and C. C. Lin

Computer Center
National Chiao Tung University
Hsin-Chu, Taiwan, R.O.C.

ABSTRACT

BRUTUS is an interactive graphic editor for IC layout. It is implemented in C language under UNIX of the VAX-11/780. BRUTUS performs 2-D geometrical functions such as rotation, translation, mirroring and is capable of zooming a specified area (called window) of a graph and thus is suitable for the manipulation of a planar graph.

The most important facility provided by BRUTUS, which makes the plot of a VLSI layout possible, is that it allows a graph of great complexity to be built up in a hierarchical approach.

BRUTUS is also capable of manipulating the CIF input file and of generating the CIF output file.

INTRODUCTION

Just as the engineering drawing in other engineering fields, the conventional way of drawing an IC layout plot is so labor-intensive, time-consuming and error-prone. An automatic graphic tool/system seems to be the only possible solution for highly complex drawing in IC design.

BRUTUS may look a little "CAESAR-like" because the functional specifications of the CAESAR system [1] were referred to at the beginning stage of the implementation of BRUTUS. It has now served as the graphic tool of generating and modifying graphic output for the VLSI CAD project held at N.C.T.U. It is also expected to facilitate students who will take part in the course of VLSI design. With BRUTUS, they can "see" what their design may look like which was simply impossible before. The instructor can check the correctness of the design submitted by students and is no longer bothered by piles of data sheets.

Moreover, students can have their layout design done in a clean and easy manner because BRUTUS allows one to construct simpler plots first (defined as cells, each with a cell name in order to be referenced later), and then to build a new cell simply by applying as many existing cells as one requires. Such a hierarchical approach means a great deal of man-hour savings in plot construction, plot modification and plot checking which in turn accelerates the design cycling and reduces the cost.

In fact, some awkwardness was found by students when using BRUTUS for their layout design. While recovering these defects, the availability of BRUTUS should be maintained simultaneously; the way we did it is included at the end of this paper.

OVERVIEW

Outlines of BRUTUS

Figure 1 shows the configuration of BRUTUS including the following:

* a Ramtek 6211 color CRT display with separate keyboard and light pen,
* an ACT I ink-inject plotter, directly coupled with the 6211 through 3 video signal cables, as a hard copier,
* a Houston CPS-15 4-pen plotter as an alternative plotting device at user's choice.

The user can interact with BRUTUS in a very easy way. All BRUTUS commands are given by simply pointing the light pen at the associated blue squares in the command table displayed at the right side of the 6211 screen. Once the command is accepted, BRUTUS reacts accordingly by modifying both the display in the plot-area and certain information about the editing status in the status table.

Figure 2 shows the picture of exactly what a user may see on the 6211 screen. The plot-area occupies about 3/4 of the screen at the left and the command table takes the rest at the right. The upper-right portion is the status table which gives the information required by the user when editing a plot of VLSI layout. There are 3 pages of command tables, each with 48 blue squares, as the mechanism for command entry.

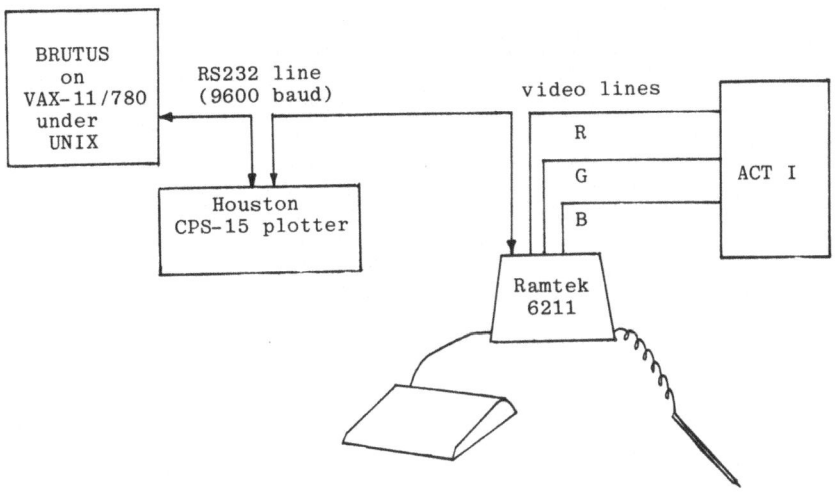

Figure 1. The BRUTUS configuration.

Construction of a Layout Plot

Before going any further, it would be a great help to understand the nature of a VLSI layout plot and the procedure through which the user must go in order to construct a plot of his layout design.

One can simply regard the layout as a plot which consists of several layers, each in turn composed of many rectangles (boxes) and, perhaps, some labels and circles. This is the nature of a VLSI layout (as shown in Figure 2) if one takes the viewpoint purely geometrically.

To construct a layout plot, one must go through the following.

(a) Invoke BRUTUS with/without a cell name which may be an existing cell to be modified or a new cell to be constructed. If the cell name being omitted, a default one, "main," will be given by BRUTUS. In either case, BRUTUS creates an edit cell of the given name which will then be the result of the invocation of BRUTUS.

(b) When constructing a new cell, one may GET existing cells from the disk file system, then ADD these cells at desired positions or one may YANK a portion of the edit cell which is internally treated as a cell named "yank" by BRUTUS, then PUT it at the desired position. In case that there were no existing cells available or suitable for use, the construction could only be done in an element by element,

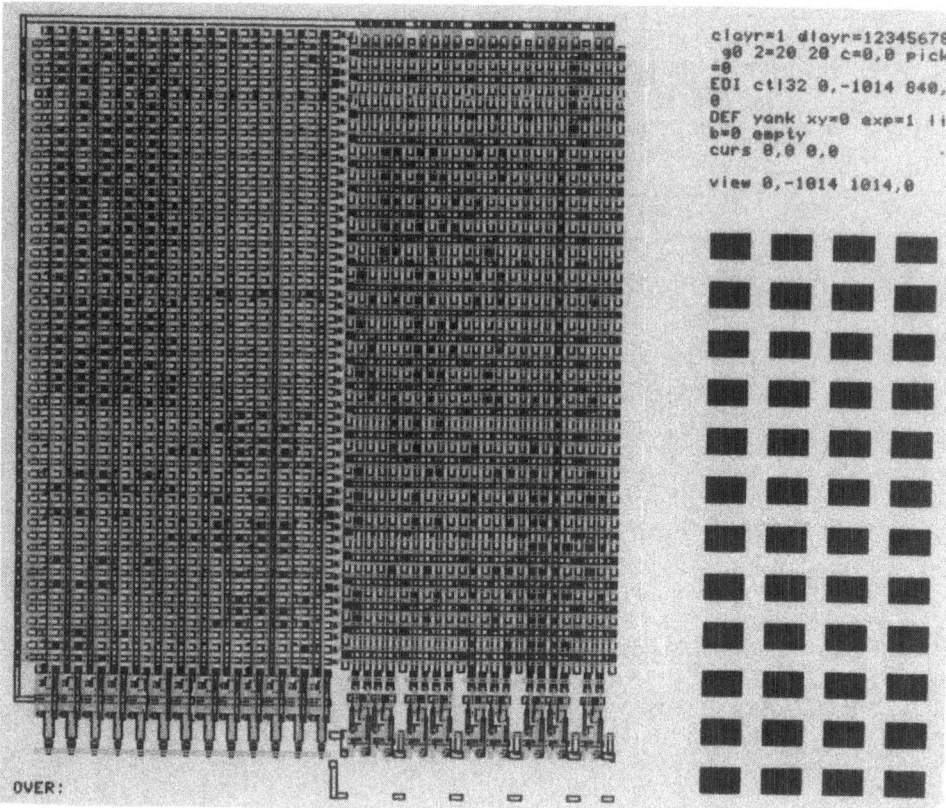

Figure 2. What a BRUTUS user exactly sees on the 6211 screen. (This layout, provided by one of the BRUTUS users, is still under modification).

layer by layer approach. Which means one has to SELECT proper layers first and ADD boxes here and there, then SELECT other layers and so forth. One of those cells used to build the edit cell can be defined as the current cell which is the default operand of most cell associated commands.

(c) When modifying the edit cell, one can either carry out step (b) to enrich the edit cell or DELETE undesired cells from the edit cell and undesired boxes from certain specified layers.

(d) To start another cell editing, exiting from the current invocation of BRUTUS and having a new invocation again is unnecessary. One simply gives the NEW command to start it. BRUTUS will create a new edit cell and still "remember" all edit cells previously manipulated.

BRUTUS   99

   (e) During editing, one may SET the size and position of the
       view window or SELECT desired layers and then DISPLAY
       particular portions/layers of the edit cell to ensure the
       correctness of the editing. One may also SET the size and
       position of the BRUTUS cursor to coordinate many editing
       commands.

   (f) Before exiting from the BRUTUS system, one should SAVE all
       edit cells kept by BRUTUS in the disk file system with the
       file names the same as those of the edit cells.

   (g) STOP the BRUTUS system.

Keywords in capital letters in the above descriptions should give one the picture of what BRUTUS can do and how it can be used, which will be of great help in reading the following section. BRUTUS also provides other commands to facilitate the graph editing as will be described in the next section.

BRUTUS FUNCTIONS

There are 3 pages of command tables in the BRUTUS system, as shown in Figure 3. All BRUTUS commands can be grouped into the following classes to perform related functions:

   * setting of BRUTUS status,
   * control of cursor,
   * control of view window,
   * manipulation of boxes,
   * manipulation of zigzags,
   * manipulation of labels,
   * manipulation of cells, and
   * others.

Setting of BRUTUS Status

The following gives explanation of every entity in the status table and by which the status setting commands should become self-explanatory.

* <u>clayr</u> & <u>dlayr</u>

BRUTUS can handle a layout of up to 7 layers (at present) and, therefore, meets the requirement of the NMOS technology. Each layer is distinguished from others by different colors and line patterns as indicated in Figure 4. One can determine the visibility of each layer on the screen by proper setting of <u>dlayr</u> (display layers) and one can also mask each layer out of certain attempt of modification by proper setting of <u>clayr</u>. Layers being opened for

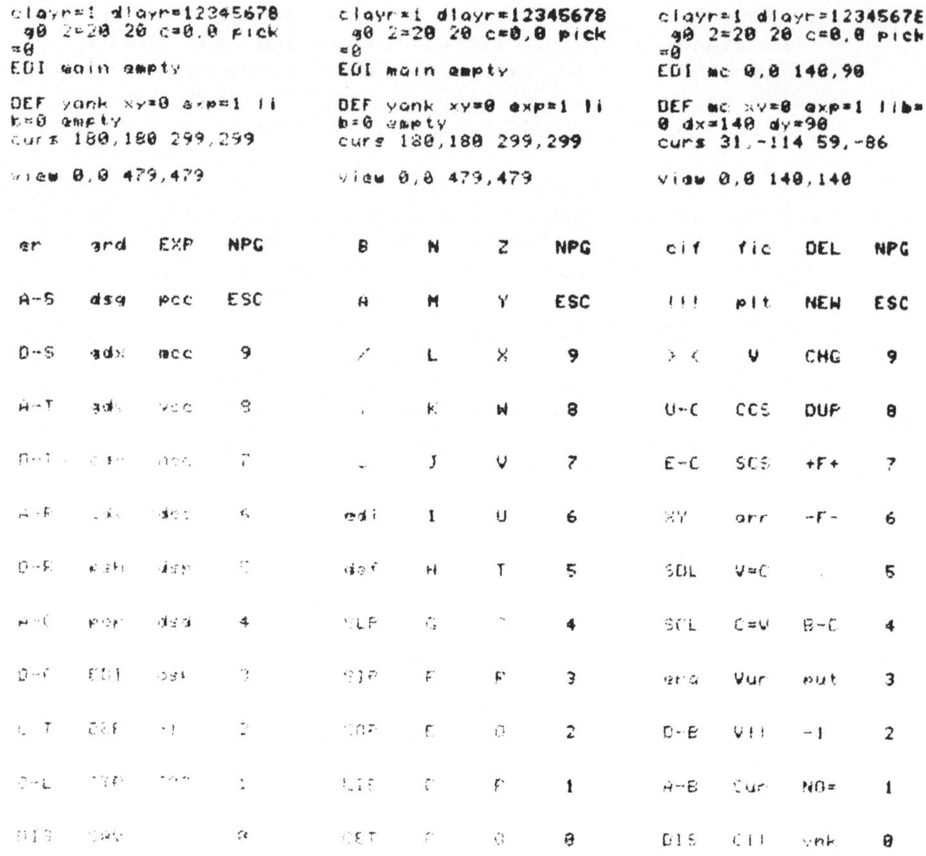

Figure 3. BRUTUS command tables (revised version).

modifications are called <u>current layers</u>. For instance, the status table of page 1 in Figure 3 indicates that all layers are visible (thus the plot will be fully displayed on the screen) and modifications can only be made on layer 1, the only current layer.

* <u>xy</u> & <u>exp</u>

When adding a current cell to the edit cell, certain transformations such as mirroring, rotation, etc. on the current cell can be performed by proper setting of <u>xy</u> as shown in Figure 5. The current cell or the yank cell can be either expanded or unexpanded when added to the edit cell. An expanded cell will be displayed in its most details while an unexpanded one appears as a box of its size only.

NMOS	ND	NP	NC	NM	NI	NB	NG
layer #	1	2	3	4	5	6	7
8 off	red line	green line	red dashed line	blue line	green dashed line	blue dashed line	red dot line
8 on	red line	green line	red face	blue line	green face	blue face	red dot line

Figure 4. Clayr and dlayr.

* <u>ecel</u> & <u>ccel</u>

   <u>Ecel</u> indicates the name and the coordinates of the lower-left and the upper-right corners of the edit cell while <u>ccel</u> indicates those of the current cell.

* <u>view</u>

   The view window can be set to any portion within, across or beyond the edit cell boundary. And the coordinates of the lower-left and upper-right corners of the window are indicated by <u>view</u>. BRUTUS takes the lower-left corner as the position of the view window.

* <u>curs</u>

   The BRUTUS cursor is a rectangle in red dashed line and <u>curs</u> indicates the coordinates of the lower-left and upper-right corners of the cursor.

0	1	2	3	4	5	6	7
no action	x - -x	y - -y	x - -x y - -y	x - y	1 & 4	2 & 4	1&2&4
	mirror vs. y-axies	mirror vs. x-axies	180° rotation	mirror vs. x=y	90° rotation	-90° rotation	

Figure 5. xy.

* SCL <u>layers</u> (select current layers)

   where <u>layers</u> can be any combination of digits 1 through 7 in the ascending order to disable the mask on specified layers.

* SDL <u>layers</u> (select display layers)

   where <u>layers</u> can be any combination of digits 1 through 8 in the ascending order to enable the visibility of specified layers. When an 8 is set in <u>layers</u>, boxes of NC, NI and NB layers will be filled with red, green and blue, respectively.

* XY <u>xy</u> (select transformation for the current cell)

   where xy can be any digit 0 through 7 to indicate the kind of transformation to be performed on the current cell or the yank cell when added to the edit cell.

* EXP

   which switches exp in the status table between 1 and 0 to indicate the expansion of the current cell.

* ER

   which switches the flag er between 1 and 0 to which BRUTUS treats an added zigzag as a closed one or not accordingly.

* ECN <u>name</u> (change the edit cell name)

   where name is the new edit cell name for the editing that follows.

* CCN <u>name</u> (change the current cell name)

   where name is the new current cell name.

Control of Cursor

   There are two kinds of cursor in the BRUTUS system. One is the BRUTUS cursor as already mentioned, the other is the 6211 cursor which is a white cross that will trace the path drawn by the light pen on the 6211 screen. The 6211 terminal will send the coordinates of the cross to BRUTUS on request.

   BRUTUS defines 6 function keys on the keyboard for the cross cursor control, namely

   Key 1: move the cross 40 units right,
   Key 2: move the cross 40 units left,

BRUTUS

    Key 3: move the cross 30 units up,
    Key 4: move the cross 30 units down,
    Key 15: move the cross top DIS square in the command table,
    Key 16: send the position of the cross to the host.

BRUTUS also provides several ways of positioning a point during the editing, denoted by + hereafter. Methods (1), (2) and (3) are used for precise positioning while method (5) is used for visual positioning. NO= is a command square in page 1 command table by which the associated x or y coordinates are retained. The coordinates' values x and y are given by pointing the light pen at digit squares in each command table.

$$+ \;=\; \begin{array}{ll} (1) & x \quad\quad y \\ (2) & x \quad\quad NO= \\ (3) & NO= \quad y \\ (4) & \text{depress Key 16} \\ (5) & \text{point the light pen at the selected position on the screen.} \end{array}$$

Commands for the control of the BRUTUS cursor are given below.

* CP   +   (change cursor position)

which changes the cursor position, i.e., the coordinates of the lower-left corner of the cursor with its size unchanged.

* CS   +   (change cursor size)

which changes the size, i.e., the coordinates of the upper-right corner of the cursor with its position unchanged.

* V > C

which moves the cursor to the position of current view window and then forms a new view window centered at the new cursor position with a size twice as large as the original one.

* SCS <u>cur</u> (save cursor)

where <u>cur</u> can be any digit 1 through 9. At most 9 cursors can be saved by BRUTUS for later use.

* CCS <u>cur</u> (change cursor)

where <u>cur</u> can be any digit 0 through 9. The cursor is changed to a previously saved one indicated by <u>cur</u>. When <u>cur</u>=0, the cursor will be the same as the boundary of the edit cell.

* VP   +   (change view window position)

which changes the position of the view window (the lower-left corner) with its size unchanged.

* VS   +   (change view window size)

which changes the size of the view window (the upper-right corner) with its position unchanged.

        * V0   set the view window the same as the boundary of the edit cell.
          V1   move the view window 1/2 the size of its width right with the size unchanged.
          V2   move the view window 1/2 the size of its width left with the size unchanged.
          V3   move the view window 1/2 the size of its height up with the size unchanged.
          V4   move the view window 1/2 the size of its height down with the size unchanged.
          V5   enlarge the size twice as large as the original view window with the center of the window fixed.
          V6   change the size 1/2 as large as the original view window with its center fixed.

* C > V

which moves the position of the view window to the cursor position with its size unchanged. The cursor size becomes 1/2 of the original size with its center fixed.

<u>Box Manipulation</u>

All box manipulation commands refer to the edit cell.

* A-B   (add boxes)

which adds boxes, on layers indicated by <u>clayr</u>, to the edit cell. The added boxes are the same as the cursor in both position and size.

* D-B   (delete boxes)

which deletes all boxes whose lower-left corners are lying within the cursor. All layers indicated by <u>clayr</u> are affected.

BRUTUS   105

* +F+ <u>ext</u>   (boxes extending)

   where <u>ext</u> is an integer varying from 1 to 4. This command
extends those boxes intersected by any edge of the cursor to the
opposite edge, as shown in Figure 6. All current layers are
affected.

* -F- <u>ext</u>   (boxes shrinking)

   which shrinks those boxes intersected by any edge of the
cursor with portions in the interior of the cursor left only, as
shown in Figure 6. All current layers are affected.

<u>Zigzag Manipulation</u>

   All zigzag manipulation commands refer to the edit cell.

* A-S <u>pt</u> ES ES   (add zigzags)

   where <u>pt</u> consists of a sequence of +'s as required. If <u>er</u>=1

ext	1	2	3	4
+F+				
-F-				

Figure 6. Results of +F+ and -F- commands.

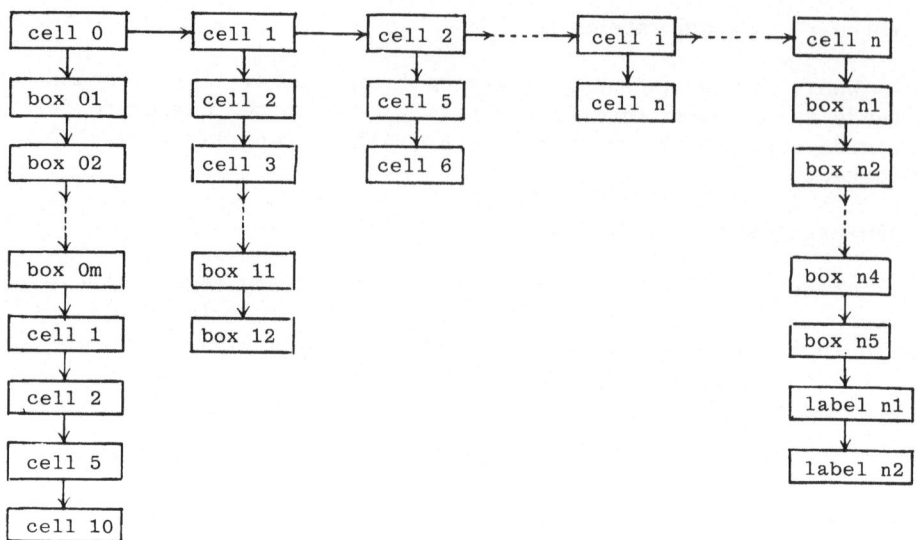

Figure 7. A typical edit cell.

* NES <u>name</u>   (create new edit cell)

where <u>name</u> is the name given for the new edit cell which must be unique in each invocation of BRUTUS. BRUTUS creates an empty edit cell to accept results of the editing that follows. All previously defined edit cells are still kept in the memory.

* CNG <u>old</u> <u>new</u>   (change cell name)

which changes the specified cell with a new name. The cell specified by <u>old</u> can either be an edit cell or a subcell. Once changed, all references to the changed one in other cell definitions or in the edit cell are changed accordingly.

* DUP <u>new</u>   (copy the current cell)

which makes a copy of the current cell with the name specified by <u>new</u>.

* GET   (get a cell from the file system)

which reads a file of the name given by <u>ccel</u> from the disk into the memory as the new current cell. BRUTUS will get all cells referenced by the new current cell if they have not been resident in the memory.

* U-C   (unexpand cells)

   which unexpands all cells of which the positions (the lower-left corners) lie in the interior of the cursor.

* E-C   (expand cells)

   which expands all cells positioned within the cursor.

* A-C   (add one cell)

   which adds the current cell, after being processed according to the entities in the status table, to the edit cell as a subcell. The added one will be positioned where the cursor is, that is, both lower-left corners of the added cell and the cursor will coincide.

* D-C   (delete cells)

   which deletes all cells lying within the cursor.

* YNK (create a yank cell)

   which creates a new cell named "yank" with all boxes which intersect with the cursor as its components. All current layers are considered. Boxes partially lying in the cursor are partitioned into several subparts (also boxes), and the yank cell includes all subparts lying completely within the cursor. A previously defined yank cell will be replaced with the new one by the next YNK commands, i.e., there is always only one cell "yank" in the BRUTUS system.

* PUT   (add the yank cell)

   which adds the yank cell, after being processed according to the entities in the status table, to the edit cell at the cursor position. One should note that if exp=0 when the PUT command has been given, he must give the CNG command to change the name of the cell "yank". Otherwise, should there be further YNK commands given, the result would follow the cell contents given by the last YNK. All current layers are affected.

* B-C   (break cells)

   which breaks all cells lying within the cursor by one level. A cell in the BRUTUS system consists of 2 major parts, one is the part of elements such as boxes, zigzags and labels, and the other is the part of subcells. As one breaks a cell by one level, the data structure is changed in the way that all its elements and its subcells become elements and subcells of the edit cell, as shown in Figure 8.

when the first ES is pointed by the light pen, BRUTUS will connect the start point and the end point and thus results in a closed zigzag. The second ES must be given to add the new zigzag to the edit cell, otherwise it will be ignored by BRUTUS. All layers indicated by clayr are affected.

* D-S   (delete zigzags)

which deletes zigzags of which every points lying within the cursor from the edit cell. All current layers are affected.

Label Manipulation

All label manipulation commands refer to the edit cell.

* A-T label   (add label)

where label is a string of characters with a length less than 30 and starts at the cursor position. The visibility of the label is up to exp in the status table. All current layers are affected.

* D-T   (delete labels)

which deletes all labels of which the starting position is lying within the cursor. All current layers are affected.

* U-T

which disables the visibility of all labels of the edit cell.

* E-T

which enables the visibility of all labels of the edit cell.

Cell Manipulation

Since BRUTUS allows a new cell to be built up hierarchically, it must have the definitions of all referenced subcells kept in the memory. Moreover, the subcells may in turn call other cells as components. Figure 7 shows a typical data structure of an edit cell in the BRUTUS system. Each cell i in the first row, except cell 0, is the head of the definition of cell i referenced by the edit cell, cell 0, or by other subcells. No mutual references are allowed.

* DEL name   (delete a cell definition)

where name can be the name of an edit cell or a subcell. The specified cell definition in the memory and all references to the deleted one in other cell definition or in the edit cell are removed.

BRUTUS

* ARR x y   (array generation)

   where x and y are digits 0 through 9 to which BRUTUS will divide the cursor into (x+1)*(y+1) partitions accordingly, and then PUTs the yank cell at each partition with the cell positioned at the lower-left corner of each partition. The user must be careful that the size of the partition is larger than the cell size.

* ERA   (erase boxes)

   which carries out YANK command (thus a yank cell is generated) and then the yanked portion displayed on the screen will be erased. Note that the effect includes a new yank cell and the disappearance of the yanked area on the display with current layers affected only.

Others
------

* CIF ES <u>int</u>

     (generate a CIF disk file)

   CIF <u>file</u> <u>int</u>

   BRUTUS generates a CIF file with the name "cif" or specified by <u>file</u> according to whether ES is given or not. The names and dimensions of all cells in the memory are recorded by the user extensions 9 and 0 in every corresponding DS & DF block of the generated CIF file. Since BRUTUS may keep several edit cells in the memory, each with several cell definitions as shown in Figure 7,

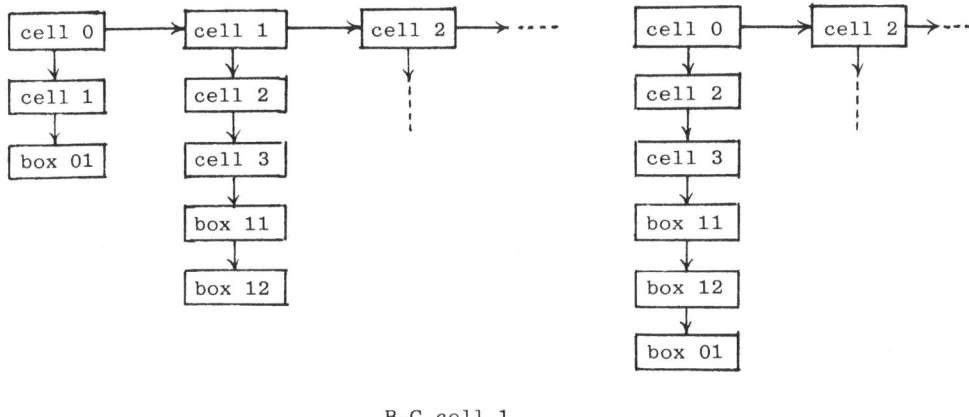

Figure 8. Break cell 1 by one level.

all cells are given a symbol number in the ascending order according to their linkage sequence for their counterparts (DS & DF blocks) in the CIF file.

If there is a need for scaling between the BRUTUS coordinates and the CIF coordinates, <u>int</u> serves this purpose.

(BRUTUS coordinates)*int=CIF coordinates.

* FIC ES <u>float</u>

    (read a CIF file)

  FIC <u>file</u> <u>float</u>

BRUTUS reads a CIF file with the name "fic" or specified by <u>file</u> according to whether ES is given or not and then creates the data structure. The name of each cell definition will be "cifn" where n is the symbol number of its counterpart (DS & DF block) in the CIF file. The value of n, however, is limited between 1 and 99. As <u>int</u> in the CIF command, <u>float</u> serves as the scaling factor.

(CIF coordinates)*float=BRUTUS coordinates.

  note: If the DS & DF block has a user extension 9 followed by a label, the corresponding cell name will be replaced by the label instead of "cifn."

* PLT <u>float</u>

which produces the layout plot on the CPS-15 pen plotter. Every 200 units of length in the layout is equivalent to 1 cm in the output plot. Since the largest plot dimension the CPS-15 can handle is about 80 cm in width and the typical size of a layout may go as high as several tens of thousands of units, a scaling factor, <u>float</u>, is needed to keep the output plot at a proper size. For instance, to have an output plot of 30*30 cm for a layout of which the size is 60000 units in each dimension, one should give a factor of 0.1.

* SAV   (save all edit cells)

which saves all edit cells kept by BRUTUS in the disk with every file name the same as its corresponding cell name.

* ><

which calculates the size of every cell in the memory.

* ASK

    which returns information concerning boxes and subcells of the edit cell.

* NP

    which selects the next page of the command table.

* DIS

    which displays the edit cell on the screen according to all settings in the status table. The command table will disappear from the screen when DIS is issued. To recover the command table when the display is completed, one simply points the light pen at an arbitrary point on the screen.

* STP

    which terminates the invocation of BRUTUS and returns control back to the UNIX Shell.

REMARKS

    The CAESAR system of the University of California, Berkeley, was the target when we started the implementation in the middle of January, 1983. As the implementation went on, we found that the deviation from our target somehow became greater and greater due to all kinds of factors. And that is the main reason why the name BRUTUS is given.

    Further implementation of BRUTUS is still going on to provide more facilities for the users.

    Figure 9 shows some IC layout designs provided by one of the BRUTUS users, Mr. Chung Hen Chow (who is now a graduate student of the Institute of Electronics, N.C.T.U.). Feedbacks returned from BRUTUS users were much stronger than expected, quite some awkwardness was pointed out by students when using BRUTUS for their layout design. Difficulties arise in that users have to explicitly specify (x,y) coordinates when positioning a desired location, and that most editing functions are cursor associated, both make the graph manipulation unnecessarily complicated and tedious. To delete a box, for example, one has to move the cursor and adjust its size to enclose the box and then give the delete command; why not just point at (or even near) the box and say, "delete it." Exact coordinate values of desired positions on the screen are often unknown to users in many circumstances.

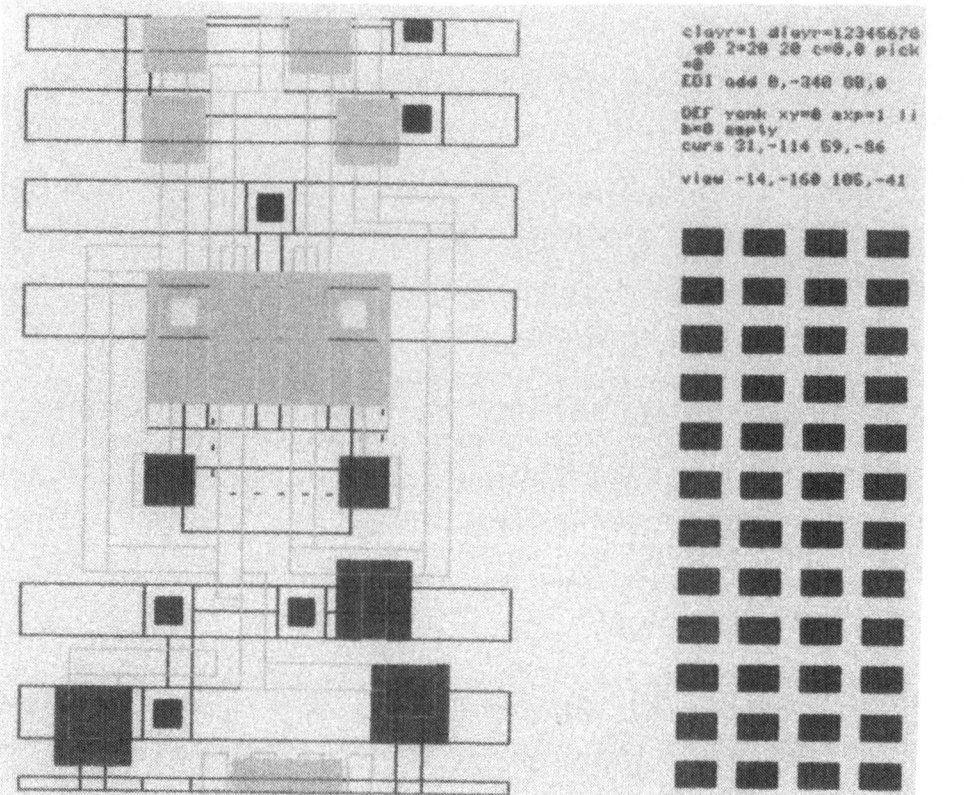

Figure 9(a). A layout design (courtesy of Mr. Chung, one of the BRUTUS users).

Commands of new approach have been designed to remedy these defects. What about those old commands? They remain in the command table while new commands are located in 2 more command tables (Figure 10). With such arrangement, we wish to give our users the feeling of steadiness and the appreciation of ongoing progress in BRUTUS.

Figure 11 shows the role played by BRUTUS in the VLSI project held at NCTU. What BRUTUS does is merely drawing, and the information transferred between BRUTUS and other modules is expressed in CIF (Caltec Intermediate Form) format [2]. BRUTUS accepts CIF for layout drawing and generates CIF for design validation.

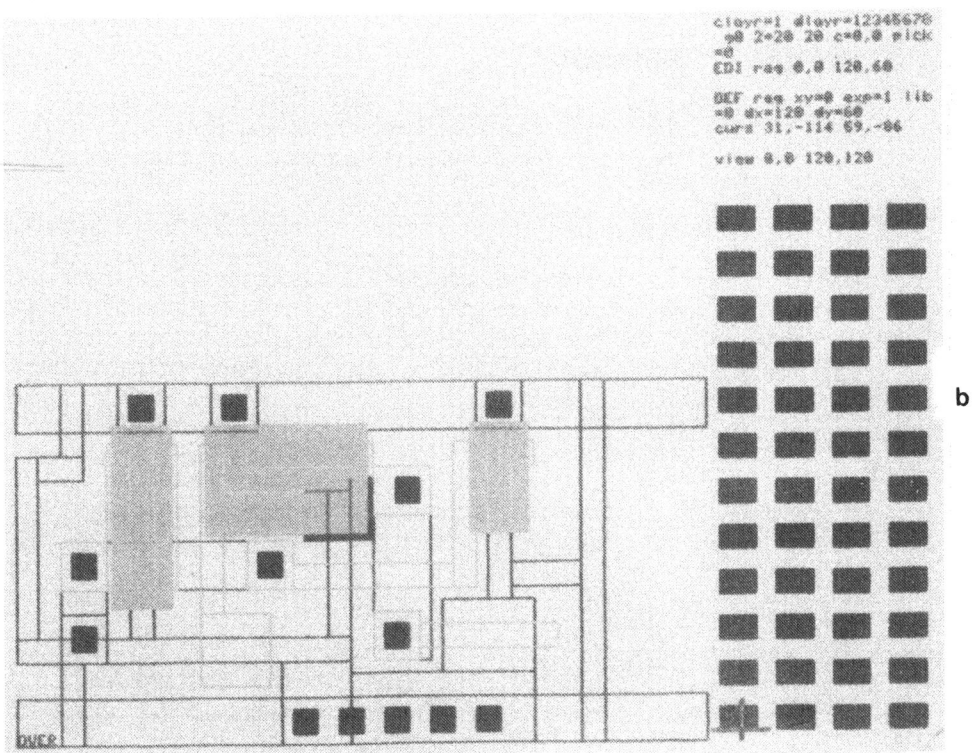

Figure 9(b). A layout design (courtesy of Mr. Chung, one of the BRUTUS users).

REFERENCES

[1] J. Ousterhout, "Editing VLSI Circuits with CAESAR," Computer Science Division, E.E. and Computer Science, University of California, Berkeley (manual of version 6 CAESAR, VLSI Tool Technical Report, 1982).

[2] R. W. Hon and C. H. Sequin, "A Guide to LSI Implementation," 2nd edition, SSL-79-7, XEROX, Palo Alto Research Center, January 1980.

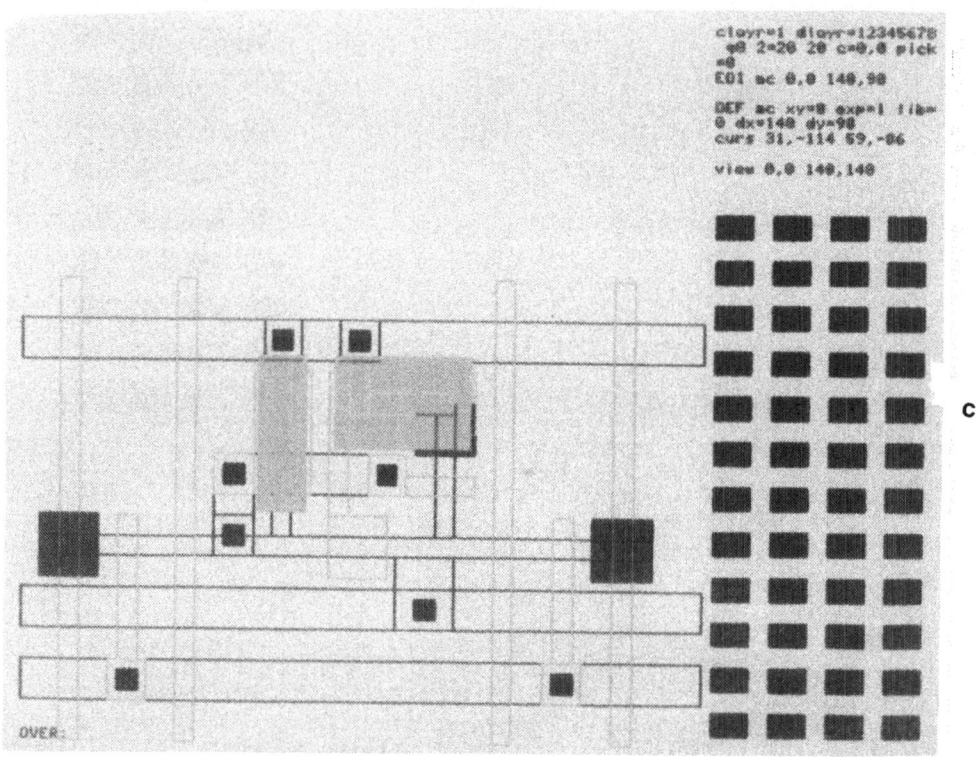

Figure 9(c). A layout design (courtesy of Mr. Chung, one of the BRUTUS users).

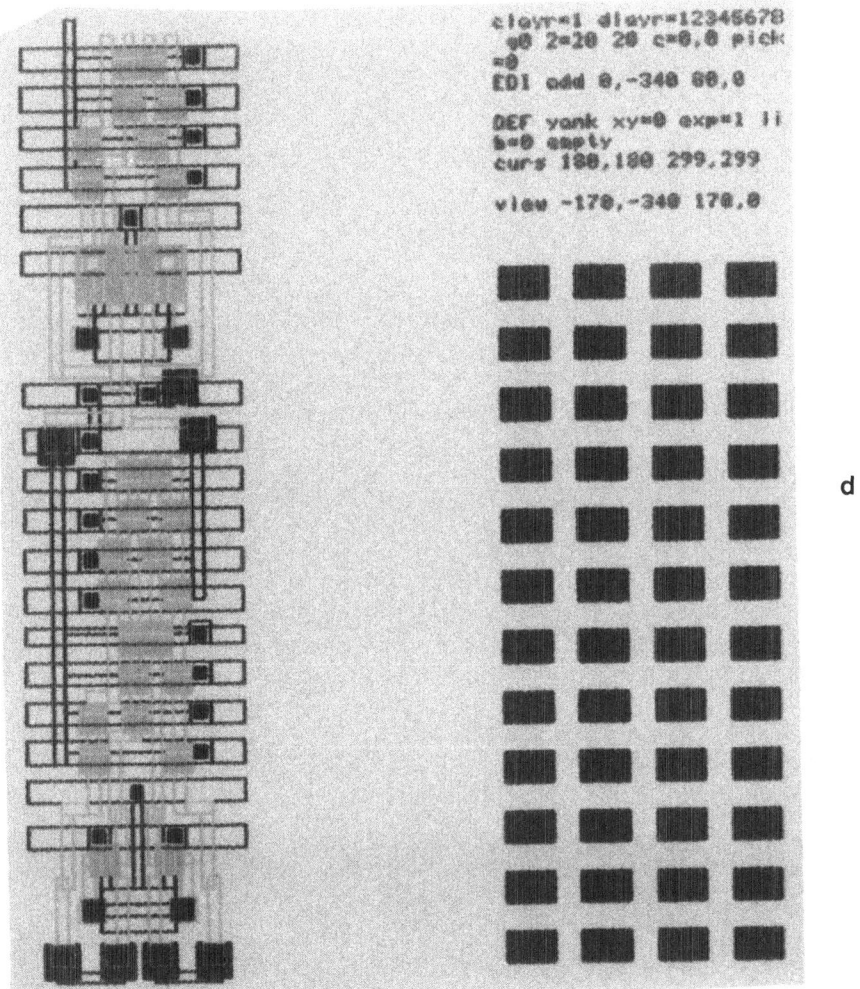

Figure 9(d). A layout design (courtesy of Mr. Chung, one of the BRUTUS users).

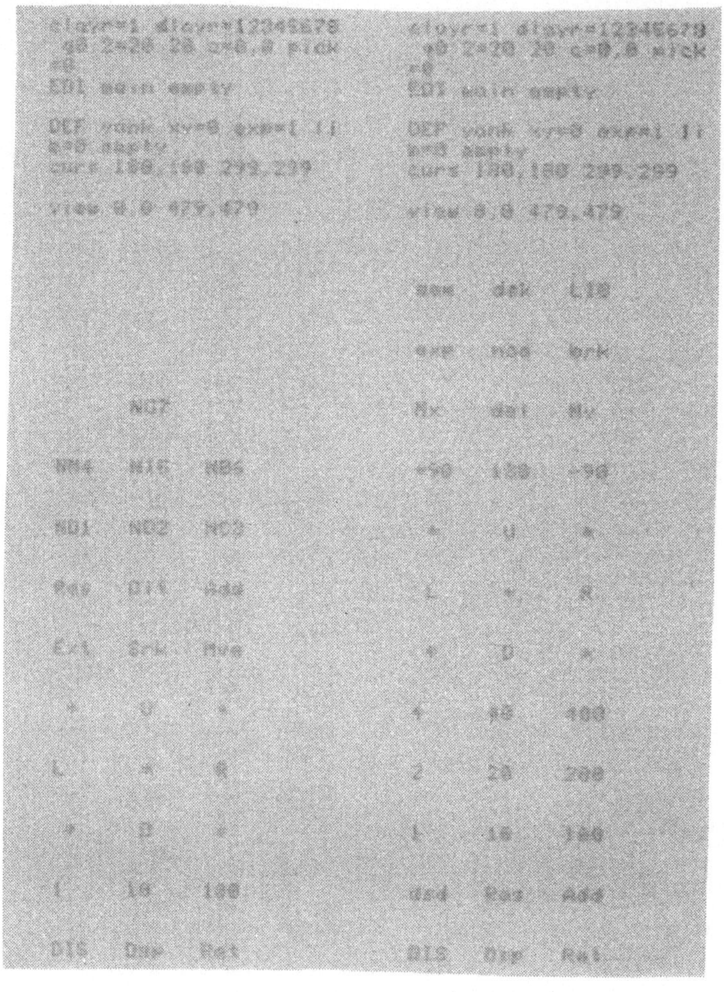

Figure 10. Two newly added command tables.

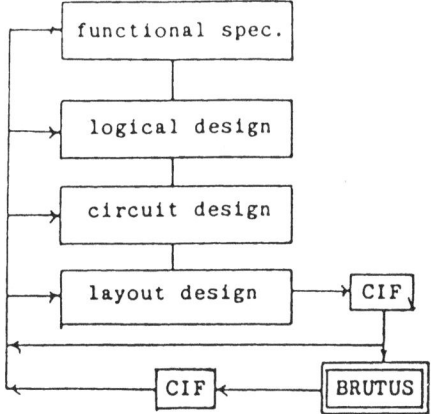

Figure 11. BRUTUS in the IC design cycle.

PLAMG: AN AUTOMATIC PLA MINIMIZER AND

GENERATOR FOR VLSI SYSTEM DESIGN

Cheng Chen, Wen Zen Shen*, Young-Li Lee, Li-Wen Shih and
Yeh Jian Liang

Dept. of Computer Engineering, National Chiao Tung
University, Hsin-Chu, Taiwan, R.O.C.
*Institute of Electronics, National Chiao Tung University
Hsin-Chu, Taiwan, R.O.C.

ABSTRACT

In this paper, we will introduce an Automatic PLA Minimizer and Generator, PLAMG, for VLSI system design, which has been designed and implemented in VAX-11/780 running under UNIX operating system. The PLAMG contains four parts. The first is Bprer which accepts the Boolean Equations as inputs and translates into equivalent truth table. The second part is product minimization which is based on MINI-algorithm. The third part is simple column folding to optimize the given PLA topologically. And finally, the reduced PLA matrix is sent to PLAGEN to generate the NMOS NOR-NOR plane layout automatically. Several examples has been tested and evaluated. The results showed that PLAMG has good performance and is a useful tool for VLSI system design. At the end of the paper, some future extensions and improvements are also discussed.

INTRODUCTION

Recently, VLSI system design has played a more important role in both semiconductor and computer fields due to the rapid progression of semiconductor technologies. Roughly speaking, a VLSI may contain from several tens of thousands to one million transistors per chip. Then according to the Mead and Conway approach [1] a good VLSI architecture in the context of this article should have one or more following properties [2]: (1) It should be implemental by only a few different types of simple cells. (2) It should have simple and regular data and control paths so that the cells can be

connected by a network with local and regular interconnections.
(Long-distance or irregular communication is thus minimized.) (3)
It should use extensive pipelining and multiprocessing. The above
three features dominate the current VLSI architecture design. For
general digital system (or digital computer) design, one of the
most irregular parts in the whole system is the random logic, either
for control logic or other functional units. Hence, PLA (Programmable Logic Array) becomes more important for VLSI system design
because of its simple and regular structure. In addition, due to
its regularity PLA provides the designer a short design cycle to
generate its layout easily and automatically [3-5]. In this paper,
we will describe one useful automatic PLA tool "PLAMG" designed and
developed in Chiao Tung University recently. The whole system is
shown in Figure 1.

As shown in Figure 1, the designer inputs his system specification by appropriate Boolean Equations. The input Boolean Equations can be parsed and translated into an equivalent truth table
form by using "Bprer" (Boolean Preprocessor). Then the equivalent
truth table will be reduced either by minterm minimization (MINI)
or by simple column folding (SCF) or both. The result is an optimized or near-optimized PLA personality matrix (PM). Finally, the
reduced PLA PM will be sent to "PLAGEN" to generate the NMOS NOR-NOR PLA layout diagram (in CIF [6]) automatically. The CIF (Caltec
Intermediate Form) can be translated and displayed at graphic terminal or plotted by using CPS-15 plotter directly [7]. In the
following sections, we will introduce and describe the basic concepts and principles of Bprer, the MINI-algorithm, simple column
folding algorithm and PLAGEN. And we also illustrate and evaluate
PLAMG by using several testing examples sampled from few to thirty
input variables and output equations. The experimental results
show that PLAMG is a useful CAD tool for VLSI system design.
Finally, some further improvements about PLAMG are also mentioned
and introduced briefly.

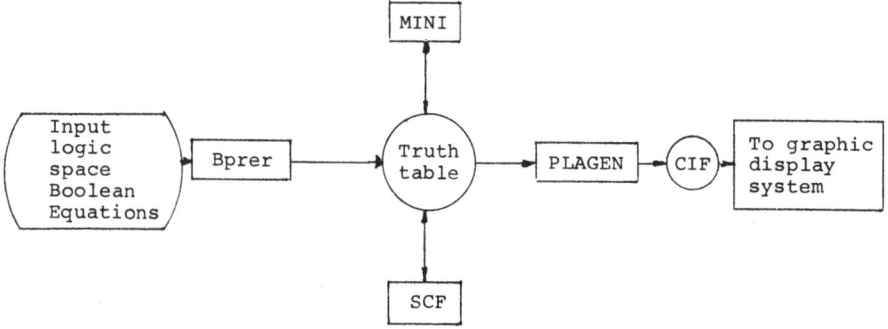

Figure 1. Overall diagram of PLAMG.

BPRER: BOOLEAN PREPROCESSOR

As the first stage of PLAMG, Bprer accepts any well-described Boolean Equations in terms of AND, OR, NOT and Don't care forms. The Bprer contains two main parts. The first part is called B-parser (Boolean parser) to parse the input Boolean Equations and to check its syntax structure. If there is any illegal form then an appropriate error message will be sent out to the designer. Otherwise, the next part of truth table converter will translate the parsed outputs into its equivalent truth table form. The input Boolean Equations have similar form as in [4]. In the beginning, the designer should specify the INORDER and OUTORDER to express the set of all input variable names and output names, respectively. And we use "+", "*", "!" and "?" to represent the "OR", "AND", "NOT" and "Don't Care", respectively.

Ex.: Consider the following input Boolean Equation form:

```
INORDER a(1), b-c, load;
OUTORDER Control (1), control (2), X1
x=[a(1)+b-c]*load;
Control (1)=[[!b-c*load]+b-c]*a(1);
Control (2)=!load+[!a(1)+b-c*a(1)]*load;
?X=!a(1)*load;
?Control (1), control (2)=!a(1)*b-c*! load;
X=! a(1)*! load;
```

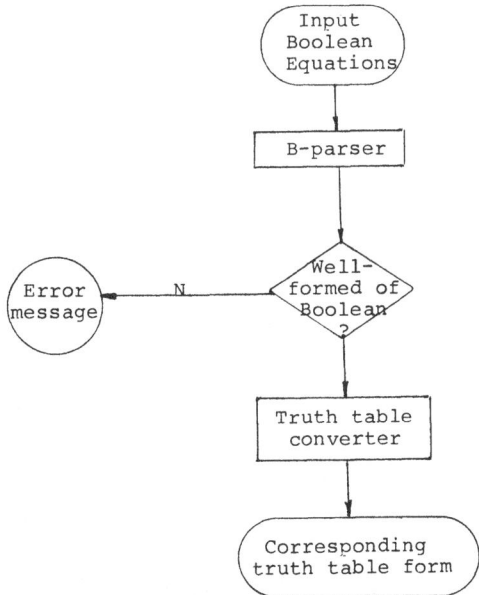

Figure 2. Simple flow of Bprer.

Here, we use ";" as the delimiter of each statement. In the appendix we give a complete example with 14 input variables and 20 outputs. Figure A(a) and A(b) are its input Boolean Equations and corresponding truth table, respectively.

PLA PRODUCT TERM MINIMIZATION

It is well known that Boolean function minimization is not a new problem. Obviously, the Quine-McClusky method is useful for single output minterm minimization [8]. As for multiple-output Boolean function minimization, the problem is not as simple and straightforward. Conversely, it is a NP-complete problem to obtain the absolute minimization according to the conventional Boolean function minimization [9]. Hence, until now, there are many heuristic algorithms being proposed to get a near-optimal solution [9-13]. Among them, MINI-algorithm [13] is the first powerful heuristic algorithm drastically different from conventional minimization algorithm [13]. In PLAMG, we base on this algorithm and develop one MINI-like algorithm to optimize the product terms of the given multiple-output Boolean Equations. Basically, MINI-algorithm has the following four operators [13]:

```
INORDER G4,G3,G2,G1,K1,K2,K3,K4,K5,K6,K7,K8,K9,K10;
OUTORDER    G4,G3,G2,G1,T0,T1,T2,T3,T4,T5,T6,T7,T8,T9,T10,T11,T12,T13,T14,T15;

T0 = !G4*!G3*!G2*G1*!K1+!G4*!G3*!G2*!G1*K1;
T1 = !G4*!G3*!G2*!G1;
T2 = !G4*!G3*!G2*!G1*K2*K3+!G4*!G3*!G2*!G1*K2*!K3+!G4*!G3*!G2*!G1*!K2*K3*!K9+
     !G4*!G3*G2*!G1*!K2*K3*K9+!G4*!G3*!G2*!G1*!K2*!K3;
T3 = !G4*!G3*G2*G1*K9+!G4*!G3*G2*G1*!K9;
T4 = !G4*G3*!G2*!G1*K5+!G4*G3*!G2*!G1*K6+!G4*G3*!G2*!G1*K7;
T5 = !G4*G3*!G2*G1;
T6 = !G4*G3*G2*!G1;
T7 = !G4*G3*G2*G1;
T8 = G4*!G3*!G2*!G1;
T9 = G4*!G3*!G2*G1*K4*K5+G4*!G3*!G2*G1*!K4*K5+G4*!G3*!G2*G1*!K4*!K5;
T10 = G4*!G3*G2*!G1*!K2*!K8*K9+G4*!G3*G2*!G1*K2*!K8+G4*!G3*G2*!G1*!K2*!K8*!K9+G4*!G3*G2*!G1*K8;
T11 = G4*!G3*G2*G1*K10;
T12 = G4*G3*!G2*!G1;
T13 = G4*G3*!G2*G1*!K9*!K10+G4*G3*!G2*G1*K9*!K10+G4*G3*!G2*G1*K10;
T14 = G4*G3*G2*!G1;
T15 = G4*G3*G2*G1;
G1 = !G4*!G3*!G2*!G1*K1+!G4*!G3*!G2*!G1*K2*K3+!G4*!G3*G2*!G1*K2*!K3+!G4*!G3*!G2*!G1*!K2*K3*!K9
     +!G4*!G3*G2*G1*!K9+!G4*G3*!G2*!G1*K5+!G4*G3*!G2*!G1*K6+!G4*G3*!G2*!G1;
     +G4*!G3*!G2*G1+G4*!G3*!G2*G1*!K4*K5+G4*!G3*!G2*G1*K4*K5+G4*!G3*G2*!G1*K2*!K8
     +G4*!G3*G2*!G1*!K2*!K8*K9+G4*!G3*G2*G1*K8+G4*G3*!G2*G1*!K9*!K10+G4*G3*G2*!G1;
G2 = !G4*!G3*!G2*G1+!G4*!G3*G2*!G1*K2*K3+!G4*!G3*G2*!G1*!K2*!K3+!G4*G3*!G2*G1*K7
     +!G4*G3*!G2*G1+!G4*G3*G2*!G1+!G4*G3*!G2*!G1*!K4*K6+G4*!G3*G2*!G1*!K2*!K8+G4*!G3*G2*!G1*K8
     +G4*G3*!G2*G1+G4*G3*G2*!G1+G4*G3*G2*!G1;
G3 = !G4*!G3*G2*!G1*K2*K3+!G4*!G3*G2*!G1*!K2*K3*!K9+!G4*G3*!G2*!G1*!K2*!K3
     +!G4*G3*!G2*G1*!K9+!G4*G3*!G2*!G1*K6+!G4*G3*!G2*G1+!G4*G3*!G2*!G1*!K2*!K8
     +G4*!G3*G2*!G1*!K2*!K8*K9
     +G4*!G3*G2*G1*K10+G4*G3*!G2*!G1+G4*G3*!G2*G1*!K9*!K10+G4*G3*!G2*G1*K10+G4*G3*G2*!G1
G4 = !G4*!G3*G2*!G1*K2*K3+!G4*!G3*G2*!G1*!K2*K3*!K9+!G4*!G3*G2*G1*!K9+!G4*G3*!G2*!G1*K5
     +!G4*G3*!G2*!G1*K7+G4*!G3*!G2*G1+G4*!G3*!G2*G1+G4*!G3*!G2*G1*!K4*K5
     +G4*!G3*!G2*G1*!K4*K3
     +G4*!G3*G2*!G1*!K2*!K3+G4*!G3*G2*!G1*!K2*!K8*K9+G4*!G3*G2*!G1*K8+G4*!G3*G2*G1*K10
     +G4*G3*!G2*!G1+G4*G3*!G2*G1*!K9*!K10+G4*G3*!G2*G1*K10+G4*G3*G2*!G1;
```

Figure A(a) Boolean Equations of 14 inputs and 20 outputs.

PLAMG 123

PLA PERSONALITY MATRIX BEFORE MINIMIZATION : NAME = PLA_14_20

```
INPUT:            OUTPUT:
00000 00000 0000  00000 00000 00000 00000

 1: 00000 ----- ----   00001 00000 00000 00000    47: 0101- ----- ----   00010 00000 00000 00000
 2: 00001 ----- ----   00001 00000 00000 00000    48: 1110- ----- ----   00010 00000 00000 00000
 3: 0001- ----- ----   00000 10000 00000 00000    49: 0001- ----- ----   00100 00000 00000 00000
 4: 0010- 11--- ----   00000 01000 00000 00000    50: 0010- 11--- ----   00100 00000 00000 00000
 5: 0010- 10--- ----   00000 01000 00000 00000    51: 0010- 10--- ----   00100 00000 00000 00000
 6: 0010- 01--- --0-   00000 01000 00000 00000    52: 0100- ----- 1---   00100 00000 00000 00000
 7: 0010- 01--- --1-   00000 01000 00000 00000    53: 0101- ----- ----   00100 00000 00000 00000
 8: 0010- 01--- ----   00000 01000 00000 00000    54: 0110- ----- ----   00100 00000 00000 00000
 9: 0011- ----- --1-   00000 00100 00000 00000    55: 1000- ----1 ----   00100 00000 00000 00000
10: 0011- ----- --0-   00010 00100 00000 00000    56: 1010- 1---- ----   00100 00000 00000 00000
11: 0100- ---10 ----   00000 00010 00000 00000    57: 1010- ----- ----   00100 00000 00000 00000
12: 0100- ----1 ----   00000 00010 00000 00000    58: 1100- ----- ----   00100 00000 00000 00000
13: 0100- ----- 1---   00000 00010 00000 00000    59: 1101- ----- ---1   00100 00000 00000 00000
14: 0101- ----- ----   00000 00001 00000 00000    60: 1110- ----- ----   00100 00000 00000 00000
15: 0110- ----- ----   00000 00000 10000 00000    61: 0010- 11--- ----   01000 00000 00000 00000
16: 0111- ----- ----   00000 00000 01000 00000    62: 0010- 01--- --0-   01000 00000 00000 00000
17: 1000- ----- ----   00000 00000 00100 00000    63: 0010- 00--- ----   01000 00000 00000 00000
18: 1001- --11- ----   00000 00000 00010 00000    64: 0011- ----- ----   01000 00000 00000 00000
19: 1001- --01- ----   00000 00000 00010 00000    65: 0100- ----1 ----   01000 00000 00000 00000
20: 1000- --0-1 ----   00000 00000 00010 00000    66: 0101- ----- ----   01000 00000 00000 00000
21: 1010- 0---- -01-   00000 00000 00001 00000    67: 0110- ----- ----   01000 00000 00000 00000
22: 1010- 1---- -0--   00000 00000 00001 00000    68: 1010- 1---- ----   01000 00000 00000 00000
23: 1010- 0---- -00-   00000 00000 00001 00000    69: 1010- 0---- ----   01000 00000 00000 00000
24: 1010- ----- -1--   00000 00000 00001 00000    70: 1011- ----- ---1   01000 00000 00000 00000
25: 1011- ----- ---1   00000 00000 00000 10000    71: 1100- ----- ----   01000 00000 00000 00000
26: 1100- ----- ----   00000 00000 00000 01000    72: 1101- ----- ----   01000 00000 00000 00000
27: 1101- ----- --00   00000 00000 00000 00100    73: 1101- ----- ---1   01000 00000 00000 00000
28: 1101- ----- --10   00000 00000 00000 00100    74: 1110- ----- ----   01000 00000 00000 00000
29: 1101- ----- ---1   00000 00000 00000 00100    75: 1010- 11--- ----   10000 00000 00000 00000
30: 1110- ----- ----   00000 00000 00000 00010    76: 0010- 01--- --0-   10000 00000 00000 00000
31: 1111- ----- ----   00000 00000 00000 00001    77: 0011- ----- --0-   10000 00000 00000 00000
32: 00001 ----- ----   00010 00000 00000 00000    78: 0101- ---1- ----   10000 00000 00000 00000
33: 1010- 11--- ----   00010 00000 00000 00000    79: 0100- ----- 1---   10000 00000 00000 00000
34: 1010- 10--- ----   00010 00000 00000 00000    80: 0111- ----- ----   10000 00000 00000 00000
35: 1010- 01--- --0-   00010 00000 00000 00000    81: 1000- ----- ----   10000 00000 00000 00000
36: 0011- ----- --1-   00010 00000 00000 00000    82: 1001- --01- ----   10000 00000 00000 00000
37: 0100- ---1- ----   00010 00000 00000 00000    83: 1000- --0-1 ----   10000 00000 00000 00000
38: 0100- ----1 ----   00010 00000 00000 00000    84: 1010- 1---- -1--   10000 00000 00000 00000
39: 0110- ----- ----   00010 00000 00000 00000    85: 1010- 0---- ----   10000 00000 00000 00000
40: 1000- ----- ----   00010 00000 00000 00000    86: 1010- ----- -1--   10000 00000 00000 00000
41: 1001- --11- ----   00010 00000 00000 00000    87: 1011- ----- ---1   10000 00000 00000 00000
42: 1001- --11- ----   00010 00000 00000 00000    88: 1100- ----- ----   10000 00000 00000 00000
43: 1010- 1---- -0--   00010 00000 00000 00000    89: 1101- ----- --00   10000 00000 00000 00000
44: 1010- 0---- -10-   00010 00000 00000 00000    90: 1101- ----- ---1   10000 00000 00000 00000
45: 1010- ----- -1--   00010 00000 00000 00000    91: 1110- ----- ----   10000 00000 00000 00000
46: 1100- ----- ----   00010 00000 00000 00000
```

Figure A(b) Truth table before minimization.

b

Disjoint sharp operation (#);
Cube expansion operation (SPE);
Cube reduction operation;
Cube reshaping operation;

Figure 3 is a simple rough flow of MINI-like algorithm to indicate the procedures of minterm minimization for multiple-output Boolean Equations. For each cube operation, there is one or more heuristic algorithms to use to select the desired cubes according to some criteria. In [13] and [14], the detailed information about these heuristic algorithms can be further described.

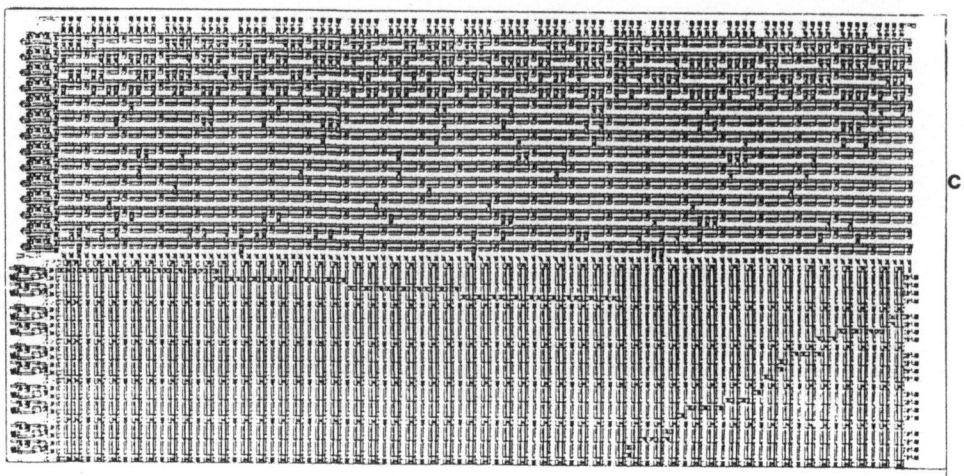

PLA PERSONALITY MATRIX AFTER MINIMIZATION : NAME = PLA_14_20

```
       INPUT:                         OUTPUT:

       00000 00000 0000               00000 00000 00000 00000
       00000 00000 1111               00000 00000 11111 11111
       01234 56789 0123               01234 56789 01234 56789

  1:   1101- ----- ----               00000 00000 00000 00100
  2:   1010- ----- ----               00000 00000 00001 00000
  3:   0010- 01--- --0-               11010 00000 00000 00000
  4:   1010- 0---- -00-               11010 00000 00000 00000
  5:   0010- 00--- ----               01000 00000 00000 00000
  6:   1001- ---1- ----               00010 00000 00010 00000
  7:   1000- --0-1 ----               00100 00000 00010 00000
  8:   0010- 11--- ----               11000 00000 00000 00000
  9:   1001- --01- ----               10000 00000 00000 00000
 10:   1010- 1---- -0--               11000 00000 00000 00000
 11:   0011- ----- --1-               00000 00100 00000 00000
 12:   0000- ----- ----               00001 00000 00000 00000
 13:   1101- ----- --00               11010 00000 00000 00000
 14:   00001 ----- ----               00010 00000 00000 00000
 15:   1011- ----- ---1               11000 00000 00000 10000
 16:   0100- ----- 1---               10100 00010 00000 00000
 17:   0100- ----1 ----               01010 00010 01000 00000
 18:   1111- ----- ----               00000 00000 00000 00001
 19:   0100- ---1- ----               10010 00010 00000 00000
 20:   1101- ----- ---1               11100 00000 00000 00000
 21:   0010- ----- ----               00000 01000 00000 00000
 22:   0011- ----- --0-               11010 00100 00010 00000
 23:   1010- ----- -1--               10110 00000 00000 00001
 24:   0111- ----- ----               10000 00000 01000 00000
 25:   -010- 1---- ----               00110 00000 00000 00000
 26:   0001- ----- ----               00100 00000 00000 00000
 27:   0101- ----- ----               01000 00001 00000 00000
 28:   1000- ----- -----              10010 00000 00100 00000
 29:   0110- ----- ----               01110 10000 00000 00000
 30:   1110- ----- ----               11110 00000 00000 00010
 31:   1100- ----- ----               11110 00000 00000 01000
```

Figure A(c) PLA layout diagram before minimization.

Figure A(d) Truth table after minimization.

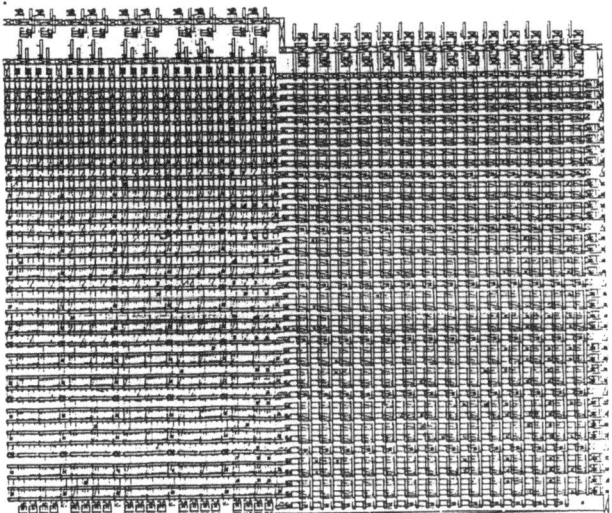

PLA PERSONALITY MATRIX AFTER MINIMIZATION : NAME = PLA_14_20

```
        INPUT:          OUTPUT:              INPUT:          OUTPUT:

        00000 0000      00000 00000          00000 0000      00000 00000
        00000 0001      00001 11111          00001 0110      01011 01000
        01234 6789      43691 12479          01252 9135      16000 78832

  1:    1101- ----      00000 00010              AND PLANE FOLDING:
  2:    1010- ----      00000 00100
 12:    0000- ----      10000 00000          4 --->12    CUTTING    ROW :  14
 18:    1111- ----      00000 00001         10 ---> 5    CUTTING    ROW :  16
 21:    0010- ----      00100 00000
 14:    00001 ----      00000 00010          7 --->11    CUTTING    ROW :   7
 29:    0110- ----      10001 00011
 27:    0101- ----      10010 00001          8 --->13    CUTTING    ROW :   6
 26:    0001- ----      01000 00001          6 ---> 9    CUTTING    ROW :   3
 28:    1000- ----      00100 01010
 24:    0111- ----      00100 10000              OR PLANE FOLDING:
 19:    0100- --1-      00100 00110
 16:    0100- ---1      00100 00101         17 ---> 3    CUTTING    ROW :   1
  9:    1001- -01-      00100 00000          6 ---> 0    CUTTING    ROW :  21
 31:    1100- ----      10110 00011          4 ---> 1    CUTTING    ROW :  12
  6:    1001- --1-      01000 00010
 30:    1110- ----      10100 01011         19 ---> 2    CUTTING    ROW :  18
 22:    00110 ----      10100 10010
 11:    00111 ----      00000 10000         14 ---> 8    CUTTING    ROW :   2
 25:    -010- ---1      00001 00011          5 --->13    CUTTING    ROW :  26
  8:    1010- 1--1      10100 00000
  5:    1010- 0--0      10000 00100         11 ---> 7    CUTTING    ROW :  24
  3:    00100 1--0      10100 00010         12 --->18    CUTTING    ROW :  28
 20:    1101- --1-      10100 00001
 15:    1011- --1-      10101 00000          9 --->18    CUTTING    ROW :  27
 13:    11010 --0-      10100 00010         10 --->15    CUTTING    ROW :  29
 17:    0100- 1---      10000 00110
  7:    1000- 10--      01000 00001
 23:    1010- -1--      00100 00011
 10:    1010- -0-1      10100 00000
  4:    10100 -0-0      10100 00010
```

Figure A(e) PLA layout diagram after minimization.

Figure A(f) Truth table after SCF.

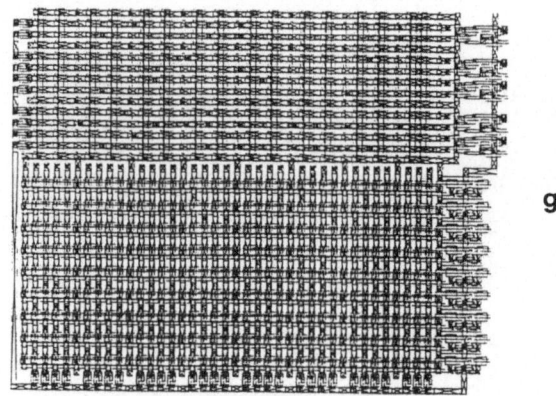

Figure A(g) PLA layout diagram after SCF
(This layout diagram is incomplete but only shows the PLA layout size after SCF).

PLA FOLDING TECHNIQUE

After the PLA logic has been minimized to reduce its product terms, the personality matrix of PLA will, in general, still be very sparse especially when the PLA matrix is large. Then the next problem is that is it possible to compact the PLA matrix topologically further? In most of the cases, the answer is definite. This is called topological optimization or PLA folding, for the given PLA matrix. Roughly speaking, folding refers to the technique by which two input/output columns or two product rows can share the same physical column or row in the layout [16]. Until now, there were many techniques proposed to do the PLA folding, such as [5,15-17]. All of those methods are based on various heuristic algorithms because the folding technique is NP-complete [15]. Recently, G. D. Hachel et al. proposed one powerful folding technique by using a simple graph theory [16]. Basically, there are two kinds of PLA folding. One is the simple column folding and the other is the simple row folding. In PLAMG, we developed a simple column folding based on the algorithm in [16,17]. The column folding problem consists of finding a set of pairs of inputs and outputs which satisfies the following two conditions: (i) Each pair of inputs/outputs must be disjoint in the product terms that depend on them. (ii) No two pairs of folded inputs/outputs imply a mutually exclusive ordering of product rows. According to [17], we can easily define a column intersection graph, $G=G(V,E)$, where V is the set of vertices of G and E is the set of edge of G, being associated to the personality of a PLA as follows. Each vertex of G is in one-to-one correspondence with a column of the personality matrix. An edge $e_k=\{v_i,v_j\}$ joins v_i and v_j, $i \neq j$, if row i and row j have "crosses" in these columns. For example, Figure 4 is a PLA form and Figure 5 is its personality matrix. Figure 6 is the column intersection graph of Figure 5. The columns

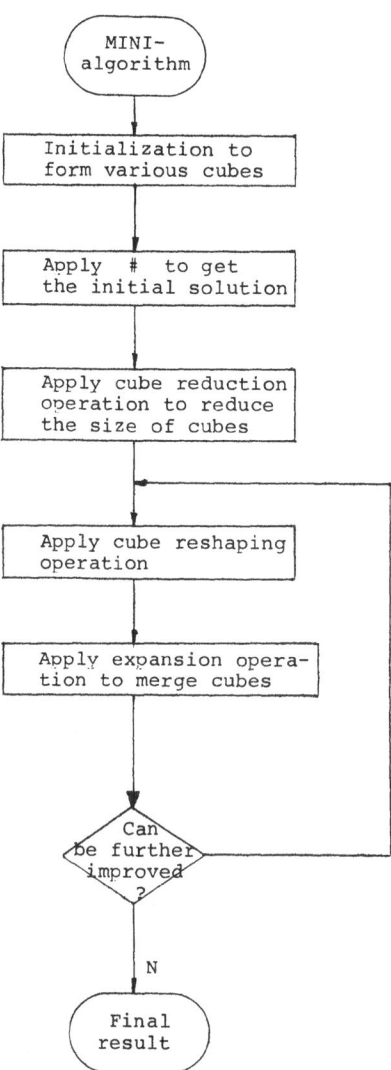

Figure 3. Overall mini-like algorithm flow.

which can be folded together according to the condition (2) described above are the ones corresponding to the set of vertices of G which have no connection edge in the column intersection graph. Now, the next problem is how to find the set of maximum possible folding pairs in the column intersection graph. By adding folding pairs as the directed edges to the given column intersection graph, we can construct a corresponding mixed graph [16]. Figure 7 is a mixed graph by adding three folding pairs (A,D), (C,B) and (f_1,f_2). By using column intersection graph and mixed graph of given PLA ckt,

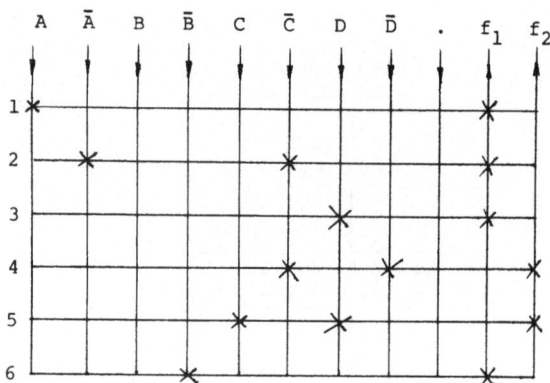

Figure 4. A PLA circuit.

we can develop a simple and straightforward heuristic algorithm to generate a set of possible maximal ordered folding pairs [14]. Figure 8 is an overall flow of this algorithm. The detailed procedures and information about this heuristic algorithm is described in [14] and [16].

PLAGEN

Because of the regularity of the PLA structure, we can generate the desired PLA layout diagram easily and automatically. According to [6], it contains the following 20 different PLA unit cells: Pla Cell, Pla Ground, Pla PXllups, Pla Connect, Pla In, Pla Clocked In, Pla OUt, Pla Clocked Out, Pla Nor Out, Pla Clocked Nor Out, Pla Hole Wires, Pla Prog Left, Pla Prog Right, Pla Prog Top, Pla Prog Bottom, Pla or Space, Pla Connect Space, Pla Ground Space, Pla

A	B	C	D	f_1	f_2
1	0	1	0	1	0
1	0	1	0	1	0
0	0	0	1	1	0
0	0	1	1	0	1
0	0	1	1	0	1
0	1	0	0	1	0

Figure 5. Personality matrix of this PLA.

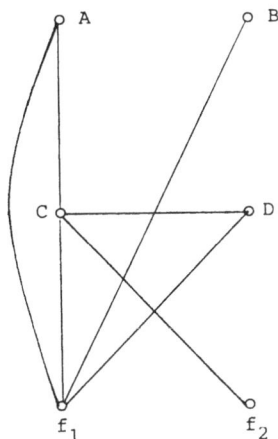

Figure 6. Column intersection graph of Figure 5.

Pullup Space, Pla Out Space. These 20 unit cells are categorized from symbol 12 to symbol 31, respectively, in [6]. Basically, one PLA layout diagram can be constructed and then generated by using these 20 unit cell symbols. At first, we generte the backbone of the given PLA ckt by using unit cell symbols 12-20 and 25-31. And then the desired PLA layout diagram will be constructed by inserting the transistors at the corresponding "minterm positions" according to the optimized PLA matrix form. Figure 9 is a four input variables/eight output with eight product terms PLA backbone diagram [6]. The generated PLA layout diagram is expressed in CIF form and

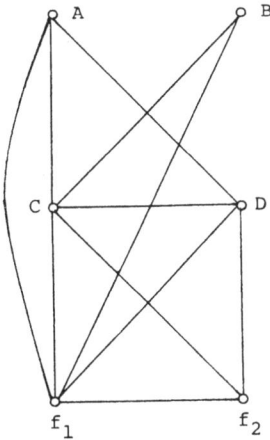

Figure 7. Mixed graph by adding three folding pairs (A,D,), (C,B) and (f_1,f_2) to the column intersection graph.

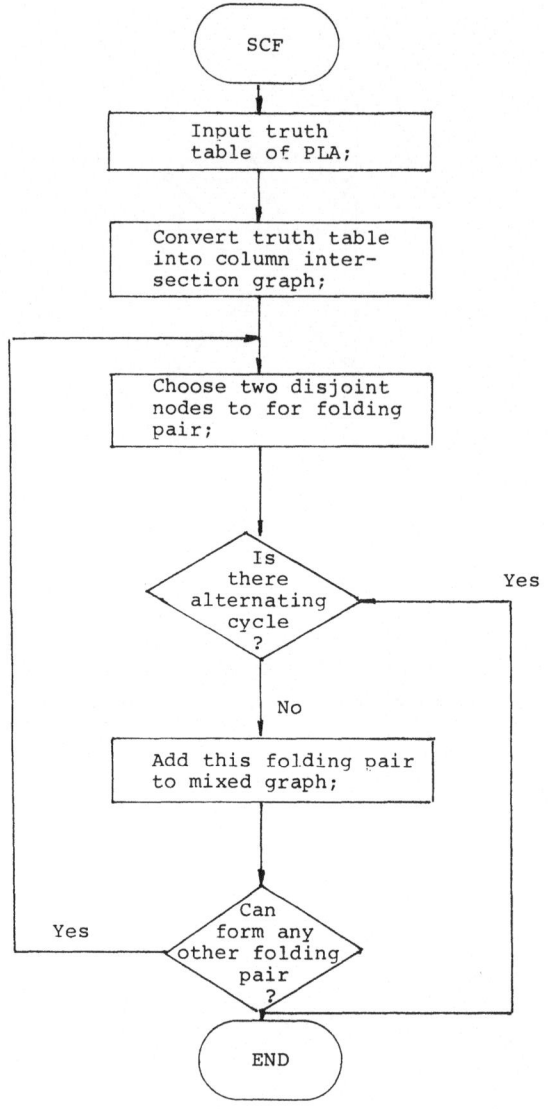

Figure 8. Overall rough flow of SCF algorithm.

can be displayed on the graphic terminal of Ramtek directly [7]. Figure 10 is an overall flow chart to show the procedure of PLAGEN roughly.

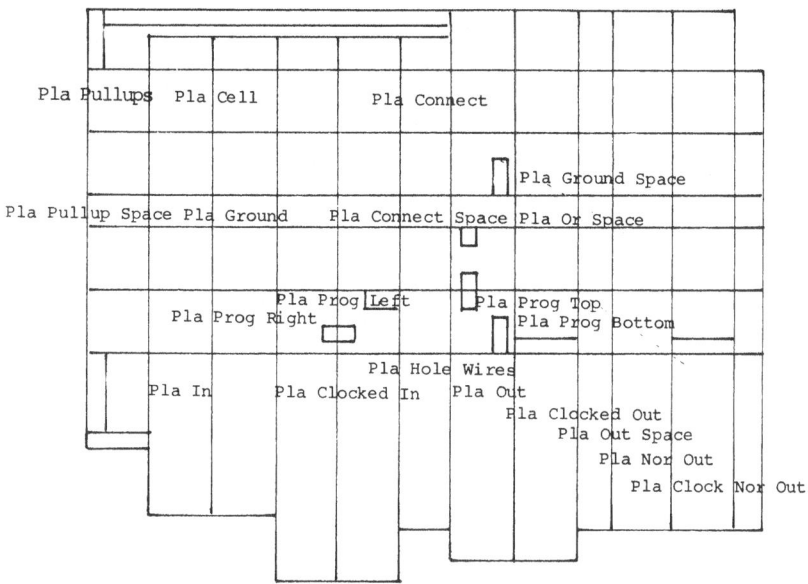

Figure 9. One PLA backbone with 4 inputs/outputs.

TEST, EVALUATION AND VERIFICATION OF SOME EXPERIMENTAL RESULTS

The whole program routines of PLAMG, including Bprer, MINI-Algorithm, SCF and PLAGEN, are implemented by using C-language and being installed in VAX-11/780 running under UNIX operating system. In our system we have tested a lot of Boolean Equations to evaluate the performances of the PLAMG. Table 1 shows the testing result of product minimization by using MINI-like algorithm. We take thirteen samples of multiple-output Boolean Equations, with input variables from four to ten. We find that there are two examples (No. 1 and No. 3) have the minimal solutions checked by using Karnaugh maps with hand calculation. And the other eleven examples also have high reduction rates with range from 8.6% to 81.4% and most of them are above 30%. Here, we define the reduction rate as follows:

Reduction rate = (Number of product terms before minimization-
                 Number of product terms after minimization)/
                 (Number of product terms before minimization)

From Table 1 we find that by using MINI-like algorithm we obtain optimal or near-optimal results in most of the cases. Among these testing examples, No. 2 and No. 4 to No. 11 are all obtained from the random number generation with the radix shown in the table [14]. And No. 12 and No. 13 are 4-bit and 3-bit binary adders, respectively, provided by the one group students in VLSI design course. Table 2 shows another eleven testing examples. All those

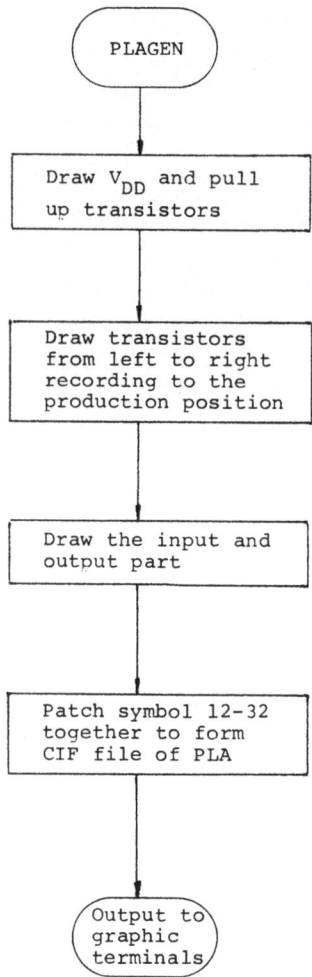

Figure 10. Overall flow of PLAGEN.

single output equations (No. 1, 2 and 4 to 8) have the results of optimal solution. These examples are obtained by using the generating method in [14,18]. And the other three examples (No. 3, 9, 10 and 11) have large output equations (above 20) but with not so many product terms. They have different results No. 10 and 11 are those Boolean Equations of control signals of pseudo machines generated from CDL (Computer Design Language) [19]. They have little reduction rates because they have no don't care terms and very scattering distribution of product terms.

Table 3 shows a set of testing examples for SCF technique. We select 9 testing examples. We find that most of the examples have

Table 1. Testing Examples and Performance Evaluation of MINI-Like Algorithm (I)

no.	number of input variable	number of output variable	random number seed	ratio of 0,1,X logic value	number of product term before minimization	number of product term after minimization	minimization ratio	optimal solution
1	4	2	--	--	23	8	--	yes
2	4	2	1467	10:10:3	15	6	60%	--
3	4	4	--	--	15	9	--	yes
4	4	7	555478	20:20:3	31	9	70.9%	--
5	5	3	478	20:20:1	24	12	50%	--
6	5	7	66745	20:20:3	31	18	42%	--
7	6	5	430952	20:20:3	31	23	25.8%	--
8	6	6	123745	10:10:3	43	28	34.8%	--
9	7	5	4925	10:10:1	53	36	32%	--
10	8	3	54398	10:10:2	23	21	8.6%	--
11	8	9	54398	10:10:2	23	17	26%	--
12	9	5	--	--	130	60	53.8%	--
13	10	6	--	--	259	48	81.4%	--

Table 2. Testing Examples and Performance Evaluation of MINI-Like Algorithm (II)

no.	number of variable	number of variable	number of product term before minimization	number of product term after minimization	minization ratio	optimal solution
1	11	1	11	6	--	yes
2	12	1	12	4	--	yes
3	14	20	91	31	66%	--
4	15	1	15	5	--	yes
5	23	1	23	8	--	yes
6	24	1	24	6	--	yes
7	28	1	28	4	--	yes
8	28	1	28	5	--	yes
9	30	1	30	9	10%	no
10	31	24	41	39	4.8%	--
11	36	42	49	49	0%	--

high column folding ratio of above 40%. Here, we define the column folding ratio as follows:

$$\text{Column folding ratio} = \frac{\text{(Number of folded columns)}}{\text{(Total number of columns before folding)}}$$

We note that the folding ratio is a relative parameter to evaluate the performance of the specified folding technique. And the result of folding is also dependent on the pattern of product terms for each case.

It is also important to verify the result of the Boolean equations after product term minimization. We take the following procedure to verify the result:

Let F, $\bar{F}$ and DC be the ON, OFF and Don't care arries of the given Boolean Equations and F' be the ON array of the result after minimization, then we check the following two cases:

Case 1: Let E1=F'-(F U DC)
If E1≠0 then F' covers some cubes in $\bar{F}$

Case 2: Let E2=F-F'
If E2≠0 then F' losses some cubes covered by F.

F' is correct if and only if E1=E2=0.

Table 3. Testing Examples and Performance Evaluation of SCF Algorithm

no.	number of input variable	number of output variable	random number seed	ratio of 0,1,X logic value	number of product term	number of folding pair	folding ratio
1	4	2	3476	1 : 1 : 2	8	3	50%
2	9	8	--	--	12	7	41.2%
3	13	21	452383	300:300:400	320	0	0%
4	14	20	--	--	31	15	44.1%
5	23	19	--	--	43	19	45.2%
6	25	18	452396	1 : 1 : 2	70	14	32.6%
7	30	20	6734	1 : 1 : 2	50	21	42%
8	31	24	--	--	39	27	49.1%
9	36	42	--	--	49	38	48.7%

The above method is simple but very useful in verifying the minimized result and also built in the PLAMG.

In the appendix, we give a complete example with 14 input variables and 20 outputs Boolean Equations shown in Figure A(a). Figure A(b) is its corresponding truth table before minterm minimization. Figure A(c) is its corresponding layout diagram. Figures A(d) & A(e) are the truth table and layout diagram after minimization, respectively. Figures A(f) & A(g) are the truth table and layout diagram after applying SCF, respectively. We find that the latter two optimizations (minterm minimization and folding technique) reduce the PLA chip area drastically.

DISCUSSIONS AND CONCLUSIONS

In this paper, we have described one automatic PLA minimizer and Generator: PLAMG for VLSI system design. The whole system has been implemented in VAX-11/780 at NCTU running under UNIX Operating System. After testing several examples, we have found that it has the following features:

(i) In VLSI system design, CAD tools play an important role in designing stage.

(ii) PLAMG provided the students in VLSI system course last semester in NCTU a useful tool to design and generate PLA layout automatically.

(iii) PLAMG has no limitations on the number of input variables and output equations. Hence, any size of PLA can be minimized and then generated.

(iv) PLAMG can be considered as one basic tool for automatic PLA design. In the future, several points in the following will be developed further.

(1) Improvement of MINI-like algorithm. This algorithm spends a lot of time to do the cube expansion by using $\bar{F}$, (where $\bar{F}$ is the set of OFF-array). Can we improve this algorithm to reduce the operation time? In [20], another approach was proposed to speed up the computation time to minimize the minterm of Boolean Equations. Faster heuristic algorithms will be proposed and developed to reduce the computing time and space during the minimization.

(2) Folding algorithms are still a hot problem to be studied to improve the topological optimization technique. Until now, many heuristic algorithms are proposed [15-17]. But a new better heuristic algorithm is necessary to be created to do the folding algorithm. In addition, the layout arrangement of folded PLA diagram will be developed and improved further.

(3) PLA partition is another new important problem. Few papers talk about this topic [3]. How to partition one large PLA size into several appropriate smaller sizes of PLAs with good chip area and fast processing time?

(4) PLA testing method is also important in reliable design of VLSI system [21,22]. In the future, we will design and develop some algorithms for PLA testing.

REFERENCES

[1] C. Mead and L. Conway, Introduction to VLSI Systems, Addison-Wesley, 1980.
[2] P. C. Treleaven, "VLSI processor architectures," IEEE Computer, pp. 33-45, June 1982.
[3] S. Kang and W. M. vanCleemput, "Automatic PLA synthesis from a DDL-P description," 18th Design Automation Conference, ACM and IEEE, pp. 391-397, 1981.
[4] H. A. Landman, R. F. Cmelik, S. Fang, and H. Hofmann, "Automatic layout of optimized PLA structures," CAD Tool, University of California at Berkeley.
[5] I. Suwa and W. J. Kubitz, "A computer-aided-design system for segmented-folded PLA macro-cells," 18th Design Automation Conference, ACM and IEEE, pp. 398-405, 1981.
[6] R. W. Hon and C. H. Sequin, "A guide to LSI implementation," Palo Alto Research Center, Xerox Corporation, 1980.

[7] K. A. Hwang, D. C. Liu and C. C. Lin, "BRUTUS: A CAESAR-like graphic editor for I.C. layout," (to be published).

[8] E. J. McCluskey, Jr., "Minimization of Boolean functions," Bell Syst. Tech. J., 35, p. 1417, 1956.

[9] J. P. Roth, "A calculus and an algorithm for the multiple-output 2-level minimization problem," Research Report RC 2007, IBM Thomas J. Watson Research Center, Yorktown Heights, New York, February 1968.

[10] Y. H. Su and D. L. Dietmeyer, "Computer reduction of two-level, multiple output switching circuits," IEEE Trans. on Computers, C-18, p. 58, 1969.

[11] J. R. Slagle, C. L. Chang, and R. C. T. Lee, "A new algorithm for generating prime implicants," IEEE Trans. on Computers, C-19, p. 304, 1970.

[12] Z. Arevalo and J. G. Bredeson, "A method to simplify a Boolean function into a near minimal sum-of-products for programmable logic arrays," IEEE Trans. on Computers, Vol. C-27, No. 11, pp. 1028-1039, November 1978.

[13] S. J. Hong, R. C. Cain, and D. L. Ostapko, "MINT: A heuristic approach for logic minimization," IBM J. Res. Develop. 18, pp. 443-458, 1974.

[14] J. L. Lee, "Design and implementation of PLAMG: A PLA minimizer and generator for LVSI system design," Master's thesis, Department of Computer Engineering, National Chiao Tung University, June 1983.

[15] J. R. Egan and C. L. Liu, "Optimal bipartite folding of PLA," 19th Design Automation Conference, ACM and IEEE, pp. 141-146, 1982.

[16] G. D. Hachtel, A. R. Newton, and A. L. Sangiovanni-Vincentelli, "An algorithm for optimal PLA folding," IEEE Trans. on Computer-Aided Design of Integrated Circuits and Systems, Vol. CAD-1, No. 2, pp. 63-77, April 1982.

[17] M. Luby, U. Vazirani, V. Vazirani, and A. Sangiovanni-Vincentelli, "Some theoretical results on the optimal PLA folding problem," Proc. 1982 International Symposium on Circuits and Systems, Rome, Italy, May 1982.

[18] D. L. Ostapko and S. J. Hong, "Generating test sxamples for heuristic Boolean minimization," IBM J. Res. Develop., 18, pp. 459-464, 1974.

[19] C. Chen, C. G. Chung, Y. Z. Horng, and J. J. Hsu, "A CAD tool for data path and control signals generation from hardware description language for digital system design," Proceedings of the International Conference on Advanced Automation," Taipei, Taiwan, R.O.C., pp. 73-81, Dec. 19-21, 1983.

[20] R. K. Brayton, G. D. Hachtel, L. A. Hemachandra, A. R. Newton, and A. L. M. Sangiovanni-Vincentelli, "A comparison of logic minimization strategies using ESPRESSO: A PLA program package for partitioned logic minimization," Proc. 1982 International Symposium on Circuits and Systems, Rome, Italy, May 1982.

[21] H. Fujiwara and K. Kinoshita, "A design of programmable logic arrays with universal tests," IEEE Trans. on Computers, Vol. C-30, No. 11, pp. 823-828, November 1981.

[22] W. Daehn and J. Mucha, "A hardware approach to self-testing of large programmable logic arrays," IEEE Trans. on Computers, Vol. C-30, No. 11, pp. 829-833, November 1981.

CAD OF TRUSS STRUCTURES BY THE INTERACTIVE SIMPLEX METHOD

Vilas Wuwongse and Mahendra Narayan Nagraj

Division of Computer Applications
Asian Institute of Technology
Bangkok, Thailand

ABSTRACT

The Interactive Simplex method which is a method developed for solving multiobjective problems in man-machine interactive mode is applied in a CAD system for optimal truss design. The method is mainly based on the Simplex method, which is one of the representative direct search methods. The techniques of minimum comparison sorting and minimum merging are applied in the method to make the requirement from the designer to be only the preference judgement of two alternatives at a time. The method is implemented as part of the CAD system. The CAD system mainly consists of optimization and graphics display subsystems. In the optimization subsystem there are Equilibrium Linear Programming and Stress Ratio method programs in addition to the Interactive Simplex method program. The graphics display subsystem shows structural models with their relevant information and converging process of the Interactive Simplex method in order to facilitate the difficulty in making decision. Applicable problems can have single or multiple objective functions. An optimal design of a bridge truss with multiple objective functions is given to illustrate the design procedure using the proposed CAD system.

INTRODUCTION

For the past two decades, much attention has been paid to solving various engineering design problems with the aid of computers. In practice there are many design problems that several requirements such as those for economics, operability, reliability, and so on have to be met. In this regard, it is more appropriate

to handle the design problem as a multiobjective one. There have
been presented various methods for solving multiobjective problems.
One of the main groups of the methods is the man-machine
interactive methods.

The interactive methods assume that the designer cannot give
global information about his preferences in advance, but can give
local preference information in a particular situation. The methods
require the designer to articulate locally approximated preference
information of given solutions or alternatives. The preference
information is then used to find a new and better solution, which
triggered new reactions from the designer. This iterative process
continues until a satisfactory solution is found or until another
termination condition becomes operative. One of the authors has
recently proposed an interactive method called Interactive Simplex
(ISP) method [1]. The ISP method is mainly based on the Simplex
method [2], which is one of the representative direct search methods. Its effectiveness has been proved through a series of numerical experiments but its practical usefulness has not yet fully
verified [1].

In the meanwhile, there have also been many attempts to combine the computing and graphics displaying capability of computers
to form so called computer-aided-design (CAD) systems to help solve
design problems. There are two major categories of CAD systems:
simulation-based and design optimization-based systems. The former
system aids the design process in the way that it predicts and
gives outputs of the design once the designer has decided the input
variables. If the designer is not satisfied with the outputs,
he/she changes some or all of the input variables and runs another
iteration. This process is repeated until a satisfactory design is
obtained. The main drawback of this system is that a satisfactory
design may not be obtained because the designer decides everything
by trial-and-error. On the other hand, the latter system provides
some mechanism to produce a best design according to the performance criteria provided by the designer. It incorporates one or
more optimization programs inside it. However, most of the optimization programs can be applied only if there is single performance
criterion and it is explicitly expressed. This restricts the range
of applicable problems of the CAD system. For example, design
problems with multiple criteria may not be able to be treated by
the system.

The purpose of this paper is to make an extensive application
of the ISP method in a design optimization-based CAD system for
truss structure design in order to enhance the system's power and
ease of use. The reason for choosing the truss structure as a
design target is that, besides being a fundamental problem in civil
engineering, it is simple yet not trivial.

CAD OF TRUSS STRUCTURES

There are 6 sections following Section 1. Section 2 defines the optimal truss design problem in a general form with multiple objective functions while Section 3 briefly describes the ISP method. Descriptions of the CAD system and the design procedure using it are given in Sections 4 and 5 respectively. Section 6 shows an illustrative example. Conclusions are given in Section 7.

PROBLEM DESCRIPTION

For given design specifications including loadings, stresses in the members, allowable member sizes and so on, the major problem of truss structure design is to determine cross sectional areas (sizes of members), coordinates of joints (geometry of the structure) and the connectivity of members in a truss system (topology of the structure) such that a design with desirable characteristics is obtained. Most of the approaches to this design problem that have been done so far take into account only one or two of the mentioned design variable subspaces, i.e., member sizes only, or geometry and member sizes, or topology and member sizes. The only approach that considers the three subspaces of design variables is the one proposed by Somekh and Kirsch [3]. However, Somekh and Kirsch have considered only the truss weight as the desirable characteristic or objective function.

In practical design not only the truss weight should be considered as the objective function. There are also other characteristics that should be taken into account, e.g., actual construction cost, shape, and so on. The design problem should therefore be formulated as a multiobjective design problem. Some of the objective functions can be expressed explicitly but some cannot. Mathematically the problem can be stated as follows:

$$\text{Maximize}_{\{X, Y, Z\}} U(f_1(X, Y, Z), f_2(X, Y, Z), \ldots, f_r(X, Y, Z)) \quad (1)$$

subject to

$$X^\ell \leq X \leq X^u \quad \text{(size constraints)} \quad (2)$$

$$Y^\ell \leq Y \leq Y^u \quad \text{(geometrical constraints)} \quad (3)$$

$$\sigma^\ell \leq \sigma \leq \sigma^u \quad \text{(stress constraints)} \quad (4)$$

and

$$D^\ell \leq D \leq D^u \quad \text{(displacement constraints)} \quad (5)$$

where

X : vector of members' cross sectional areas
Y : vector of members' coordinates of joints
Z : number of members
σ : vector of stresses
D : vector of displacements
$f_1, f_2, \ldots, f_r$: r distinct objective functions
U : designer's preference function
ℓ : superscript denoting lower bound
u : superscript denoting upper bound

U is a function of the objective functions f_k's. It cannot be defined in advance in an explicit form, but should be identified through a sequence of decision process. f_k's are defined by different scales, and there exist trade-off relationships among them. Some of f_k's may not be able to be expressed in mathematical formula or in numerical data. An example of such objective functions is aesthetic character of a structure.

INTERACTIVE SIMPLEX METHOD

The ISP method is briefly described here since a detailed discussion on it has already been made elsewhere [1].

The main concern of interactive methods for multiobjective design problems is how to extract information about U in equation (1) from the designer and how to use it to find the most preferable design. The methods can be divided into two groups according to how information about U is extracted. One group extracts cardinal information and the other extracts ordinal information. Ordinal information is generally easier to extract than the cardinal information. The ISP method belongs to the latter group.

The ISP method is based on the Simplex method [2], and is developed by applying the Merge Insertion algorithm of Ford and Johnson [4] to sort the vertices of initial or contracted simplexes and the Binary Merging algorithm of Hwang and Lin [5] to merge a reflection or expansion or reduction vertex into an ordered set of simplex vertices. This modification results in a good property that the preference information required by the method is only the comparing of two alternatives at a time. In other words, the designer has only to make pairwise comparisons of alternatives throughout the search process for a best alternative. The procedure of the method can be briefly described as follows:

Step 1 [MACHINE]   Set up a finite initial set of alternatives.

Step 2 [MAN]   Sort members of the alternative set over a preference by repeating pairwise comparisons.

Step 3 [MACHINE] Estimate an alternative according to the mechanism of the Simplex method.

Step 4 [MAN] If the estimated alternative is satisfactory, terminate the interaction process. Otherwise, merge it with the ordered alternative set over a preference by repeating pairwise comparisons. Go to Step 3.

The ISP method is incorporated into a CAD system to solve (1). The detail of the CAD system is described in the next section.

ACTUAL CAD SYSTEM

As Somekh and Kirsch [3] have pointed out, it is not practical to solve the design problem by a single optimization method and treatment of all the design variables at the same time leads to slow convergence. Therefore, the solution of the design problem will be implemented using multilevel optimization and the concept of a separate design variable space. Different optimization methods may be used for different levels and design variable subspaces.

The problem (1) can be stated as a two-level optimization problem as follows:

$$\text{Maximize}_{\{Y\}} U(f_1(X^*, Y, Z^*), \ldots, f_r(X^*, Y, Z^*)) \qquad (6)$$

subject to

$$X^* \text{ and } Y^* \text{ minimize } W(X, Y) \qquad (7)$$

and equations (2) ~ (5)

where W is the weight of the structure.

The lower level problem (7) is to find member number (topological variables) and member sizes that minimize the structure weight, given fixed coordinates of joints (geometrical variables). In this problem, member number and member sizes are treated as separate design variable subspaces, and Equilibrium Linear Programming and the Stress Ratio methods [6,7] are applied to the two subspaces, respectively. The higher level problem (6) is a preference optimization one. It is to find members' coordinates of joints that optimize the designer's overall preference function. The ISP method is applied to this problem.

The CAD system can be divided into three parts:

1. Optimization subsystem,
2. Graphics display subsystem and
3. Filing subsystem.

Optimization Subsystem

Three optimization programs are implemented in this subsystem.

Equilibrium Linear Programming program [6]. This program yields optimal topology and member sizes for a structure with given geometry, such that the structure has minimum weight. Also, unnecessary members are deleted and member sizes which give minimum weight are selected. However, the solution does not satisfy the compatibility conditions and may violate the stress limits.

The input to this program is a structure with geometrical variables generated by the ISP method and with initial topology and member sizes given by the designer. When the program execution is over, the output, i.e., a structure with optimal topology, is shown to the designer using the graphics display subsystem.

Stress Ratio method program [7]. The input to this program is the structure obtained from the Equilibrium Linear Programming method. The output is a structure with optimal topology and optimal member sizes which do not violate the stress constraints and yield nearly minimum weight. Before entering this program, the designer has a choice of adding or deleting additional members from the optimal topological structure. This flexibility is provided, because addition or deletion of a few members from the structure may be necessary to make it structurally stable.

Interactive Simplex Method program. This is used for obtaining a structure with optimal geometry. The other two methods can be thought as acting as suboptimization methods. The input variables of this program are coordinates of joints of members. The output is a structure with satisfactory geometry and with optimal topology and member sizes determined by the other two optimization programs. Associated with each output structure are various kinds of information about its attributes such as weight, differences in members' cross sectional areas, maximum displacements of nodes. All this information can be shown on graphics display terminal together with the structure. The ISP method requires the designer to make certain times of pairwise comparisons of outputs with all or some of this information provided. Using this information of pairwise comparisons, the ISP algorithm tries to estimate a new set of geometrical variables which will yield a structure with better attributes. The algorithm will continue until a satisfactory structure is obtained.

Graphics Display Subsystem

This subsystem consists of a structure display program used to show structures along with relevant information about their attributes, and a converging display program used to show the converging process of the ISP method. Tektronix graphics terminal 618 has been used for this purpose. The designer can use this subsystem to

- set initial structures and make pairwise comparisons for the ISP method.
- add or delete members to make structures stable or satisfy construction limitations.
- see the converging process of the ISP method to decide whether he/she should terminate the searching process.
- see the final optimal structure with all attributes and variables.

Filing Subsystem

This subsystem is used to store and provide all necessary data. Examples of the data stored in this subsystem are problem parameters provided by the designer and data of the finally obtained optimal structure.

DESIGN PROCEDURE

The main design procedure follows the algorithm of the ISP method with the Equilibrium Linear Programming and Stress Ratio methods being used whenever a new structure is generated. The procedure starts after the designer gives necessary information about the design problem. The steps are as follows:

1. Initializing

In this step, the designer is asked questions about the structure to be designed such as:

- number of geometrical variables,
- initial coordinates of joints in the structures,
- topology of the structure,
- initial member sizes for all the members,
- joints loaded and their loads,
- constraints on stress in member,
- modulus of elasticity and density of the material to be used.

All of these interactions are done via an alphanumeric terminal. After obtaining answers to all these questions, the resultant structure is shown graphically to the designer for verification.

2. Obtaining initial simplex

- Treating the initial values of the geometrical variables (whose number is supposed to be n) as coordinates of the centroid of a simplex, n+1 vertices of the simplex, with n coordinates for each, are generated.
- Using the same topology and member sizes as the ones given by the designer for the initial structure, n+1 structures are generated.
- Applying the Equilibrium Linear Programming method, and then the Stress Ratio method, an optimal structure is obtained for a vertex. Repeat this step for the whole n+1 vertices.
- Ask the designer to make pairwise comparisons among these n+1 structures. As a result, a simplex with an ordered set of vertices (each of which corresponds to one of the n+1 structures) is obtained.

3. Applying the ISP method

After obtaining an initial simplex, the rest of the procedure follows the algorithm of the ISP method until a satisfactory design is obtained.

The above procedure will be illustrated by an example in the next section.

ILLUSTRATIVE EXAMPLE

Let us consider an example of designing a symmetrical bridge truss loaded successively at the five fixed joints of the roadway as shown in Figure 1. The arch has five joints, but only the first three joints (joints 12, 11, 10) need to be specified since the structure is symmetrical. Moreover, the middle joint (joint 10) has a fixed horizontal coordinate. Therefore, if four coordinate values of the first two joints (joints 12, 11) and vertical one of the middle joint are specified, a bridge truss structure is obtained. In other words, the number of geometrical variables in this example is five.

The design objective functions or criteria are assumed to be structure weight, stress-ratio efficiency, maximum displacements, and differences in member cross sectional areas. The last objective function is used as an indirect indicator of actual construction cost since in practice it is cheaper to order members with the same cross sectional area than with different ones. In the example members are divided into three groups: horizontal, vertical, and

CAD OF TRUSS STRUCTURES

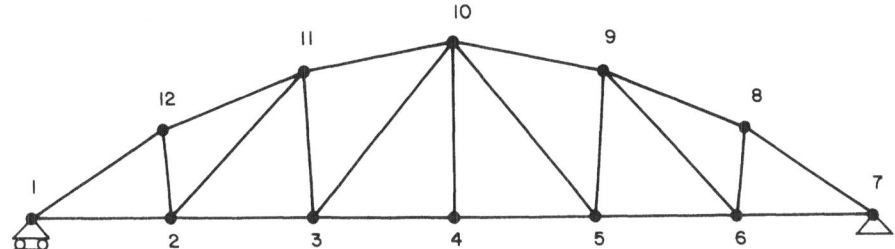

Figure 1. A symmetrical bridge truss structure.

slanted members. The difference between maximal and minimal cross sectional areas in each group is calculated and displayed to the designer together with other information.

Figure 2 shows how the designer interacts with the CAD system when he/she is asked to specify parameters of the design problem. The inputs from the designer are underlined in the figure. The interaction is done through an alphanumeric terminal. After

```
GIVE THE LENGTH OF THE BRIDGE TO BE DESIGNED
30
THIS BRIDGE IS DIVIDED INTO 6 EQUAL PARTS AT BOTTOM
THE TOP JOINTS ARE SYMMETRICAL AROUND THE CENTER ONE
SO THAT GIVEN THE FIRST THREE JOINTS, OTHER TWO CAN BE KNOWN
THE BOTTOM JOINTS ARE DETERMINED BY THE LENGTH OF THE BRIDGE
THEREFORE, GIVE THE X AND Y COORDINATES OF FIRST 2 JOINTS
AND Y COORDINATE OF THIRD JOINT (THE X COORDINATE OF THIRD
JOINT IS FIXED AND IS HALF THE LENGTH OF BRIDGE)
5.0, 3.33, 10, 5.33, 6.33
GIVE NO. OF JOINTS, NO. OF MEMBERS & ELASTICITY MODULUS
12, 25, 2.1E11
GIVE NUMBER OF JOINTS AT WHICH LOAD IS PRESENT
5
FOR EACH LOADED JOINT, GIVE JOINT NO., LOAD IN X AND Y DIRECTION
2, 0, -6.E5
3, 0, -6.E5
4, 0, -6.E5
5, 0, -6.E5
6, 0, -6.E5
   .
   .
   .
THE DEGREE OF INDETERMINACY OF NEW STRUCTURE IS 4
SO FOR THE PURPOSE OF ANALYSIS 4 MEMBERS ARE TO BE MADE REDUNDANT
THEREFORE GIVE MEMBER NOS. WHICH WOULD BE MADE REDUNDANT,
ONE AT A TIME
19, 21, 22, 24
```

Figure 2. Dialogues between the designer and the CAD system in the problem specification stage.

obtaining necessary information, a model of the structure with
optimal topology and member sizes is displayed on the graphic terminal and further questions are asked on the alphanumeric terminal
as shown in Figure 3. An initial simplex is then generated with
this structure as its centroid and with appropriate size. After
this stage, all the designer has to do is to compare two structures
at a time. An example of one of the pairwise comparisons made in
the first iteration is shown in Figure 4. To help the designer
make decision, necessary information is provided on the terminals.
After 15 iterations a satisfactory structure shown in Figure 5 is
obtained. A display on converging process of the example is shown
in Figure 6. The system will provide detailed information of the
satisfactory structure if the designer requires.

CONCLUSIONS

In practical engineering design problems, there are usually
multiple criteria or objective functions to consider. It is therefore natural to treat the problems as multiobjective design problems. This paper has presented a CAD system which incorporates as
its part an interactive multiobjective optimization method. As a
result, the CAD system can be used to solve single as well as multiple objective design problems. As the system provides relevant
information to the designer and requires from him/her modest information, mostly pairwise comparison judgement of alternatives, it is
easy to use from the designer's point of view. A bridge truss
structure design problem has been taken up to illustrate how the
CAD system works. In fact this approach of incorporating the
interactive optimization method into a CAD system is not good only
for the truss design CAD system. It can also be applied to other
CAD systems. This possibility is being under investigation and the
result will be reported later.

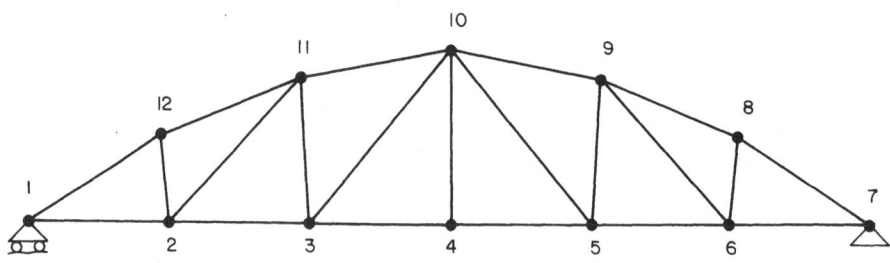

DO YOU WANT TO ADD SOME MEMBERS? (Y/N)
N

Figure 3. A structure after applying the Equilibrium Linear
Programming and the Stress Ratio methods.

CAD OF TRUSS STRUCTURES

WEIGHT   12156.6 KG
EFFICIENCY   100.0 %
MAX X DISPLACEMENT .018 M
MAX Y DISPLACEMENT .039 M

DIFFERENCE OF MAX AND MIN SECTIONAL AREAS
HORIZONTAL MEMBERS   0.00014 SQ M
VERTICAL MEMBERS   0.01260 SQ M
SLANTED MEMBERS   0.00719 SQ M

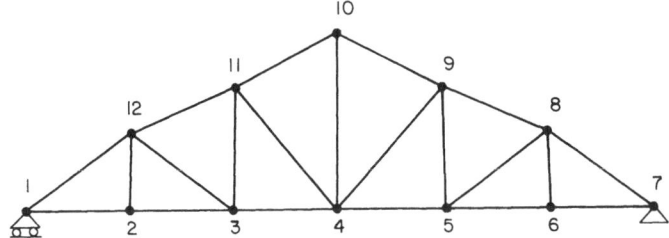

WEIGHT   11569.3 KG
EFFICIENCY   99.9 %
MAX X DISPLACEMENT .018 M
MAX Y DISPLACEMENT .034 M

DIFFERENCE OF MAX AND MIN SECTIONAL AREAS
HORIZONTAL MEMBERS   0.00076 SQ M
VERTICAL MEMBERS   0.00396 SQ M
SLANTED MEMBERS   0.00358 SQ M

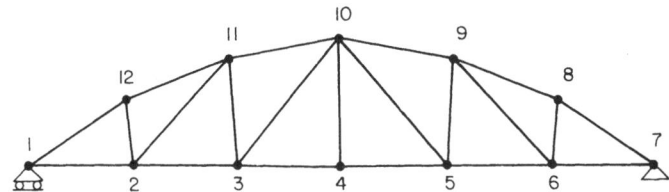

UPPER STRUCTURE			LOWER STRUCTURE		
JOINT NO.	CO-ORDINATES X	Y	JOINT NO.	CO-ORDINATES X	Y
10	15.00	6.45	10	15.00	6.04
11	10.12	5.45	11	9.71	5.04
12	6.54	3.45	12	4.71	3.04

WHICH STRUCTURE DO YOU PREFER?
INPUT   1 FOR UPPER STRUCTURE
    2 FOR LOWER STRUCTURE
    3 IF BOTH ARE INDIFFERENT
<u>2</u>

Figure 4. A comparison judgement of two structures in the first iteration.

REFERENCES

[1] V. Wuwongse, "Multiobjective optimization by interactive pairwise comparison," Doctoral dissertation, Dept. of Systems Science, Tokyo Inst. of Technology, 1982.

[2] J. A. Nelder and R. Mead, "A simplex method for function minimization," Comput. J., Vol. 7, pp. 308-313, 1965.

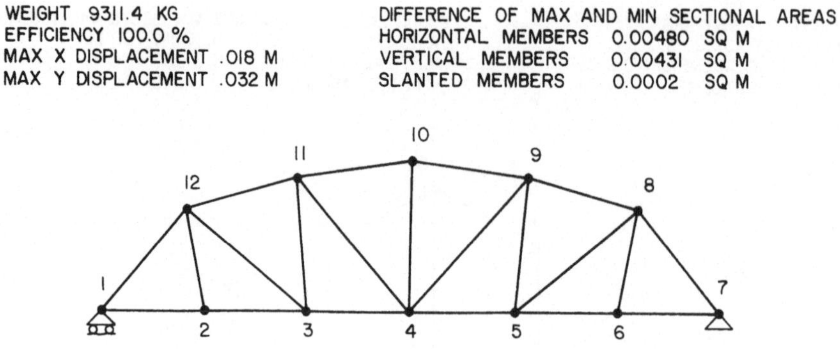

WEIGHT 9311.4 KG
EFFICIENCY 100.0 %
MAX X DISPLACEMENT .018 M
MAX Y DISPLACEMENT .032 M
DIFFERENCE OF MAX AND MIN SECTIONAL AREAS
HORIZONTAL MEMBERS 0.00480 SQ M
VERTICAL MEMBERS 0.00431 SQ M
SLANTED MEMBERS 0.0002 SQ M
1
2
3
4
5
6
7
8
9
10
11
12

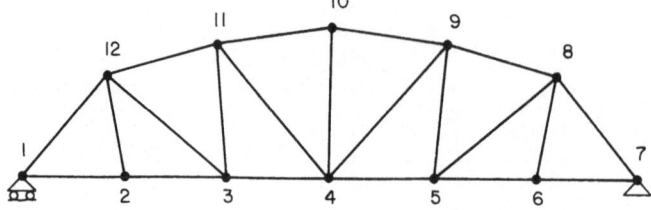

Figure 5. A satisfactory structure.

X_{12}
Y_{12}
X_{11}
Y_{11}
Y_{10}
10
5
1
5
10
15
ITERATIONS

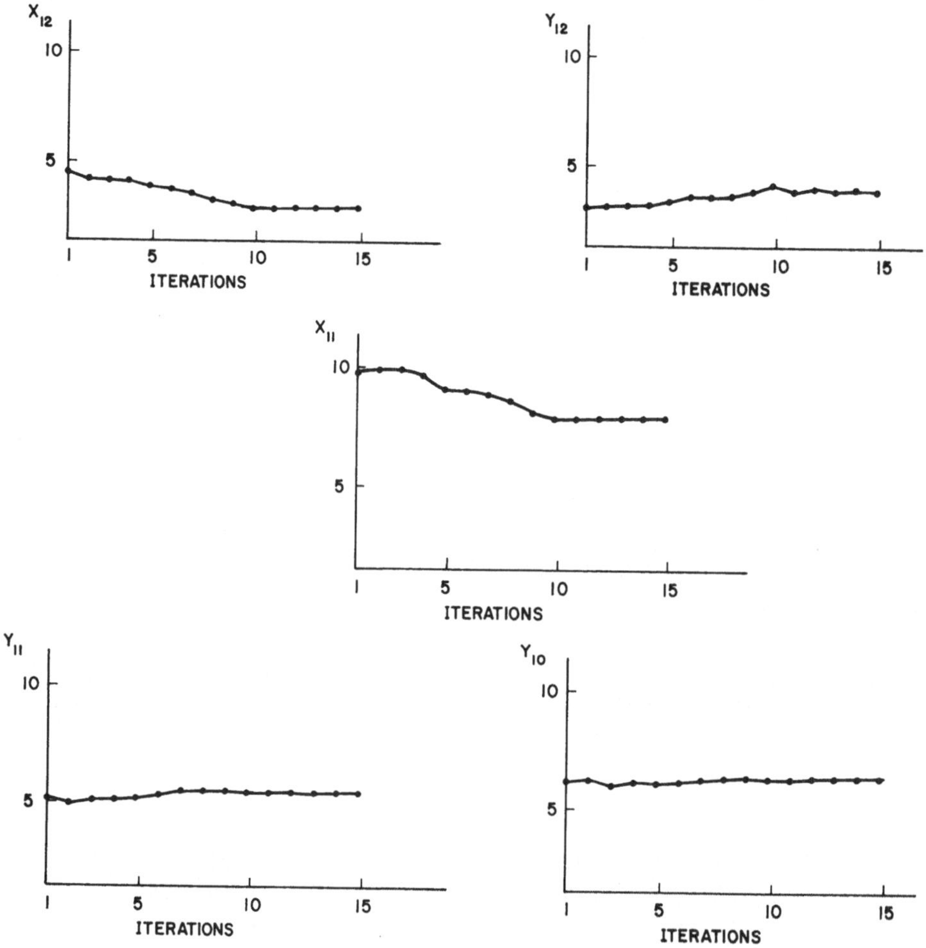

Figure 6. The display of converging process.

[3] E. Somekh and U. Kirsch, "Interactive optimal design of truss structures," CAD, Vol. 13, pp. 253-259, 1981.
[4] L. R. Ford and S. M. Johnson, "A tournament problem," Am. Math. Monthly, Vol. 66, pp. 387-389, 1959.
[5] F. K. Hwang and S. Lin, "A simple algorithm for merging two disjoint linearly ordered sets," SIAM J. Computing, Vol. 1, pp. 31-39.
[6] K. F. Reinschmidt and A. D. Russel, "Applications of Linear Programming in Structural Layout and Optimization," Comput. Struct., Vol 4, pp. 855-869, 1974.
[7] R. H. Gallagher and O. C. Zienkiewiez, Optimum Structural Design, Theory and Applications, John Wiley, pp. 19-26, 1973.

LOW COST CAD/CAM SYSTEM FOR MACHINE PARTS

W. B. Ngai and Y. K. Chan

Department of Computer Science
Chinese University of Hong Kong
New Territories, Hong Kong

ABSTRACT

A CAD/CAM system known as the MODCON(1,2,3) has shown to be useful for producing machine parts such as rocker-lever, connecting-rods, etc. In spite of the usefulness of the MODCON system, it suffers the inherent disadvantage of a batch system and therefore does not interact with a MODCON user wishing to design a machine part. An interactive CAD/CAM system is now proposed which is conceptually similar to the MODCON system. The proposed system is implemented on a microprocessors based machine which includes a central processor, a numerical data processor and another processor for I/O. This low-cost/high-power machine enables a designer to interactively design a machine part to his satisfaction before a N.C. paper tape is produced for manufacturing purpose.

INTRODUCTION

There are a variety of CAD/CAM systems for the design and manufacturing of geometric components in the mechanical and industrial engineering areas. These type of geometric components are usually referred to as machine parts. Some better known CAD/CAM systems for the design and manufacturing of machine parts, based on the "primitive" and "shape operation" approach, are BUILD(4), PADL(5), ROMULUS(6) and MODCON. Of these sytems, the MODCON is more oriented towards the manufacturing aspect rather than the design aspect. Currently, the MODCON offers no interactive design capability for a designer to interactively design a machine part and then subsequently to manufacture the machine part using a N.C. machine. Nevertheless, the MODCON system has shown to be capable of producing a variety of complex machine parts based on drawing specifications. The other disadvantage of the MODCON system is that it was implemented as a batch system on an expensive mainframe which is beyond the financial resource of small and medium companies. Therefore, if the MODCON system were to be enhanced to include interactive design feature, inexpensive dedicated computer or a computer with time-sharing O/S and graphics Input-Output peripherals should be chosen. On a cost to effectiveness basis, a Seiko 9500 is chosen for implementing the proposed interactive CAD/CAM system. The Seiko 9500 is a single user system which is a microprocessors based system consisting of INTEL 8086, 8087, 8088. The 8086 is the central processor, while as the 8087 and 8088 are used for number crunching and I/O, respectively. The system also consists of two mini-floppy disc drives giving a total of 1.2 M-bytes of formatted capacity. Interactive design of machine part using the proposed system can be achieved by the graphics capability of the Seiko 9500 which has a raster-scan CRT with 512 * 480 display resolution in 8 colors. This paper demonstrates how low cost and yet poerful CAD/CAM system can be achieved with acceptable response time and throughput.

SOLID PRIMITIVES IN THE PROPOSED SYSTEM

The proposed system is a low-cost system which aims to provide genuine design as well as manufacturing capability in order to produce machine part by N.C. milling. The approach of the proposed system is very similar to the MODCON system which allows a machine part to be built from a collection of individual components. The components are previously created from the "PRIMITIVEs" available from the proposed system. The synthesis of components to form a machine part is referred to as the "MERGE" operation. The components which are created from a set of primitives in the system are referred to as the "PRIMITIVE" operation. The primitives available in the proposed system such that components are created from "PRIMITIVE" operation and then "MERGEd" to form a machine part are

CONE+, ELLIPSOID+, ARC-BLOCK+, PRISM+, CURVE+, TOROID+

The "+" signs appended at the end of the primitive identifiers indicates that these primitives are used for creating solid components which protrudes from a flat surface. The solid components created can then be "MERGEd" together to form a complex machine part lying on a flat surface. As the proposed system is designed to cater for manufacturing machine parts by forging and casting, the primitives in the proposed system are considerably more complex in physical characteristics than the primitives used in other CAD/CAM systems. The more complex physical characteristics of a truncated cone specified from CONE+ is obviously illustrated from the basic data in Figure 1. The basic data include "RE" and "α" usually referred to as edge radius and draft angle. Typically, the

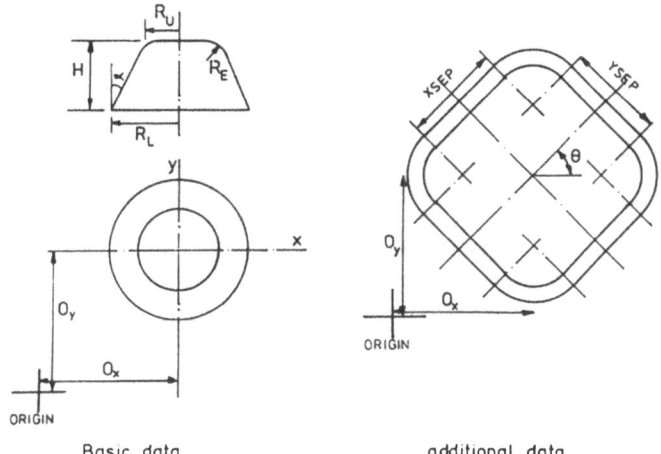

Figure 1. Characteristic data for primitive CONE+.

edge radius is of the order 3 mm in order to eliminate any sharp edges which might exist in a component. Typically, the draft angle is approximately 5-7 degrees in order to avoid any vertical faces. During forging operation to produce a machine part the existence of either sharp edge or vertical face can cause die crack and therefore must be avoided by using the parameters "RE" and "α". On the other hand, primitives which are more complex in their physical characteristics enables a variety of shapes to be created from the same

Photo 2. Typical CONE and its variants.

"PRIMITIVE" and therefore covers a wider application area. This advantage is clearly demonstrated by reference to the CONE+ primitive. Figure 1 illustrates the shape, basic data and additional data for a typical truncated cone. The lower right hand side component of Photo 2 is obtained by using typical basic data of a truncated cone without additional data. By assigning other values of the basic data, three other components in Photo 2 are obtained and these are hemisphere, sharp circular cone and vertical cylinder when viewed from the right hand side to the left hand side of Photo 2. By using typical basic data for specifying a truncated cone as well as different values for additional data, other components are also shown in Photo 2. All these components in fact protrudes from a flat surface and that the resolution for the components in Photo 2 can be improved and indeed the resolution for the components manufactured by N.C. milling can be specified by the designer using the proposed system. The specification for defining the resolution is dealt with in more detail in the next section.

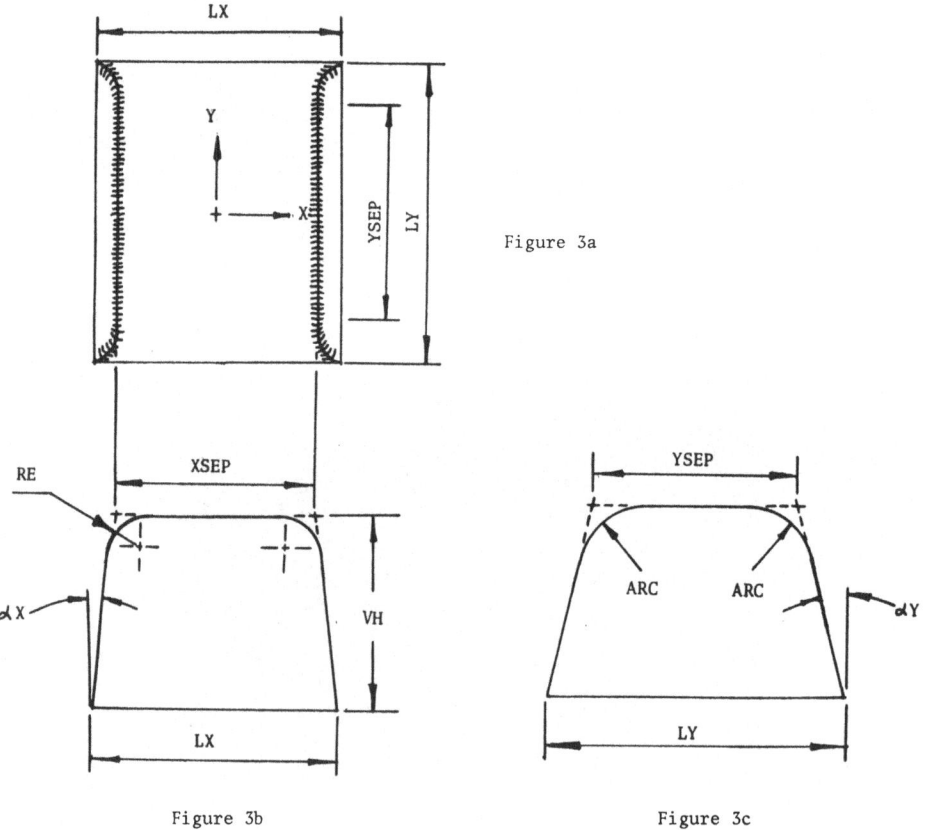

Figure 3a

Figure 3b

Figure 3c

Figure 3. Physical characteristics of an ARC-BLOCK+ component.

From time to time it is necessary to manufacture machine parts with cavity embedded onto some other solid components. For this reason, the proposed system will also have "PRIMITIVEs" for creating impression/impressions on a flat surface but will not be dealt with in here.

PHYSICAL CHARACTERISTICS AND EXAMPLE FROM ARC-BLOCK+

One of the "PRIMITIVEs" in the proposed system is referred to as ARC-BLOCK+. Components, created from the ARC-BLOCK+, have been successfully used with other components to form a complex connecting-rod [1,2,3]. The geometry of a typical component created from ARC-BLOCK+ can be best illustrated by referring to Figure 3a, 3b and 3c which contain the plan view and the two side views. Figure 3b illustrates one of the parameters "RE" which is usually known as edge radius and has the same significance as the "RE" in Figure 1. Figure 3a contains two sequences of short curve lines which indicates the visual effect of edge radius along the edges which would otherwise be sharp. A typical ARC-BLOCK+ component is illustrated in Photo 4. Photo 5 illustrates the N.C. cutter-path for machining the component in Photo 4. In Photo 5 it can be readily seen that the flat top of the component is made possible by the cutter-path indicated by the green rectangular closed contours. The edge radius near the flat top is cutted by the white contours as indicated by the parameters "RE" in Figure 3b. The "ARC" radius is cutted by the white and red cutter contours when referred to Figure 3c. The bottom contours in Photo 5 illustrates the cutter-path for milling the sloping side faces when referred to the parameters "αX" and αY" in Figures 3b and 3c. An isometric view of the cutter-path is also

Photo 4. Model of an ARC-BLOCK+ component.

Photo 5. Multi-view of ARC-BLOCK+ component cutter paths.

shown in Photo 5, from which the presence of edge radius running along the edges can be easily observed.

CUTTER-PATH FOR N.C. MILLING AN ARC-BLOCK+ COMPONENT

Figure 6 shows the variety of cutters that can be used for manufacturing components as well as machine parts. Figure 6a illustrates a typical cutter while as Figures 6b and 6c illustrate an end mill cutter as well as a ball ended cutter. In Photo 5, it can be seen that a component specified from the ARC-BLOCK+ is actually separated into 4 regions for N.C. milling. These 4 regions are indicated by the green contours for cutting the flat top, the white contours for cutting the edge radius near the flat top, the red contours for cutting the ARC radius and the bottom green contours for cutting the sloping faces. The contours illustrated in Photo 5 in fact are cutter paths for N.C. milling the component shown in Photo 4. The physical characteristics of the cutter used is shown in Figure 6a. The cutter path calculation for N.C. milling the flat top is relatively easy and is related to the cutter parameters T1, T2 and the ARC-BLOCK+ parameters XSEP and YSEP. When calculating the vertical cutter path separation for the edge radius near the flat top, the resolution for cutting can be specified as "EPSEDG" by the user of the proposed system and this is illustrated in Figure 7. When referring to Figure 7, it can be seen that the

LOW COST CAD/CAM SYSTEM

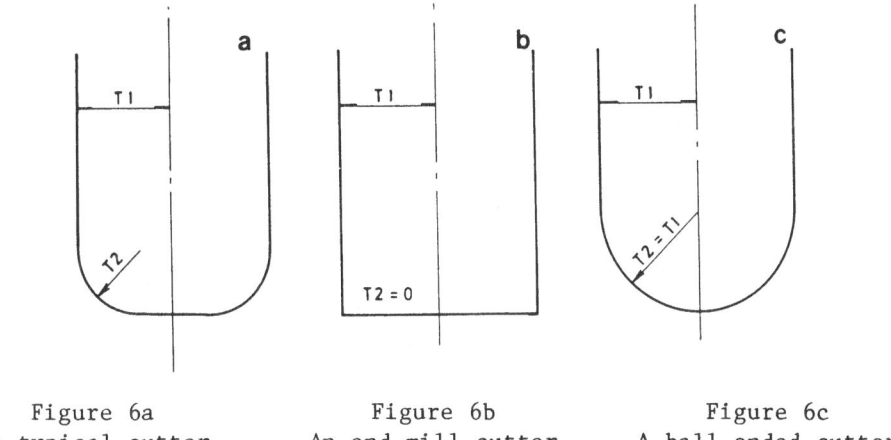

      Figure 6a   Figure 6b   Figure 6c
A typical cutter   An end mill cutter   A ball ended cutter

specification for "EPSEDG" in fact implicitly determines the vertical separation of two discrete vertical distance of two cutter passes denoted by "Z1" and "Z2". Photo 5 shows that 7 vertical cutter paths are used to produce the edge radius near the flat top. Figure 8 illustrates how vertical cutter path separation is calculated, when given "ARC" radius of the ARC-BLOCK+ component, cutter characteristics T1 and T2 and a user-specified parameter "EPS" which governs the surface finishing of the component. In Photo 5 it can be seen that 10 vertical cutter paths are used to cut the lower part of the "ARC" radius. The upper part of the "ARC" radius has been cutted at an earlier stage during the cutting for edge radius. Figure 9 shows the cutting of one of the four sloping faces bounding the ARC-BLOCK+ component. Again the resolution for cutting is chosen by the user in terms of "EPS". The "EPS" parameter has the same meaning and significance for both Figure 8 and Figure 9. From Figure 9, it can be seen that the vertical cutter path separation is calculated when given the angle of the sloping face, "EPS", T1 and T2. In Photo 5 it can be seen that 15 vertical cutter paths are used for cutting the ARC-BLOCK+ component in Photo 4. In practice, since the edge radius is in the order of 3 mm and therefore a large value of "EPSEDG" may mean that only a couple of vertical cutter paths are employed for cutting the edge radius region. This is not acceptable as the 3 mm edge radius will not be truly represented and it is found that at least five vertical cutter paths are necessary. From experience it is found that the parameter "EPSEDG" should be around 0.01 mm for typical cutter size T1 and T2. Again, from experience, "EPS" should be chosen to be around 0.1 mm for typical cutter in order that satisfactory surface finishing is to be obtained. T1 and T2 are typically of 1/8 inch radius and 1/16 inch radius. In fact, these recommended values for "EPSEDG", "EPS", T1 and T2 are applicable

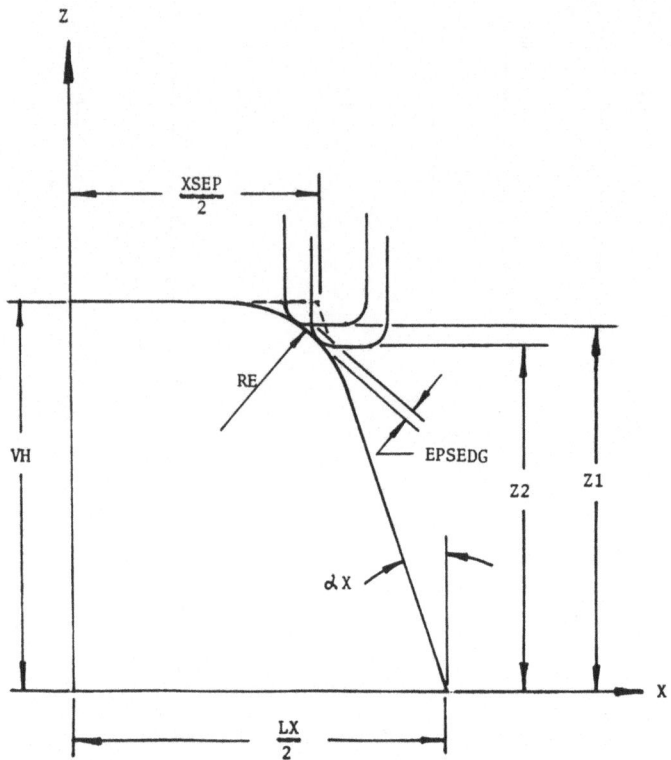

Figure 7. Vertical cutter path separation at "RE" region based on "EPSEDG".

for components specified from other "PRIMITIVEs" in order to ensure suitable N.C. machining surface finishing.

AN ARC-BLOCK+ COMPONENT AND TWO PRISM+ COMPONENTS

Photo 10 shows a component that is obtained from the ARC-BLOCK+ primitive. It can be seen that Photo 10 actually represents a horizontal cylinder. In general, a horizontal cylinder can be obtained by choosing appropriate parameters for the ARC-BLOCK+ primitive such that LY=2*ARC, YSEP=LY and ARC=VH indicated in Figures 3a and 3c. The cutter paths for N.C. machining the horizontal cylinder is shown in Photo 11. The presence of edge radius and draft angle to avoid die crack can be easily seen in Photo 11. Photos 12a and 12b show the plan view and the perspective view of a component from the PRISM+ primitive. It is called PRISM+ as it resembles a prism lying on one of its four sloping sides rather than lying on its base. In some ways the PRISM+ can be considered as an ARC-BLOCK+ with tapering on one end and having a sloping face

LOW COST CAD/CAM SYSTEM

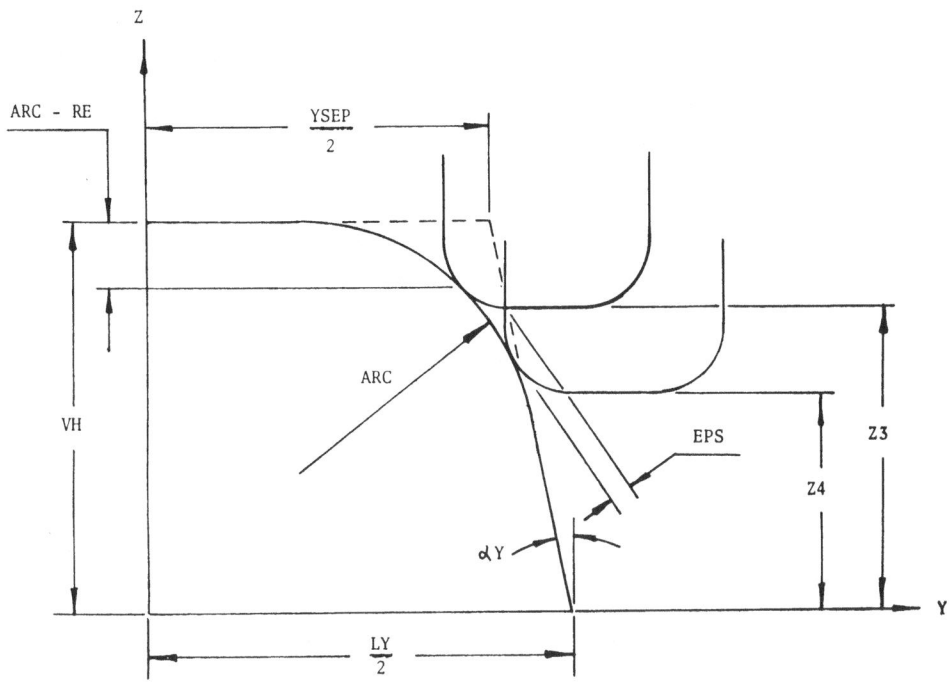

Figure 8. Vertical cutter path separation at "ARC" region based on "EPS".

rather than a horizontal flat face. Photos 13a and 13b show another component from the PRISM+. Photo 14 shows the cutter paths for N.C. machining the component shown in Photos 13a and 13b.

A CRANK-SHAFT FROM ARC-BLOCK+ AND PRISM+ COMPONENTS

Based on the horizontal cylinder shown in Photo 10 and a component from PRISM+ shown in Photos 13a and 13b, an attempt was made to produce a crank-shaft based on these components. The operation involved in producing the crank-shaft is to orient the necessary components in the appropriate positions and then to "MERGE" them. In this case, the meaning of "MERGE" is in fact to reduce the overlapping cutter paths. Photo 15 shows the final cutter paths after merging cutter paths from the cylinder components and the PRISM+ components. The green and white colors in Photo 15 are horizontal cylinders from ARC-BLOCK+ component while as the reddish color are PRISM+ components. The actual physical crank-shaft produced by N.C. machining is shown in Photos 16 and 17.

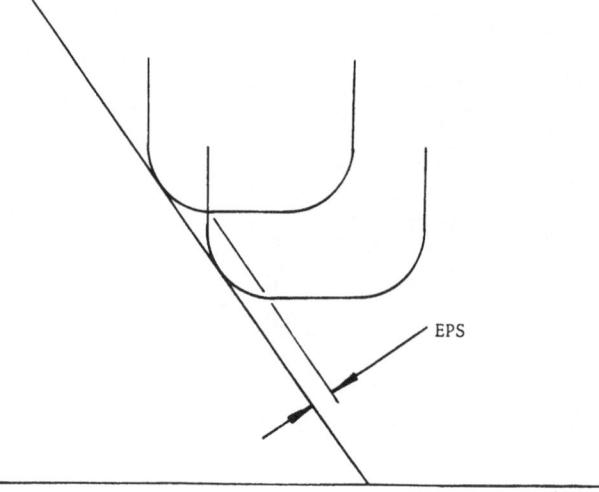

Figure 9. Vertical cutter path separation at sloping face based on "EPS".

CONCLUSION AND FURTHER WORK

With results so far obtained from the proposed system, it can be seen that a user can verify the cutter path for individual component as well as to verify the cutter path for a machine part before generating a paper tape for N.C. machining. With graphics capability the proposed system will, at a later stage, enable a user to interactively design components based on the available "PRIMITIVEs" and then drag individual components into their respective positions in the X-Y plane and then "MERGE" this collection of components to form a complete machine part. At this moment in time, only interpreted BASIC with graphics extension is available and the computing time for evaluating cutter path for ELLIPSOID+ components can take up to 2 minutes on the Seiko 9500. This clearly is not acceptable in terms of interactive design. However, the Seiko 9500 does have a FORTRAN compiler without graphics extension and hands-on experience showed that FORTRAN generated object code for CPU bound programs actually execute approximately 20 times faster than the equivalent BASIC version. With PASCAL and FORTRAN soon to be available with graphics extension, the proposed system will then be truly interactive in terms of design response time as well as throughput. Apart from the design and manufacturing capability of the proposed system, the system will be further enhanced to include features such as volume calculation, center of gravity calculation and moment of inertia about a given axis. In short this paper illustrates how a true CAD/CAM system can be implemented on a low-cost high-performance microprocessors based hardware to produce results previously expected from hardware systems such as

Photo 10. Cylinder from ARC-BLOCK+.

mainframe and super minis. From an industrial user point of view the proposed system genuinely represents low-cost machine part CAD/CAM automation.

REFERENCES

[1] Y. K. Chan and W. A. Knight, "MODCON: A system of the CAM for dies and moulds," in Proc. Computer Aided Design Conference, Brighton, England, pp. 361-369, April 1980.
[2] Y. K. Chan, "A perspective view of the MODCON system," in Proc. 18th Design Automation Conference, Nashville, TN, pp. 179-188, June 1981.
[3] Y. K. Chan, "Computer aided manufacturing of machine parts by volume buildings," in Proc. Intergraphics '83, Tokyo, Japan, Session B3, pp. 1-19, April 1983.
[4] I. C. Braid, "Designing with Volumes," Cambridge University Press, 1978.
[5] C. M. Brown, "PADL-2: A technical summary," IEEE Computer Graphics and Applications, pp. 69-84, March 1982.
[6] W. A. Carter, "ROMULUS - The design of a geometric modeller," Publication p-80 CM 01 by CAM-I, Inc., November 1979.

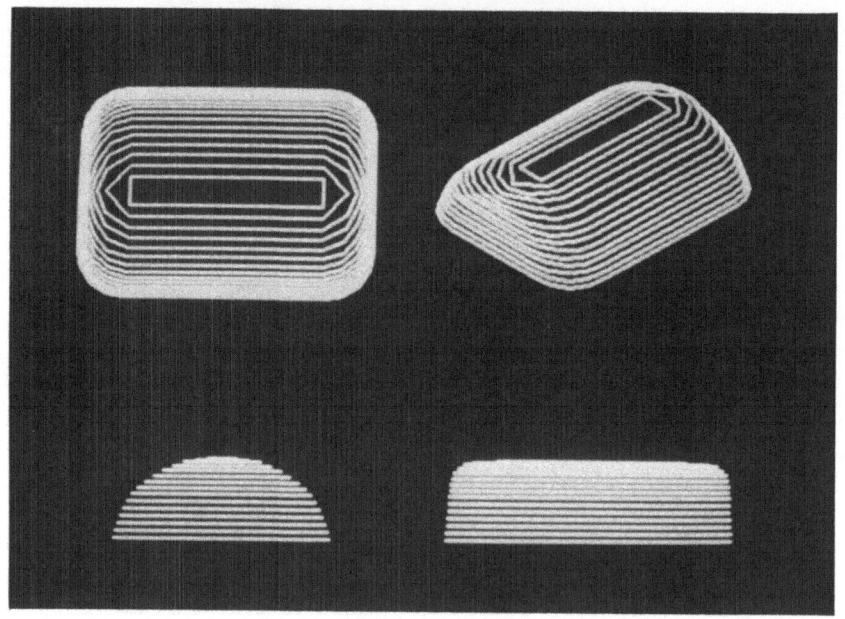

Photo 11. Multi-view of CYLINDER cutter paths.

LOW COST CAD/CAM SYSTEM 165

Photo 12a,b. Model of a PRISM+ component.

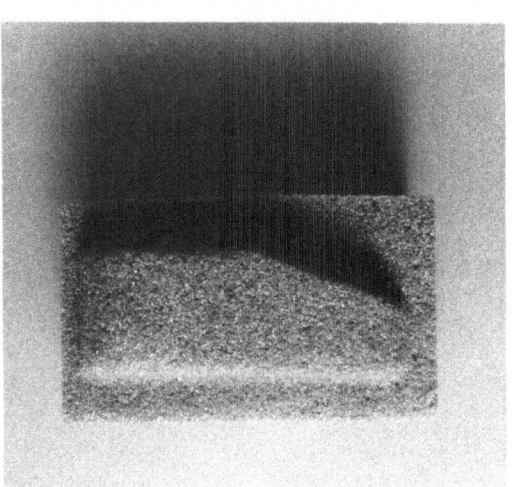

Photo 13a,b. Model of another PRISM+ component.

LOW COST CAD/CAM SYSTEM

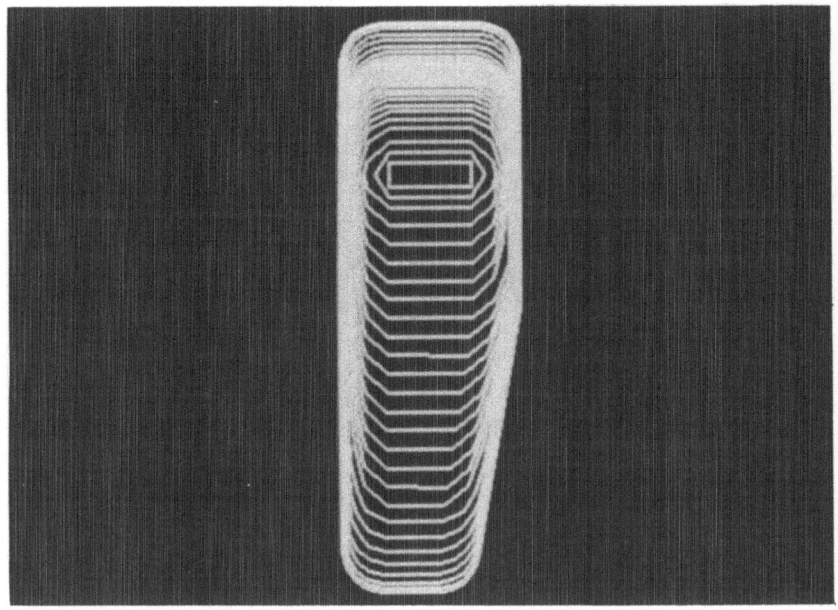

Photo 14. Cutter paths for a PRISM+ component.

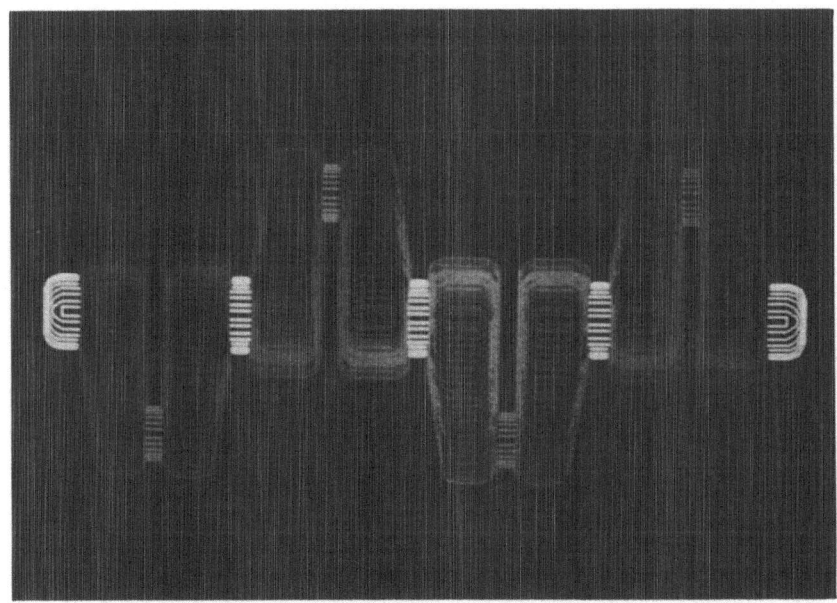

Photo 15. Cutter paths for a crank-shaft.

Photo 16. Top view of a crank-shaft model.

Photo 17. Perspective view of a crank-shaft model.

COMPUTER-ASSISTED METHODS FOR DEFINING 3D GEOMETRIC STRUCTURE OF

MECHANICAL PARTS

Z. Chen and D. B. Perng

Institute of Computer Engineering
National Chiao Tung University
Hsinchu, Taiwan, R.O.C.

ABSTRACT

In a mechanical CAD/CAM system the 3D geometric structure of parts have to be specified. There are two classes of methods for specifying the geometry of the parts, i.e., 2D-oriented and 3D-oriented. In the first class we examine the methods of representing the mechanical part by an engineering drawing with the top, front, and side views in orthographic projection. We shall propose a system which automatically interprets the engineering drawing by employing image processing techniques and derives the 3D geometry of the part from the 2D views.

Next, the class of 3D-oriented methods for part geometric modeling is reviewed. In this class we shall provide a system to facilitate the user to creat and edit a 3D part, with the aid of both wire frame and shaded displays.

INTRODUCTION

In a mechanical CAD/CAM system a crucial step is to obtain the correct 3D geometry of a mechanical part. There are two classes of methods for specifying the part geometry, i.e., 2D-oriented and 3D-oriented. The methods in the first class use the 2D view which is a flat representation of the part, while the methods in the second class use the 3D view which provides the solid figure of the part. In our system we shall consider these two types of methods. More specifically, we shall propose two ways of defining the geometry. One is based on the 2D orthographic views and the other is based on the 3D solid views.

To recognize the 3D figure of a part from its 2D representation is generally not easy [1-3]. There are two major types of 2D representation for showing the true geometry of the part, namely, orthographic and perspective views. In mechanical manufacturing it is more often to use orthographic views to depict the part.

Ejiri et al. [4], Idesawa et al. [5], and Lafue [6] addressed automatic interpretation schemes using the top, front, and side views. They mainly dealt with plane-faced objects, i.e., polyhedra. They are different in their assumptions about the ways that the edges of planar faces are represented in the orthographic views. The projected edge can be invisible [4], or shown in a dashed line [5], or explicitly stated [6]. However, they are more or less similar in their methods of deriving the part. Namely, they first found a set of possible vertices based on the given three views, then they tested the condition for the existence of possible edges and faces. Finally, they determined which faces could embody subparts. A theoretical treatment built upon formal geometric definitions and concepts of algebraic topology can be found in [7,8]. The drawback in these methods is that spurious vertices and edges will result. The elimination of spurious vertices and edges has to be provided in these methods. Recently, Aldefeld [9] proposed a method for automatic recognition of the 3D structures of parts whose constituent elementary objects are of the uniform-thickness type. In this method he first found all elementary loops of edges and selected one as a base silhouette according to some evaluation function. Then he tested the hypothesis that the 3D part obtained via a translational sweeping of the base silhouette is a legitimate part. Although the method can reconstruct polyhedral and curved objects, yet these objects have to be of the uniform-thickness type.

We shall propose a new method for recovering the 3D solid figure from the 3-view orthographic drawing without facing the foregoing problem of spurious vertices and edges. Furthermore, our method will not only deal with parts of the uniform-thickness type. However, we shall consider the class of parts whose orthographic views consist of only line segments, arcs, and circles; besides, no arcs or circles in any two views can correspond to the same part. Thus, doubly curved parts are excluded. In addition, we shall make an assumption that the given engineering drawing corresponds to a single object. In other words, we are not dealing with an assembly of multiple objects.

Our system is designed to read the 2D geometric structure from the 3-view engineering drawing via a TV camera. Furthermore, it outputs the reconstructed 3D part in two graphic displays, i.e., wire frame and color shading.

Next, we shall consider the 3D-oriented method for specifying the geometry of the mechanical part. The 3D solid modeling has received increasing attentions [10-19]. There are various kinds of schemes for representing the 3D part geometry, including

(1) Pure Primitive Instancing Schemes
(2) Spatial Occupancy Enumeration Schemes
(3) Cell Decompositions Schemes
(4) Sweeping Schemes
(5) Constructive Solid Geometry Schemes
(6) Boundary Data Schemes

The advantage of using the 3D-oriented approach is due to the fact that the human being is more effective with the 3D picture than with the 2D picture. In our system we shall emphasize on the 3D-oriented method which relieves much of the burden related to geometrical manipulations in the part design from the user.

In Section II we shall define a data structure for describing the 3D geometry of the mechanical parts. Section III is about the automatic interpretation of 2D geometric structure contained in the orthographic views of a part. Based on the extracted 2D geometric relations we shall provide an algorithm for recovering the 3D geometry of the part in Section IV. Some computer simulation results will be used to illustrate the process of our algorithm. These results are given in Section V. A convenient 3D-oriented method for the part design will be described in Section VI and some experimental results will be given in Section VII. Conclusions and suggestions in Section VIII will wind up this paper.

3D GEOMETRIC STRUCTURE OF PARTS

The 3D geometric structure of parts derived from either 2D orthographic views or 3D solid views will be described by the hierarchical representation shown in Figure 1. The specific data structure used to record these geometric entities is given in Figure 2. Notice that the curved surface of the part have a field indicating the solid/void flag which is needed in the shaded display computation, while the planar surface do not need the solid/void flag.

Based on the above 3D data structure two graphic display modes will be provided, wire frame display and shaded display. The wire frame display simply utilizes the face tables, while the shaded display is more complicated. Only visible faces will be displayed with different colors in the shaded display mode. The algorithm for hidden face removal used here is the depth-sorted algorithm [20]. The actual projection used for the display is an oblique projection. The portions of two partially occluded faces and their

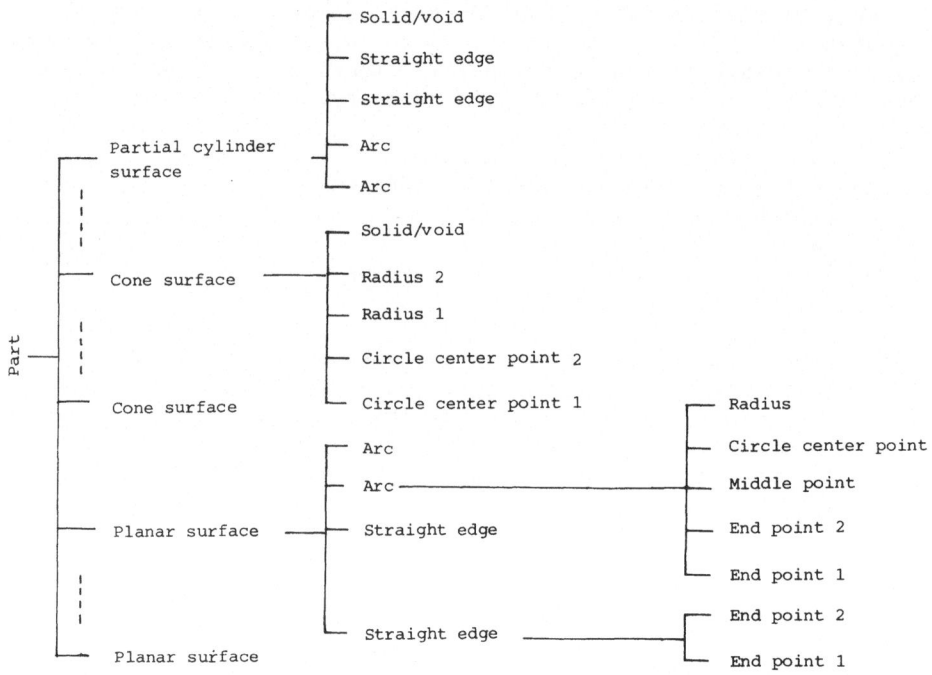

Figure 1. A hierarchical representation of a 3D part.

projected shapes have to be calculated. The whole process is carried out automatically by computer. For a detailed implementation, please refer to [21].

AUTOMATIC INTERPRETATION OF GEOMETRIC STRUCTURE OF 2D VIEWS

The 3-view engineering drawing will be contained in a rectangle box with the view axes parallel to the rectangle sides, as shown in Figure 3. The top view appears in the upper left corner of the box, the front view in the lower left corner, and the side view in the lower right corner. In the upper right corner a scale of one unit length is provided.

The visible and invisible edges of a part are represented by solid and dashed lines, respectively. The engineering drawings under consideration follow the general rules in the field of engineering drawing except for two minor modifications. The first one is depicted in Figure 4. Here the intersection line of a planar face and a cylindrical face, indicated by the line segment, $\overline{ab}$, must be added in the side view. The second modification is shown in Figure 5 where the center lines are drawn for the two silhouettes of a cylinder, as seen in Figures 5(b) and 5(c). To simplify

(a) Vertex table

vertex index	coordinate		
	x	y	z

(b) Straight edge table

edge index	pointers to two end points		pointers to two faces	
	Head	Tail		

(c) Arc table

arc index	coordinates of circle center			radius of circle	pointers to two end points and a middle point		
	x	y	z		Head	Tail	Middle

(d) Face table (ax+by+cz+d=0)

face index	hyperplane parameters				pointers to the constituent straight edges and/or arcs		
	a	b	c	d			
						>0, to stright edge <0, to arcs	

(e) Generalized cone surface table

cone index	coordinates of upper circle			coordinates of lower circle			radius of upper circle
	x	y	z	x	y	z	

radius of lower circle	solid/void flag

(f) Partial cylinder surface table

partial cylinder index	pointers to the two associated arcs	pointers to the two associated straight edges	solid/void flag

Figure 2. The 3D data structure of a part.

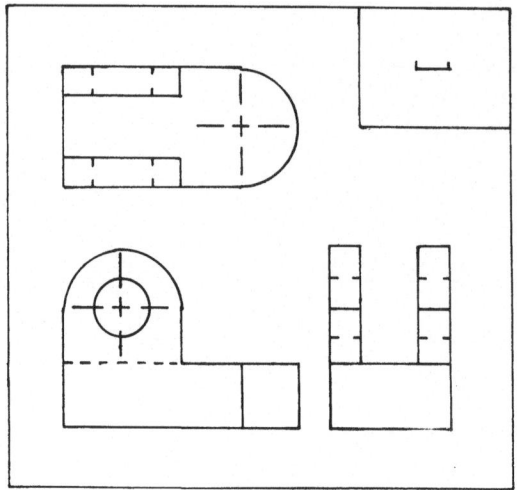

Figure 3. The engineering drawing used in the system.

the image processing work of reading the input drawing, the center line for the lateral silhouette must be removed, as indicated in Figure 5(d); the two center lines in Figure 5(c) will be retained for extracting the center point of the circle.

Next, we shall describe image processing techniques used to extract the 2D structure of the 3-view engineering drawing. First of all, the engineering drawing is inputted into a computer via a TV camera digitizer. Here a digitized picture of the engineering drawing contains 256 x 256 pixels with 256 grey levels. The picture can be easily thresholded and thinned to obtain a binary picture.

Now image processing techniques can be applied to extract solid/dashed line segments, arcs, circles and circle center points in the three views. The geometric relations between these entities are also to be derived simultaneously. We define a hierarchical

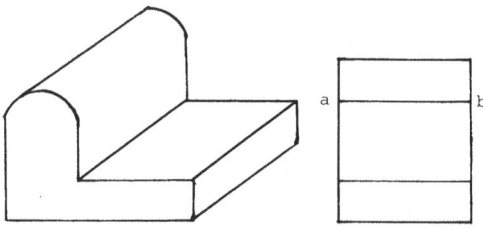

Figure 4. A curved object and its side view.

COMPUTER-ASSISTED METHODS   175

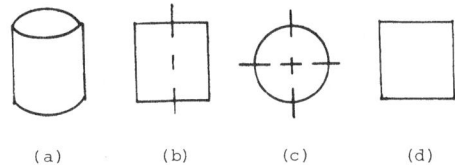

Figure 5. A cylinder and its side and top views.

representation of the 3-view engineering drawing as given in Figure 6 and a specific data structure in Figure 7.

In the binary picture the first step is to find the maximum and minimum x,y coordinate values of the three rectangle boxes enclosing the orthographic views as well as the unit length of the scale in the upper right corner of the picture. The actual value of the unit length is obtained by examining the best divisibility of bounding rectangle side lengths by the scale length. We shall

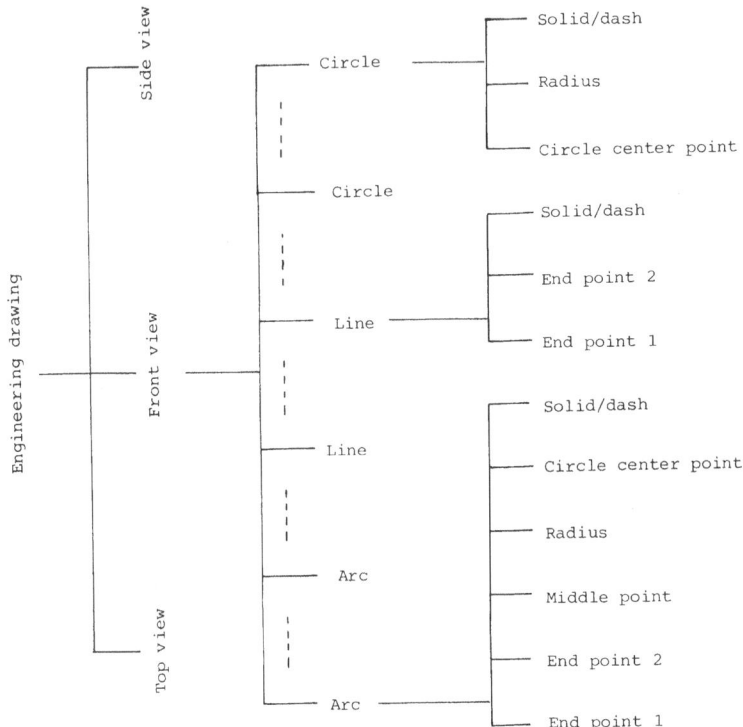

Figure 6. The hierarchical representation of an engineering drawing.

(a) Vertex table

vertex index	coordinates		pointers to edges that contain this vertex
	x	y	
			- - - -

(b) Straight edge table

edge index	solid/dashed flag	pointers to two end points		pointers to the middle points
		Head	Tail	
				- - - -

(c) Arc (curved edge) table

arc index	solid/dashed flag	coordinates of circle center		radius
		x	y	

pointers to two end points and a middle point		
Head	Tail	Middle

(d) Circle table

Circle index	solid/dashed flag	coordinates of circle center		radius
		x	y	

Figure 7. The 2D data structure of each of the three orthographic views.

use the scale length to partition each bounding rectangle into an array of grids.

To derive the 2D geometric structure of the orthographic views, we only need to check the binary picture at all grid points. Because there may be slight impression in the drawing lines, we shall examine the small square box centered at each grid point. There are several situations in which picture lines may appear in the square box, as shown in Figure 8. In the following we outline (1) PROCEDURE

COMPUTER-ASSISTED METHODS   177

CREATION_OF_VERTICES, (2) PROCEDURE CREATION_OF_STRAIGHT_EDGES and
(3) PROCEDURE CREATION_OF_CURVED_EDGES for deriving the 2D geometric
structure in each of the three orthographic views.

(1) PROCEDURE CREATION_OF_VERTICES

 {Let the grid points be numbered from 1 to total_grid_number}.
BEGIN
 FOR grid# = 1 to total_grid_number
  BEGIN
   FIND any binary picture points in the square box
    centered at the grid point;
   IF no binary picture points are found
    THEN no vertex at the grid point
   ELSE
    BEGIN
     IF the binary picture points constitute more
      than one straight line
     THEN
      BEGIN
       FIND the intersection point;
       PUT the intersection point in the
        vertex table
      END
     ELSE
      no vertex exists at the grid point
    END
  END
END {CREATION_OF_VERTICES}

(2) PROCEDURE CREATION_OF_STRAIGHT_EDGES

 {Let the vertices be numbered from 1 to total_vertex_number}
BEGIN
 FOR vertex# = 1 to total_vertex_number
 {To check whether all emanating vectors at the current
 vertex are marked (i.e., traced)}
 BEGIN
  WHILE any unmarked emanating vector at the vertex DO
   BEGIN
    FIND the terminal point along the emanating vector;

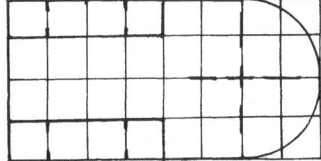

Figure 8. Grid points and imprecise vertices.

```
          IF  the terminal point is another vertex
          THEN
            BEGIN
              RECORD  a solid line between the two vertices;
              MARK  the emanating vector;
              FIND  other vertices in between these two
                    vertices;
              IF  found
                THEN
                  ENTER  the pointer fields of the vertex
                         table and the straight edge table
                         accordingly
                ELSE  {not found}
                  no vertices in between
            END
          ELSE  {The terminal point is not a vertex along a
                 solid line}
            BEGIN
              FIND  the terminal point on the extended
                    dashed line;
              IF  the terminal point is another vertex
              THEN
                BEGIN
                  RECORD  a dashed line between the two
                          vertices;
                  MARK  the emanating vector;
                  FIND  other vertices in between these
                        two vertices;
                  IF  found
                    THEN
                      ENTER  the pointer fields of the
                             vertex table and the straight
                             edge table accordingly
                    ELSE  {not found}
                      no vertices in between
                END
              ELSE  {The terminal point is not a vertex on
                     a dashed line}
            END
          {To check whether the current vertex is a circle
           center point}
          IF  the emanating vector consists of a sequence of
              long and short dashed lines
            THEN
              CALL PROCEDURE CREATION_OF_CURVED_EDGES
            ELSE
              PRINT "error in the drawing"
        END
      END
END {CREATION_OF_STRAIGHT_EDGES}
```

(3)   PROCEDURE CREATION_OF_CURVED_EDGES

      BEGIN
        {To find the radius with the given circle center point}
        FIND the radius of a curved edge;
        IF a circle is found
         THEN {To create the circle table}
           BEGIN
             RECORD the circle center point and the radius;
             DETERMINE the solid/dashed flag of the circle;
             RECORD the solid/dashed flag
           END
         ELSE {To create the arc table}
           BEGIN
             RECORD the circle center point and the radius;
             DETERMINE the two end points and the middle point of
                    the arc;
             FIND the solid/dashed flag;
             RECORD the two end points, the middle point and the
                    solid/dashed flag
           END
      END {CREATION_OF_CURVED_EDGES}

3D RECONSTRUCTION OF PARTS FROM 2D VIEWS

   Now we begin to describe an algorithm for finding a unique 3D
structure which corresponds to the given 2D data. The algorithm
consists of three major stages. The first stage is a decomposition
procedure. It recursively resolves the part under consideration
into a number of possible subparts which constitute the part under
geometric operations of union and difference. The second stage is
a reconstruction procedure which recovers the solid figures of any
cylinders, partial cylinders, and polyhedra. The last stage is a
composition procedure which pieces together all subparts to form
the final part. The algorithm is given in the form of a number of
procedures.

(1)   PROCEDURE 2D_TO_3D_RECONSTRUCTION

      BEGIN
         CALL PROCEDURE SUBPART_DECOMPOSITION
         CALL PROCEDURE SUBPART_RECONSTRUCTION
         CALL PROCEDURE SUBPART_COMPOSITION
      END {2D_TO_3D_RECONSTRUCTION}

(2)   PROCEDURE SUBPART_DECOMPOSITION

      BEGIN
         subpart# = 1

```
            WHILE  any set of views needed for decomposition check      DO
              BEGIN
                WHILE  any view of the set needed for decomposition
                       check     DO
                  BEGIN
                    WHILE  any isolated subview in the view      DO
                      BEGIN
                        MARK  and  REMOVE  the isolated subview and its
                              corresonding two other views;
                        STORE  these subviews as a new set of views for
                              further decomposition check;
                        subpart# = subpart# + 1
                      END
                    {All isolated subviews in the current view have been
                     extracted}
                  END
                {The current set of views has been processed}
              END
         END {SUBPART_DECOMPOSITION}

(3)    PROCEDURE SUBPART_RECONSTRUCTION

       BEGIN
         WHILE  any set of 3_views needed for reconstruction      DO
           BEGIN
             arc# = 0;
             WHILE  any view of the set contains any circle or
                    arcs     DO
               BEGIN
                 WHILE  any circle on the contour of the view      DO
                   BEGIN
                     CALL PROCEDURE CIRCLE_RECONSTRUCTION
                   END
                 WHILE  any arc on the contour of the view      DO
                   BEGIN
                     CALL PROCEDURE ARC_RECONSTRUCTION
                   END
               END
             {The circles or arcs in the current set of views have
              been processed
             WHILE  any straight edges left on the contour of the
                    current set of views      DO
               BEGIN
                 CALL PROCEDURE POLYHEDRAL_RECONSTRUCTION
               END
           {The current set of 3_views has been reconstructed}
           END
         {All sets of 3_views have been reconstructed}
       END  {SUBPART_RECONSTRUCTION}
```

COMPUTER-ASSISTED METHODS 181

(4) PROCEDURE CIRCLE_RECONSTRUCTION

```
BEGIN
  IF  circle is partitioned into arcs
    THEN
      BEGIN
        WHILE   any arc needed for reconstruction     DO
          BEGIN
            CONSTRUCT   a partial cylinder using the partial
                        circle as the base by translational
                        sweeping;
            subpart# = subpart# + 1
            STORE   edge-face relationships
          END
          subpart# = subpart# - 1
      END
    ELSE
      BEGIN
        CONSTRUCT   a cylinder using the circle as the base by
                    a translational sweeping;
        STORE   edge-face relationships
      END
END  {CIRCLE_RECONSTRUCTION}
```

(5) PROCEDURE ARC_RECONSTRUCTION

```
BEGIN
  CONSTRUCT   a partial cylinder using the partial circle as
              base by translational sweeping;
  STORE   the edge-face relationships;
  REPLACE   the convex arc by its bounding rectangle sides or
            the concave arc by the two lines obtained by
            extending its two connected lines until they meet;
  arc# = arc# + 1
END  {ARC_RECONSTRUCTION}
```

(6) PROCEDURE POLYHEDRAL_RECONSTRUCTION

```
BEGIN
  IF  a non-rectangular contour exists in any view
    THEN
      SELECT   such a non-rectangular contour as the sweep base
    ELSE
      SELECT   the contour of any view as the sweep base;
      CONSTRUCT   a block S with the sweep base by
                  translational sweeping;
  WHILE   any line L in any view which has no corresponding
          spatial line in the block S     DO
    BEGIN
```

```
            FIND  a set of one or two cutting planes such that the
                  line L will be formed after the set of plane
                  cutting;
            WHILE  any set of cutting planes     DO
              BEGIN
                CUT  the current block according to the set of
                     cutting planes to obtain an updated block S;
                FIND  the projected lines of the cutting planes in
                      the three views;
                MARK  the lines as being processed;
                STORE  the edge-face relationships;
              END
          END
      WHILE  arc#1 > 1      DO
        BEGIN
          RESTORE  the modified arcs in the three views;
          arc# = arc# - 1
        END
  END {POLYHEDRAL_RECONSTRUCTION}

(7)  PROCEDURE SUBPART_COMPOSITION

     BEGIN
        IF  subpart# > 1
          THEN
            BEGIN
              SELECT  the first subpart as the composite subpart;
                      {The first subpart is a solid subpart}
              FOR  subpart index = 2 to subpart#
                BEGIN
                  IF  the current subpart is contained by the
                      composite part
                    THEN
                      BEGIN
                        PERFORM the DIFFERENCE  operation by
                        subtracting the current subpart from the
                        composite subpart;
                        OBTAIN  the edge-face relationships of the
                                resultant subpart;
                        REPLACE  the composite subpart by the
                                 resultant subpart
                      END
                    ELSE  {The subpart is not contained by the
                           composite subpart}
                      BEGIN
                        PERFORM  the UNION operation of the current
                                 subpart and the composite subpart;
                        OBTAIN  the edge-face relationships of the
                                resultant subpart;
```

COMPUTER-ASSISTED METHODS

                        REPLACE the composite subpart by the
                                 resultant subpart
                END
         END
                VERIFY the correctness of the final composite
                        subpart
         END
      ELSE
         VERIFY the correctness of the subject
   END {SUBPART_COMPOSITION}

We shall use two examples to illustrate how our algorithm works. The first example is shown in Figure 9. The decomposition process is applied to the front view. The subview related to the inner circle is extracted as indicated in Figure 10. Next the decomposition process is applied to the side view of Figure 10(a) to extract another isolated circle. This leads to the decomposition result given in Figure 11. The subparts corresponding to Figure 10(b) and Figure 11(b) can be reconstructed easily by invoking the reconstruction process. They represent two cylinders. The reconstruction process for the 3-views given in Figure 11(a) is as follows. We find a circle in the side view and an arc in the front view. The subview containing the circle is extracted as shown in Figure 12. It can be easily decided that there exists a cylinder. And the subview containing the arc can be shown to correspond to a

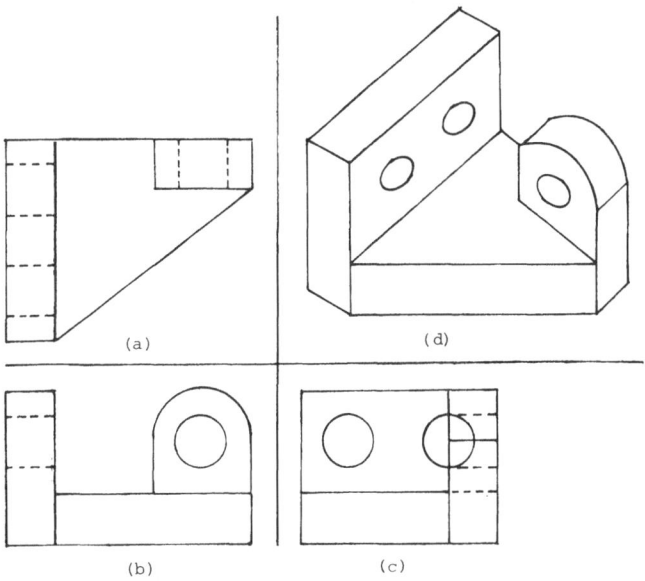

Figure 9. The engineering drawing for example 1.

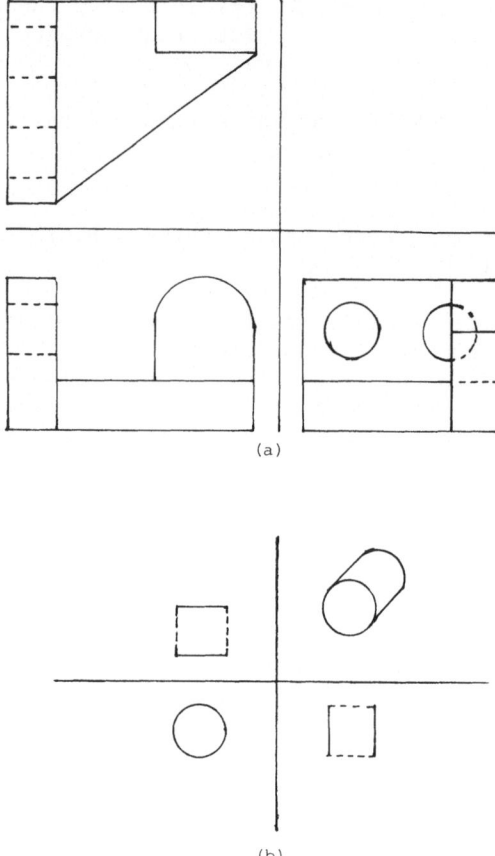

Figure 10. The decomposition and the reconstruction of subpart 1.

partial cylinder as shown in Figure 13(b). At this stage the remaining three views are given as Figure 13(a). This set of views becomes a polyhedral reconstruction problem. By using the translational sweeping and plane cutting, the subpart reconstructed is given in the upper right corner of Figure 13(a). The final process is to compose the subparts given in Figures 10(b), 11(b), 12(b), 13(a) and 13(b). The final part is given in Figure 9(d).

The second example is given in Figure 14, which is borrowed from [9]. The decomposition process is applied to the orthographic views recursively. Then each set of subviews can be reconstructed. Finally, the mechanical part is obtained by the composition process. The intermediate results are given in Figures 15-17. The part shown in Figure 14(d) can be finally obtained.

COMPUTER-ASSISTED METHODS

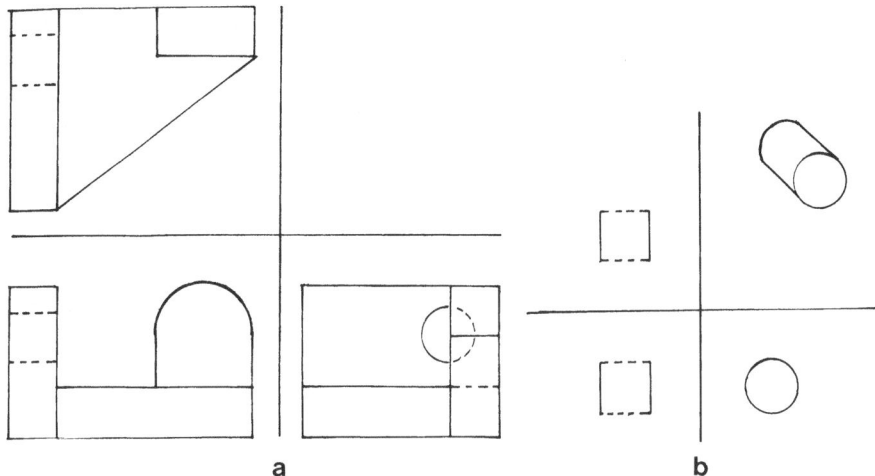

Figure 11. The decomposition and the reconstruction of subpart 2.

SIMULATION RESULTS OF RECOVERY OF 3D PART GEOMETRY FROM 2D VIEWS

Two different engineering drawings are digitized and processed to produce two binary pictures in our laboratory. These binary pictures are loaded into a VAX-11/780 computer. The graphic display terminal used is a Ramtek 6211 color display device.

The first engineering drawing that corresponds to a polyhedron is given in Figure 18. First we examine the outmost silhouette of the side view and find that it is not a rectangle, so we choose the silhouette as the sweep base. The solid obtained by a translational sweeping for a length of 4 units in the direction perpendicular to the base is given in Figure 19. Next, we detect a dashed line $\overline{a_s b_s}$ in the left hand side of the side view which is not able to be interpreted with the solid constructed so far. From the other two views we can find two cutting planes abcd and abgh which both contain the spatial line $\overline{ab}$. Therefore, we use these two planes to cut the solid to obtain a new solid shown in Figure 20. Similarly, another two cutting planes can be found to explain the uninterpreted solid lines $\overline{u_t v_t}$ and $\overline{v_t w_t}$ in the top view. The final solid is shown in Figure 21. At this stage all 2D data can be interpreted with the 3D figure of the solid constructed so far and vice versa. The process of 2D to 3D reconstruction is thus accomplished.

The second example involves a solid with curved surfaces. The 2D input data is the one given previously in Figure 3. In the front view we detect an isolated circle which stands for two void cylinders in the solid. Next, there is an arc in the front and top views, respectively. Each of these arcs induces a partial cylinder. Then a process is invoked to replace these arcs by the three

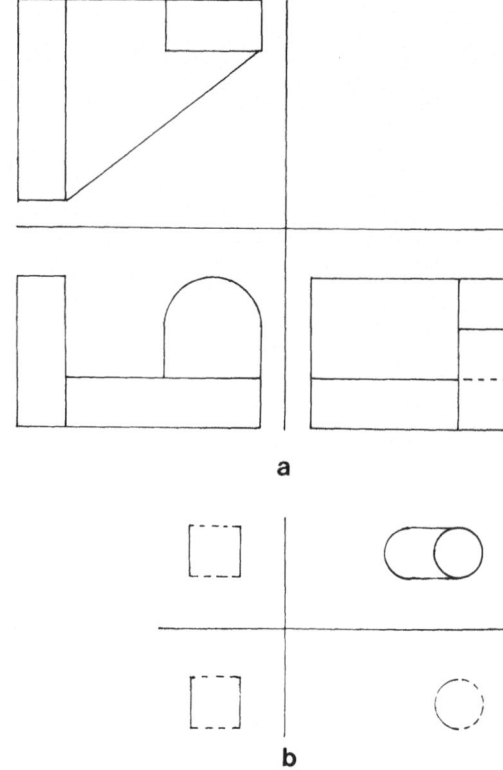

Figure 12. The decomposition and the reconstruction of subpart 3.

enclosing rectangle sides. The remaining problem reduces to be a polyhedral reconstruction problem. As illustrated in the first example, we can use translational sweeping and plane cutting operations to obtain the solution polyhedron. Finally, by the composition procedure, we can obtain the actual solid shown in Figure 22.

A 3D GRAPHICS APPROACH TO PART GEOMETRIC MODELING

Now we propose a 3D approach for defining the geometry of mechanical parts. Basically, our approach is similar to other current existing methods [11-15]. However, we will include new means to facilitate the user to design and manufacture 3D mechanical parts. The main features of the system include (1) geometric modeling of the 3D object, (2) interactive process for creating and

COMPUTER-ASSISTED METHODS 187

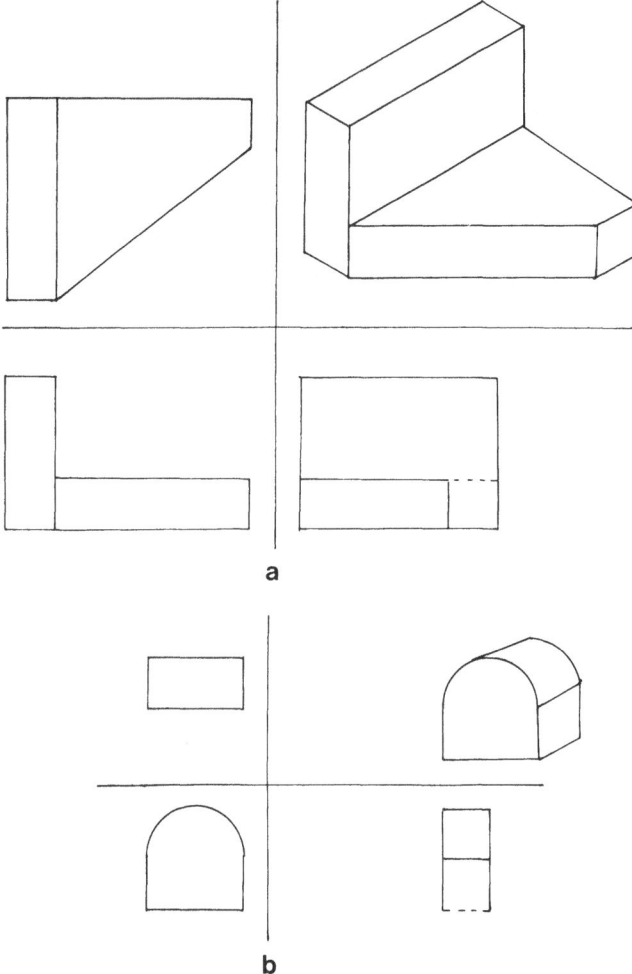

Figure 13. The reconstruction of the remaining views after the extraction of isolated subviews.

editing the 3D object, (3) 3D graphic display, and (4) automatic generation of APT program of the 3D object.

The layout of our system is illustrated by Figure 23. In this system the following rules are adopted:

(a) the name of an object have a maximum of five alphanumeric characters starting with an alphabet, e.g., ABC, A23.

(b) the name of a point is provided by the system automatically; the names range from P0 to P99.

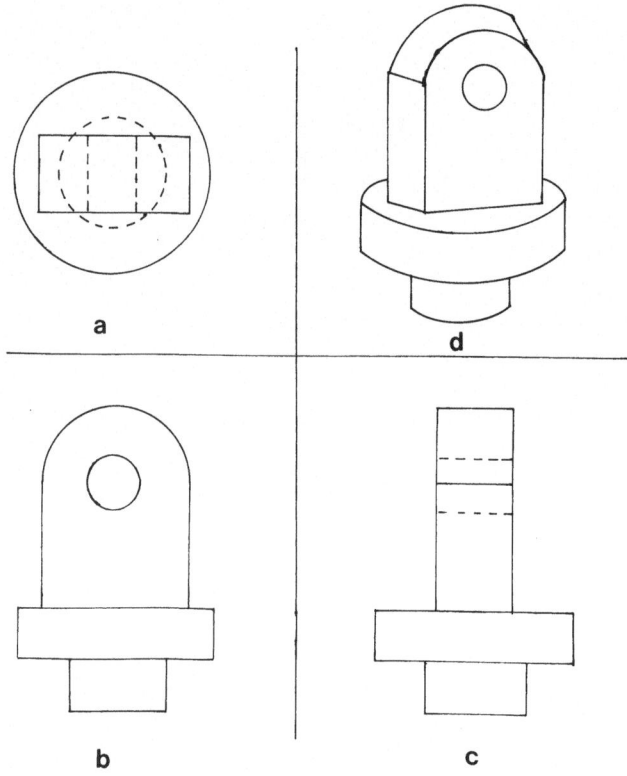

Figure 14. The engineering drawing for example 2.

(c) the name of a <u>line</u> segment is defined by the two end points. For instance, $\overline{P1,P2}$ defines a line segment with P1 and P2 as the end points.

(d) the name <u>of a plane</u> is defined by three points. For instance, $\overline{P1,P2,P3}$ represents a plane determined by the three points P1, P2, and P3.

We use the Constructive Solid Geometry (CSG) method to generate objects. The system provides six types of primitive objects, namely, (1) block, (2) generalized cone (including cylinder), (3) wedge, (4) partial cylinder, (5) sector, and (6) fillet. These primitive objects are given in Figure 24.

Now we begin to briefly introduce the functions of all system modules. When the user selects the module PO for generating primitive objects, the system will first prompt a list of six primitive objects for the user to choose and name. Subsequently, the system

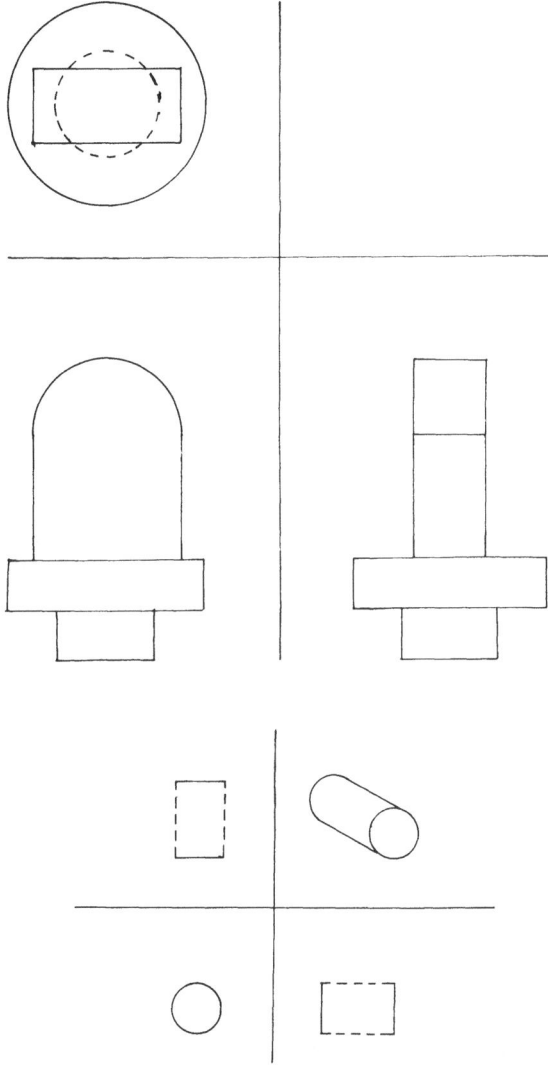

Figure 15. The decomposition and the reconstruction of subpart 1.

requires the user to specify the object dimension. At this point the primitive object has been completely defined, the system can output the graphic display of the object with feature points of the object being designated with point names generated by the system itself. On the other hand, if the user needs additional points, he can define point locations and requests point names by invoking the module DP.

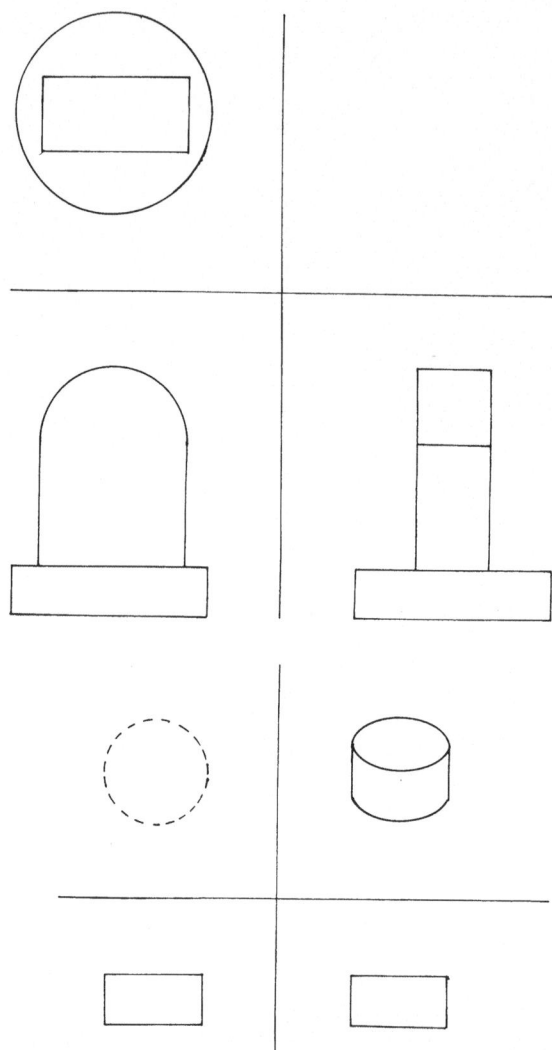

Figure 16. The decomposition and the reconstruction of subpart 2.

When the user wants to perform geometric operations on two objects, he needs only to specify at most three pairs of corresponding points between the two objects. The necessary translations and/or rotations will be carried out by the system instead of the user; thus the user does not need to keep track of the actual coordinate values of the objects. In the process of specifying the correspondence between points, some new points are often needed. We shall illustrate this fact in Figure 25. Blocks B1 and B2 are to form a new block, called B3, under the union operation. The

COMPUTER-ASSISTED METHODS

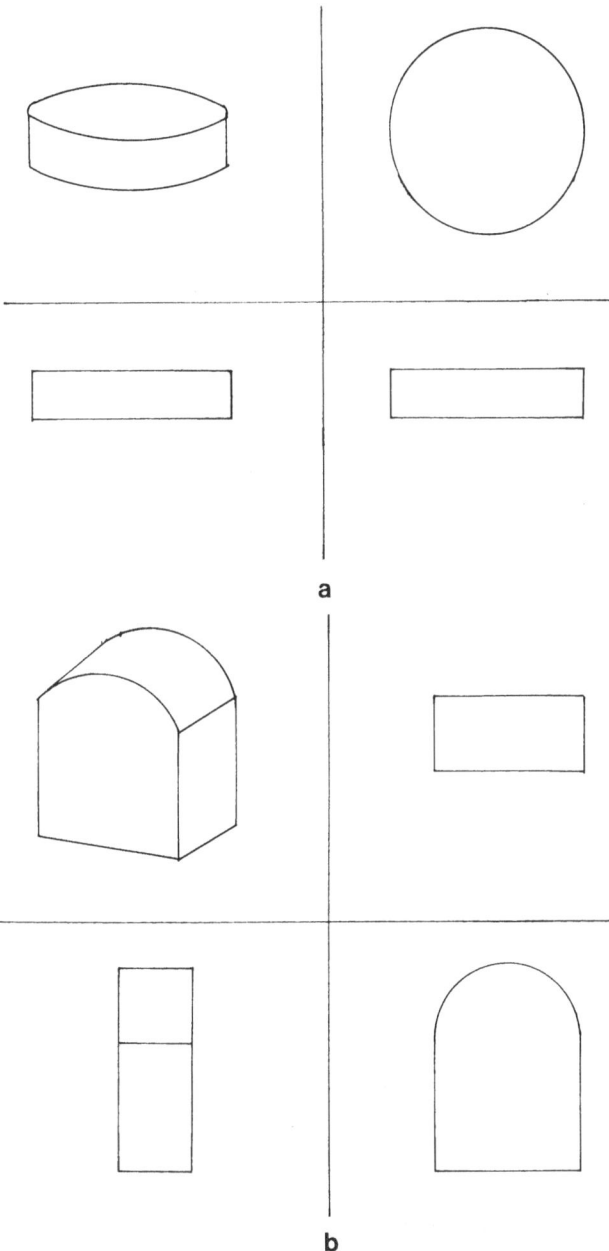

Figure 17. The reconstruction of the remaining views after the extraction of isolated subviews.

primitive points P1, P2 of block B2 correspond to primitive points P5, P6 of block B1. A new point P9 on line segment $\overline{P5,P8}$ must be created to match the primitive point P4 of block B2. The creation of a new point is done by invoking the module DP, as mentioned before.

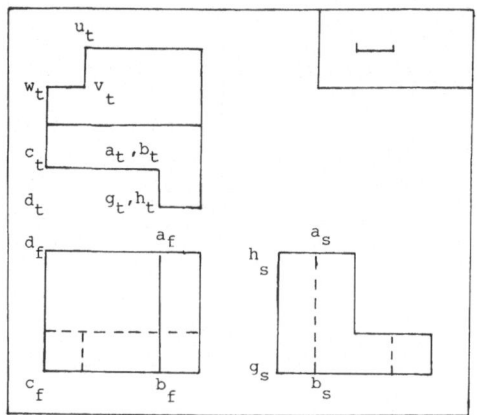

Figure 18. The engineering drawing of a polyhedron.

There are two ways in creating a new point in the module DP, namely,

(1) a new point is defined on a line segment. In this case the user indicates the distance between the new point and any one of the two end points.

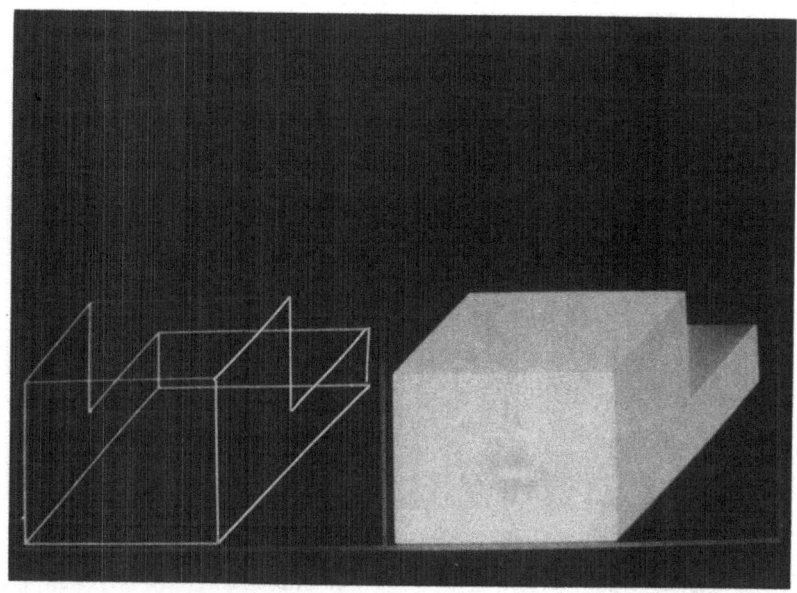

Figure 19. The solid obtained by a translational sweeping.

COMPUTER-ASSISTED METHODS

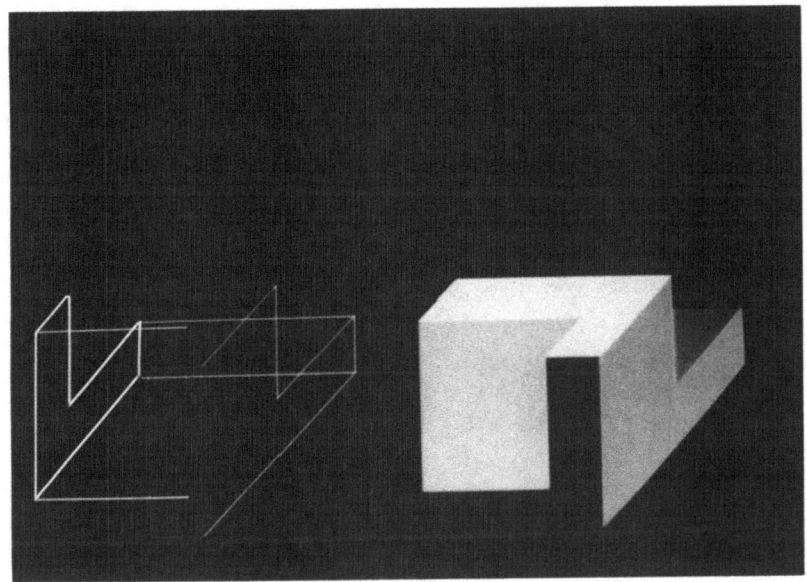

Figure 20. The resultant solid after two plane cutting.

(2) a new point is defined on a plane. Here the distances between the new point and any two line segments on the plane must be specified.

Next, we shall introduce the module GO for specifying the geometric operation. Here only three types of operation are implemented, i.e., (1) union, (2) difference and (3) assembly. In the union operation the coincident line segments are removed, while they are unchanged in the assembly operation. On the other hand, a difference operation is to remove a smaller object from a larger object with the former lying entirely in the latter. We avoid the complicated operation of intersection by employing the six primitive objects which are not mutually independent if the intersection operation is included.

The module GD is used to display the wire frame and color shading of the 3D object created. And the module APT can be called to produce the APT part program of the object. The functions of the rest of system modules are obvious from their names.

EXPERIMENTAL RESULTS OF PART DESIGN BY THE 3D GRAPHICS APPROACH

The overall system design is given in Figure 26. The system is implemented on a VAX 11/780 computer under VMS/3.0. The programs

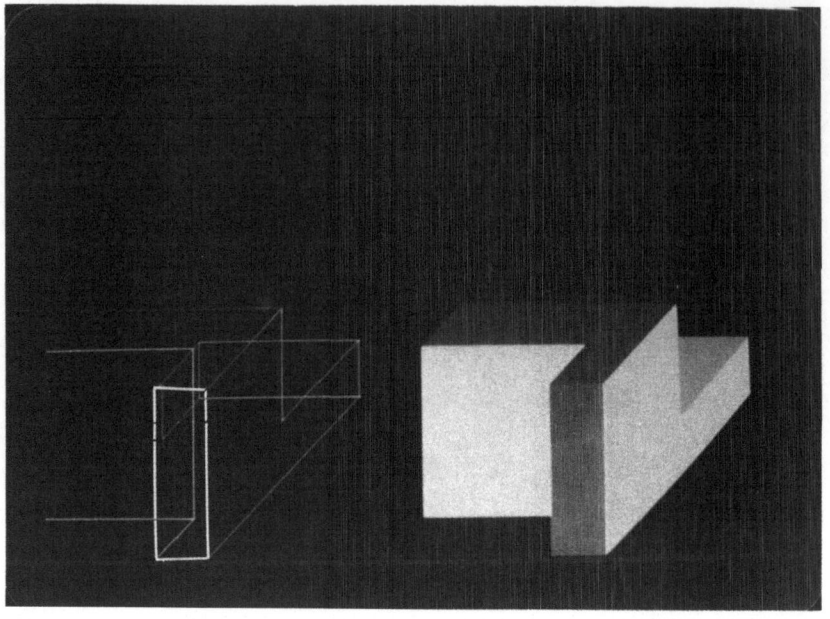

Figure 21. The final solid obtained.

are written in FORTRAN 77; the graphic terminal is Ramtek 6211. We shall illustrate the important functions of the system with three experiments.

Experiment 1: The primitive objects, block and wedge, are created which are displayed in the form of wire-frames. Notice that all feature points of the two objects are supplied by the system automatically. These two objects are composed under the union operation. The point correspondence is given as

(1)   Point P1 of the wedge corresponds to point P6 of the block,
(2)   Point P3 of the wedge corresponds to point P9 of the block,
(3)   Point P4 of the wedge corresponds to point P5 of the block.

The graphic displays of wedge, block and the composed object are shown in Figure 27.

Experiment 2: Perform the difference operation by subtracting a partial cylinder from a block. The graphic display of the resultant object is shown in Figure 28.

COMPUTER-ASSISTED METHODS 195

Figure 22. The final solid obtained.

Experiment 3: More complicated objects are possible to be derived from the blocks, fillets, cylinders, and partial cylinders through a sequence of union and difference operations. Three such objects are displayed in Figure 29.

On the other hand, after the 3D mechanical part is created by our 3D graphics method, one can invoke the module APT to generate

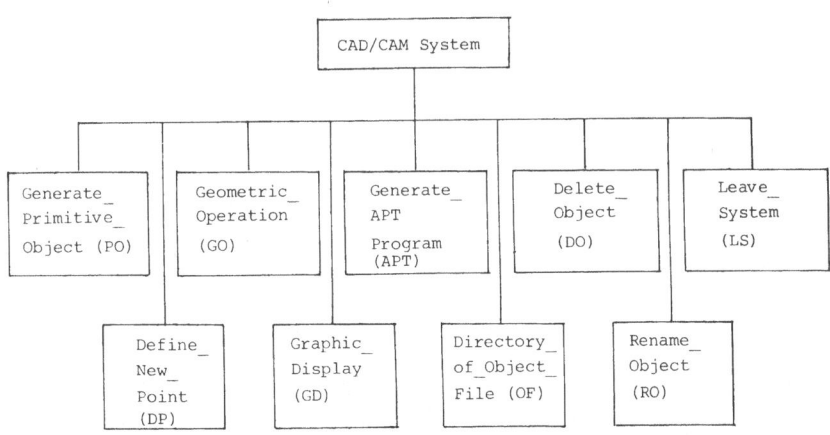

Figure 23. The layout of system modules.

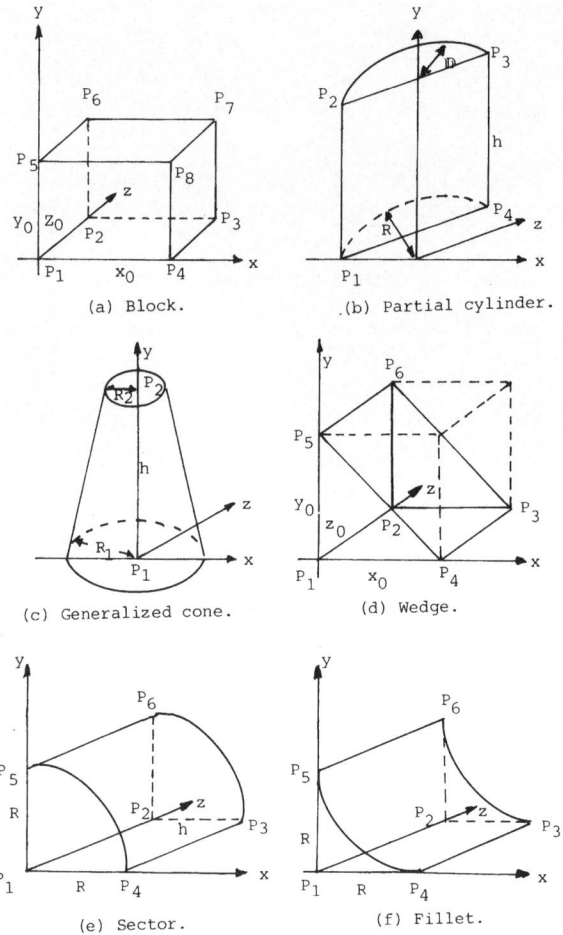

Figure 24. Six types of primitive objects.

the APT part program. The system then outputs two separate lists of geometry statements and motion statements of the APT part program. This type of experiment is omitted here. For more information one can refer to [22].

CONCLUSIONS

We have proposed an automatic system that can read the 2D data directly from a 3-view orthographic drawing via a picture digitizer. Furthermore, the system will interpret the 2D data to derive the 3D geometry of a part corresponding to the given engineering drawing.

COMPUTER-ASSISTED METHODS

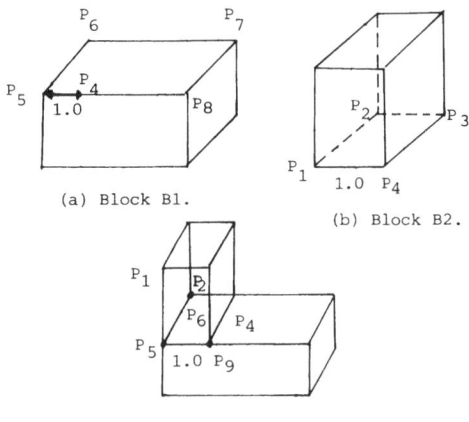

(a) Block B1.   (b) Block B2.

(c) Composed block B3.

Figure 25. A union operation.

The class of objects handled by our system comprises arbitrary polyhedra, curved objects derivable by translational sweeping, and a combination of these two objects under union and difference operations. However, at present we restrict the above combination to be carried out with the sweeping direction parallel to the three coordinate axes. We shall remove this restriction by extending our system to be able to recognize ellipses in the orthographic views. Furthermore, we are also investigating the possibility of handling other classes of objects.

On the other hand, we also propose a 3D graphical method for 3D mechanical part design and production. The desirable features of the system include:

(1) An interactive question-answering process to generate the 3D objects.

(2) Geometric operations are specified by the point pairs supplied by the user. The necessary coordinate transformations are automatically done by the system.

(3) 3D graphic display helps the user visualize the creating and editing of 3D objects.

(4) APT part program for a 3D mechanical part created is automatically generated by the system.

The following topics are suggested for the future research:

(1) The inclusion of the intersection operation,
(2) The tolerance and dimensioning problem,
(3) Graphic data base for integrated CAD/CAM and management information system.

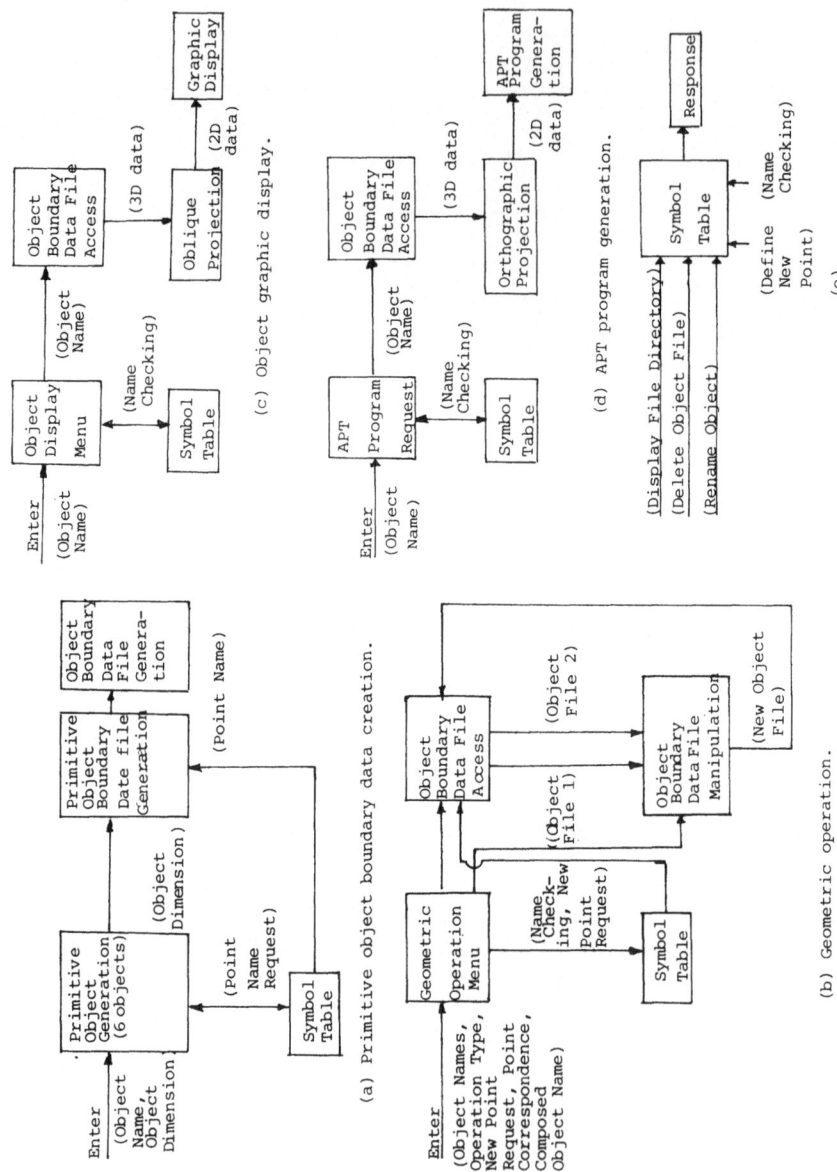

Figure 26. The overall system design.

COMPUTER-ASSISTED METHODS

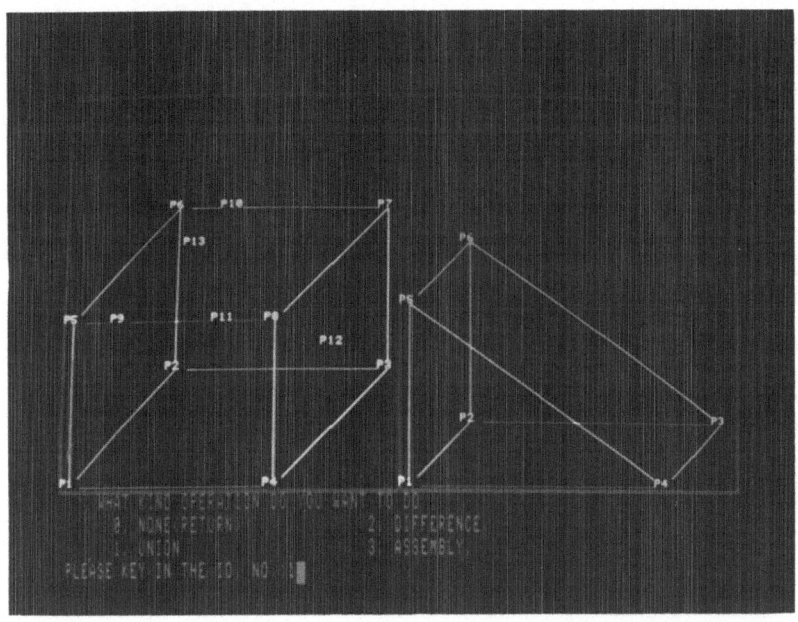

Figure 27. Experiment 1.

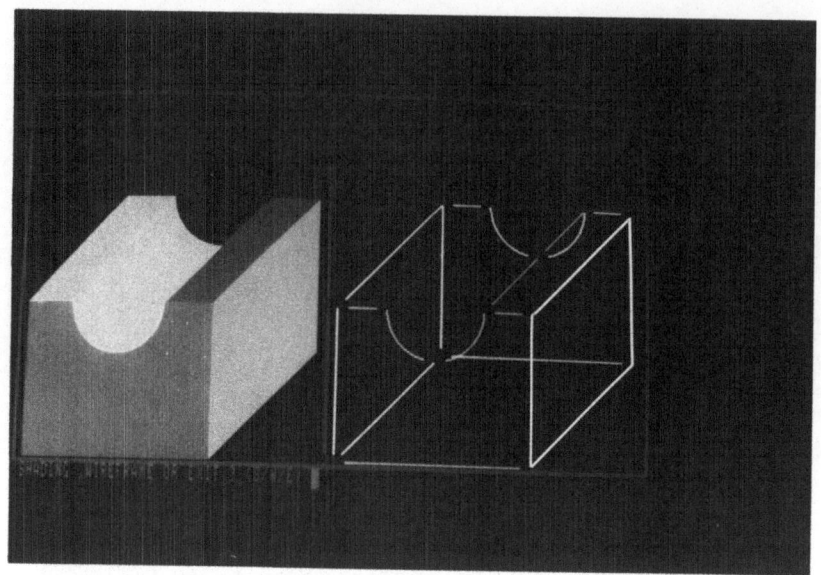

Figure 28. Experiment 2.

REFERENCES

[1] T. Kanade, "Recovery of the three-dimensional shape of an object from a single view," Artificial Intelligence 17, pp. 409-460, 1981.

[2] J. K. Udupa, "Determination of 3D shape parameters from boundary information," Computer Graphics and Image Processing 17, pp. 52-59, 1981.

[3] S. Lobregt, P. W. Verbeek, and F. C. A. Groen, "Three-dimensional skeletonization: principle and algorithm," IEEE Trans. on Pattern Analysis and Machine Intelligence, No. 1, pp. 75-77, Jan. 1980.

[4] M. Ejiri, T. Uno, H. Yoda, T. Goto, and K. Takeyasu, "A prototype intelligent robot that assemblies object from plan drawings," IEEE Trans. on Computer, vol. C-21, No. 2, pp. 161-170, 1972.

[5] M. Idesawa, T. Soma, E. Goto, and S. Shibata, "Automatic input of line drawing and generation of solid figure from three-view data," in Proc. Int. Comput. Symp., vol. 2, pp. 304-311, 1975.

[6] G. Lafue, "Recognition of three-dimensional objects from orthographic views," in Proc. SIGGRAPH, pp. 103-108, 1976.

[7] M. A. Wesley and G. Markowsky, "Fleshing out projections," IBM J. Res. Develop., vol. 25, No. 6, pp. 934-954, 1981.

[8] G. Markowsky and M. A. Wesley, "Fleshing out wire frames," IBM J. Res. Develop., vol. 24, No. 5, pp. 582-597, 1980.

COMPUTER-ASSISTED METHODS 201

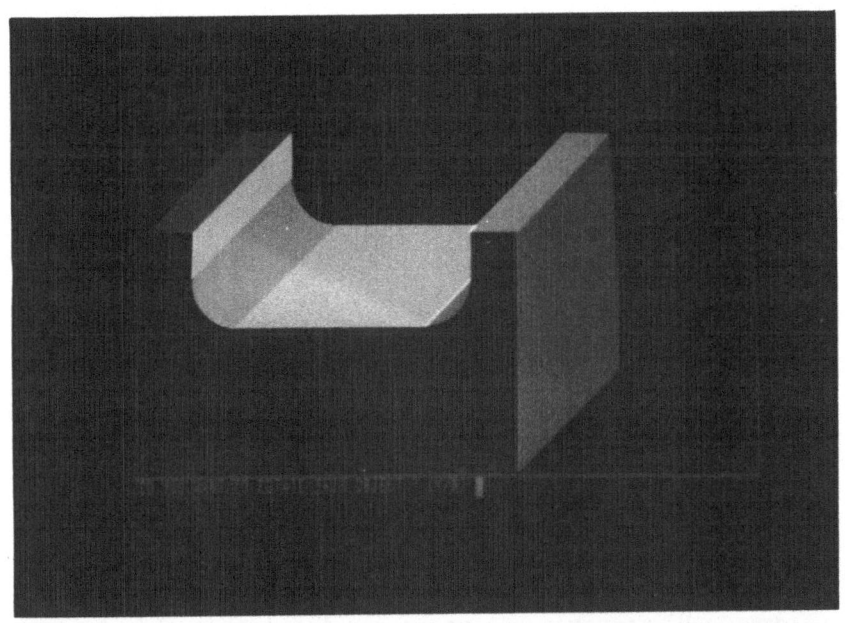

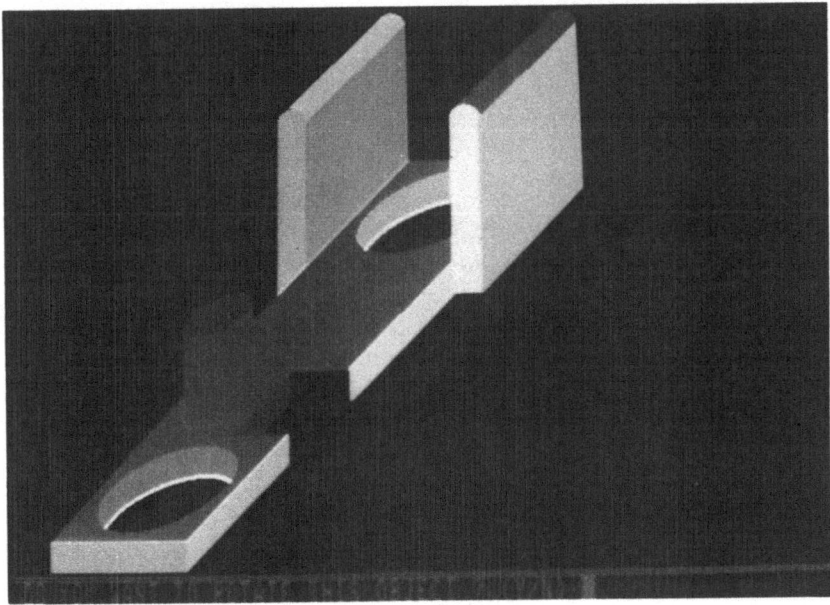

Figure 29. Experiment 3.

[9] B. Aldefeld, "On automatic recognition of 3D structures from 2D representations," Computer Aided Design, vol. 15, No. 2, pp. 59-64, March 1983.

[10] A. A. G. Requicha and H. B. Voelcker, "Solid modeling: a historical summary and contemporary assessment," IEEE CG & A, pp. 9-23, March 1982.

[11] A. A. G. Requicha and H. B. Voelcker, "An introduction to geometric modeling and its applications in mechanical design and production," The University of Rochester, Rochester, NY, Tech. Rep., Chapter 5, 1977.

[12] A. A. G. Requicha, "Representations for rigid solids: theory, method, and systems," Computing Surveys, vol. 12, No. 4, pp. 437-465, Dec. 1980.

[13] J. W. Boyse and J. E. Gilchrist, "GMsolid: interactive modeling for design and analysis of solids," IEEE CG & A, pp. 27-40, March 1982.

[14] R. Hillyard, "The build group of solid modelers, "IEEE CG & A, pp. 43-52, March 1982.

[15] A. Baer, C. Eastman, and M. Henrion, "Geometric modelling a survey," CAD, vol. 11, No. 5, pp. 253-272, Sept. 1979.

[16] I. C. Braid, "The synthesis of solids bounded by many faces," Commun. Assoc. Comput. Mach., pp. 209-216, April 1975.

[17] M. A. Wesley, T. Lozano-Perez, L. I. Lieberman, M. A. Lavin, and D. D. Grossman, "A geometric modeling system for automated mechanical assembly," IBM J. Res. Develop., pp. 64-74, Jan. 1980.

[18] D. D. Grossman, "Procedural representation of 3D objects," IBM J. Res. Develop., pp. 582-589, Nov. 1976.

[19] H. B. Voelcker and A. A. G. Requicha, "Geometric modeling of mechanical parts and processes," Computer, pp. 48-57, Dec. 1977.

[20] J. D. Foley and A. Van Dam, Fundamentals of Interactive Computer Graphics, Massachusetts: Addison-Wesley, 1982.

[21] S. J. Joe, "3D reconstruction and graphic display of mechanical parts from orthographic drawings," Master's thesis, Inst. of Comput. Eng., National Chiao Tung University, Taiwan, R.O.C., 1983.

[22] C. Y. Lo, Z. Chen, and D. B. Perng, "A CAD/CAM system for mechanical part design and production," in Proc. Int. Conf. Adv. Automation, pp. 266-272, 1983.

COMPUTER-AIDED FINITE ELEMENT ANALYSIS: INTERFACING SYMBOLIC AND NUMERICAL COMPUTATIONAL TECHNIQUES

Paul S. Wang

Department of Mathematical Sciences
Kent State University
Kent, Ohio 44242   U.S.A.

ABSTRACT

The design and implementation of a software system for generating finite elements and related computations are described. Exact symbolic computational techniques are employed to derive strain-displacement matrices and element stiffness matrices. Methods for dealing with the excessive growth of symbolic expressions are discussed. Automatic FORTRAN code generation is described with emphasis on improving the efficiency of the resultant code. The generated code is used in conjunction with a FORTRAN-based finite element analysis package.

INTRODUCTION

In recent years we have seen increasing interest in using computer-based symbolic and algebraic manipulation systems for computations in both linear and nonlinear finite element analysis. Application areas include the symbolic derivation of stiffness coefficients [1,5], the reduction of tedium in algebraic manipulation, the generation of FORTRAN code from symbolic expressions [6,7], etc. The strength of this approach lies in the combined use of symbolic and numerical computational techniques. The potential benefits and usefulness of such an approach in finite element analysis as well as other scientific research areas are evident. However, several problems need be solved before this approach can become widely accepted and practiced for finite element work:

i. the efficiency of the symbolic processor and its ability to handle the large expressions associated with practical problems,

ii. the interface between a symbolic system and a finite element system on the same computer, and

iii. the inefficiencies that are usually associated with automatically generated code.

The MACSYMA system [8] is a sophisticated computer system for performing exact symbolic mathematical computations. It provides many high-level commands for symbolic computations such as differentiation, matrix multiplication and matrix inverse. MACSYMA is developed at the Massachusetts Institute of Technology and has recently been made available on the Digital Equipment Corp. VAX computers. It is a family of affordable high-performance minicomputers with a very large address space. The version of MACSYMA on the VAX under the UNIX operation system [9] is known as VAXIMA [4]. The author participated in the design and implementation of the MIT MACSYMA system.

We describe our ongoing research on the design and implementation of a finite element generator. The functionalities of this generator depends critically on interfacing the symbolic processing capabilities of VAXIMA and the numerical programs on the same computer.

Functional Specifications and Design

The finite element generator (Generator) as a software system should provide a set of predefined functionalities and be constructed to exhibit prescribed characteristics. These functional specifications and design goals will then guide the detailed program design and implementation of the software system.

From a functional point of view, the Generator will perform the following:

(1) to assist the user in the symbolic derivation of finite elements;

(2) to provide routines for a variety of symbolic computations in finite element analysis, including linear and nonlinear applications especially for shells [2];

(3) to provide easy to use interactive commands for most common operations;

(4) to allow the mode of operation to range from interactive manual control to fully automatic;

(5) to generate, based on symbolic computations, FORTRAN code in a form specified by the user;

(6) to automatically arrange for generated FORTRAN code to compile, link and run with FORTRAN-based finite element analysis packages such as the NFAP package [3];

(7) to provide for easy verification of computational results and testing of the code generated.

In providing the above functions attention must be paid to making the system easy to use, modify and extend. Our initial effort is focused on the isoparametric element family. Later the system can be extended to a wider range of finite elements.

GENERATION OF ELEMENT STIFFNESS MATRICES

Symbolic processing can play an important role in the generation of element matrices, especially for higher order elements. As an example, we shall describe the automatic processing leading to the derivation of the element stiffness matrix [K] in the isoparametric formulation. From user-supplied input information such as the element type, the number of nodes, the nodal degrees of freedom, the displacement field interpolation polynomial and the material properties matrix [D], the generator will derive the shape function, the strain-displacement matrix [B] and the element stiffness matrix [K].

The computation is divided into five logical phases (Figure 1), each is implemented as a LISP program module running under the VAXIMA system. Aside from certain interface considerations, these modules are largely independent and can therefore be implemented and tested separately. This allows different people to work on different modules at the same time. After the modules are individually tested then they can be integrated and verified together. Any problems can be isolated to a module and fixed quickly. Detailed description of these phases follows.

Phase I: Define Input Parameters

The task of this phase is to interact with the user to define all the input names, variables, and values that will be needed later. The basic input mode is interactive with the system prompting the user at the terminal for needed input information. While the basic input mode provides flexibility, the input phase can be tedious. Thus, we also provide a menu-driven mode where

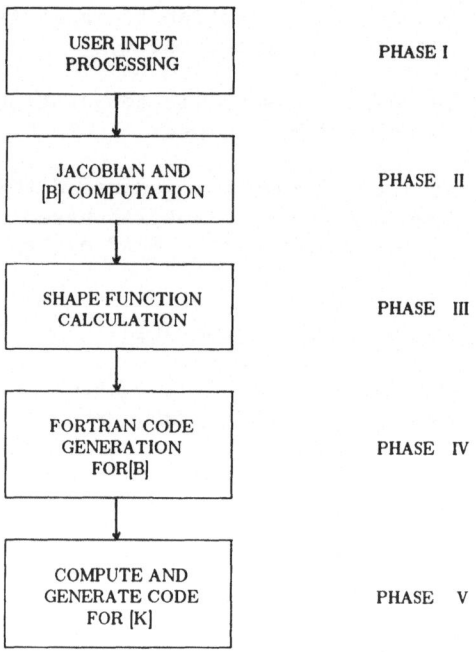

Figure 1. Automatic generation of [B] and [K].

well-known element types together with their usual parameters values are predefined for user selection. A fully user-friendly input phase is a goal of our system.

The input handling features include:

(1) free format for all input with interactive prompting showing the correct input form;

(2) editing capabilities for correcting typing errors;

(3) the capability of saving all or part of the input for use later;

(4) the flexibility of receiving input either interactively or from a text file.

Phase II: Jacobian and [B] Matrix Computation

The strain-displacement matrix [B] is derived from symbolically defined shape functions in this phase. Let n be the number of nodes then

$$H = (h_1, h_2, \ldots, h_n)$$

is the shape function vector whose components are the n shape functions h_1 through h_n. The value for the shape functions will be derived in a later phase. Here we simply compute with the symbolic names. Let r, s and t be the natural coordinates in the isoparametric formulation and HP be a matrix

$$HP = \begin{matrix} H,r \\ H,s \\ H,t \end{matrix}$$

where H,r stands for the partial derivative of H with respect to r. The Jacobian J is then

$$J = HP \cdot [x,y,z]$$

where x stands for the column vector $[x_1, \ldots, x_n]$, etc. Now the inverse, in full symbolic form, of J can be computed as

$$J^{-1} = \frac{invj}{det(J)}$$

By forming the matrix (invj · HP) we can then form the [B] matrix.

Phase III: Shape Function Calculation

Based on the interpolation polynomials and nodal coordinates the shape function vector H is derived and expressed in terms of the natural coordinates r, s and t in the isoparametric formulation. Thus, the explicit values for all h_i and all their partial derivatives with respect to r, s and t needed in HP are computed here.

The input handling module, the derivation of the shape functions, the [B] matrix and its corresponding FORTRAN code were done by Mr. Young as his master's thesis project [10].

Phase IV: FORTRAN Code Generation for [B]

A fortran subroutine for the numerical evaluation of the strain-displacement matrix [B] is generated for use with the NFAP package. This NFAP package is a large FORTRAN based system for linear and nonlinear finite element analysis. It is developed and made available to us by P. Chang of the University of Akron. It has been modified and made to run in FORTRAN 77 under UNIX.

Assignment statements will be generated to define the components of the shape function vector H and the various partial derivatives. These variables are then used in assigning values to the FORTRAN array corresponding to [B]. FORTRAN code generation is controlled by a "template" file. The template file is a skeleton FORTRAN program with special instructions for code generation.

Details of the FORTRAN code generator will be discussed later.

Phase V: Compute and Generate FORTRAN Code for [K]

The inverse of the Jacobian, J, appears in [B]. By keeping the inverse of J as INVJ/det(J), the quantity det(J) can be factored from [B] and, denoting by [BJ] the matrix [B] thus reduced, we have

$$[K] = \int\int\int_{-1}^{1} \frac{[BJ]^T \cdot [D] \cdot [BJ]}{det(J)} \, dr \, ds \, dt$$

The determinant of the Jacobian involves the natural coordinates. This makes the exact integration in the above formula difficult. We elect to evaluate det(J) at r=s=t=0 and factor it out of the integral. The resulting integrand involves only polynomials in r, s and t which are readily integrated. We believe this approximation is reasonable and the resulting symbolic expression for [K] will still be more accurate than results obtainable from numerical quadrature.

To avoid accumulating large amounts of symbolic expressions, we generate the stiffness matrix [K] one entry at a time. Namely, the integrand matrix is not formed all at once, instead each entry is computed and integrated individually. Furthermore, the FORTRAN code of each [K] entry is output after it is computed. Usually [K] is symmetric and only the upper triangular part need be computed. Although there is a general purpose integration routine available in VAXIMA, we use a specially designed integration program to gain speed and efficiency. The integration is organized to combine common subexpressions and produce compact and efficient FORTRAN code. Details on this later.

THE FORTRAN CODE GENERATOR

A set of LISP programs have been written for generating FORTRAN code based on symbolic mathematical expressions derived in VAXIMA. This code generator runs under VAXIMA and generates FORTRAN code by following a given "template" file (Figure 2). A template file contains FORTRAN statements any of which may contain an "active part" which is one or several correct VAXIMA statements enclosed in "{" and "}". A typical template is shown in Figure 3.

To invoke the FORTRAN generator the following VAXIMA command is used.

GENFORTRAN (template-file-name, output-file-name)

FINITE ELEMENT ANALYSIS

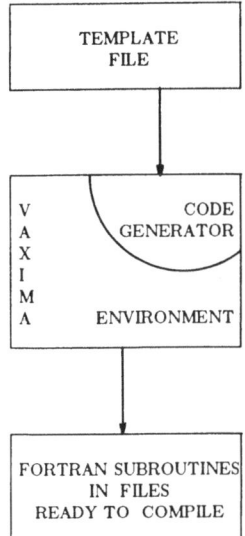

Figure 2.   FORTRAN code generation.

GENFORTRAN reads the given template file and passes all characters into the output file without modification except for the active parts contained in "{ " and "}". The VAXIMA commands in the active part will be executed by the Generator in the order given, and any resultant output will be directed to the output file as well. Thus, in the above example, we can see where the strain-displacement matrix [B] and the stiffness matrix [K] are being generated. The output file will be template with all active parts replaced by generated code. This output file will therefore be a syntactically correct FORTRAN file ready to compile.

Comment lines are copied into the output file without checking for any active parts. This allows the use of {,} inside comment lines. Recursive invocation of GENFORTRAN from within a template is allowed. This makes it easy to generate a set of related subroutines under the control of GENFORTRAN. This arrangement is very flexible and works quite well for our purposes.

When the code generator is fully developed, it will have the ability to generate code in separate sessions rather than all at once. This means partially generated code can be completed later without regenerating what's already been done. When substantial amount of code is generated automatically, this feature will become important. To help organize the code generated, multiple output files can be automatically used to contain groupings of subroutines.

```
              SUBROUTINE STIFF2(ST,B,YZ,C,THICK,R,S,ND,DK)
              IMPLICIT REAL*8 (A-H,O-Z)
              DIMENSION ST(1),B(4,1),YZ(2,1),C(4,1),SK(18,18)
       C
       C   DK = 1 STIFFNESS matrix calculations
       C   DK = 2 STRAIN-DISPACEMENT matrix calculations
       C
              IF ( DK .EQ. 1 ) GO TO 500
       C Here are some active statements enclosed in {}
       { f77(det=detj), bfortran() }
              RETURN
        500   CONTINUE
       C kfortran generates code for [K]
       { kfortran(b,dx,[r,s]) }
              KL = 0
              DO 510 I = 1 , ND
               DO 510 J = 1 , ND
                KL = KL+1
                ST(KL) = SK(I,J)*THICK
        510   CONTINUE
              RETURN
              END
```

Figure 3. A typical template.

EXPRESSION GROWTH AND EFFICIENCY OF FORTRAN CODE

Previous work in using systems such as MACSYMA for finite element computation uses user-level programs which do not allow much control over the exact manner in which computations are carried out. As a result, the ability of handling realistic cases in practice is very restricted because of expression growth, a phenomenon in symbolic computation when intermediate expressions become too large for efficient manipulation.

We use LISP programs with direct access to internal data structures. Thus, it is possible to construct programs which will avoid expression growth in the intermediate computations as much as possible. Therefore, our programs are better suited to handle practical problems with efficiency. Let us illustrate the control of expression growth by the [K] matrix computation.

First of all, the [B] matrix in the integrand is computed with "unevaluated" symbols to keep things small. Thus, a typical nonzero entry of [B] looks like

$$hr_2 hs \cdot y - hr \cdot y\, hs_2$$

FINITE ELEMENT ANALYSIS

where hr_2 = the second component of H,r and $y = [y_1,\ldots,y_n]$. But these symbols remain unevaluated at this point. As stated before, we proceed to generate the entries of [K] one at a time to keep expressions small. In doing this, we apply the formula,

$$[K] = \sum_i \sum_j \iiint_{-1}^{1} [B]^T{}_i [D]_{ij} [B]_j \, dr \, ds \, dt,$$

to collect terms with respect to $[D]_{ij}$. In the above formula, $[B]^T{}_i$ denotes the ith row of [B] transpose, etc. Coefficients thus obtained are kept in unexpanded form on a list which is consulted for duplicates whenever a new coefficient is generated. This identifies common (duplicate) subexpressions in different entries of [K] and keep the resulting FORTRAN code compact and more efficient. When a new coefficient is formed then it is evaluated and expressed as a polynomial involving the natural coordinates r, s and t. Now a special-purpose integration routine is used. The integration result for each coefficient is converted into FORTRAN code and assigned to a temporary FORTRAN variable. Thus, a typical section of the code for [K] may look like the following.

```
      t30 = -((16*y3-4*y2-12*y1)*y4+(-12*y2-4*y1)*y3
     1 + 4*y2**2+8*y1*y2+4*y1**2)/3.0
      t31 = ((16*x3-4*x2-12*x1)*y4+(-12*x3+4*x2+8*x1)
     1 *y2+(-4*x3+4*x1)*y1)/3.0
      t32 = ((16*x4-12*x2-4*x1)*y3+(-4*x4+4*x2)*y2
     1 +(-12*x4+8*x2+4*x1)*y1)/3.0
      t33 = -((16*x3-4*x2-12*x1)*x4+(-12*x2-4*x1)*x3
     1 +4*x2**2+8*x1*x2+4*x1**2)/3.0
      k(5,7) = 4*(d6*t33+d3*(t32+t31)+d1*t30)/detk
      k(5,8) = 4*(d5*t33+d6*t32+d2*t31+d3*t30)/detk
```

In the above, t30, etc. are the temporary variables and d1, d2, etc. are entries of the material properties matrix [D]. Without the techniques mentioned here, the code for each single [K] entry will require 5 to 8 continuation cards (for a plane 4-node element).

Experiments on the VAX-11/780 with the NFAP package together with code for the [B] and [K] matrices generated show that there is a 10% CPU time savings with the above described simplification for the 4-node plane element. The savings will be much greater for larger problems. Among other things, we are currently studying ways to further simplify the expressions for the t's. For three-dimensional elements with many degrees of freedom, controlling the size of the intermediate and the resulting code will become critical. Among the techniques we are exploring are: generating functions and subroutines to simplify the code, using any

symmetries that may be present in the given problem, and
systematically identify and collect subexpressions.

ACKNOWLEDGMENT

Work reported herein has been supported in part by the U.S.
National Aeronautics and Space Administration under Grant NAG 3-298
and in part by the U.S. National Science Foundation under Grant
MCS-82-01239.

REFERENCES

[1] M. M. Cecchi and C. Lami, "Automatic generation of stiffness matrices for finite element analysis," Int. J. Num Meth. Eng., 11, pp. 396-400, 1977.
[2] T. Y. Chang and K. Sawamiphakdi, "Large Deformation Analysis of Laminated Shells by Finite Element Method," Comput. Structures, Vol. 13, 1981.
[3] T. Y. Chang, "NFAP - A Nonlinear Finite Element Analysis Program, Vol. 2 - User's Manual," Technical Report, College of Engineering, University of Akron, Akron, Ohio, 1980.
[4] J. K. Foderaro and R. J. Fateman, "Characterization of VAX Macsyma," Proceedings, ACM SYMSAC'81 Conference, Aug. 5-8, Snowbird, Utah, pp. 14-19, 1981.
[5] A. R. Korncoff and S. J. Fenves, "Symbolic generation of finite element stiffness matrices," Comput. Structures, 10, pp. 119-124, 1979.
[6] A. K. Noor and C. M. Andersen, "Computerized symbolic manipulation in nonlinear finite element analysis," Comput. Structures, 13, pp. 379-403, 1981.
[7] A. K. Noor and C. M. Andersen, "Computerized symbolic manipulation in structural mechanics-progress and potential," Comput. Structures, 10, pp. 95-118, 1977.
[8] MACSYMA Reference Manual, Version Nine, the MATHLAB Group, Laboratory for Computer Science, M.I.T., Cambridge, MA, 1977.
[9] UNIX Programmer's Manual, Vol. I and II, Seventh Edition, Bell Telephone Laboratories, Inc., Murray Hill, NJ, 1979.
[10] P. Y. Young, "Automatic finite element generator," master's thesis, Dept. Mathematical Sciences, Kent State University, Kent, Ohio, 1983.

A THEORETICAL PROPOSAL FOR A CASD SYSTEM EXTENDING JACKSON'S METHOD FOR PROGRAM CONSTRUCTION

C. J. Lucena, R. C. B. Martins, P. A. S. Veloso,[+] and D. D. Cowan[§]

[+]Departamento de Informática,
 Pontifícia Universidade Católica,
 Rio de Janeiro, Brazil
[§]Department of Computer Science
 University of Waterloo
 Waterloo, Ontario, Canada

ABSTRACT

This paper presents a new programming method called the data transform programming method. In particular, we present a specialization of data transform programming to deal with file processing applications. Direct comparison is made with Jackson's approach by the presentation of uniform solutions to problems that cannot be solved through his basic method.

The new method consists of the application of data transformations to the abstract problem statement, following the formal notions of problem reduction and problem decomposition. Data transformations are expressed in programming terms through a basic set of data type constructors. The method reduces the original problem to a set of sub-problems that can be solved through the direct application of Jackson's method. It produces a solution which is correct by construction.

INTRODUCTION

As computer costs go down the use of computer assistance in the process of problem solving increases. In fact, Computer Assisted Design (CAD) is rapidly catching up in most technological areas. Very recently, as the software development process became better known, a new area has been receiving widespread attention:

Computer Assisted Software Design (CASD). Most of the work in the CASD area can be roughly classified into two categories: systems to support the activity of programming-in-the-large (systems level programming) and systems to aid the process of programming-in-the-small (module level programming).

The present work describes a methodology that can be used as a basis for a CASD system for module level programming.

It has been observed that many of the changes in typical data processing applications, often called file processing programs, are caused by the changes in the structure of the data to be processed or to be output as the result of processing and by the accompanying actions which must occur to reflect these changes in the structure of the input/output data. Hence, if a program or system of programs can be designed to reflect the structure of the data that is being processed, then modifications to the data might be more easily reflected in the modifications of the program necessitated by these changes.

The above ideas were captured by experienced practitioners who have formulated programming methodologies that have considerably influenced today's programming practices in industry. The work of Jackson [1], Warnier [2], and Yourdon and Constantine [3] are often quoted as some of the most important in this area.

Data transform programming deals with the class of problems that can be solved by the basic Jackson method. It can also solve, through a uniform approach, problems that Jackson can handle only through major departures from his basic method.

The formalization of data transform programming was made possible through the association of the notion of data abstraction to file processing programming and through the utilization of formal definitions for concepts such as program decomposition and program reduction borrowed from the areas of logic and problem solving.

In order to put the original Jackson basic method on a more formal basis, Hughes [4] establishes a correspondence between the class of programs available to treatment by his method and the formal language concept of generalized sequential machine. It turns out that Jackson's basic method gives rise to transformations which are gsm computable (in the sense that the required transformation can be performed by a generalized sequential machine).

That, of course, explains why Jackson's basic method cannot solve backtracking problems (multiple passes over the input) and problems that he calls structure clash problems. Jackson solves the latter problems by using ad hoc solutions and the technique of program inversion (preparation of a program to be used, for the same function, as a subroutine to another program).

Cowan and Lucena [5], by introducing a new factor (abstract levels of specification for data and program and the subsequent implementation thereof in terms of more concrete levels of abstraction) into Jackson's method have solved the sorting problem to illustrate how the exercise of thinking abstractly about a problem can lead to novel solutions, or solutions which were thought to be unavailable due to shortcomings of a given method. The informal notion of data flow design by Yourdon and Constantine [3], together with the formal notion of problem solving by Veloso and Veloso [6] were instrumental for the formulation and improvement of the original ideas in Cowan and Lucena [5].

In previous papers ([7] and [8]) we have formulated the data programming method and applied it to the sorting problem (unsolvable by the basic Jackson method). Here we concentrate on other examples proposed by Jackson to illustrate the shortcomings of his method. These other examples are particular cases of the structure clash problem. The telegram problem illustrates a boundary clash situation, the system log problem is an example of a multithreading problem and the matrix transposition problem illustrates an ordering clash.

We try to make this work self-contained by reviewing Jackson's method and Yourdon and Constantine's data flow design methodology.

The data transform method is presented through some concepts in problem solving theory and theory of data types associated with informal arguments.

The regularity of the solutions enabled by the data transform method, which can be verified through the comparison among the various examples, should convince the reader about the viability of automating many of its aspects. We envision a computer-assisted software-design (CASD) approach based on the method. In the conclusion of the paper we reference an appendix (Appendix III) where we define informally the set of routines we have been using to support automatically the application of the method.

THE JACKSON METHOD

Following the formalization of the method by Hughes [4] in terms of generalized sequential machines (gsm's), we outline a correspondence between the class of programs developed by using the basic Jackson method and gsm computable functions between regular languages.

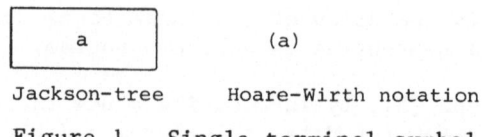

Figure 1. Single terminal symbol.

Jackson pointed out that input (and output) may often be regarded as possibly infinite languages (over some primitive set of data values). A Jackson tree provides a finite representation of such a language and can be used to represent only a regular language. One can show therefore that Jackson trees are an alternative notation of regular expressions as are data type definitions using the types sequence, cartesian product and discriminated union [8]. In fact, all the notations above are capable of representing only restricted forms of regular expressions if we assume the usual conventions concerning the priority of the regular expression operations "*", "U" and "." (if we do not make this assumption and proceed strictly according to the formal definitions of these operations, there is an exact one-to-one correspondence).

We proceed to outline the four notations below by defining how a given regular expression is represented in the other two notations.

i) A single terminal symbol a is represented in Figure 1.

ii) A regular expression $(\alpha_1,\ldots,\alpha_n)$, which is a concatenation of regular expressions, is represented in Figure 2.

iii) An expression $(\alpha_1 \cup \alpha_2 \cup \ldots \cup \alpha_n)$, which is the union of regular expressions, is represented in Figure 3.

iv) An expression $\alpha*$, which is the iteration of a regular expression, is represented in Figure 4.

The reader should note that bracketing of regular expressions with respect to the same operation has been abandoned in the above

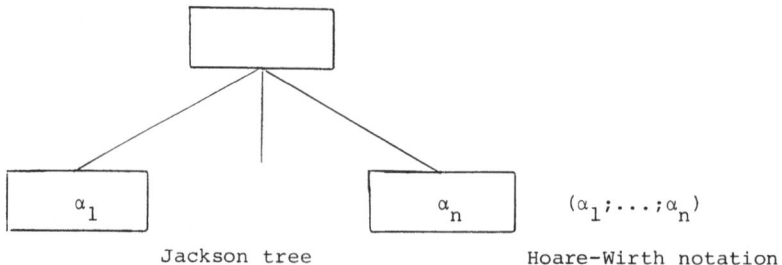

Figure 2. A regular expression.

CASD SYSTEM EXTENDING JACKSON'S METHOD

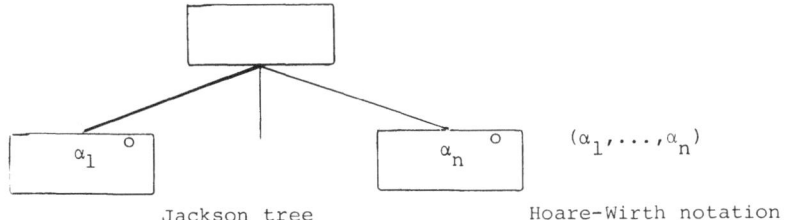

Figure 3. The union of regular expression.

notations to create more compact notations. This reduces the size of our various notations without loss of correctness. Note also that this corresponds to generalizing the usual binary operation "." and "U" to a family of operations with any finite number of arguments.

Now, the method requires that correspondences are identified between the specification of the input and that of the output in terms of correspondences between substructures in the two specifications. This is done in a bottom-up fashion so that the translation of a node in the tree (or graph) depends only on its descendants (i.e., the sub-expression or subtree defined by the node).

The correspondence effectively defines a desired translation between the nodes of the input specification tree and those of the output specification tree.

As it was proven in [4], for a given characterization of gsms it can be shown that Jackson's basic method gives rise to transformations which are gsm computable.

DATA FLOW DESIGN

Data flow design has been proposed by Yourdon and Constantine [3] as a program or programming system design methodology. As we

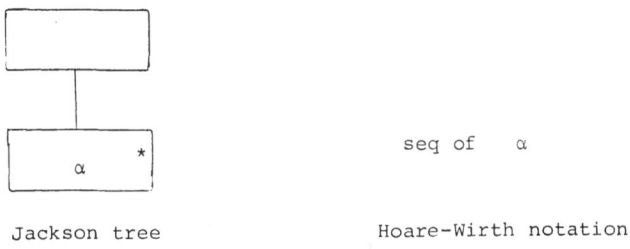

Figure 4. Iteration of a regular expression.

did with the Jackson method we will now outline the central
concepts in data flow design.

The purpose of the methodology is to identify the primary
processing functions of the system, the high-level inputs to those
functions, and the high-level outputs. It then creates high-level
modules within the hierarchy to perform each of these tasks:
creation of high-level inputs, transformation of inputs into high-
level outputs and the processing of those outputs. Clearly, data
flow design is an information flow model rather than a procedural
model.

Like other information flow models, transform analysis makes
use of a graph model of computation. It is called a data flow
graph. The nodes of the graph are called transforms. Each node
represents a data transformation (to be accomplished later by a
module or a program) from one data representation to another. The
data elements are represented by labelled arrows connecting the
nodes. Figure 5 shows a transform with a single input stream and
a single output stream.

A transform may require (or accept) elements of more than one
input data stream in order to produce its outputs. An asterisk "*"
or a disjunction symbol "+" indicates simultaneous requirements of
data elements or a mutually exclusive situation, respectively.

Yourdon and Constantine [3] have proposed a data flow design
methodology composed essentially of the following four steps. (in
what follows we have tried to capture the central aspects of each
step).

Step 1: Consists of the statement of the problem as a data
flow graph; the authors recommend that the designer should be concerned first with the "main" data paths dealing with primary
inputs.

Step 2: Identification of the initial and final data elements.
Initial and final data elements are those high-level elements of
data which are furthest removed from physical input and output,
respectively.

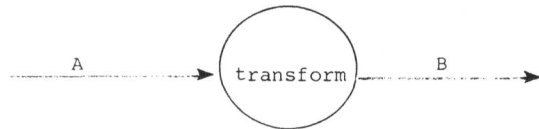

Figure 5. A transform.

Step 3: This step is further subdivided into four parts. Having identified the initial and the final data elements of the system:

   i) specify a "main" module which, when activated, will perform the entire task of the system by calling upon subordinates.

   ii) for each initial data element feeding a central transform an initial module is specified as an immediate subordinate to the main module.

   iii) for each final data element emerging from any central transform a subordinate final module is defined which will accept the final data element.

   iv) for each central transform or functionally cohesive[1] composition of central transforms, we specify a subordinate transform module which will accept from the main module the appropriate input data and transform it into the appropriate output data.

Yourdon and Constantine note at this point that there is a simple (usually one-to-one) correspondence between the initial data flow graph and the module diagram that can be associated to it.

Step 4: This step consists of the factoring of the initial, final and transform modules until the ultimate physical input and output are reached as well as the detailed transform modules detected during the analysis of the problem (those for which it is not possible to state a transform with any clearly discernible sub-tasks).

Yourdon and Constantine point out that the objective of transform analysis is to make the program structure reflect the structure of the problem statement as a data flow graph.

THE DATA TRANSFORM METHOD

The General Method

Programs solve problems. According to Veloso and Veloso [6] a problem is a structure $P = \langle D,O,q \rangle$ with two types, where the

---

[1] Because of the goal of the present work, we have omitted many aspects of the method being described, such as cohesiveness, coupling, functional strength, etc., which are not directly related to the subject of this paper.

elements of D are the problem data, the elements of O are the
solutions (outputs) and q is a binary relation between D and O.

A program P solves a problem if P defines a total function
between D and O such that

$$(\forall d_\epsilon : D) \; q(P(d),d) \tag{1}$$

holds. To derive a program through the data method consists of,
given an input specification for $d \epsilon D$ and an output specification
for $o \epsilon O$, to construct a program P such that formula (1) holds.

Certain data-directed design approaches, such as Jackson's,
proceed as above by trying to find at the beginning of the derivation
process a direct mapping between the input data structures and
the output data structures (a mapping from a representation of $d \epsilon D$
to a representation of $o \epsilon O$). As it was pointed out in section 2,
for some situations it is not possible to solve some problems
through Jackson's basic method (problems which are not gsm computable).
The data transform method proposes a canonical form for the
expression of programs that include trivially problems which are
solvable through the Jackson basic method and that is amenable to
simple transformations which lead to solutions to problems which
are not Jackson solvable.

The data transform method starts by expressing the abstract
notions of $d \epsilon D$ and $o \epsilon O$, instead of trying to look for data representations
for these two entities. This approach, of course,
became a standard procedure in many programming methodologies but
is not very common in the context of data-directed programming.
The strategy for program derivation through the data transform
method consists of applying the concept of problem reduction and
decomposition while using Hoare's general data type construction
mechanisms (section 2 and [9]). Problem reduction and decomposition
are applied in a way which will leave us with a set of Jackson
solvable problems in hand. In the process of decomposing the problem
the method bears some similarity with Yourdon and Constantine's
data flow design.

We say a problem $P_1 = \langle D_1, O_1, q_1 \rangle$ is a reduction of $= \langle D, O, q \rangle$
and write $P \rightarrow P_1$ if we can define a unary function insert, ins:
$D \rightarrow D_1$ and a unary function retrieve, retr: $O_1 \rightarrow O$ such that the
program defined by

$$P(d) = retr(P_1(ins(d))) \tag{2}$$

solves P when P_1 solves P_1.

In Figure 6 we illustrate this situation: Note that

q is a subset of D×O
q_1 is a subset of $D_1 \times O_1$
P is a solution to p (a total function between D and O)
P_1 is a solution to p_1 (a total function between D_1 and O_1)

and that the functions ins and retr need to be defined in such a way that the composition expressed in (2) is satisfied.

The first step of the data transform method consists of defining D_1 and O_1 as the cartesian product of D and O; ins such that $ins(d) = (d, o_0)$ for some $o_0 \in O$; $retr(d, o_n) = o_n$. In other words, the reduction through ins and retr makes use of the data type constructor cartesian product (record) which is one of the three basic constructors proposed by Hoare [9]. Intuitively, it avoids the problem of structure clashes between the input and output spaces which sometimes occur when the basic Jackson method is directly applied. The input and output data of P_1 have now, trivially, the same structure (independently of any chosen representations for d and o). Figure 7 further clarifies the previous considerations.

This first step is clearly an intermediate step in the reduction process and is basically motivated by the existence of the structure clash type of problems in a data-directed programming type of solution. A trivial case, in practice, would be the one for which it is possible to define compatible data structures for d and o. That is, a situation in which P is Jackson solvable.

The method requires a second step whenever P_1 is not a simple problem, but requires for instance, modularization or the treatment of backtracking or recursive situations.

The second step of the data transform method consists of defining a new reduction $P_2 = \langle D_2, O_2, q_2 \rangle$ of P_1. In this step we

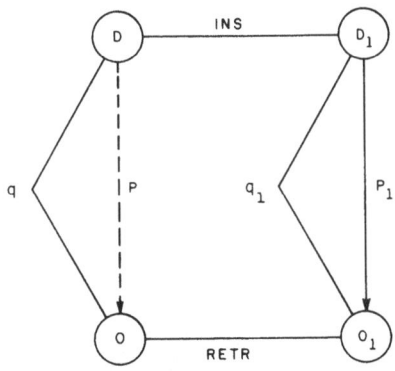

Figure 6. $P^r \to P_1$.

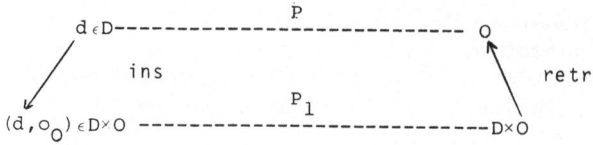

Figure 7. Data transform.

will make use of the sequence (file) data type constructor. We will define D_2 as D_1^*; O_2 as O_2^* and the function ins from D_1 to D and retr from O_1 to O as being, respectively, the functions make and last which have the normal meaning of these operations when applied to sequences, that is,

    make: builds a unitary sequence from a given argument
    last: returns the last element of the sequence

Figure 6 would now be replaced by the situation pictured in Figure 8.

The diagram in Figure 8 can now be expanded in the following way, Figure 9.

The outcome of this step is a program P_2 which we want to decompose into simpler programs. Let us be more precise about what we mean by decomposition [7]. If we take the problem $P_2 = \langle D_2, O_2, q_2 \rangle$, a n-ary decomposition Δ of P_2, $P_2 \uparrow \Delta$, consists of

    i) n functions $decmp_i$: $D_2 \to D_2$, i=1,...,n;

    ii) a (n+1) ary function merge: $D_2 \times O_2^n \to O_2$;

    iii) a unary function immd: $D_2 \to O_2$

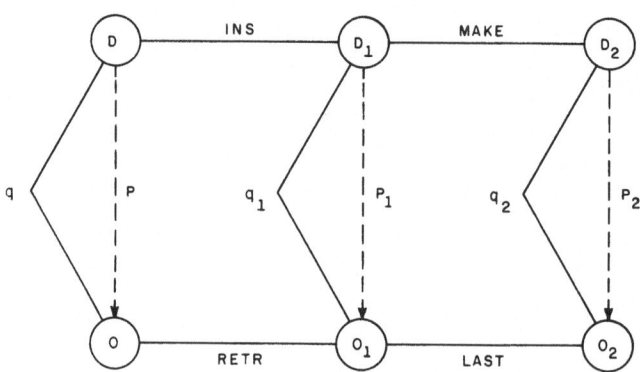

Figure 8. $P \; \Gamma \to P_1 \Gamma \to P_2$.

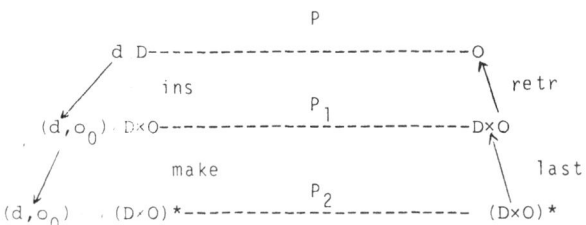

Figure 9. Data transform.

iv) a unary relation easy $\subseteq D_2$

We call items (i) to (iv) a good n-ary decomposition of P_2 if

$$P_2(d_2) = \begin{cases} \text{immd}(d_2) & \text{if easy}(d_2) \\ \text{combine}(d_2, \text{sol}_1[\text{decmp}_1(d_2)], \ldots \\ \qquad \ldots, \text{sol}_n[\text{decmp}_n(d_2)]) & \text{otherwise} \end{cases} \qquad (3)$$

where sol stands for the part of the solution of P_2 contributed by each decomposition. Intuitively, if the problem is simple (easy), that is, Jackson solvable, decomposition is not necessary and we have a direct (immd) solution. Otherwise, the solution for P_2 is obtained through the combination (combine) of the solutions (sol's) to the programs P_{2_1}, P_{2_2}, ..., P_{2_n} which correspond to the solutions. The decomposition process is guided by a data flow design type of analysis while we try to identify as many gsm solvable problems as possible. If one or more of the identified programs are not Jackson solvable, steps 1 and 2 and decomposition are applied to all programs at hand and applications of steps 1 and 2.

THE DATA TRANSFORM METHOD FOR FILE PROCESSING PROGRAMMING

We are mainly interested here in an important specialization of the data transform method to deal with file processing programming. These problems are identified in association with the data transform method as problems for which the inputs for P are always entities of the general type (files) and as problems for which the constituent programs of P_2 (obtained by decomposition) are always similar, in the sense that a while statement can drive a copy of them by changing the necessary inputs through its parameter.

The program schema below defines the family of programs (in the sense of [10]) that can be obtained by the data transform method as specialized for file processing programming, when we have one application of the first step of the followed by one application of the second step.

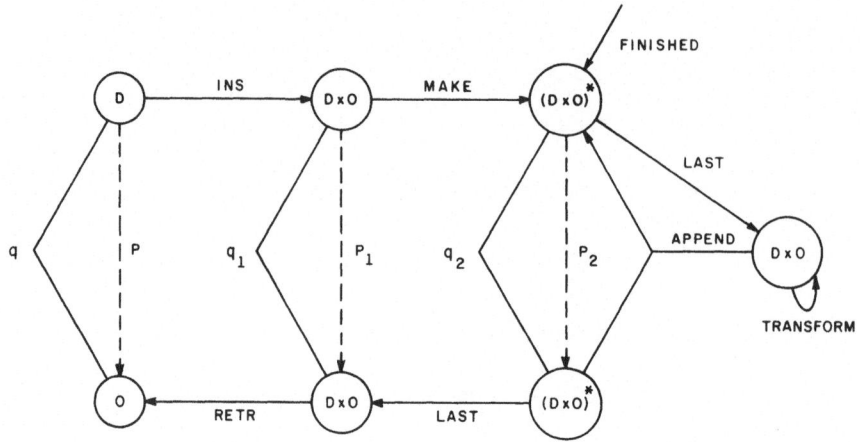

Figure 10. Diagram for file processing problems solution by the Data Transform Method.

The notation used in Figure 11 is Pascal-like. The programs that constitute Schema are presented in the order of their derivation, therefore violating a Pascal syntax rule.

From now on, any standard function not defined in the text is explained in the glossary of functions in Appendix I.

In the program schema the selectors i and r simulate the function ins and retr and the symbol Λ stands for the null sequence. The program schema only creates an instance of the input data to allow the application of the method.

The function update for the class of file processing problems has been defined as

$$\text{update}(x_3) = \text{append}(x_3, \text{transform}(\text{last}(s_3)))$$

where transform is a function from D×O to D×O which contributes to the solution of the problem. Refer to definition of $P_2(d_2)$ in equation (3).

The function append has the usual meaning of the operator with the same name, normally associated to the type sequence that is

append: $(D \times O)^* \times (D \times O) \rightarrow (D \times O)^*$

append $((p_1,\ldots,p_n),p) = (p_1,\ldots,p_n,p)$

In Figure 12 append is Pascal-like and belongs to the program schema.

```
Program schema;
    type D = seq of objects₁;
    type O = objects₂;
    type D·O = record i:D;
                      r:O
                      end;
    type(D·O)* = seq of D·O;
    var x,d:D;
    var y,o:O;

    begin
        x · copy(d);
        P ;
        o · copy(y)
    end {schema} .
Procedure P;
    var x₁,y₁:D·O;
    begin
        x₁.i · x; x₁.r · Λ;
        P₁;
        y · y₁.r
    end {P} ;
Procedure P₁;
    var x₂,y₂:(D·O)*;
    begin
        x₂ · make(x₁);
        P₂ ;
        y₁ · last(y₂)
    end {P₁} ;
Procedure P₂;
    var x₃:(D×O)*;
    begin
        x₃ · x₂;
        while not finished (x₃) do
        x₃ · update (x₃);
        y₂ · x₃
    end    {P₂};
```

Figure 11. Program schema for file processing programming through the data transform method.

Finally, to complete the program schema we present in Figure 13 a skeleton of the function transform.

```
Function update(x_3:(D×O)*):(D×O)*;
var x_4:D×O;
    y_3:(D×O)*;
begin
    y_3 ← x_3;
    x_4 ← last(x_3);
    x_4 ← transform(x_4);
    update ← append(y_3,x_4)
end {update} ;
```

Figure 12. Function update for file processing programming.

A Correctness Criterion for the Method

We define initially the termination condition for the program schema displayed in Figure 11. We have:

i) $update(x_3) = append(x_3, transform(last(x_3)))$

ii) $\forall x_3 \in (D \times O)^*$, $smllr(transform(x_3).i, x_3.i)$

iii) smllr is a well-founded relation in $D \times D$ such any $d \in D$ is in a finite smllr chain starting at Λ:

$smllr(\Lambda,d_1)$ $smllr(d_1,d_2)$... $smllr(d_n,d)$ (that is usual for file processing program)

iv) $last(x_3).i = \Lambda \leftarrow finished(x_3) = true$

Transform and finished must be specified so as to satisfy the above conditions. We can now state the partial correctness condition for the class of programs.

```
Function tranform(x_4:D×O):D×O;
var x_5,x_6:D;
    y_5,y_6:O;
    result: object_2;
begin
    x_5 ← x_4.i;
    y_5 ← x_4.r;
    Process(result)
    y_6 ← append(y_5,result)
    transform ← recombine(x_6,y_6)
end {transform}
```

Figure 13. Function transform.

v) $\forall x_3 \in (D \times 0)^*$, $\text{finished}(x_3) \Rightarrow q_2(x_3, \text{make}(d.i, \Lambda))$

vi) $\forall x_3 \in (D \times 0)^*$, $q_2(x_3, \text{maked}(d.i, \Lambda)) => q(\text{last}(x_3).r, d)$

Intuitively, the relation smllr guarantees that in each step the transform function contributes some more for the solution of the problem. The smllr relation, which is a well-founded relation, characterizes the empty element as a distinguished element that will necessarily be reached to accomplish the termination of the program.

Condition (v) guarantees that when the program stops x_3 is the solution of the problem for which the input is obtained from d by the application of ins and make and condition (vi) ensures that the reduction from the original problem P to P_2 is good, i.e., that the element from x_3 obtained by the application of retr and last is the solution to the original problem with input d.

THE TELEGRAM ANALYSIS PROBLEM

The classical telegram analysis problem, often used as an example of structure clash, boundary clash in Jackson's [1] terminology, has been defined in his book (page 155) as follows.

"An input file on paper tape contains the texts of a number of telegrams. The tape is accessed by a "read block" instruction, which reads into main storage a variable-length character string delimited by a terminal EOB character: the size of a block cannot exceed 100 characters, excluding the EOB. Each block contains a number of words, separated by space characters; there may be one or more spaces between adjacent words, and at the beginning and end of a block there may (but need not) be one or more additional spaces. Each telegram consists of a number of words followed by the special word "ZZZZ"; the file is terminated by a special end-file block, whose first character is EOF. In addition, there is always a null telegram at the end of the file, in the block preceding the special end-file block: this null telegram consists only of the word "ZZZZ". Except for the fact that the null telegram always appears at the end of the file, there is no particular relationship between blocks and telegrams: a telegram may begin and end anywhere within a block, and may span several blocks; several telegrams may share a block.

The processing required is an analysis of the telegrams. A report is to be produced showing for each telegram the number of words it contains and the number of those words which are oversize (more than 12 characters). For purposes of the report, "ZZZZ" does not count as a word, nor does the null telegram count as a telegram."

We will define a program TELEGRAM that will create an instance of the data that will be used for the application of the reductions and decompositions that will take us to our canonical form.

```
Program Telegram
   type D = seq of Telegrams;
        O = seq of Telegram-analysis;
        (D×O) = record i:D;
                       r:O
                end;

        (D×O)* = seq of (D×O);
   var x,d:D;
        y,o:O;
        begin
            x ←  copy(d);
            P ;
            y ←  copy(y)
        end {Telegram}.
```

The solution of the problem follows exactly the same steps presented in the presentation of the method, up to the point where we need to define the programs Transform and Process.

The change in the Transform function is minor, and the program can be expressed as follows:

```
Procedure Transform(x₄:D×O):D×O;
   var x₅,x₆:D;
        y₅,Y₆:O;
        report:telegram-analysis;
        begin
            x₅ ← x₄.i;
            y₅ ← Y₄.r;
            Process(report);
            y₆ ← append(y₅,report);
            Transform ← recombine(x₆,y₆)
        end {Transform};
```

We are now going to derive Process. According to the data transform method we need Process to be gsm solvable or decomposable in gsm solvable programs. Recall that the method makes use of the notion of data abstraction. In particular, Process will deal with sef of telegrams. It means, in practice, that we are focusing on the concept of a Telegram instead of reasoning at the block "level" as Jackson does.

The core of the program Process, which is dealing with the cartesian product of the sequence of telegrams with the sequence of

CASD SYSTEM EXTENDING JACKSON'S METHOD

telegram analysis can be represented graphically by the following picture (Figure 14).

To implement Process, it is necessary to scan the tape, block by block. Within each block Process must analyze word by word and compute each one for report purposes. When finding the end of a telegram before the end of a block, Process places the rest of the block as the first block in the output tape. The processing of words through this approach involves no prediction and therefore Process is gsm solvable.

One possible schematic version for Process is the following:

Procedure Process (report: Telegram Analysis);
   begin
      $x_6 \leftarrow \Lambda$;
      get (first block in x_5)

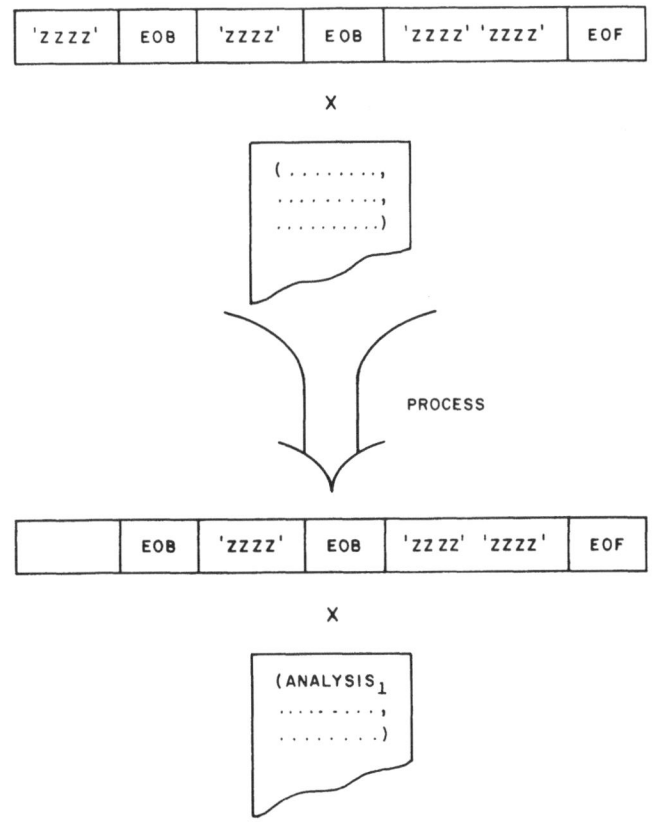

Figure 14. Program process.

```
            if first word in block is 'ZZZZ'
            then report ← Λ
            else
                begin
                    initialize report;
                    while telegram not empty do
                        begin
                            while telegram not empty and block not
                            empty    do analysis of a word in
                            report;
                            while (block not empty)    do
                            construct the first block in x₆;
                            get (another block in x₅)
                        end;
                    while    x₅ not empty    do
                        begin
                            append (x₆,block from x₅);
                            get (block in x₅)
                        end
                end
end {Process};
```

We need now to characterize the predicate finished. It takes the following form:

$$\forall x_3 \in (D \times O)^*, \text{finished}(x_3) \leftrightarrow \text{length}(\text{last}(x_3).i) = 0$$

That, of course, is so because we are dealing with a standard file processing problem, as defined by the data transform method. We reach this standard form for the termination procedure because the final problem reduction (cartesian product) leaves us with the input data to be processed as the first component of the product.

The input data is always reduced (each execution of Process has at least one operation get) and saved and therefore the program terminates when the input part of product is empty.

For the proof of correctness of the program we proceed as described at the end of section 2, after verifying the inner simple details of the operations "initialize report" and "analysis of words" in the Process program.

Although the previous level of decomposition may be considered satisfactory, a reader could possibly feel more confortable with a further decomposed solution. We will illustrate this possibility by decomposing Process one more time. Process can be decomposed into three sub-problems. The first, get-telegram, reads the input tape block by block (x_5) and within each block it looks for the word "ZZZZ". Once "ZZZZ" is found, the rest of the block which is being processed is appended to the output tape (x_6).

CASD SYSTEM EXTENDING JACKSON'S METHOD

The second sub-problem, get-tape, reads the rest of the input tape (x_5) and transfers its contents to the output tape (x_6).

The third sub-problem, analysis, makes the analysis of one telegram. The input of this sub-program is a telegram which consists of a sequence of words, and the output is a report about the analysis of one telegram, which can be in turn recognized as a file processing problem (and therefore further reduced).

The new version of Process becomes:

```
Procedure Process;
    type T: seq of words;
        var t₁: T;
            r₁: telegram-analysis;
          begin
                Get-telegram;
                Get-tape;
                Analysis
          end {Process};

Procedure Get-telegram;
        begin
              x₆ ← Λ;
              get (first block in x₅);
              if first word is 'ZZZZ'
              then r₁ ← Λ
              else
                   begin
                         t₁ ← first word;
                         while telegram not finished    do
                              begin
                                    while (block and telegram not
                                    empty)    do
                                         if word is not ZZZZ'
                                         then t₁ ← append (t₁,word)
                                         else telegram finished;
                                    while (block not empty)         do
                                         x₆ ← rest of block;
                                    get (another block)
                              end
                   end
        end  Get-telegram;

Procedure Get-tape;
        begin
              while (x₅ not empty)     do
                   begin
                         x₅ ← append (x₆,block in x₅)
                         get (block in x₅)
```

```
                    end
        end {Get-tape};

Procedure Analysis;
    type T: seq of words;
         R: telegram-analysis;
         (T×R):record i:T;
                      r:0
               end
         (T×R)*:seq of (T×R)*;
    var x',t_1:T;
        y',r:R;
        x_1',y_1':T×0;
        x_2',y_2',x_3':(T×0)*;

Procedure P_2';
    begin
         x_3' ← x_2';
         while (last(x_3').i ≠ 0)       do
              x_3 ← update_T(x_3');
         x_2' ← x_3'
    end {P_2'};

Procedure P_1';
    begin
         x_2' ← make(x_1');
         P_2';
         y_1' ← last(y_2)
    end {P_1'};

Procedure P';
    begin
         x_1'.i ← x';
         x_1'.r ← Λ;
         P_1';
         y ← y_1'.r
    end {P'};

begin
    x' ← copy(t_1);
    P';
    r ← copy(y')
end {Analysis};
```

Note that for the sake of clarity we have redefined the types T and R. Update$_T$ could be defined in such a way that each of its executions would perform the required analysis of one word of a telegram.

THE SYSTEM LOG PROBLEM

The System Log Problem is a structure clash problem which Jackson classifies as a multi-threading problem. The problem as stated by Jackson is the following (page 160) of [1].

"A time-sharing system collects information about system usage. This information consists of records, one for each log-on, log-off, program-load and program-unload. When a user of the system logs on, he is allocated a unique job-number for that session: one user would receive two different job-numbers if he logged on two different occasions. The system ensures that no user can log on unless the terminal is free (that terminal), and that he cannot log off unless he has previously logged on. Further, he is allowed only one active program at any one time: he must unload that program before he can load another or load the same program again.

The collected information is written to magnetic tape. The records contain the following information:

log-on record: code "N"; job-number; time of logging on;
log-off record; code "F"; job-number; time of logging off;
program-load record: code "L"; job-number; program-id; time of loading;
program-unload: code "U"; job-number; program-id; time of unloading.

The records are written in strict chronological sequence."

The instance of the data that will be used for the transform application of the reductions and decompositions used by the data method will be generated by the following SYSTEMLOG program.

```
Program Systemlog;
   type D = seq of job records;
        O = seq of job reports;
        (D×O) = record i:D;
                       r:O
                end;
        (D×O)* = seq of (D×O);
   var x,d:D;
       y,o:O;
       begin
            x ← copy(d);
            P ;
            y ← copy(y)
       end {Systemlog}.
```

As in the previous section we will skip the steps of the method that takes us from the first step to the decomposition when Transform need to be defined. We need now to specify the programs Transform and Process. As before, the change needed by the Transform function is trivial.

```
Procedure Transform (x₄:D×O):D×O;
    var x₅,x₆:D;
        y₅,y₆:O;
        report: job report;
    begin
            x₅ ← x₄.i;
            y₅ ← x₄.r;
            Process
            y₆ ← append(y₅, report);
            Transform ← recombine(x₆,y₆)
    end {Transform};
```

We shall look now for a Process program which is gsm solvable at this level of decomposition. We use the abstraction that the input tape is a sequence of job records (seq of job records). We are therefore overlooking details such as the fact that there are various types of records associated to the same job (log-on records N, log-off records F, program-load records L, and program-unload records U) and that the records referring to the same job are not contiguous in the tape. All the same, we will deal with the problem, one job record at a time (the same way we looked at a telegram at a time in the previous example). We will produce, as before, a gsm solvable program. Figure 15 illustrates the procedure just described in a graphical form.

The core of the Process program scans the tape collecting and processing information about one job, while constructing an output tape with the information about this job suppressed.

```
Procedure Process;
    begin
        x₆ ← Λ;
        get (first register in x₅);
        get (another register from x₅);
        while (x₅ not empty)    do
            begin
                if job-number of reg.=job-number of first reg.
                then
                    analysis of job-report
                else
                    x₆ ← append(x₆,register);
                get (another register from x₅)
            end
    end {Process};
```

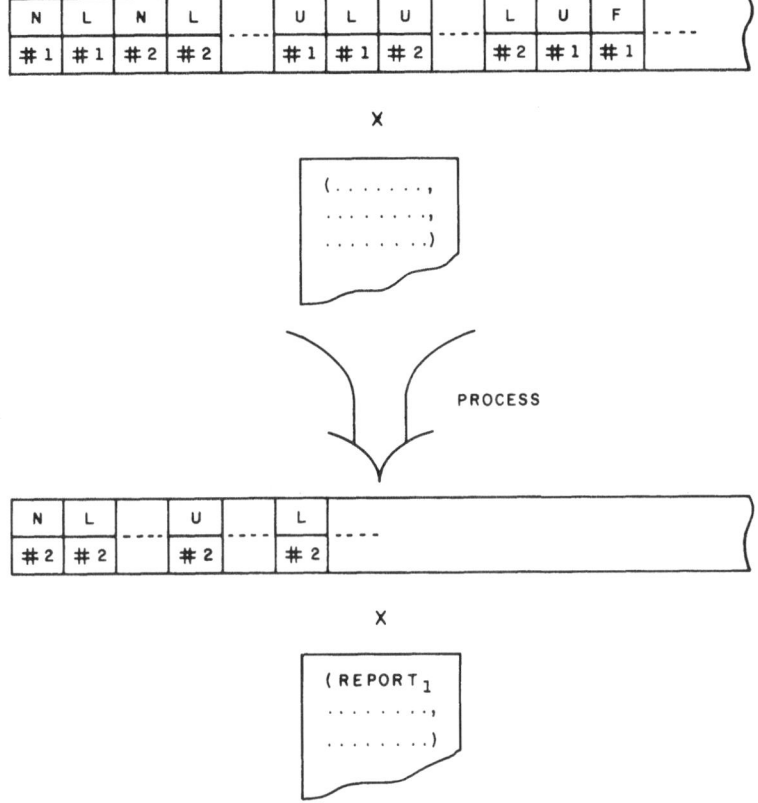

Figure 15. A gsm solvable program.

THE MATRIX TRANSPOSITION PROBLEM

This last example completes the number of structure clashes problems presented in Jackson. In his terminology this problem is called an ordering clash problem.

We will state the problem in the following simple way. One tape contains the elements of an m x n matrix recorded by line. The transpose problem will display the elements of the same matrix by column. In other words, we will find the transpose A^T of matrix Z.

As before, we will define a program TRANSPOSE that will create an instance of the data that will be used for the application of the reductions and decompositions that will take us to our canonical form.

```
Program Transpose;
   type D = seq of Aobjects;
        O = seq of Aobjects;
        (D×O) = record i:D;
                       r:O
                end;
        (D×O)* = seq of (D×O);
   var x,d:D;
       y,o:O;
       begin
            x ← copy(d);
            P ;
            y ← copy(y)
       end {Transpose}.
```

The nucleus of the Process problem looks very much like the systemlog basic problem with the difference that when scanning the input tape for a column we can determine exactly where each element of the column is located. Figure 16 illustrates the approach taken in this case.

One possible schematic version for process is the following:

```
Procedure Process;
   begin
        get (first element in x_5); column first element of x_5;
        x_6 ← Λ
        get (x_5);
        while (x_5 not empty)      do
             begin
                  if (element x_5 is in the same column as the
                      first) then
                         column ← append (column,element)
                  else
                         x_6 ← append(x_6,element);
                  get (one element of x_5)
             end
   end {Process};
```

CONCLUSIONS

We have presented in this paper the data transform programming method and applied it to the solution of some classical programming problems. The choice of the examples was meant to compare clearly our approach with Jackson's method, since his method cannot solve directly the problems we have dealt with.

The present is an updated version of previous papers [7,8], extended through the presentation of additional examples. Still,

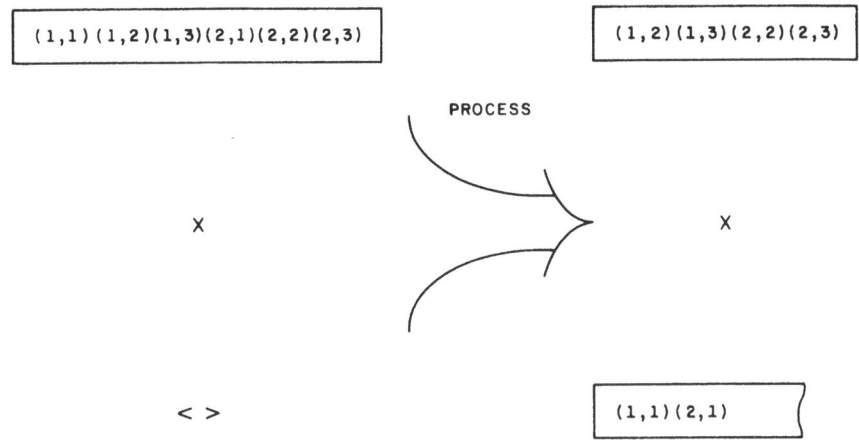

Figure 16. An ordering clash problem.

many interesting developments of the present work are in sight. The idea of automating the method follows the motivation of after researches. The work by Coleman, Hughes and Powell [11] and Logrippo and Skuce [12] follow this general direction although they are restricted to Jackson's basic method.

We believe, as [13], that for a large, long-lived software project, the existence of an accurate, readable model or specification, such as the one produced by the data transform method, can be as important as the existence of an efficient implementation of it. By using the set of procedures described in the Appendix III, a programmer can produce such a model with the assistance of a processing environment. We are presently working on a friendly human-interface which will guide the programmer on the application of the method while trying to make the details of using the procedures transparent to the user. Besides, we are developing a refinement procedure that will allow us to arrive to an efficient version for the solution at hand through a set of well-defined program transformations [14].

APPENDIX I

Glossary of Functions

copy — copies the arguments to produce another instance of the type

first — exhibits the first element of a sequence, that is, first $(<l_1,l_2,\ldots,l_n>) = l_1$ and the original sequence is not changed

get   - exhibits and removes the first element of a sequence

make   - exhibits the last element of a sequence, i.e., last $(\langle l_1, l_2, \ldots, l_u \rangle) = l_n$

tail   - constructs a sequence by removing the first element of the original sequence, i.e., tail $(\langle l_1, l_2, \ldots, l_n \rangle) = \langle l_2, \ldots, l_n \rangle$

recombine - constructs an ordered-pair from two given elements, i.e., recombine$(l_1, l_2) = l_1; l_2)$

APPENDIX II

```
Program Telegram;
   type D = seq of Telegram;
        O = seq of Telegram-analysis
        (D×O) = record i:D
                       r:O
                end;
        (D×O)* = seq of (D×O);
   var x,d,x₅,x₆:D;
       y,o,y₅,y₆:O;
       x₁,y₁,x₄:D×O;
       y₃,x₃,x₂,y₂:(D×O)*;
       report:Telegram-analysis;

Procedure Process;
    begin
        x₆ ← Λ;
        get (first block in x₅);
        if first word in block is 'ZZZZ'
        then report
        else
            begin
                initialize report;
                while telegram not empty      do
                    begin
                        while telegram not empty      and
                              block not empty         do
                              analysis of a word      in
                              report;
                        while (block not empty)       do
                              construct the first block in
                              x₆;
                        get (another block in x₅)
                    end;
                while x₅ not empty      do
                    begin
```

                                    append $(x_6,\text{block from }x_5)$;
                                    get (block in x_5)
                        end
            end
    end {Process};

Procedure Transform:$D\times 0$;
    begin
        $x_5 \leftarrow x_4.i$;
        $y_5 \leftarrow x_4.r$;
        Process;
        $y_6 \leftarrow \text{append}(y_5,\text{report})$;
        Transform $\leftarrow \text{recombine}(x_6,y_6)$
    end {Transform};

Procedure Update:$(D\times 0)*$;
    begin
        $y_3 \leftarrow x_3$;
        $x_4 \leftarrow \text{last}(x_3)$;
        $x_4 \leftarrow \text{transform}(x_4)$;
        update $\leftarrow \text{append}(y_3,x_4)$
    end {update};

Procedure P_2;
    begin
        $x_3 \leftarrow x_2$;
        while length(last(x_3).i) $\neq$ 0   do
            $x_3 \leftarrow \text{update}(x_3)$;
        $y_2 \leftarrow x_3$
    end {P_2};

Procedure P_1;
    begin
        $x_2 \leftarrow \text{make}(x_1)$;
        P_2 ;
        $y_1 \leftarrow \text{last}(y_2)$
    end {P_1};

Procedure P;
    begin
        $x_1.i \leftarrow x$;
        $x_1.r \leftarrow \Lambda$;
        P_1;
        $y \leftarrow y_1.r$
    end {P};
    begin
        $x \leftarrow \text{copy}(d)$;
        P ;
        $o \leftarrow \text{copy}(y)$
    end {sort}.

APPENDIX III

COPY
    FORMAT: Variable1 ← COPY (Variable2)
    FUNCTION: COPY Variable2 to Variable 1

GET
    FORMAT: Variable1 ← GET (Variable2)
    FUNCTION: Transfer the first component of Variable2 to Variable1

INSERT
    FORMAT: Variable1 ← INSERT (Variable2)
    FUNCTION: Constructs an ordered pair and the type of just component is the same of Variable2 and the second component is empty

RETRIEVE
    FORMAT: Variable1 ← RETRIEVE (Variable2)
    FUNCTION: Puts in Variable1 the second component of Variable2

FIRST
    FORMAT: Variable1 ← FIRST (Variable2)
    FUNCTION: Copies the first element of Variable2 into Variable1

LAST
    FORMAT: Variable1 ← LAST (Variable2)
    FUNCTION: Copies the last element of Variable2 into Variable1

MAKE
    FORMAT: Variable1 ← MAKE (Variable2)
    FUNCTION: Constructs a unitary sequence from Variable2

RECOMBINE
    FORMAT: Variable1 ← RECOMBINE (Variable2,Variable3)
    FUNCTION: Constructs an ordered pair, the first component Variable1 is Variable2 and the second component is Variable3

APPEND
    FORMAT: Variable1 ← APPEND (Variable ,Variable2)
    FUNCTION: Constructs a new sequence from an old one appending to it another element

FINISHED
    FORMAT: Variable1 ← FINISHED (Variable2)
    FUNCTION: Variable1 is true if Variable2 is empty and false otherwise.

REFERENCES

[1] M. A. Jackson, Principles of Program Design, London: Academic Press, 1975.

[2] J. D. Warnier, Logical Construction of Programs, New York: Van Nostrand Reinhold, 1974.

[3] E. Yourdon and L. L. Constantine, Structured Design: Fundamentals of a Discipline of Computer Program and System Design, Yourdon Press, 1978.

[4] J. W. Hughes, A Formalization and Explanation of the Michel Jackson Method of Program Design, Software-Practice and Experience, Vol. 9, 1979.

[5] D. D. Cowan, J. W. Graham, J. W. Welch, and C. J. Lucena, A Data-Directed Approach to Program Construction, Software-Practice and Experience, Vol. 10, 1980.

[6] P. A. S. Veloso and S. R. M. Veloso, "Problem Decomposition and Reduction: Applicability, Soundness, Completeness," Trappl, R., Klir, J., Pichler, F. (eds), Progress in Cybernetics and Systems Research, Vol. VIII, Hemisphere Publ. Co., 1980.

[7] C. J. Lucena, R. C. B. Martins, P. A. S. Veloso, and D. D. Cowan, "The data transform programming method - an example for processing problems," Proceedings of the 7th International Conference of Software Engineering, IEEE Computer-Society, 1983.

[8] C. J. Lucena, R. C. B. Martins, P. A. S. Veloso, and D. D. Cowan, "A theoretical proposal to a CASD system extending the Jackson's method, 1983 International Conference on Advanced Automation, Taiwan, Republic of China, 1983.

[9] C. A. R. Hoare, "Notes on Data Structuring," in Dahl, O. J., Dijkstra, E. W., Hoare, C. A. R., Structured Programming, Academic Press, 1972.

[10] D. L. Parnas, "Designing software for ease of extension and contraction," IEEE Trans. on Software Engineering, Vol. SE-5, No. 2, 1979.

[11] D. Coleman, J. W. Hughes, and M. S. Powell, "A method for the syntax directed design of multiprograms," IEEE Trans. on Software Engineering, Vol. SE-7, No. 2, 1981.

[12] L. Logrippo and D. R. Skuce, "File structures, program structures, and attributed grammers," Technical Report TR82-02, Computer Science Department, University of Ottawa, 1982.

[13] T. E. Cheatham, G. H. Holloway, and J. A. Townley, "Program refinement by transformation," Proceedings of the 5th International Conference on Software Engineering, 1981.

[14] R. C. B. Martins, "Data Transform Method," Doctoral Thesis (in Portuguese), Computer Sciences Department, PUC/RJ, Rio de Janeiro, Brazil, 1984.

Robotics

COORDINATING MULTIPLE ROBOT ARMS TO INCREASE PRODUCTIVITY

John Roach and Paul Montague

Virginia Polytechnic Institute and State University
Department of Computer Science
Blacksburg, VA 24061   U.S.A.

ABSTRACT

Robotics research has mainly concentrated on low-level kinematics, dynamics, and path control problems. Intelligent decision making will also be required for truly effective robots. Planning for tasks requiring multiple arm coordination, including human/robot cooperative efforts, is becoming increasingly important. This paper reports progress toward a theory and implementation of collaborative action. An execution environment for coordinated parallel action is also presented.

INTRODUCTION

Robot arms in most commercial installations today are programmed for pick and place tasks. In the factory of the future, robot arms will work together and with human beings to achieve tasks far more complex than pick and place operations allow. This paper describes research exploring the coordinated use of robot arms. The tasks chosen, simple blocks problems, illustrate the capabilities achievable by coordinating arms. Coordination is not created by manual training sequences as in pick and place operations. Rather, intelligence is required to create and execute plans for arm coordination.

We have referred to "coordination" of robot arms above. Coordination in this case means the purposeful movement of several arms in parallel to achieve a common end. Coordination saves time for completing a task, makes a task easier and therefore more amenable to solution for intelligent robots, or makes a task possible that one arm could not otherwise achieve. In all cases, productivity will be increased.

One abstraction of the multiple robot arm problem is concurrent computing. Each robot arm may be viewed as an independent process, and its actions must be coordinated with other processes to achieve common goals. By implication, solutions described in this paper should apply to coordinated computing, and techniques of coordinated computing [6] might apply to help solve problems of multiple robot arms. Other implications include the need to have an execution environment for multiple cooperating processes, the need to handle "deadlock" [3], and the need for scheduling operations to achieve optimal utilization of resources. The techniques developed here should also apply to planning collaborative actions of multiple mobile robots.

Other problems for coordinating robot arms include control problems such as compliant motion in which the movement of each arm is constrained by the other, path planning, space intersection problems, as well as all the usual problems for an individual robot arm. This paper will concentrate, however, on the concurrent computing problems of multiple robot arms.

OVERVIEW

Several design difficulties face the builder of a multiple robot arm system: should the system make high level plans, to what level should planning proceed, how should parallel actions be represented, how does the system know that multiple arms should be used? The system reported in this paper derived from the work on STRIPS-like systems [5], [12], and [11] but with a generalization for explicit parallel planning. A traditional STRIPS operator and world description are shown in Figure 1; a plan devised from this kind of description is also included. Figure 2 gives an example plan output from the planning system described here. Plans for coordinated robot arms are represented as graphs somewhat like PERT charts. Unlike PERT charts, however, these plan graphs include synchronization primitives that specify simultaneous movement.

Once a plan for coordinated action has been produced, actual execution of the plan commences. We have built a tool for creating parallel processes on a VAX/VMS 11/780 to manage the supervision of execution for each arm. The parallel processes communicate via a message passing facility. The main execution process schedules work for each subprocess in control of an arm. The entire execution system amounts to something like an operating system for robot arms.

COORDINATING MULTIPLE ROBOT ARMS

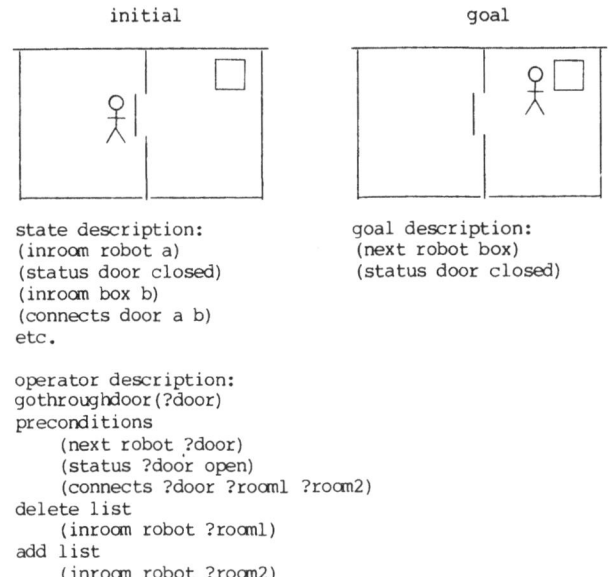

Figure 1. STRIPS operator and world description.

The robot operating system consists of four major sections: the Planner, the Execution Monitor (Execmon), and two processes to control the two arms as well as low-level control components. Each section will comprise at least one process in Hccom. The linkages between sections are represented graphically in Figure 3.

THE NEED FOR COORDINATED ACTION

A program making plans for robot arms needs to know what one arm can or cannot achieve. The knowledge people use to decide how

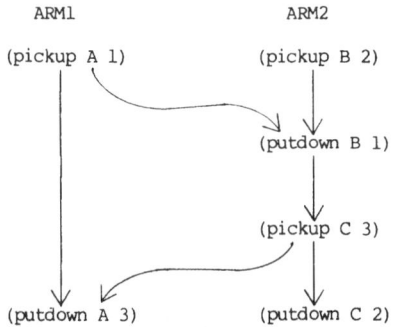

Figure 2. An example plan output.

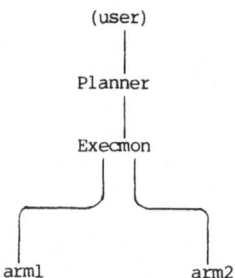

Figure 3. Linkages in Hccom.

many hands are needed for different tasks lies in the realm of intuitive physics [2], [7], and [8]. People have a large storehouse of knowledge about physics. We know when objects will fall without additional support, when they will slide, or when they will roll. Robots need this knowledge too. We have formalized a number of rules applicable for simple roll, slide and tumble determination [9]. Some of this particular intuitive physical knowledge will be useful in a multi-arm world, but additional intuitive physical knowledge is needed. Listed below are classes of problems for which rules of intuitive physics are needed in the multiple robot arm domain.

First are situations where one hand performs an action while another arm anchors, steadies, or holds up the object being acted upon. A nail will not stand up by itself while we hammer it (the rule formulated for objects falling can predict this). Other examples not predictable by the falling rule include holding a pad still while we write on it, pushing peas onto a fork with a knife, pushing the two parts of a seat belt together, or holding two boards in alignment while we put in a screw. One hand may prevent interference from the environment for another hand's action. For example, when we take a book down from a shelf, we often use our other hand to hold back the books on either side to keep them from falling as a result of the friction between the book covers. This class of action, where one hand aids the other in a passive way, is "symbiotic" in nature.

A second kind of two-hand action involves performing similar movements at different places, usually on the same object. For example, using both hands to lift an object from both sides so that we do not tip the object over. Similarly, when steering a car with two hands, the motion of either hand is reflected in the other one. These are "compliant" actions.

A third class concerns both hands actively working toward the same goal but with different coordinated actions. Knitting, collating papers, twisting a Rubik's cube, pouring something into a bowl while stirring what is already in the bowl are all examples of "co-operative" activity. Some motions co-operate by opposition, as when one hand twists a lid in one direction and the other hand twists the jar in an opposite direction to increase torque.

THE DESIGN OF THE PLANNING SYSTEM

In this section, we shall discuss the design of the planner for fundamental coordination problems. The system reported here is being programmed in Virginia Tech Prolog/Lisp [10].

The basic form of the planner is a backward-chaining system with rules to maintain parallel action paths in the graph. Our planner somewhat resembles the work of Dreussi [4] and differs from previous planners [5], [11], [12], [14], [15], and [16] in that it reasons not primarily from the standard "operator" definitions, with precondition/delete/add sets, but from a graph that expresses knowledge of causality. This graph -- the Precondition Constraint Graph (PCG) -- has nodes representing conditions that may hold in the environment and edges containing information about actions that may cause another condition to become true. The planner solves problems by setting up a chain of intermediate states, as in most STRIPS-like systems, but it does so by reasoning from a fund of knowledge primarily about those states, not about actions. This reflects the human tendency to think in "snapshots" and also facilitates reasoning about actions that can be used to achieve more than one condition.

As an example, consider the simple Precondition Constraint Graph given in Figure 4. This is the causal graph for a simple blocks world. The nodes are represented as condition predicates; the <> brackets denote secondary conditions, used to differentiate

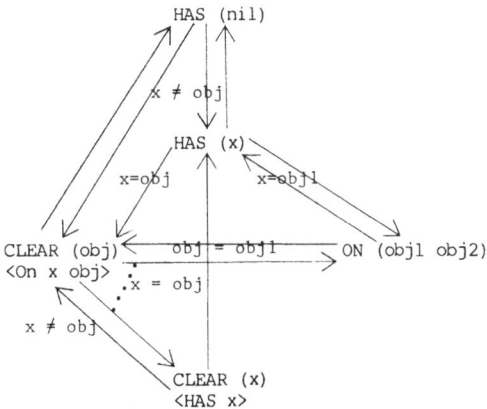

Figure 4. Sample PCG.

syntactically similar predicates. Thus, in the blocks world our planner knows that clearing an object that is in its hand is different from clearing an object that has another object sitting on top of it.

The actual PCG nodes are considerably more complex. They include information used in selecting variable bindings and resources required to achieve the condition. Likewise, the edges contain more than just the binding constraints shown in the figure. They also give the system its knowledge of actions and limitations on actions and secondary state changes related to the one represented by the edge. Additional constraints such as timing could be included as part of the graph.

Planning Examples

In this section, we show how the planner uses the Precondition Constraint Graph of Figure 4 to solve blocks world problems. Figure 5 contains an overview of the algorithm. Consider a problem of shuffling blocks (Figure 6). The goal predicates for the problem are (on B 1) (on C 2) (on A 3).

Initially, each goal is solved independently of the other goals. From the Precondition Constraint Graph, the goal (on block position) can be achieved using a PUTDOWN operator if (has arm block) and (clear position) are simultaneously true. Neither subgoal is true in the initial world, however. An arm can have a block if (pickup arm block) is executed which from the PCG has preconditions (has arm nil), (on block position), and (clear block). See Figure 7.

At this point, assignments can be made for variables in the subgoals and operators. The positions of blocks can be determined

```
                    Planning Algorithm

    1. frontier <- current goals
    2. repeat until frontier = nil or no changes
           a. mark all frontier nodes that are true
              in the world
           b. for each unmarked frontier node
                   i. find its entry in the Precondition
                      Constraint Graph (PCG)
                  ii. create new nodes from edges leading
                      out of the PCG
                 iii. check constraints on edges
           c. check for duplicate nodes in the plan graph
                   i. merge duplicate nodes
                  ii. do not merge duplicate nodes if a loop
                      is formed
```

Figure 5. Overview of the algorithm.

COORDINATING MULTIPLE ROBOT ARMS 251

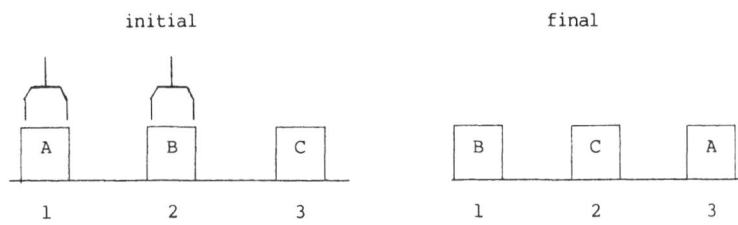

Figure 6. Block shuffling.

from the initial state of the world; for example, (on b ?p1) becomes (on b 2). At this point no robot arm(s) has been assigned to solve the goal. The assignment of arms to optimize expenditure of energy, to minimize time to completion and so forth amounts to a theory of resource usage. Optimizing resource usage has long been a topic of study in the operating systems field of computer science; studies show that complete optimization requires a tremendous expenditure of computing resources. Heuristics designed to approximate optimal mechanical resource usage but which take far less computing time are therefore preferable. We have not emphasized the formulation of efficient heuristics, but the assignment techniques adopted have, in all examples, worked out very intuitive, reasonable allocations of arms in plans. In the current plan, arm2 is assigned to pick up B and arm1 is assigned to pick up A. Since there are two arms and three blocks, this leaves no arm assigned to pick up C, so that part of the plan must be taken care of later.

The plan graph at this point is shown in Figure 8. In the figure, boxes surround sets of conditions belonging together to form a context of statements that must be true before an operation leading out of that context can be performed. Additional conditions that are made true for a context by operations leading into a

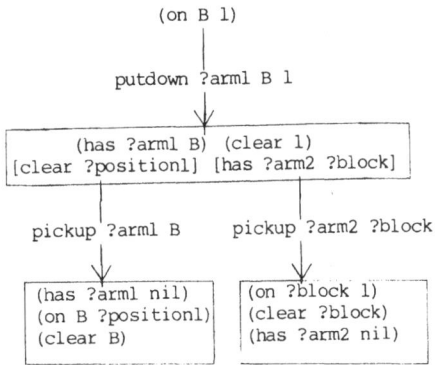

Figure 7. Initial plan graph involving block B.

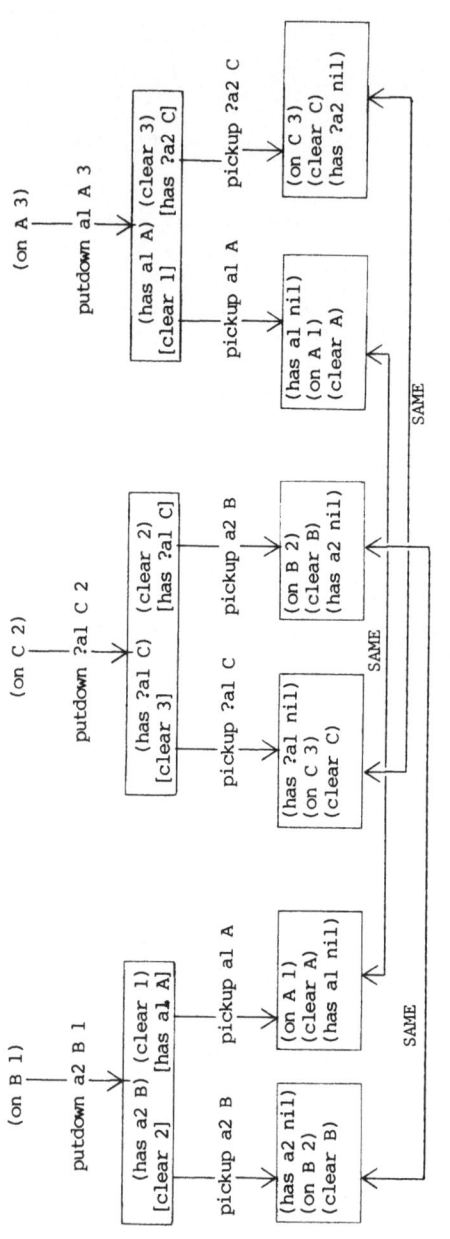

Figure 8. Plan graph.

context but which are not essential preconditions are enclosed by square brackets.

Step 2c of the planning algorithm checks for contexts that are the same. In Figure 8 we have indicated the duplication of contexts by connecting similar contexts by arrows labelled "same." The planner collapses these contexts showing that one action such as (pickup arm1 A) can achieve several conditions such as (clear 1) and (has arm1 A) at the same time. The result is shown in Figure 9 where almost all conditions on the frontier are satisfied in the initial world.

Only the goal (has ?arm nil) is not satisfied in Figure 9. The operation to make an arm hold nothing is (putdown ?arm ?object ?place); the preconditions for this operation are (has ?arm ?object) and (clear ?place). One of the contexts in the plan graph already satisfies these conditions with ?arm = arm2, ?object = B and ?place = 1. A connection is therefore made between the context with goal (has ?arm nil) and the context that satisfies the goal. See Figure 10. Other contexts at the same level of the graph would seem to satisfy the same goal, but a careful examination of the plan graph shows that an infinite loop would be formed in the graph if the goal context were connected to the other contexts that would satisfy the goal. If there were genuine choices of contexts to solve a goal, heuristics for optimizing resources would once again be employed.

The planning technique described thus far does not include the need for concerted action by both arms. If object B were too heavy to lift for one arm, a different plan would be created. A constraint equation on the PICKUP operator would indicate the need for a second arm. Simultaneous action links would be added to the graph and additional preconditions would have to be added to the contexts dealing with object B. The additions are shown in Figure 11. The (has ?a2 B) precondition and the operator needed to achieve it (pickup ?a2 B) cause the deletion of the (has ?a2 nil) precondition for the (pickup ?a2 A) action. [This is determined from a deletion graph not shown here.] The preconditions of (pickup ?a2 A) must precede the preconditions of (pickup ?a2 B). The preconditions for (putdown ?a2 A ?position) are (has ?a2 A) and (clear ?position) which are achieved by (pickup ?a2 A) as shown in Figure 12. This example shows the generation of plan graphs with simultaneous actions.

A different problem and its associated plan graph are shown in Figure 13.

At this point the plan graph looks somewhat mysterious. No means of traversing the graph to extract an ordering of actions is apparent: a program to linearize the graph is needed.

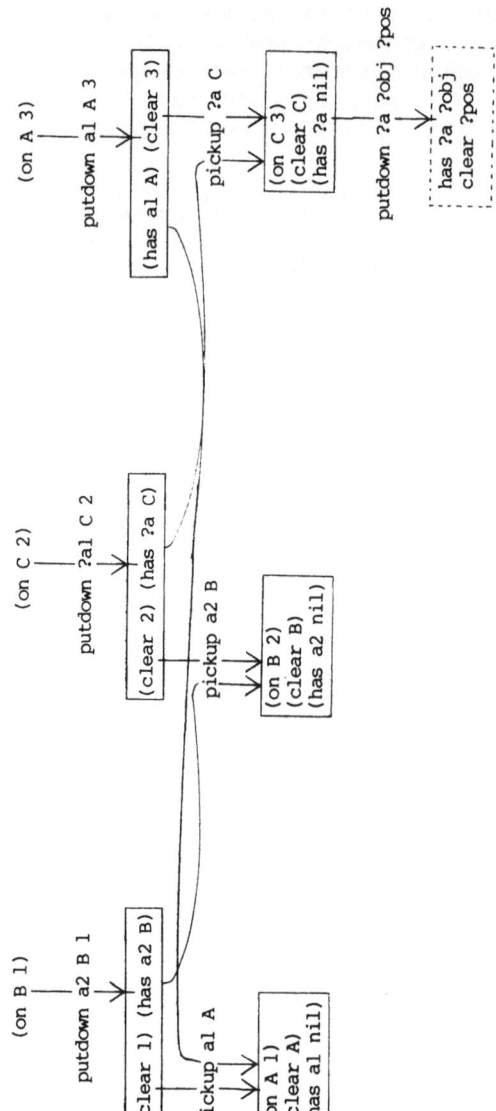

Figure 9. Intermediate result of planning algorithm.

COORDINATING MULTIPLE ROBOT ARMS

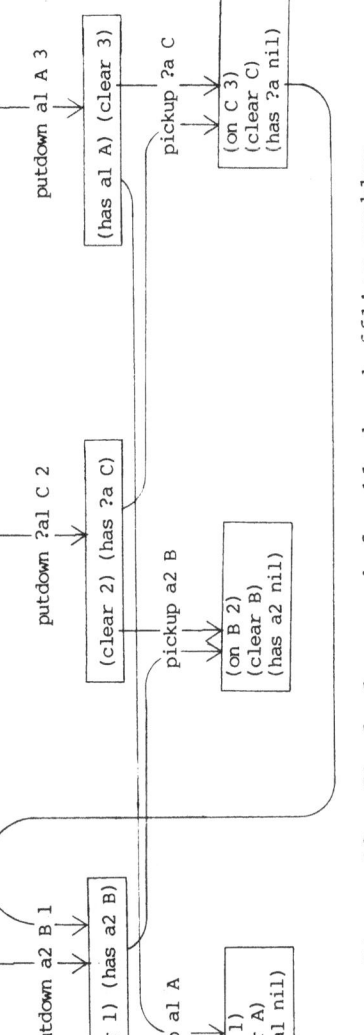

Figure 10. Final plan graph for blocks shuffling problem.

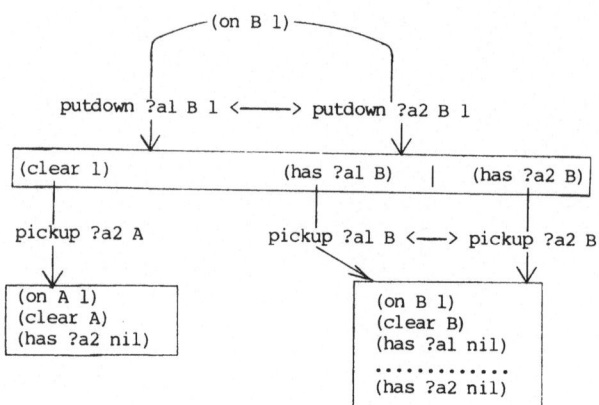

Figure 11. Addition of a second arm.

The Linearizing Algorithm

Figure 14 presents an outline of the algorithm for linearizing plan graphs. Arcs in the plan graph represent dependencies, so linearization requires that we find the deepest dependencies in the graph. Depth from the final state (goal context) is counted by marking each arc in the graph. Starting at the nodes of the graph

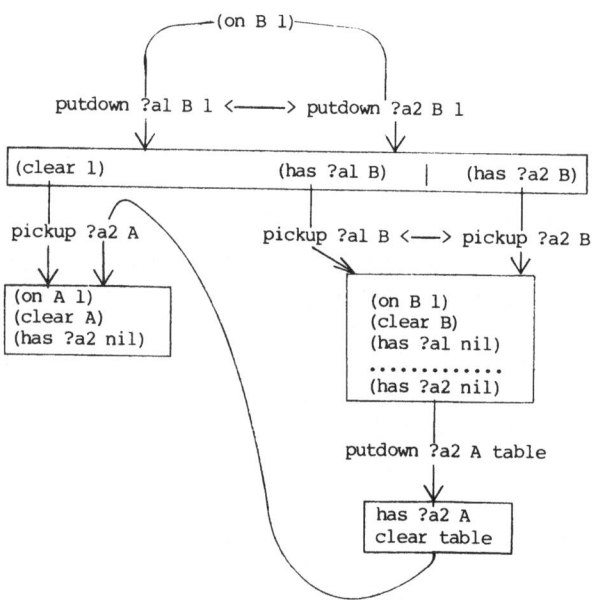

Figure 12. Generation of plan graphs with simultaneous actions.

COORDINATING MULTIPLE ROBOT ARMS

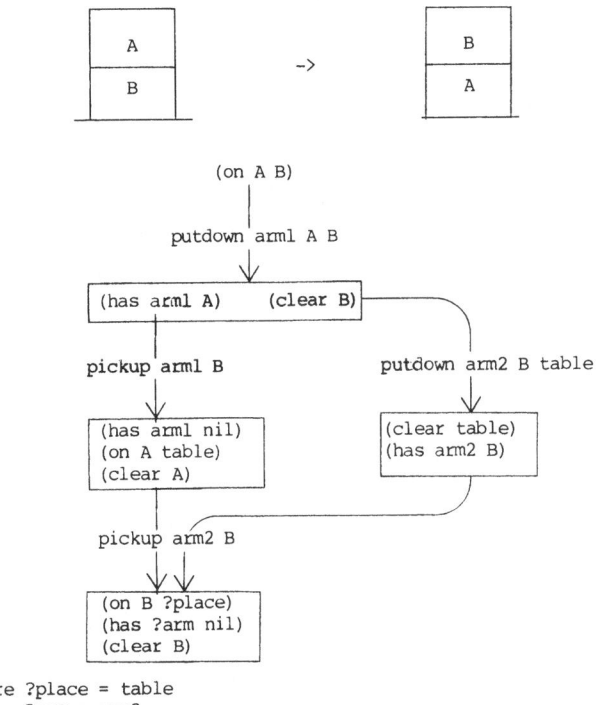

Figure 13. Plan graph for a different problem.

without any successors, the linearizing algorithm works backwards toward the goal context, always following the arc backward with the greatest depth marking when there is a choice. Figure 15 presents an actual computer run of the linearizing algorithm.

The plan graph generation and linearization technique have been tried with many problems, including Sussman's non-linear blocks problem [13] and the closed door problem [12]. In all cases, this system generates its own ordering of actions automatically and

Linearizer Algorithm

Repeat until all nodes of the graph have been visited.

1. Current set <- deepest nodes in graph marked true.
2. Follow "deepest" edges out to nodes, putting actions into linearized plan graph. Create new current set.
3. Create any dependency arc or parallel arc in the linearized plan graph as needed.

Figure 14. Outline of linearizer algorithm.

```
Example partial input to the linearizer

((problem_goals (gcl gc2 gc3 gc4 gc5) ))
((contexts (gcl gc2 gc3 gc4 gc5 ncl nc2 ... nc10) ))
((edges (el e2 e3 e4 e5 e6 e7 e8 e9 ... e18) ))
((gcl (mark nil) (depth 0) (nodes (gl)) (outedges (el)) (inedges nil) ))
((gc2 (mark nil) (depth 0) (nodes (g2)) (outedges (e2)) (inedges nil) ))
((ncl (mark nil) (depth 3) (nodes (nl n2)) (outedges (e6 e7)) (inedges (el e18))))
((el (mark nil) (action (putdown otherarm b 1)) (depth 1) (from gl) (to ncl) ))
((e6 (mark nil) (action (pickup otherarm b) (depth 2) (from nl) (to nc7) ))
((nl (mark nil) (has otherarm b) (outedge e6) (nc ncl) ))
((gl (mark nil) (on b l) (outedge el) (nc gcl) ))


The linearized plan is:

(onearm ((G2 (pickup onearm a)
             (preds nil)
             (succs (G4 G6)))
                    (G6 (putdown onearm a 5)
                        (preds (G2 G5))
                        (succs (G7)))
                    (G7 (pickup onearm d)
                        (preds (G6))
                        (succs (G8 G11)))
                    (G11 (putdown onearm d 3)
                         (preds (G7 G9))
                         (succs nil))))
(otherarm ((G3 (pickup otherarm b)
               (preds nil)
               (succs (G4)))
                     (G4 (putdown otherarm b 1)
                         (preds (G2 G3))
                         (succs (G5)))
                     (G5 (pickup otherarm e)
                         (preds (G4))
                         (succs (G6 G8)))
                     (G8 (putdown otherarm e 4)
                         (preds (G7 G5))
                         (succs (G9)))
                     (G9 (pickup otherarm c)
                         (preds (G8))
                         (succs (G10 G11)))
                     (G10 (putdown otherarm c 2)
                          (preds (G9))
                          (succs nil))))
```

Figure 15. Five block shuffle problem.

finds a solution that assigns parallel actions to arms in a very intuitive fashion.

AN OPERATING SYSTEM FOR MULTIPLE ROBOTS

An execution environment is needed to control the motions of arms as dictated by plans. The problems encountered in robotics require considerable computation; multiple robot arms compound this difficulty. A robot operating system for multiple arms requires concurrent computation to achieve parallel action and sufficient computational power. The following sections report the construction of an execution environment employing multiple processes.

Hccom, a Parallel Processing System for Hc

Hccom is an extension of the Virginia Tech Prolog/Lisp system [10] to allow concurrent processing. Hccom provides facilities to set up subprocesses structured as a tree running under the user's hc command-level process, for the user to communicate with these subprocesses, and for the subprocesses to communicate among themselves. This section explains the use of Hccom.

Subprocesses may be started by the com$run command. (All commands unique to Hccom are prefixed by "com$".) Com$run sets up the internal structures necessary to keep track of the newly created process. Once a subprocess is created it is invisible to the user except through messages it may send to the highest-level hccom process. This means that all actions executed by the subprocess, including message-passing, must be coded inside the file loaded by the com$run command.

There are 10 commands in the hccom language:

(com$init)	Enter process into hccom communication system
<return>	Hccom process id is returned (nil on failure)
(com$exit)	Remove process from hccom communication system
<return>	t (success) or nil (failure)
(com$send <to> <message> [<priority>])	Send message to one or more processes
<to>	Hccom id of process to send <message> to
<message>	S-expression to send as message.
<priority>	Optional priority, high=1, low=10, default=5
<return>	t (success) or nil (failure)
(com$receive [<from>])	receive message from process
<from>	Optional hccom id to receive from (default is any process in the group).
<return>	(<from> . <message>)
(com$wait)	Return if messages available, else wait.
<return>	t (success) or hand the process.
(com$run <cmd> [<input>])	Run command to start subprocess
<cmd>	Name of file to run (usually "hc.exe")
<input>	File to be used as standard input to <cmd> file
<return>	Local system process id.
(com$whom)	Return list of hccom process id's in group
<return>	List of hccom id's in group or nil (failure)
(com$cancel)	cancel any messages intended for current process.
<return>	t (success) or nil (failure)

```
(com$post
 <message>)            Post status of current process
   <message>           S-expression to be posted as status
   <return>            t (success) or nil (failure)
(com$status
 [<id>])               Read status posted by process
   <id>                Optional hccom process id (default is self)
   <return>            S-expression status or nil (failure)
```

These commands form the basis of a message passing, multi-processing robot operating system. Each arm is assigned a separate process with an overlord process called Execmon to coordinate the execution of the arms.

An Example Parallel Plan to be Executed

A phenomenological study of human behavior as in [1] shows that human beings are normally carrying out several activities in parallel at the same time. Robot planning and executing studies of the past have ignored difficulties with representing, deriving and executing parallel planning structures. This means that the robot programs were naturally limited to single actor systems.

A partial ordering over a set determines which elements of the set precede other elements. If the set elements are actions, then the partial ordering on the set tells which actions must be executed before others. Partial ordering is routinely used by builders to construct PERT charts. The partial ordering method of PERT chars, however, can only account for actions preceding one another. A new kind of relation between actions that indicates simultaneity is needed. Allowing another kind of ordering relation and a different link type in the parallel action graph results in the simplest representation method.

The following example generalizes the blocks world problems described earlier to a generic assembly task where the arms must move a large part (part1) into the work area. They must both cooperate in doing this because the part is too heavy for one arm to lift. Arm1 pulls a pin out of a box, puts it in place in part1, and arm2 pulls part2 out of another box and pushes it down over the pin which arm1 has just inserted.

A diagrammatic representation of the plan is shown in Figure 16.

The Robot Execution Monitor

Execmon is responsible for overseeing the execution of the plans output by the planner. It is, therefore, the heart of the robot operating system. Its first responsibility is to pick up the

COORDINATING MULTIPLE ROBOT ARMS

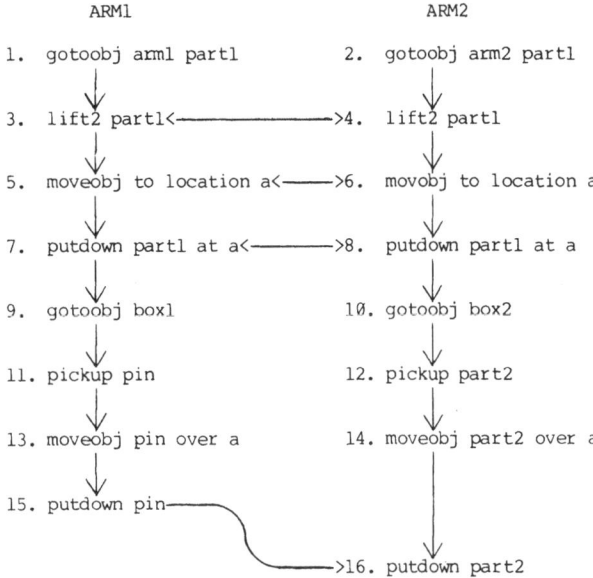

Figure 16. Diagrammatic representation of the plan.

plan parts. Once it has collected the two arms' portions of the plan and entered them in its data space, Execmon uses a startup routine to set arm1 and arm2 running and entered into the hccom system. An initialization process allows processes to exchange their id numbers. Once this bookkeeping is completed, Execmon begins its real work.

To keep track of the current state of execution of the plan, Execmon maintains records of the plan nodes that each arm is currently working on. As messages are received from the arms reporting successful completion of nodes, assertions detailing the successful execution of the plan are remembered.

The heart of Execmon is a message processing loop. It can receive three basic message types: "shutdownnow", a process name and number, or the message that an arm has completed executing a node's action. The shutdown message triggers the end of execution, normally when the plan has successfully been completed.

When Execmon receives the message that a node has been executed, it first checks whether the arm that sent the message has completed its plan. If it has, the planner is notified. If Execmon finds nodes still to be executed, it examines the next node to check whether it is ready to be executed or whether other conditions (being achieved by another arm, for example) are not yet fulfilled. If all the 'precondition' nodes have been achieved, it

checks whether all the 'parallel' nodes are ready. If they are,
Execmon sends out the goals to be executed to the appropriate
arm. Then a "go" message is broadcast.

    Execmon continues processing messages in a loop until the plan
is completely executed or it receives an error message that calls
for processing to halt.

Trace of System Performance

    This section presents part of a trace of the robot execution
monitor. The plan being executed was presented in Figure 16.
Execmon receives the process numbers for the subprocesses "arm1"
and "arm2" (colons ":" precede echoes of messages received). Once
the plans for each arm have been loaded, Execmon starts plan processing. To execute a parallel action with two arms, Execmon sends
the goal to each arm, waits for a (readyarm armi) message from each
arm, and then broadcasts a (go) message. Shown below is a trace of
the first part of the plan:

```
   ((first node (gotoobj part1) sent to arm1))
   ((first node (gotoobj part1) sent to arm2))
:     ((message (arm1 nodedone) received))
   ((lift2 part1) sent to arm1)
      (1 (gotoobj arm1 part1) sent to planner)
:     ((message (readyarm arm1) received))
:     ((message (arm2 nodedone) received))
   (((lift2 part1) sent to arm2))
   (((2 (gotoobj arm2 part2) sent to planner))
:     ((message (readyarm arm2) received))
   ((go signal sent to (arm1 arm2)))
:     ((message (arm2 nodedone) received))
              .
              .
              .
```

Arm2's final section in Figure 16 cannot be performed until an
action -- "putdown arm1 pin" -- has been completed by arm1.
Execmon does not, therefore, send out the node to arm2 immediately.
When Execmon verifies that arm1's action is done, it tells the
planner that arm1 is finished and sends the waiting node to arm2.

```
:(arm2 nodedone)
        :(arm1 nodedone)
        (15 (putdown arm1 pin a)...)
        sent to planner, arm1done
(putdown arm2 part2 a)
sent to arm2
(14 (moveobj arm2 part2 a)...)
sent to planner, continuing
```

:(arm2 nodedone)
(16 (putdown arm2 part2 a)...)
sent to planner, arm2done

Both arms have finished their portions of the plan; the timing performance of the system may be summed up in Figure 17. This run was timed on a VAX/VMS 11/780 and shows that the execution environment does not achieve synchronous starting motion for both arms. The time difference in execution is caused by the time-shared environment of the VMS operating system. Without running at real-time priority (devoting the entire VAX to running the robot arms), it is not possible to improve the synchronization. The architecture of the system built here allows the separate arm controllers to be moved to dedicated machinery that could monitor and control each arm. True synchronization will require a hardware solution; a single time-shared computer is insufficient.

CONCLUSION

In this paper, we have presented a system that can plan and execute parallel actions with multiple robot arms. The theory presented, however, is not restricted to multiple robot arms. A good theory of planning should transfer to other domains. We are current working on the application of this theory to planning concurrent computations.

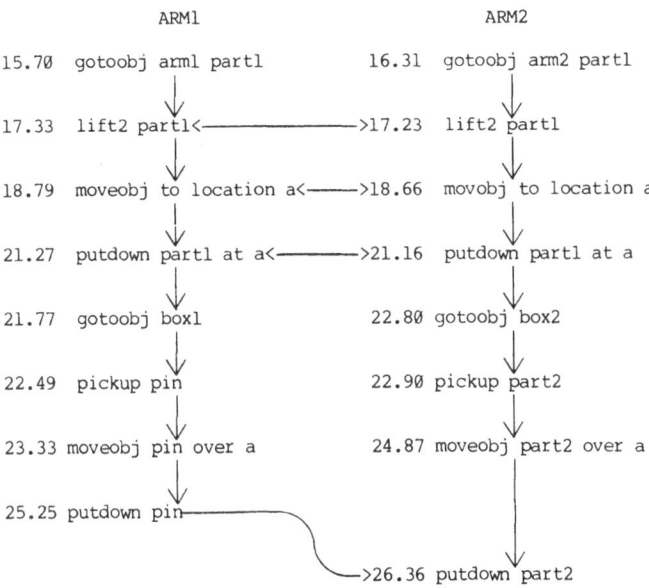

Figure 17. Timing performance of the system.

A full implementation of intelligent robotics clearly requires considerable parallelism with distinct functions running on separate hardware passing messages back and forth. The abstract problem that must be solved if we are to have intelligent robots is distributed coordinated computing.

ACKNOWLEDGMENT

The research reported here was supported in part by the National Science Foundation, Grant MCS - 8210194, Coordinating Robot Arms.

REFERENCES

[1] R. G. Barker and H. F. Wright, The Midwest and Its Children, Row, Peterson and Co., Evanston, IL, 1954.
[2] J. deKleer, "Causal and ieleological reasoning in circuit recognition," Artificial Intelligence Laboratory, Technical Report 529, M.I.T., Cambridge, MA, 1979.
[3] E. Dijkstra, "Cooperating sequential processes," Technical Report EWD-123, Technological University, Eindhoven, The Netherlands, 1965.
[4] J. Dreussi, "The detection and correction of errors in problem-solving systems," Ph.D. dissertation, Dept. of Computer Science, University of Texas, Austin, TX, 1982.
[5] R. Fikes and N. Nilsson, "STRIPS: A new approach to the application of theorem proving to problem solving," Artificial Intelligence Journal, vol. 2, no. 3-4, 189-208, Winter 1971.
[6] R. Filman and D. Friedman, Coordinated Computing, McGraw Hill, New York, 1984.
[7] D. Gentner and A. Stevens, Mental Models, Lawrence Erlbaum Associates, Hillsdale, NJ, 1983.
[8] P. Hayes, "The Naive Physics Manifesto," in Expert Systems in the Micro Electronics Age, Donald Michie, ed., Edinburgh University Press, Edinburgh, Scotland, 1979.
[9] J. Roach and T. Dean, "Intuitive physics for robots," in preparation.
[10] J. Roach and G. Fowler, "The HC Manual: Virginia Tech Prolog," Technical Report, Dept. of Computer Science, Virginia Polytechnic Institute and State University, 1983.
[11] E. Sacerdoti, A Structure for Plans and Behavior, Elsevier, New York, 1977.
[12] L. Siklossy and J. Dreussi, "An efficient robot planner which generates its own procedures," Proc. 3IJCAI, Stanford, CA, 423-420, 1973. Reprinted in Tutorial on Robotics, C. Lee, R. Gonzalez, and K. S. Fu, eds., Silver Spring, MD: IEEE Computer Society Press, 1983.

[13] G. Sussman, "A computational model of skill acquisition," Ph.D. dissertation, M.I.T., Cambridge, MA, August 1973, AI-TR-297.
[14] A. Tate, "INTERPLAN: A plan generation system that can deal with interactions between goals," Memo MIP-R-109, Machine Intelligence Research Unit, University of Edinburgh, 1974.
[15] S. Vere, "Planning in Time: Windows and durations for activities and goals," IEEE PAMI, vol. PAMI-5, no. 3, 246-266, May 1983.
[16] D. Warren, "WARPLAN: A system for generating plans," Memo No. 76, Dept. of Computational Logic, School of Artificial Intelligence, University of Edinburgh, 1974.

DESIGN OF A COMPUTER AIDED ROBOT DESIGN SYSTEM

Linfu Cheng

Department of Electrical and Computer Engineering
University of Miami
Coral Gables, FL   33124   U.S.A.

ABSTRACT

Utilization of robots has recently received considerable attention from the engineering community; however, the state-of-the-art design and control of robots leave much to be desired. The sheer high dimensionality of robot models and the large geometrical variations in system parameters have made the study and control of robots an extremely complex task. As robots become lighter, more intelligent, more maneuverable and affordable, the modelling of robots necessary for the design and control of such systems presents a higher level of complexity.

To further the technologies of robot design and control, automation through a suitable computerized design system is essential. This article describes the design of a unified computer aided robot design system that will aid in robot modelling, specification of body parameters and control algorithms, as well as simulation in a controlled environment. While this system, called MADSIM, aims to increase the productivity of designers or researchers, it also provides the necessary structures so that new techniques in CAD or future concepts in robot systems can be accommodated. By incorporating extensibility, maintainability, and transportability into the MADSIM architecture, the design also serves to achieve the equally important goal of improving productivity of MADSIM programmers/designers.

INTRODUCTION

Robots, particularly the industrial robot or IR, will play an ever important role in the automation of industry [1]. They are most likely the solution to the manufacturing productivity, quality and cost dilemma.

The state-of-the-art technology in the design and control of robots, however, remain to be primitive. The main difficulties arise primarily from the complexity of the robot body modelling, including the high dimensionality and nonlinearity, and the associated control algorithms.

To further the robot design technology, aids provided by a suitable computerized design automation system are essential. The technology of computer aided design (CAD) has been successfully applied to various areas and are most beneficial in the design of large complex systems such as electrical power system, building structural system, and computer VLSI system. In the area of robot design, there is already partial automation in the design aids and some have been successfully reported [2]. An integrated computer aided robot design (CARD) has been lacking.

A CARD system will also be of tremendous value in robot education and training. The availability of a CARD system provides a very flexible method in learning various aspects of robot mechanisms. This is particularly attractive as robot mechanisms are made available without the commitment of real production machinery.

Overall, the CARD contributes to the state-of-the-art robot design technology by increasing the design productivity and product quality at lower cost.

This article describes the architecture of a CARD system called MADSIM (for Manipulator Design and Simulation System) which aids researchers and, to a lesser degree, the educators in the various aspects of manipulator design. This MADSIM system architecture furnishes a convenient, flexible and friendly environment in which a modeller, a robot body designer, or a control specialist can utilize, individually or collectively, towards the synthesis of robot body and its ancillary parts, the design of control algorithms as well as the coding of control software.

Several features of modern software engineering are appropriately applied to the design of MADSIM, among them are the extensibility, maintainability, transportability and simplicity. The ultimate goal of this MADSIM endeavor is to help increase the productivity of not only the robot designers but also that of the MADSIM developers.

Far from providing a full-featured CARD system, the MADSIM represents a sound starting effort towards the final goal of automating the robot design process.

RELEVANT WORK

Early efforts in computer aided manipulator design dated back to the 60's when single analysis methods were applied to spatial linkage systems [6]. Batch processing, command driven method, standalone simulator and analysis software to aid the design soon followed. Before long, interactive design with graphic capability became common practices [6,7]. All these use computer graphics terminals, rely on design data bases, and are useful in evaluating existing robot manipulators. Recent endeavors in simulating robot activities to aid robot design have also been reported [8,9,10].

Various efforts are also evident in the simulation of different existing industrial robots as placed in the manufacturing environment. PLACE (McAuto) is one such system [40]. Among the most ambitious is the CAM-I project [9] that attempts to automate the plant robot design and to finally automate the manufacturing process.

As geometric modelling plays an important role in 3-D mechanical systems of which a robotic system is, most systems designed for geometric modelling also have the capability in the representation and manipulation of robot body and its kinematics. Notable examples are GMSolid [23], IBM Solid [or GDP/GRIN) [22], PADL [25], and many others.

Simulation of robot at some levels is fairly common. A notable example is EMULA [26] which emulates a robot at the manipulator or behavior level.

In all these endeavors, geometric modelling, data base, and simulation strategies have been of primary concern. While these systems represent important efforts towards more efficient, flexible and thus productive design, they suffer several drawbacks. First, they are independently developed and thus are not compatible with each other, which leads to awkward system integration or extension and thus are in general not conducive to full automation in the design. Secondly, most of these systems are not developed with the maximum productivity of designer users as the main objective. They are generally designed to meet the functional specifications.

No serious efforts have been reported on a formal integrated environment which emphasizes full design automation and promotes designer productivity in the whole robot design process. This type of system ideally consists of natural language user interface, centers on common design data base, and is modular and transportable as well. The need for such a design setting, however, is clear [9,13].

Work in computer aided design [3] and in software engineering associated with CAD [26,35] has been extensive. Many projects are of special relevance and can provide exemplary guidelines for any attempt to automate the design of robots. Two such successful examples are Hewlett-Packard's CAEE [34] and Bell Labs' BELLCAD [36].

ROBOT DESIGN: METHODOLOGY AND REQUIREMENTS

A robot or manipulator is a computer controlled mechanical linkage, consisting of arms for reaching, hands for manipulation, and wheels or legs for mobility. All of its capabilities and intelligence come from the control computer and its associated mechanisms. These characteristics of robot systems lend the design of the whole robot system to computerization.

To better understand and more effectively design a robot system, it is customary to regard the system as a hierarchy of several level machines. Some common levels are:

(1) goal level - This is in the area of artificial intelligence. Only the desired goal such as 'pick the pin' is specified.

(2) behavior level - This is the area of intelligent control. Trajectories are specified.

(3) motion level - This is the area of physical and geometrical modelling. Sensor information is utilized and appropriate actuator signals are issued. Stepwise motion needs to be detailed.

Virtual Robot Machines

A virtual robot machine can be thought to exist at each level. A virtual machine accepts a collection of instructions which form a language. The common robot/manipulator programming languages such as WAVE (SRI [19]), VAL (Unimation [20]), and AML (IBM [22]), in general, represent the behavior level machine. A pseudo goal level language and the machine is AUTOPASS [5]. Other virtual robot machines can be defined similarly according to the various robot languages that exist currently [32] or will appear in the future.

The MADSIM as described in the next section aims to emulate a machine at the motion level. A body modeller can specify the models using differential equations. Modellers for the transducers or sensors can conduct their work similarly. A control specialist may wish to lay out the control laws. And the designers would probably conduct a simulation.

Robot Design Requirements

Several levels of robot system design are well defined [1,7] and each of these includes several aspects. Briefly, they are: (1) goal, (2) behavior, and (3) motion.

Each of these levels requires several aspects of design details and they include at least the following:

1. Modelling and synthesis of body
2. Control algoritohms and controller design
3. Programming

At the functional level which we are particularly interested in, the above consists of many detailed steps such as:

a. 3-D modelling of robot body mechanisms
b. Functional specification of actuators and transducers
c. Specification of control laws
d. Simulation of motion/function
e. Optimization of parameters to meet specification

As the robot models are typically nonlinear and time-varying with large variations, model representations and the associated operations have to provide for these modelling requirements. To make the MADSIM project manageable, however, there will be constraints and assumptions although the underlying structures are flexible enough to permit extension. A minimum requirement could include the capability to specify the models of body and controls in terms of differential equations (or Laplace transforms) with certain nonlinearities such as saturation, backlash and hysteresis. Piecewise linearization should also be allowed.

Detailed analyses of robot body and the ancillary parts will not be furnished in the MADSIM because of its potential complexity in the algorithms and the associated tremendous amount of programming effort. Various other systems are available for these purposes.

SIR: A VIRTUAL ROBOT

To understand the design of a robot and to illustrate the concept of a virtual robot machine, we introduce the SIR or the Simple Instructional Robot for short.

SIR Machine

SIR is a conventional sensor-based multiply-linked robot manipulator. It has a body made of rigid members, a set of internal sensors capable of measuring joint angles or positions, an internal wrist sensors for measuring static forces at the last joint, and two external sensors that provide SIR with positional information about objects in the environment.

Logically, SIR is the robotic machine at the physical level that consists of the robot body, its sensors, actuators and the necessary electronics and microprocessors. It may correspond to the main control microcomputer that provides the overall central control and monitoring functions for the entire robot manipulator.

SIR can be thought of as a conventional von Neumann computer that sequentially executes on a finite set of instructions stored in the control memory and appropriate for the physical level of robot functions. It has a set of registers particular to the robot as described above and performs operations suitable for its normal functions.

This SIR machine may have an organization defined as follows in Figure 1.

SIR is a data acquisition system as it collects a large amount of data for control and monitoring purposes. It is a process control system as it commands a large complex mechanical plant dynamically engaged in high-speed maneuver. It is obviously a real-time system as all operations and decision makings have to be made in time to assure proper machine functions.

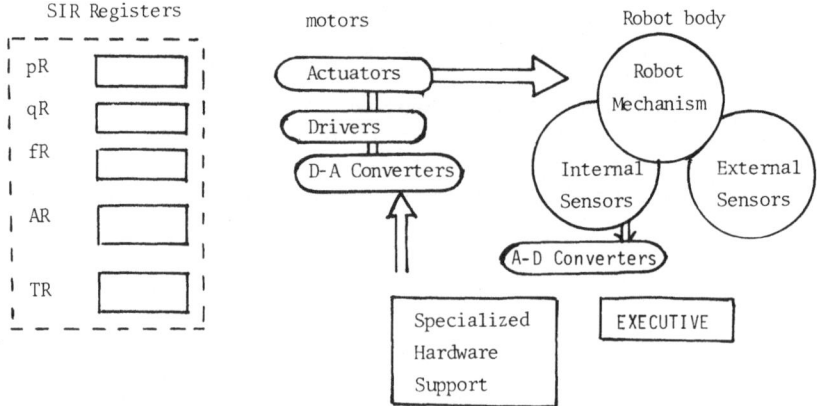

Figure 1. Organization of SIR machine.

COMPUTER AIDED ROBOT DESIGN SYSTEM

<u>SIR Registers</u>

Several sets of registers exist corresponding to all important parameters that characterize the SIR robot. They include:

(i) $pR(1)$, $pR(2)$, ..., $pR(6)$ (Generalized Positions Register Set) --- This set of registers indicate the generalized position for the last joint of SIR, namely three positions and three orientations. The quantities are expressed in absolute values.

(ii) $qR(1)$, $qR(2)$, ..., $qR(6)$ (Generalized Joint Variable Register Set) --- This represents the joint angles or displacements in absolute values.

(iii) $fR(1)$, $fR(2)$, $fR(3)$ (Force Register Set) --- This represents the three orthogonal forces as measured at the wrist.

(iv) $AR91)$, $AR(2)$, ..., $AR(6)$ (A Matrix Register Set) --- This keeps records of current values of A matrices as per [18].

(v) $TR(1)$, $TR(2)$, ..., $TR(6)$ (T Matrix Register Set) --- This keeps records of current values of T matrices as per [18].

Depending on implementation, these registers may correspond to real processor registers, emulated system memory units, common area, or even subroutine parameters.

<u>SIR Instructions</u>. SIR instructions are used to effect the common control and management of SIR registers and thus the operations of SIR. Some of these instructions are as follows:

(i) RESETQ --- Set all q's (the joint control variables) to known predetermined values.

(ii) SETQ_ABS Q1 ... Q6 --- This instruction moves all q's to the specified values.

(iii) SETQ_REL Q1 ... Q6 --- same as the previous except that q's are expressed in values relative to the existing values.

(iv) GET_P P1 ... P6 --- As the generalized positions are determined by the EXECUTIVE or SIR machine and are set by the associated controller and the physical robot mechanism, the SIR machine has to query this generalized position.

(v)   GET_F F1 ... F3 --- Get the force reading.

(vi) SET_TIME T1 T2 --- This instruction alters the timer parameters that drive the EXECUTIVE. It may be noted that SIR is a real-time computer system in which all tasks are time dependent and are prioritized.

(vii) Other conventional instructions for data manipulations and program controls.

SIR System Requirements. To implement SIR entails a good balanced commitment in both hardware and software so as to meet the required system performance in true real-time response and heavy numerical computations.

Software provides the sophistication needed for coping with the system complexity inherent with the system model and the resultant control requirement. To handle the strenuous real-time and computation-intensive operations as evidenced by the complex robot model and control algorithms, the software needed to manage the SIR robotic system has to be well-structured and properly managed by a well-behaved control operating software. Inasmuch as SIR operates in the Executive and all application must be conducted in clock time, the SIR software system is thus real-time and multitasking, driven by a clock on priority basis.

Hardware provides the raw power necessary to meet the requirement for intensive numerical computations. As a computer controlled system, SIR has to monitor system variables, to compute, to reason (sort of), and to control all actuating mechanisms at the right precise moments. The processors have to be powerful and fast enough so that they can compute and handle large amounts of information in time. Special processors such as floating point arithmetic, array, or high-performance processors are needed. Multiple processors and distributed processing appear to furnish a practical configuration for the computing needs of SIR.

SIR Implementation. Implementation of SIR ranges from direct customized processor via microcoding, to interpretation using conventional processor via native assembly language, and to common library of subroutines embedded high-level languages, provided that they meet the speed and memory requirement.

Customized processor is well within the economical reach of today's technologies. Though it offers the desired speed and features, it may not be cost-effective. Interpretation via conventional processors is high in speed and is relatively convenient to develop. The subroutine library approach may well be the most expedient method to obtain fast operational systems because the common popular high-level programming languages and systems are the primary tools in which to develop the software. Despite the fact that this approach is less machine efficient, it is viable as today's processors are fast and the use of multi-processing is practical to obtain designer efficiency at the expense of machine efficiency.

In normal implementation using subroutine library, the EXECUTIVE is a program resident on a conventional microcomputer that emulates and executes the SIR machine.

CARD SYSTEM REQUIREMENTS

Generally, a CAD system consists of two major parts: the modelling and the computer graphics. The modelling provides the necessary framework to represent and manipulate the various aspects of robot system design process. Modelling portion contains suitable mechanisms for representing various data structures in the CARD system. The graphics portion communicates with users these aspects in a friendly and coherent fashion in order to facilitate the design process. A wide variety of data are communicated via the graphics subsystem, among them are primarily graphic data for the object or components of the robot system. For a robot arm, these include all the parameters regarding links, joints, etc., their dimensions, weights, types, and bounds of these parameters.

Three software components are of particular importance in an integrated computer system such as CARD system; they are language subsystem, an integrated data base and user interface. The language system handles all communication between CARD systems and the user interface subsystem. The integrated data base system serves as a central control outpost [15]. User interface is of utmost importance as it directly affects CARD users' or designers' productivity.

Users communicate with the CARD system normally in an interactive fashion, but occasionally in the batch mode. The underlying structure of the CARD system is through procedural calls.

Graphics and Languages

Efficient and natural communications between users and programs are necessary for all design activities both inside and outside the CARD system. These communications are conducted through suitable forms of languages. Procedural languages permit formal definition and manipulation of physical models as well as various transfers and transformations of other data or information. On the other hand, the nonprocedural languages refer to query languages in the forms of menus, commands, or question-and-answer types standard in interactive user-program communications.

Graphic (particularly the colored) display and various graphic input devices represent much more natural, friendly and wider communication media. In addition, the graphic capabilities to effectively communicate three-dimensional objects and their operations pertinent in robotic environment are of extraordinary value to any CARD system.

The study of robot manipulation deals intensively with the relationship between robots and manipulators. Graphical representations of the manipulator and its environment so as to provide the necessary visualization and graphics animation for the dynamimcs of robot manipulation are highly conducive to designer/user productivity.

Integrated Data Base

All automation systems are data-oriented. Data base has long been proposed and established to be a basis for data processing [27] and has been well recognized as crucial in the CAD/CAM integration [28]. Various efforts have been reported [29]. Various types of data including the design details, design rules or guidelines, documentations, management control, and others are typically stored and managed in the data base.

Data-drive automation has triggered a new industrial awareness, and it is evident that an integrated data base provides the key to total integration of various computer aided engineering and automation.

User Interface

Designer efficiency and design quality are greatly enhanced if user-program communications are effectively conducted. Graphics subsystem provides the enhanced communication bandwidth and visualization. The associated (graphics) language subsystem furnishes the required semantic, syntactical, and lexical aspects in communication format and protocol. Use of innovative input devices such as special function keys, light pen, tablet digitizer and the now ubiquitous mouse all serve to increase the effect of interactive user program communications.

Following are considered sound guidelines for these subsystems [30]:

1. Know the users: from novice to the experienced
2. Structure the display
3. Give feedbacks
4. Accommodate errors and allow backup
5. Minimize memorization and maintain consistency

Both menu and command user-machine interfaces must be adopted in order to support all users from novice to the experienced. Nonetheless, as the majority of the users of CARD are designers or specialists who have some exposure to the art of computer usage and also are well motivated, the system can be tilted towards the command approach in which the users keep the initiative and may get the assistance or help on demand only.

The use of natural language and graphic icons [39] will enhance the ease and the friendliness to use the CARD system. Various other disciplines including perceptual and cognitive psychologies may be needed to implement a successful user-computer interface.

In addition, there is one area that is of particular importance to the functioning of a CARD system, namely solid modelling.

Solid Modelling

As the robot body system and its operation are characterized by three-dimensional quantities, these measurements and characterization are needed to describe robots' behavior effectively. A CARD system should therefore provide facilities to represent and to manipulate various three-dimensional parameters and attributes. In addition to the standard geometrical quantities, some physical measurements such as the centers of gravity and the weights must also be accommodated.

As a full-fledged general purpose solid modelling system is complex and expensive to build, to assimilate and to operate [23], a standalone CARD system cannot support extensive and detailed system analysis. It suffices to incorporate less sophisticated and yet specialized solid modelling systems that are adequate and efficient to characterize robotic systems. Utilization of existing solid modelling systems is very practical in obtaining early demonstration capability.

CAD of CARD

The MADSIM is to be developed, coded, tested, and throughout the majority of its lifespan maintained and upgraded. The productivity and quality of MADSIM can and must be improved by applying existing software technologies [26]. To allow orderly development, upkeep, and evolution, it is imperative that the program structures can be reconfigured and the associated data managed under program control. The success of YACC [34] and other software systems that utilize software design automation techniques affirms the need for similar consideration on the MADSIM.

MADSIM ARCHITECTURE AND DESIGN OBJECTIVES

The single most important goal in undertaking MADSIM projects is to prototype on integrated CARD system encompassing major steps in the robot system design which is complete with modelling, design (parameters specification), analysis, simulation and performance optimization. To achieve this goal, we set out to target the following as the MADSIM system design objectives:

  (i) transportability - This feature allows the majority of MADSIM codes to be transportable to other hardware and programming environments. It would prolong the life of this project and enlarge users' base.

  (ii) extensibility - This capability allows the system to grow as the CAD technology and robot design methods evolve.

  (iii) maintainability - This ensures the controllability of MADSIM project and its smooth modification as need for upgrade arises.

  (iv) simplicity - This makes possible early completion and facilitates development. Coherence and uniformity in software structure and in system usage should be enforced, and this will have positive effect towards simplicity for the system design.

It might be worth mentioning that real-time response is not among the design objectives. This permits the designers to surpass the financial limitation. This poses no real hindrance to the final performance as meeting the transportability objective necessitates the separation of hardware dependency from the architecture. Indeed, the device independence that comes with the transportability enables the incorporation of extra hardware into the existing system.

Transportability of MADSIM will be obtained through several levels: (1) High-level language (HLL) level via highly portable languages such as FORTRAN, Pascal or ADA, (2) Algorithms level by structuring the software system into device dependent and independent portions.

Just like any other CAD system, the MADSIM can be decomposed into at least three major parts: (a) analysis and synthesis methods for robot design, (b) software and hardware system architecture, and (c) user interface. The software portion can be further broken into four areas:

  (1) Modelling/Design/Simulation language subsystem
  (2) Design data base management subsystem

COMPUTER AIDED ROBOT DESIGN SYSTEM   279

(3) Design graphics subsystem
(4) Configuration control

MADSIM interfaces with users and other program through some forms of languages. In general, command languages or menu operations or question-and-answer dialogues are used to communicate with interactive users, and procedural languages are native to the MADSIM Executive and are used to interface with programs.

To maintain the independence of MADSIM from user interface, host computer or other units and to preserve modularity, portability, and efficiency, attention will be given to the following areas: (1) MADSIM and user-programmer interface, (2) MADSIM and host-computer interface, (3) Language systems, and (4) Configuration management and support software control.

Transportability of MADSIM allows MADSIM to be embedded in a variety of host configurations, ranging from a single processor to a network of processors. The software structures of MADSIM with its associated hardware may be summarized graphically in Figure 2.

In the diagram, MML is the manipulator modelling language used for interactive users to model the robot. MDL is the manipulator design language used for designers to select or specify robot model parameters. MSL is the manipulator simulation language used to conduct the simulation and possibly the optimization. An effort has been made to consolidate these languages so that a unified user interface can be accommodated [33].

To ensure that the whole MADSIM system is manageable, extensive solid modelling will not be used. However, it would be beneficial that MADSIM is able to exchange robot model information with some general purpose solid modelling systems such as MOVIE.BYU and TEMPLATE. It is important that this MADSIM specialized solid

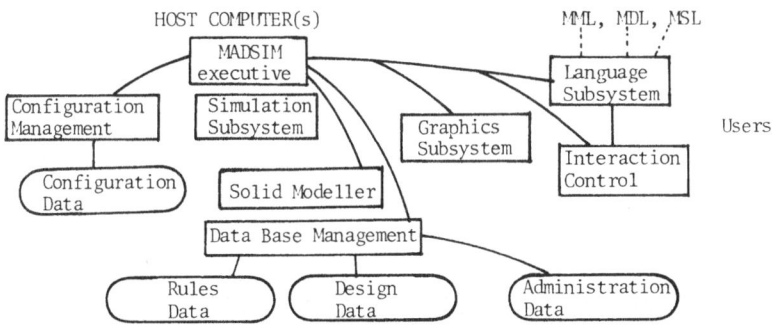

Figure 2. MADSIM system architecture.

modelling subsystem has good rapport with finite element modelling facility in which detailed analysis can be performed.

MADSIM IMPLEMENTATION: CONSIDERATIONS AND PLANS

Software development is a complex process requiring extraordinary investment in manpower and time. It is thus very important to fully take advantage of modern software design and development methodologies and practices. The ultimate goals for the MADSIM software architecture and its implementation are (1) higher productivity for all major activities that are involved throughout its life cycle and (2) better quality of the resultant products. Of the system requirements and design objectives that are described in the previous sections, some are fundamental and must be met at all reasonable cost. Some, however, are cosmetic and may be realistically put off to latter stage for implementation if this can simplify the early phases of design and it does not impose unduly constraint on the whole project.

Throughout the lifetime of the MADSIM project, three types of persons are involved and their productivities should be properly enhanced. They are (1) MADSIM designer, developer, or programmer who has the responsibility in the design, construction and testing of MADSIM System from conceptual level to prototype and to release, (2) System manager who maintains, updates and administers the MADSIM system; and (3) User or robot designer who uses MADSIM in order to automate the many aspects of robot design. To reach the goal of higher productivity for the first two types of person, it is imperative that features such as software modularity and hierarchy be incorporated into MADSIM system. Mechanisms for the monitoring, triggering and modification of data structures which control and configure the MADSIM environment must be provided.

As for the higher productivity of robot designers, the user interface must be handled efficiently and flexibly by allowing non-rigid natural formats in user-computer dialogues and by giving MADSIM system manager the capability to vary and the dialogue formats. In addition, the MADSIM system implementation should also strive to attain fast prototyping. This is desirable as it demonstrates the CAD project's feasibility and usefulness as well as boosts the confidence in its early completion.

With the above capabilities delineated for MADSIM initially, we can set forth to implement the MADSIM system by applying the modern software design methodology and utilizing the software support environment. The software architecture as laid out in previous sections is amenable to these design formalisms.

Several efforts in Computer Aided Engineering have been successful and should give us a guideline, if not a blueprint, in implementing MADSIM. Of particular interest to the implementation of CARD system are Hewlett-Packard's CAEE (Computer Aided Electrical Engineering) [35], Bell Lab's Designer's Workbench [34] and BELCAD [37]. HP's CAEE is closer to the MADSIM in terms of project complexity and system requirement and thus warrants the utilization of CAEE's software models in the MADSIM system. The program control and configuration mechanism are driven by a set of system files including command file, configuration file, and message files. The message file is used for user-computer communication. Command file contains a finite-state machine (FSM) which determines the system operation and, in particular, the modes and behavior of user-computer interaction. The interrelationship between the system components (inside and outside the system 'shell') are determined and managed by the configuration file. All three system files can be changed.

Excluding the system files, the CAEE system can be broken into two parts: the system utility shell, and the applications program. The 'shell' consists of procedures and utility programs, and it shields from the intelligence-oriented applications program all machine or data dependent details. Applications program is that part of the CAEE which handles the needed target problem analysis, simulation, and other operations.

The utilitarian approach in which the system shell takes advantage of the plethora of utility routines that are available for a given operating system and additional custom-written routines is appropriate and practical for several reasons [38]. One important fact is that most existing operating system such as UNIX and VMS on DEC's VAX are abundant with reliable routines. It may not be very efficient in CPU time and not all that suitable for CAD purpose; its very existence and popularity with programmers certainly make the utility routines a solid foundation for bringing up the MADSIM system. It obviously is conducive to faster development and earlier prototyping.

<u>MADSIM Implementation</u>

The structure of MADSIM will be similar to that of HP's CAEE system with some additional features catering to special needs of robot design. The design requirement and capabilities as outlined in the above can be satisfied under the general architecture of CAEE. Minor implementation aberrations and sophistication via further structure decomposition are needed.

The graphic functions are mainly performed by the utilities or procedures in the system shell and partially by the applications programs. The graphic functions manage the 3-dimensional display and handle user computer interactions. The MADSIM data base which includes details of design, rules and some project administration will be managed by a set of utilitarian or procedural shell functions. This set of shell functions as a unit will function as the needed integrated data base management.

User interface of MADSIM will be managed at the front end by the application programs which are strongly supported by the graphics and data base shell functions. On top of the data base exists a layer of software which provides the structures for the modelling of robot system including the robot body, homogeneous transformations, dynamics equations, sensor and actuator characteristics, tool descriptions, and a simple workspace. Both geometrical and physical quantities as well as the related symbolic data have to be managed.

Graphically, the MADSIM as discussed in the above may be described in Figure 3.

Generality needed to implement the MADSIM software architecture as outlined in previous sections is not compromised by adopting the above implementation structure. Though it may obscure the clearer desirable MADSIM structure partition as expounded before, the shell facilities in the implementation is general enough to permit inner structuring. The integrated data base, the graphics and the user-computer dialogue handling can all be carved out from the system shell and thus creating newer layers or modules of software units. Generally, extension of these facilities either in the shell or in the applications is fully appropriate.

In addition to the modular and hierarchical structures that have to be built in MADSIM, certain mechanisms must also be in place inside the MADSIM system at source code level that allows direct modifications and monitoring of the MADSIM system software. The 'frame' concept [38], for instance, proves to be a valid

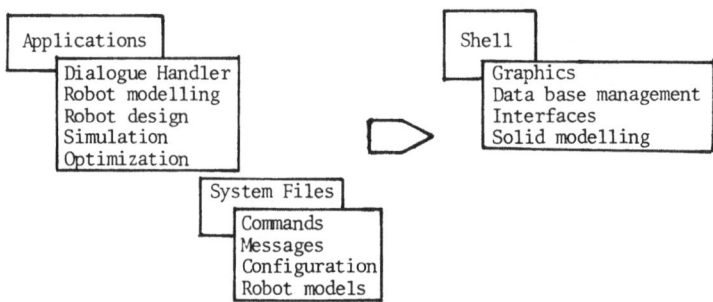

Figure 3. Organization of a MADSIM implementation.

vehicle to improve the productivity in source code generation and modification.

Robot Model Definition and Languages

How can the robot model and its associated data enter the MADSIM data base? Two mechanisms are common, namely via a procedural language, and through interactive graphics. As each method has its own merits, both will be provided in MADSIM. Interactive graphics approach is handled by the dialogue handling mechanism and the graphics display facility. The language approach allows formal definition of robot hardware and software using formal language. As with all cases of CAE system, data or model definitions are predominant and are more important than the data manipulation aspects. This is so prevalent that the language subsystems are characteristically similar to the data base query languages. In the context of a formal high-level language, this means high percentage of declaration of data types and structures and thus low portion of execution statements. This is all evident in robot languages such as AML [22], AL [4], and AUTOPASS [5]. Whereas, language processing falls under the general applications program, the system shell provides the more primitive language interface, including both procedural and nonprocedural interfaces.

MADSIM System Files

To ensure proper system operations, MADSIM utilizes four major textual system files: (i) command file, (ii) configuration files, (iii) message files, and (iv) robot modelling file. The first three files are similar to those in HP's CAEE and the fourth one is the main media for holding the various data defining the robot. The management primitives for the creation, retrieval and alteration of robot models are provided in a set of customizmed utilities that are built on top of the system shell utilities or procedures.

Multiple Level Robot Machines in MADSIM

In order to make MADSIM a useful tool in robot design, it is necessary that multiple levels of robot machines are allowed to exist in MADSIM system. For practical purpose, only the functional or motion level language robot as typified by the SIR system as previously discussed will be represented. The addition of the functional level to MADSIM will not unduly increase the complexity as these two levels are rather close and differ principally in the forms. At a later time, a behavior level robotic machine can be implemented on top of this motion level SIR machine in a separate unit in the applications program of MADSIM. Robot machine at a still higher level can be added at a later stage by expanding the then well-structured MADSIM.

CONCLUSIONS

With the robots becoming more intelligent and complex, the associated tasks of robot design more and more complicated, and the benefit of optimizing the robot performance or reducing the cost all the more important, the role of design automation is evidently becoming more crucial. While CAD techniques serve to increase the productivity of robot design and the quality of final product, the success and thus its contribution of a CARD system depend on the productivity and quality of the very CARD system itself.

Just like the design of robot itself, the design requirements and techniques as well as tools for CAD will evolve. To ensure that the CARD system can also evolve as robot design techniques and CAD technologies progress, it is imperative to utilize the state-of-the-art techniques in software technologies which generally facilitate CAD system development and promote user/designer efficiency. The proposed MADSIM project represents a timely effort along this direction.

REFERENCES

[1] J. Birk and R. Kelley, Workshop on the Research Needed to Advance the State of Knowledge in Robotics, NSF Report, 1981.
[2] R. P. Paul, "Robots, models, and automation," IEEE Computer, July 1979.
[3] Special Issue of Computer Aided Design, IEEE Proceedings, October 1981.
[4] R. Finkel, R. H. Taylor and R. C. Bolles, "AL, a programming system for automation," Stanford Artificial Intelligence Laboratory Memo AIM-243, Stanford University, Stanford, CA, 1974.
[5] L. I. Lieberman and M. A. Wesley, "AUTOPASS: An automatic programming system for computer controlled mechanical assembly," IBM J. Res. & Dev., 21, pp. 321-333, 1977.
[6] N. Orlandia, D. A. Calahan and M. A. Chace, "A sparsity-oriented approach to dynamic analysis and design of mechanical systems," ASME J. Engineering for Industry, August 1977.
[7] W. B. Heginbotham, M. Dooner, and K. Chase, "Rapid assessment of industrial robot performance by interactive computer graphics," Proc. Ninth ISIR, 1979.
[8] A. Liegeois, A. Fournier, M. J. Aldon and P. Borrel, "A system of computer aided design of robots and manipulators," Proc. Tenth ISIR, 1980.
[9] J. Johnson, "Pushing the state of the art," Datamation, Feb. 1982.

[10] S. Derby, "Simulation motion elements of general purpose robot arms," Int. J. of Robotic Research, Vol. 2, No. 1, pp. 3-12, 1983.
[11] G. Dodd and L. Rossol, Computer Vision and Sensor Based Robots, Plenum Press, New York, 1979.
[12] T. Meyer, "An evaluation system for programmable sensory robots," IBM J. Res. & Dev., 25, 6, 1981.
[13] B. I. Soroka, "A program for computer aided robot design," Proc. ASME Second Int'l. Computer Engineering Conference, San Diego, CA, August 1982.
[14] C. B. Besant, Computer Aided Design and Automation, John Wiley, New York, 1980.
[15] M. Kutcher, "Automating it all," IEEE Spectrum, May 1983.
[16] F. Bliss and G. M. Hyman, "Selecting and implementing a turnkey graphics system," IEEE Computer Graphics and Applications, April 1981.
[17] J. F. Engleburger, Robots in Practice, Avenbury Publishing Co., London, 1980.
[18] R. Paul, Robot Manipulators, MIT Press, 1981.
[19] R. Paul, "WAVE: A model based language for manipulator control," The Industrial Robots, 4, 1, 1977.
[20] User's Guide to VAL, Unimation, Inc., June 1980.
[21] R. Finkel, R. Taylor, R. Bolles, R. Paul and J. Seldmen, "An overview of AL, a programming language for automation," Proc. 4th Joint Conference on Artificial Intelligence, pp. 758-765, 1975.
[22] A Manufacturing Language Reference, IBM, 1983.
[23] R. D. Tilove, "Extending solid modelling systems for mechanism design and kinematic simulation," IEEE CG&A, May-June 1983.
[24] W. Fitzgerald, F. Gracer and R. Wolfe, "GRIN: Interactive graphics for modelling solids," IBM J. Res. & Dev., 25, 4, July 1981.
[25] "An Introduction to PADL," Production Automation Project Technical Memorandum, University of Rochester, Dec. 1974.
[26] A. I. Wasserman and L. A. Belady, "Software engineering: The turning point," IEEE Computer, Sept. 1978.
[27] J. Meyer, "An emulation system for programmable sensory robots," IBM J. Res. & Dev., 25, 6, Nov. 1981.
[28] CODASYL Data Base Task Group Report, ACM, April 1971.
[29] W. D. Beeby, "The heart of integration: a sound data base," IEEE Spectrum, May 1983.
[30] J. Encarnacao and F. I. Krause, File Structures and Data Base for CAD, North-Holland, 1982.
[31] J. D. Foley and A. Van Dam, Fundamentals of Interactive Computer Graphics, Addison-Wesley, 1982.
[32] S. Bonner and K. Shin, "A comparative study of robot languages," IEEE Computer, Dec. 1982.
[33] L. Cheng, "A manipulator design language," 20th ISMM Symposium of Mini-Micro Computer Applications, Cambridge, MA, July 1982.

[34] S. C. Johnson, "YACC - Yet Another Compiler Compiler," Comp. Sci. Report No. 32, Bell Labs, July 1975.
[35] C. L. Leath and S. J. Ollanik, "Software architecture for the implementation of a computer aided engineering system," Proc. 30th IEEE/ACM Design Automation Conference, pp. 137-142, 1983.
[36] R. A. Friedenson, J. R. Breiland and T. J. Thompson, "Designer's workbench: Delivery of CAD tools," Proc. 19th IEEE/ACM Design Automation Conference, pp. 15-22, 1982.
[37] F. W. Day, "Computer aided software engineering," Proc. 20th Design Automation Conference, pp. 129-136, 1983.
[38] T. J. Thompson, "A utilitarian approach to CAD," Proc. 19th IEEE/ACM Design Automation Conference, pp. 23-29, 1982.
[39] K. N. Lodding, "Iconic interfacing," IEEE Computer Graphics and Applications, 3, 2, pp. 11-24, 1983.
[40] R. Waterbury, "Factory simulation: testing automations' what if's," Assembly Engineering, July 1983.
[41] V. D. Hunt, Industrial Robotics Handbook, Industrial Press, New York, 1983.

A CAD TOOL FOR THE KINEMATIC DESIGN OF ROBOT MANIPULATORS

Chi-haur Wu

Department of Electrical Engineering and Computer Science
Northwestern University
Evanston, IL 60201 U.S.A.

ABSTRACT

In order to fully develop a CAD tool for the kinematic design of a robot manipulator, the relationship between the designed errors in the kinematic parameters and the resultant errors in the robot's position and orientation at the World coordinates (the working coordinates of a robot) has to be mathematically modeled. Since the control basis for all robot manipulators is the relationship between the Cartesian coordinates of robot's end-effector and the joint coordinates. The correct relationship between two connective joint coordinates of a robot manipulator is defined by four link parameters; one is the joint variable and the others are geometric values. Hence, the fidelity of robot manipulator's position and orientation in the real world depends on the accuracy of these four link parameters of each joint. In this paper, a linear analytical model between the six errors in the position and orientation of the World coordinates and the four kinds of kinematic errors is developed. Based on this model, the error envelopes in the position and orientation of the World coordinates due to any combination of these four kinds of kinematic errors can be easily determined. Hence, this model can be used as a CAD tool to optimize the kinematic parameters of a robot manipulator and to increase the accuracy of designed manipulator. Furthermore, a simple calibration scheme based on this model has been developed to correct the kinematic errors of present robots.

INTRODUCTION

In the sixties, industrial robot was introduced. It was a universal transfer machine and could be programmed to perform many different tasks. Since computers have been added to robots, robots' flexibility and productivity have been increased. Although a great deal of sophistication has been achieved in specifying positions, in control and in programming, the robot accuracy is still an unsolved problem. At present, robot manufacturers used different calibration scheme to minimize the manufacturing errors of their robots. However, these calibration schemes need heavy fixtures mounted on robot and require a very time-consuming process. Also, the fixtures and the algorithms are different for different types of robots. These techniques are not practical and incur the extra cost of the expensive fixtures. Hence, there is a need to develop an efficient method for the process of calibration. Although a calibration method with a kinematic control scheme can minimize the robot's errors, the best way to improve the accuracy of a robot manipulator is still to design a robot with minimum errors. In order to achieve this goal, there is a need to develop a kinematic CAD model which can describe the problem of robot's accuracy. With this CAD model, the kinematics of the designed robot can be optimized to meet the standards of required accuracy.

A robot manipulator is a position-oriented mechanical device. A serial link manipulator consists of a sequence of mechanical links connected together by actuated joints. The relationship between two connective joint coordinates is well-defined by a homogeneous transformation matrix [1]. This matrix is determined by four kinds of link parameters, also called kinematic parameters; one is joint variable and the others are geometric parameters. At present, all the robot manipulators are the open-loop linkage control. The control basis is a relationship between the Cartesian coordinates of the end-effector and the joint coordiantes. Hence, the accuracy of robot position and orientation to the real world depends on the accuracy of the four link parameters of each joint. Due to the nature of these kinematic parameters, the Cartesian errors can be grouped into two categories: (1) the Cartesian errors due to the position accuracy of the joint variables, (2) the Cartesian errors due to the dimensional errors of the geometric kinematic parameters. Waldron [3] has developed a model in a general form for the first category. As to the second category, he also described it in the general case but without any mathematical formulation. Based on his developed model, a computer program [4] which can generate and plot the surfaces of the position accuracy of a robot manipulator has also been developed; however, this work only considered the errors of first category and assumed that there are no dimensional errors in the geometric parameters. In order to fully understand the impact of these kinematic errors on robot

A CAD TOOL FOR THE KINEMATIC DESIGN

accuracy, a detailed mathematical error model to describe the above impact is necessary.

In author's previous paper [2], a linear explicit mathematical error model for the four kinds of kinematic errors and the six Cartesian errors of the end-effector has been formulated. However, from the practical point of view, the representations of Cartesian errors are more like imaginary errors in the space; and they are not so easily understandable as the errors represented in the working World coordinates of a robot manipulator. Hence, in order to fully develop a CAD tool for the kinematic design of robot manipulators, the error model between the position and orientation errors in the World coordinates and the kinematic errors needs to be mathematically derived. In this paper, a reversed approach compared to the approach in [2] is applied, and a linear error model is also obtained. Based on this linear error model, the error envelopes of the position and orientation in the World coordinates caused by the kinematic errors can be easily generated. Thus, the kinematic parameters of the designed robot manipulator can be optimized to increase the accuracy of a robot manipulator. In addition, a simple calibration scheme has also been proposed to improve the accuracy of present robots.

KINEMATICS

In order to find the impact of kinematic errors on robot's position and orientation in the real world, an explicit mathematical model is needed. In order to derive this model, a thorough understanding of robotic kinematics is necessary. Although there are several ways to represent the robotic kinematics, the homogeneous transformation [1,6] is still the most popular method in the robot industry.

For an N degrees-of-freedom manipulator, there will be N joints and N+1 links. The relationship between the joint coordinate frames i-1 and i can be well represented by a homogeneous transformation matrix A_i [1,6] and

$$A_i = \begin{bmatrix} C\theta_i & -S\theta_i C\alpha_i & S\theta_i S\alpha_i & \ell_i C\theta_i \\ S\theta_i & C\theta_i C\alpha_i & -C\theta_i S\alpha_i & \ell_i S\theta_i \\ 0 & S\alpha_i & C\alpha_i & r_i \\ 0 & 0 & 0 & 1 \end{bmatrix} \quad (1)$$

where S and C refer to SINE and COSINE functions and θ_i, r_i, ℓ_i, α_i are the kinematic parameters of the ith joint. These parameters are shown in Figure 1. For a revolute joint, θ_i is the joint

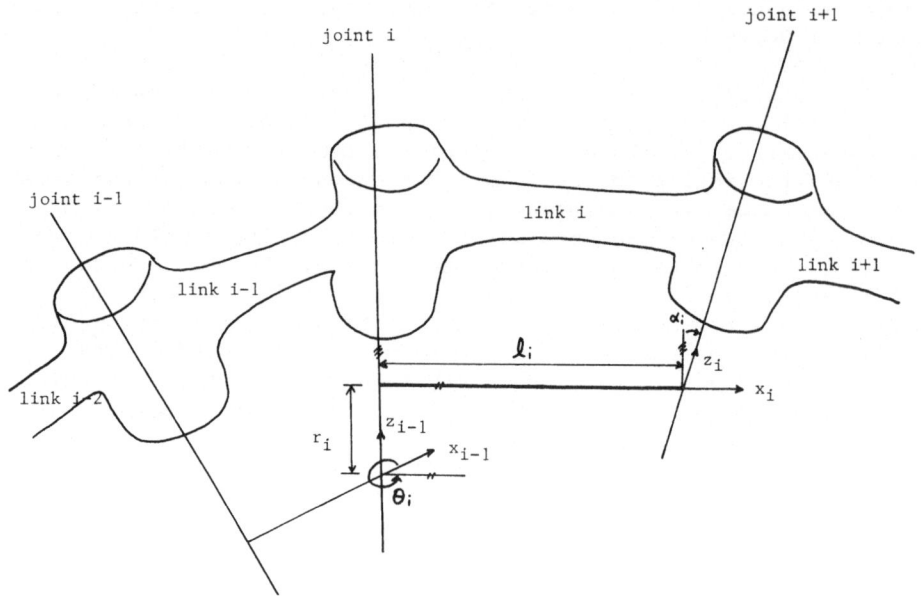

Figure 1. Link coordinates and parameters θ_i, r_i, ℓ_i, α_i.

variable. As to a prismatic joint, r_i is the joint variable and $\ell_i = 0$. After defining the joint variable, the other three kinematic parameters are the fixed geometric values. For convenience, A_i can be represented by a 3 by 3 rotational matrix R and a 3 by 1 positional vector $\underline{p}$ as follows

$$A_i = \begin{bmatrix} R & \underline{p} \\ \underline{0} & 1 \end{bmatrix} \quad (2)$$

where $\underline{0} = [0\ 0\ 0]$.

With the definition of homogeneous transformation matrix A_i, the motion of the end of an N degrees-of-freedom manipulator with respect to the base coordinates can be represented as

$$T_N = A_1 * A_2 * \ldots\ldots * A_{N-1} * A_N \quad (3)$$

where "*" represents the matrix multiplication. As to the motion of joint coordinate frame i with respect to the base coordinates can be represented as

A CAD TOOL FOR THE KINEMATIC DESIGN

$$T_i = A_1 * A_2 * \ldots\ldots * A_i \;. \tag{4}$$

For simplicity, we will assume that the base coordinates is the working World coordinates of a robot for the rest of the paper. Hence, T_i represents the motion of joint coordinate frame i with respect to the World coordinates.

Same as Equation (2), the matrix T_i can also be represented by the following form

$$T_i = \begin{matrix} R_i & \underline{p}_i \\ \underline{0} & 1 \end{matrix} \tag{5}$$

and when i = 0, T_0 = I (the identity matrix).

BACKWARD DIFFERENTIAL CHANGES BETWEEN TWO COORDINATE FRAMES

In [2,5,6], the forward differential changes between two coordinate frames has been formulated. In this section, the relationship of backward differential changes between two coordinate frames will be similarly formulated.

Given a small change δT_2 in the position and orientation of coordinate frame 2, there will be a corresponding small change δT_1 in coordinate frame 1. If the relationship between two coordinate frames represents by a transformation T_1^2 then the relationship between the differential changes can be represented as [7]

$$\delta T_1 * T_1^2 = T_1^2 * \delta T_2 \tag{6}$$

and

$$\delta T_1 = T_1^2 * \delta T_2 * (T_1^2)^{-1} \;. \tag{7}$$

The differential error matrices δT_i, i=1,2, can be represented in the following form [6] by ignoring the higher order terms

$$\delta T_i = \begin{matrix} 0 & -\delta z_i & \delta y_i & dx_i \\ \delta z_i & 0 & -\delta x_i & dy_i \\ -\delta y_i & \delta x_i & 0 & dz_i \\ 0 & 0 & 0 & 0 \end{matrix} \tag{8}$$

where $\underline{d}_i = [dx_i \; dy_i \; dz_i]^t$ and $\underline{\delta}_i = [\delta x_i \; \delta y_i \; \delta z_i]^t$ are the small translational and rotational changes at coordinate i; and the superscript 't' represents the matrix transpose.

If T_1^2 and δT_2 were known then the components of δT_1 can be solved analytically by following two equations

$$\underline{\delta}_1 = R_t * \underline{\delta}_2 \tag{9}$$

$$\underline{d}_1 = R_t * \underline{d}_2 + \underline{p}_t \times \underline{\delta}_1 \tag{10}$$

where R_t is the rotational matrix and $\underline{p}_t$ is the position vector of T_1^2, respectively, and 'x' is the cross product of two vectors.

JOINT DIFFERENTIAL CHANGES DUE TO THE KINEMATIC ERRORS

Since the correct relationship A_i between joint coordinate frames i and i-1 is determined by its four kinematic parameters, $\theta_i, r_i, \ell_i, \alpha_i$. If there are errors in these kinematic parameters, there will be a differential change dA_i between these two joint coordinate frames. Thus, the accurate relationship between these two joint coordinate frames could no longer be correctly defined by the transformation A_i, and the true relationship will be equal to $A_i + dA_i$. This differential change dA_i can be estimated by the following linear differential equation

$$dA_i = \frac{\partial A_i}{\partial \theta_i} \Delta \theta_i + \frac{\partial A_i}{\partial r_i} \Delta r_i + \frac{\partial A_i}{\partial \ell_i} \Delta \ell_i + \frac{\partial A_i}{\partial \alpha_i} \Delta \alpha_i \tag{11}$$

where $\Delta \theta_i, \Delta r_i, \Delta \ell_i$ and $\Delta \alpha_i$ are the small error changes in the kinematic parameters.

By differentiating Equation (1) with the kinematic parameters, following results can be obtained

$$\frac{\partial A_i}{\partial \theta_i} = D_\theta * A_i \; , \tag{12}$$

$$\frac{\partial A_i}{\partial r_i} = D_r * A_i \; , \tag{13}$$

$$\frac{\partial A_i}{\partial \ell_i} = D_\ell * A_i \; , \tag{14}$$

$$\frac{\partial A_i}{\partial \alpha_i} = D_\alpha * A_i \tag{15}$$

where

$$D_\theta = \begin{matrix} 0 & -1 & 0 & 0 \\ 1 & 0 & 0 & 0 \\ 0 & 0 & 0 & 0 \\ 0 & 0 & 0 & 0 \end{matrix} \tag{16}$$

$$D_r = \begin{matrix} 0 & 0 & 0 & 0 \\ 0 & 0 & 0 & 0 \\ 0 & 0 & 0 & 1 \\ 0 & 0 & 0 & 0 \end{matrix} \tag{17}$$

$$D_\ell = \begin{matrix} 0 & 0 & 0 & C\theta_i \\ 0 & 0 & 0 & S\theta_i \\ 0 & 0 & 0 & 0 \\ 0 & 0 & 0 & 0 \end{matrix} \tag{18}$$

and

$$D_\alpha = \begin{matrix} 0 & 0 & S\theta_i & -r_i S\theta_i \\ 0 & 0 & -C\theta_i & r_i C\theta_i \\ -S\theta_i & C\theta_i & 0 & 0 \\ 0 & 0 & 0 & 0 \end{matrix} \tag{19}$$

Based on above results, Equation (11) can be rewritten as

$$dA_i = (D_\theta \Delta\theta_i + D_r \Delta r_i + D_\ell \Delta\ell_i + D_\alpha \Delta\alpha_i) * A_i . \tag{20}$$

By defining an error matrix transform δA_{i-1} with respect to joint coordinate i-1 and

$$dA_i = \delta A_{i-1} * A_i \tag{21}$$

then

$$\delta A_{i-1} = D_\theta \Delta\theta_i + D_r \Delta r_i + D_\ell \Delta\ell_i + D_\alpha \Delta\alpha_i \tag{22}$$

and, mathematically, δA_{i-1} can be solved as

$$\delta A_{i-1} = \begin{bmatrix} 0 & -\Delta\theta_i & S\theta_i\Delta\alpha_i & C\theta_i\Delta\ell_i - r_i S\theta_i\Delta\alpha_i \\ \Delta\theta_i & 0 & -C\theta_i\Delta\alpha_i & S\theta_i\Delta\ell_i + r_i C\theta_i\Delta\alpha_i \\ -S\theta_i\Delta\alpha_i & C\theta_i\Delta\alpha_i & 0 & \Delta r_i \\ 0 & 0 & 0 & 0 \end{bmatrix} \quad (23)$$

As a result, δA_{i-1} has the same form as Equation (8) with three translational errors $\underline{d}^A_{i-1}$ and three rotational errors $\underline{\delta}^A_{i-1}$ caused by the kinematic errors. By defining the following three new vectors

$$\underline{m}^1_{i-1} = [0 \quad 0 \quad 1]^t , \quad (24)$$

$$\underline{m}^2_{i-1} = [C\theta_i \quad S\theta_i \quad 0]^t , \quad (25)$$

$$\underline{m}^3_{i-1} = [-r_i S\theta_i \quad r_i C\theta_i \quad 0]^t , \quad (26)$$

the translational and rotational errors at joint coordinate i-1 due to the kinematic errors at joint i can be expressed in the following linear error equations

$$\underline{d}^A_{i-1} = \underline{m}^1_{i-1} \Delta r_i + \underline{m}^2_{i-1} \Delta\ell_i + \underline{m}^3_{i-1} \Delta\alpha_i \quad (27)$$

$$\underline{\delta}^A_{i-1} = \underline{m}^1_{i-1} \Delta\theta_i + \underline{m}^2_{i-1} \Delta\alpha_i .$$

From the results of above two equations, the error transformation δA_{i-1} can be easily obtained. Hence, the correct relationship between joint coordinates i and i-1 can be expressed as

$$A_i + dA_i = (I + \delta A_{i-1}) * A_i \quad (29)$$

where I is the identity matrix.

KINEMATIC ERROR MODEL OF AN OPEN-LOOP ROBOT MANIPULATOR IN WORLD COORDINATES

The position accuracy of an open-loop, N degrees-of-freedom robot manipulator in the real world depends on the accuracy of four kinematic parameters of every joint. In the previous section, the accurate relationship between joint coordinates i and i-1 due to the kinematic errors is represented by Equation (29). Hence, for an N degrees-of-freedom manipulator, the accurate position and orientation of robot's end-effector with respect to the World coordinate due to the 4N kinematic errors can be expressed as

$$T_N + dT_N = (A_1 + dA_1) * (A_2 + dA_2) * \ldots * (A_N + dA_N) \qquad (30)$$

where dT_N represents the total differential changes at the end of a robot manipulator due to these 4N kinematic errors.

By expanding Equation (30) and ignoring the higher order differential changes, dT_N can be obtained as

$$dT_N = \sum_{i=1}^{N} [A_1 * \ldots * A_{i-1} * \delta A_{i-1} * A_i * \ldots * A_N] . \qquad (31)$$

Using the definition of T_i matrix in Equation (4), above equation can be rewritten as

$$dT_N = \left\{ \sum_{i=1}^{N} \left[T_{i-1} * \delta A_{i-1} * (T_{i-1})^{-1} \right] \right\} * T_N . \qquad (32)$$

By defining an error matrix transformation δT_w with respect to the World coordinates and

$$dT_N = \delta T_w * T_N , \qquad (33)$$

then the error transformation at the World coordinates due to the kinematic errors can be obtained from Equation (32) as

$$\delta T_w = \left\{ \sum_{i=1}^{N} \left[T_{i-1} * \delta A_{i-1} * (T_{i-1})^{-1} \right] \right\} \qquad (34)$$

where δT_w has the same form as Equation (8) and contains three translational errors $\underline{d}^w$ and three orientational errors $\underline{\delta}^w$.

Based on the backward differential relations, Equations (7), (9) and (10) the three translational errors $\underline{d}^w$ and three rotational errors $\underline{\delta}_w$ at the World coordinates can be expressed by the following two equations

$$\underline{\delta}^w = \sum_{i=1}^{N} (R_{i-1} * \underline{\delta}^A_{i-1}) \qquad (35)$$

$$\underline{d}^w = \sum_{i=1}^{N} [(R_{i-1} * \underline{d}^A_{i-1}) + (\underline{p}_{i-1} \times (R_{i-1} * \underline{\delta}^A_{i-1}))] \qquad (36)$$

where R_{i-1} is the rotational matrix and $\underline{p}_{i-1}$ is the positional vector of T_{i-1} defined in Equation (5); $\underline{d}^A_{i-1}$ and $\underline{\delta}^A_{i-1}$ are the six error components of δA_{i-1}.

By substituting Equation (27) and (28) into Equation (35) and (36), the position and orientation errors at World coordinates can be solved as the linear function of the 4N kinematic errors; and

$$\underline{\delta}^W = \sum_{i=1}^{N} [(R_{i-1} * \underline{m}^1_{i-1}) \Delta\theta_i + (R_{i-1} * \underline{m}^2_{i-1}) \Delta\alpha_i] \qquad (37)$$

$$\underline{d}^W = \sum_{i=1}^{N} \{[\underline{p}_{i-1} \times (R_{i-1} * \underline{m}^1_{i-1})] \Delta\theta_i + (R_{i-1} * \underline{m}^1_{i-1}) \Delta r_i$$

$$+ (R_{i-1} * \underline{m}^2_{i-1}) \Delta\ell_i + [(\underline{p}_{i-1} \times (R_{i-1} * \underline{m}^2_{i-1}))$$

$$+ (R_{i-1} * \underline{m}^3_{i-1})] \Delta\alpha_i \} . \qquad (38)$$

For easy expression, above linear results can be grouped into following compact equation

$$\begin{matrix} \underline{d}^W \\ \underline{\delta}^W \end{matrix} = \begin{matrix} W_1 \\ W_2 \end{matrix} \Delta\underline{\theta} + \begin{matrix} W_2 \\ 0 \end{matrix} \Delta\underline{r} + \begin{matrix} W_3 \\ 0 \end{matrix} \Delta\underline{\ell} + \begin{matrix} W_4 \\ W_3 \end{matrix} \Delta\underline{\alpha} \qquad (39)$$

where

$$\Delta\underline{\theta} = [\Delta\theta_1 \ldots \Delta\theta_N]^t , \quad \Delta\underline{r} = [\Delta r_1 \ldots \Delta r_N]^t ,$$

$$\Delta\underline{\ell} = [\Delta\ell_1 \ldots \Delta\ell_N]^t , \quad \Delta\underline{\alpha} = [\Delta\alpha_1 \ldots \Delta\alpha_N]^t ,$$

and $\Delta\theta_i$, Δr_i, $\Delta\ell_i$, $\Delta\alpha_i$ are the errors in the kinematic parameters of the ith joint for $i = 1,2,\ldots,N$; and W_1, W_2, W_3 and W_4 are all a 3 by N matrix whose components are the functions of N joint variables, $\underline{q} = [q_1 \, q_2 \, \ldots \, q_N]^t$ where $q_i = \theta_i$ for a revolute joint and $q_i = r_i$ for a prismatic joint. The ith column of W_1, W_2, W_3 and W_4 can be expressed as follows

$$W_1^i = \underline{p}_{i-1} \times (R_{i-1} * \underline{m}^1_{i-1}) \qquad (40)$$

$$W_2^i = R_{i-1} * \underline{m}^1_{i-1} \qquad (41)$$

A CAD TOOL FOR THE KINEMATIC DESIGN

$$W_3^i = R_{i-1} * \underline{m}_{i-1}^2 \tag{42}$$

$$W_4^i = \underline{p}_{i-1} \times (R_{i-1} * \underline{m}_{i-1}^2) + (R_{i-1} * \underline{m}_{i-1}^3). \tag{43}$$

Observed from the results of Equation (39) that all four kinds of kinematic errors will cause translational errors at the World coordinates, and only two kinds of kinematic errors will cause rotational at the World coordinates. Hence, the errors in the two rotational kinematic parameters have the dominant effect on robot's accuracy. Since the matrices W_1, W_2, W_3 and W_4 are the functions of joint variables such that the six World errors will vary for different joint positions. Equation (39) describes that the impact of the kinematic errors on the accuracy of a robot manipulator in its World coordinates can be represented by a linear equation. Due to these errors, the correct position and orientation of a robot manipulator will be equal to

$$T_N^c = T_N + dT_N = (I + \delta T_w) * T_N. \tag{44}$$

ERROR ENVELOPES OF ROBOT'S POSITION AND ORIENTATION

In order to find out the error envelopes of robot's position and orientation in the real world, a stochastic approach will be applied. A reasonable assumption is that these 4N kinematic errors are 4N independent Gaussian random variables with zero mean. The standard deviation of each kinematic error is the manufactured error of the kinematic parameter or the error in calibrating the zero position of the joint variable. Hence, these four kinds of kinematic errors are four independent, N-variables Gaussian random vectors with zero mean and variances V_θ, V_r, V_ℓ and V_α, respectively. These variances are all an N-by-N diagonal matrix whose components are the standard deviations of kinematic errors.

Due to the property of Gaussian distribution and the relation in Equation (39), $\underline{d}^w$ and $\underline{\delta}^w$ are also Gaussian random vectors with zero mean and variances as

$$\begin{aligned}V_d &= E[(\underline{d}^w - E[\underline{d}^w])(\underline{d}^w - E[\underline{d}^w])^t] \\ &= W_1 V_\theta W_1^t + W_2 V_r W_2^t + W_3 V_\ell W_3^t + W_4 V_\alpha W_4^t,\end{aligned} \tag{45}$$

$$\begin{aligned}V_\delta &= E[(\underline{\delta}^w - E[\underline{\delta}^w])(\underline{\delta}^w - E[\underline{\delta}^w])^t] \\ &= W_2 V_\theta W_2^t + W_3 V_\alpha W_3^t\end{aligned} \tag{46}$$

where V_d and V_δ are 3 by 3 matrices and whose components are the functions of the joint variables.

Based on the above results, a stochastic model described the relationship between the kinematic errors and the error envelopes of robot's position and orientation has been obtained. From this model, the error envelopes of any robot's task can always be predicted by knowing the standard deviations of the kinematic errors.

The trivariable Gaussian density function of $\underline{d}^w$ and $\underline{\delta}^w$ are of the form

$$f(dx^w, dy^w, dz^w)$$
$$= (2\pi)^{-3/2} \|V_d\|^{-1/2} \exp\{-.5[(\underline{d}^w)^t \|V_d\|^{-1}(\underline{d}^w)]\} \quad (47)$$

and

$$f(\delta x^w, \delta y^w, \delta z^w)$$
$$= (2\pi)^{-3/2} \|V_\delta\|^{-1/2} \exp\{-.5[(\underline{\delta}^w)^t \|V_\delta\|^{-1}(\underline{\delta}^w)]\} . \quad (48)$$

With the knowledge of the error standard deviation of each kinematic parameter, the error envelopes of a robot manipulator in the World coordinates can be easily obtained from Equation (47) and Equation (48).

The error envelopes of three independent translational World errors and three independent rotational World errors can also be obtained by rotating the axes of V_d and V_δ into their eigenvectors, i.e., their principal axes. After this transformation, six independent, zero mean, normal random variables $\{v_i; i=1,\ldots,6\}$ with standard deviation $\{\sigma_i; i=1,\ldots,6\}$ can be obtained. v_1, v_2 and v_3 are on the principal axes of V_d; v_4, v_5 and v_6 are on the principal axes of V_δ. The density function of v_i is of the form

$$f(v_i) = (1/(2\pi)^{-1/2} \sigma_i) \exp\{-v_i^2/2\sigma_i^2\} \quad (49)$$

and σ_i is the function of the joint variables.

In order to find out the error bound of robot's trajectory, a statistical method can be applied. Since the probability of $\|v_i\| \leq U$ is that

$$\mathrm{Prob}(\|v_i\| \leq U) = \int_{-U}^{U} f(v_i)\, dv_i$$
$$= 2\,\mathrm{erf}(U/\sigma_i) \quad (50)$$

where erf(·) is the error function of Gaussian distribution and which is defined as

$$\mathrm{erf}(x) = (1/(2\pi)^{-1/2}) \int_0^x \exp(-y^2/2)\, dy . \quad (51)$$

A CAD TOOL FOR THE KINEMATIC DESIGN 299

With the knowledge of the upper bound of the error probability that

$$\text{Prob } (\|v_i\| \leq U) \leq C, \tag{52}$$

then, from the table of erf($\cdot$), there is a constant B satisfied both Equation (50) and Equation (52). In addition, the lower bound of the standard deviation σ_i can be obtained by the following equation

$$\sigma_i \geq U/B. \tag{53}$$

In order to maintain the robot's accuracy within the error bound C in the World coordinates, robot's trajectory at any time must satisfy Equation (53).

In general, based on above technique and the developed error model, the error envelopes of robot's position and orientation can be generated, and the trajectory bound can also be found. Another usage of this error model is that the error envelopes of any variables which is the function of the six errors in the World coordinates, such as error volume, can also be generated. These results are very useful for the kinematic design of a robot manipulator because the designed kinematic parameters can be optimized to satisfy the required accuracy.

CALIBRATION FOR KINEMATIC ERRORS

The accuracy of an open-loop robot manipulator in the real world depends on the accuracy of its kinematics. Due to the manufacturing errors, a designed robot always has kinematic errors in its kinematic parameters. With these errors, a calibration process of a robot manipulator is necessary such that the impact of these kinematic errors can be minimized. In the general case, there are 4N kinematic errors needed to be calibrated in any N degrees-of-freedom manipulator. However, in a special case where a robot manipulator has been designed with precision points on each joint, then the calibration of joint variables can always be achieved by moving each joint to its mechanical precision points. Thus, there are only 3N kinematic errors needed to be calibrated.

Current robot manufacturers do not have the mathematical model to represent these errors. Therefore, they use the primitive measurement schemes to measure these kinematic errors. The disadvantages of the methods are machine dependence and incurring more cost. In addition, due to the tedious calibration procedure, a calibration expert is required to do the job. Hence, the development of an efficient universal calibration scheme is needed.

In this section, a universal calibration scheme will be proposed. The proposed method is based on the developed kinematic error model and using the method of linear regression to solve the kinematic errors. If there are NK kinematic errors then we need at least NK equations. The proposed procedures will be as follows. First, the joint variables can be roughly calibrated by moving the joints to their precision points. Secondly, by moving the robot to a known precise position T_N^c in the World coordinates and calculate the actual robot position T_N from its kinematic parameters, a group of six errors in the World coordinates can be obtained be comparing the actual robot position with this precise known position from the relation of Equation (44). Third, calculate the four matrices W_1, W_2, W_3 and W_4 in our error model from the joint positions and 3N geometric parameters. Hence, for one precise known position in the World coordinates, there are six equations for solving these NK kinematic errors. By moving the robot to P precise positions where NK $\leq$ 6P, these NK kinematic errors can be solved analytically.

After above procedure, a set of new accurate kinematic parameters can be obtained by compensating with these solved kinematic errors. After repeating the procedure several times, these kinematic errors should converge to a minimum.

DESIGN OF A ROBOT MANIPULATOR

Although a calibration scheme can correct the kinematic errors, these errors will cause the joint solutions of a robot manipulator much harder to solve. Hence, the best way to improve the accuracy of a robot manipulator is to design it properly to ensure that the robot has the minimum errors. The results of a proper design cannot only increase the robot's accuracy, but also simplify the calibration method.

In order to preserve the fidelity of a robot manipulator in the real world, the design of the kinematic parameters has to be optimized. In this section, we will discuss how to minimize the errors of an open-loop manipulator in the World coordinates.

Observed from the results of the developed kinematic error model that three World translational errors $\underline{d}^w$ are dominated by the errors in the two rotational kinematic parameters $\underline{\theta}$ and $\underline{\alpha}$ due to the error terms consist of the position vector $\underline{p}_{i-1}$ of the T_{i-1} matrix; and the World rotational errors $\underline{\delta}^w$ are only affected by these two kinds of parameters. Thus, if the kinematic parameters $\underline{\theta}$ and $\underline{\alpha}$ have very high precision then the World errors of an open-loop robot manipulator can be reduced to a minimum.

In order to design a robot manipulator with high accuracy, our error model can be applied as a CAD tool to optimize the kinematic parameters and increase the robot's accuracy. Since the error envelopes of a robot manipulator depend on the standard deviations of kinematic errors such that, with the constraint of the World errors to certain error bounds, the maximum manufacturing error tolerances of the designed kinematic parameters can be decided. By setting the upper bound of the error envelopes for the designed manipulator, different sets of manufactured error tolerances can be tested and the World error envelopes of the whole working space of the manipulator can be generated. By this procedure, the maximum manufacturing tolerances of the kinematic parameters can be obtained. In addition, if the standard deviations of the kinematic errors are known then the accuracy of the designed manipulator can be tested for different set of kinematic data. Based on this method, the optimized kinematic data for the designed manipulator can be obtained such that the requirements of robot's accuracy can be satisfied.

CONCLUSION

The accuracy problem of a robot manipulator has long been one of the principal concerns of robot control. Although in author's previous paper [2], a linear error model between the Cartesian errors of the end-effector and the kinematic errors has been developed and the kinematic data can be optimized. However, for the purpose of easy understanding and user friendly of a CAD tool, a kinematic error model which is represented in the World coordinates is necessary.

In this paper, a simple linear error model between the kinematic errors of a robot manipulator and the errors of robot's position and orientation in the World coordinates has been developed in a straightforward manner from the kinematic equations. The method is based on the backward differential relation between two coordinates. By using the error transformation matrix, the errors at each joint can be transformed back to the World coordinates of the manipulator.

Based on this model, a simple calibration scheme has been proposed to improve the accuracy of present robots. In addition, the error envelopes of a robot manipulator in World coordinates caused by any combination of the four kinds of kinematic errors has been generated through a stochastic approach. Thus, this error model can be used to minimize the errors of an open-loop manipulator by proper design of the kinematic parameters. For example, by knowing the manufacturing error tolerances, the error envelopes of a robot manipulator in the World coordinates for the designed kinematic parameters can be generated by this model. If the generated error envelopes were larger than the required accuracy then the designed kinematic parameters can be redesigned to satisfy the requirements. Hence, we can always determine that whether a robot manipulator with the designed kinematic parameters will meet the accuracy requirements. In general, the developed model can be used as a CAD tool to optimize the kinematic parameters of a robot manipulator and increase the robot accuracy.

REFERENCES

[1] J. Denavit and R. S. Hartenberg, "A kinematic notation for lower-pair mechanisms based on matrices," ASME J. Appl. Mech., pp. 215-221, June 1955.

[2] C. H. Wu, "The kinematic error model for the design of robot manipulator," 1983 American Control Conference, San Francisco, CA, June 22-24, 1983.

[3] K. J. Waldron, "Positioning accuracy of manipulators," Proc. of NSF Conf., held at University of Florida, pp. 111-141, Feb. 1978.

[4] A. Kumar and K. J. Waldron, "Numerical plotting of positioning accuracy of manipulators," Mechanism and Machine Theory, Vol. 16, No. 4, pp. 361-368, 1981.

[5] C. H. Wu, "Force and position control of robot manipulator," Ph.D. thesis, Purdue University, Dec. 1980.

[6] R. Paul, "Robot Manipulators: Mathematics, Programming, and Control," The MIT Press, 1981.

[7] B. E. Shimano, "The kinematic design and force control of computer controlled manipulators," Ph.D. thesis, Stanford University, March 1978.

Computer Vision and Image Processing

A GRAPH-THEORETIC APPROACH TO 3-D OBJECT RECOGNITION AND ESTIMATION OF POSITION AND ORIENTATION

E. K. Wong and K. S. Fu

School of Electrical Engineering
Purdue University
West Lafayette, IN   47907   U.S.A.

ABSTRACT

   This paper describes a new technique for modeling and recognition of 3-D objects. 3-D objects are modeled as graphs where the nodes correspond to object vertices and the branches correspond to object edges. Allowable junction types at each object vertex are assigned as the property at the corresponding node of the model graph. Allowable junction type combinations of the neighboring vertices at a given vertex are established as constraints required at the corresponding node of the model graph. The 2-D projections of an object are modeled as subgraph isomorphisms of the model graph. Recognition is by searching if a 2-D projection graph is a subgraph isomorphism of the model graph, and each node in the projection graph also satisfies the constraints at the node of the model graph matched. Techniques are described for the estimation of position and orientation of a 3-D object based on the proposed scheme. Experiments are conducted to evaluate the performance.

INTRODUCTION

   In machine recognition of 2-dimensional patterns and shapes, features are extracted and subject to statistical classification; or primitives are extracted and subject to syntax analysis. The former approach is called statistical or decision-theoretic pattern recognition [1] and the later approach is called syntatic pattern recognition [2]. Recognition techniques which have been well developed in both approaches cannot be applied directly to the recognition of 3-dimensional objects or patterns. The difficulty lies in the fact that the 2-D projections of a 3-D object is viewpoint dependent. Features or primitives which are invariant to rotation, translation, scaling and perspective transformations cannot be easily extracted. Other approaches must therefore be sought in the recognition of 3-D objects.

One approach is based on using the silhouette of a 3-D object as a basis in recognition. This method is effective only in the recognition of classes of simple objects, in which objects can be discriminated from each other by their silhouettes alone. Another approach is based on multiple-view representations [3,4] of a 3-D object. The object's line structure is characterized in terms of a variety of two-dimensional features [5]. Shape and feature matching techniques, invariant to rotation, translation and scaling are used for identification [3,4,6]. This approach requires stringent positioning of the object with respect to the camera and therefore is not suitable for general purposes. Other efforts have been directed towards extracting spatial relationships and 3-D features from an object's line drawing projections [7,8,9,10].

In all recognition schemes, the computer stores canonic models of the 3-D objects to be recognized. The process of recognition is to match the input 2-D image against the stored models for the best fit. The canonic models stored by the computer can be categorized as 2-D or 3-D in nature. 2-D models are characterized by sets of silhouettes, set of characteristic views [5], set of symbolic sentences [11] and multiple-view representation [3,4]. 3-D geometric models are characterized by (1) volumetric representations based on spatial occupancy, cell decomposition or constructive solid geometry; (2) surface representations utilizing bicubic patches, quadric-surface patches and others; and (3) generalized cylinder representations [12]. However, not all of the modeling schemes are suitable for recognition; especially 3-D geometric models such as those mentioned above. The intrinsic difficulty lies in the fact that a 3-D model takes on an infinite number of possible 2-D projections.

To resolve the infeasible task of matching an input 2-D projection against an infinite number of 2-D projections of a 3-D model, various other approaches have been attempted. One approach is characterized by the classical work of Roberts [13]. In Roberts' work, line drawing of the 2-D image is first extracted. A list of two-dimensional features is associated with each model. For example, in the line drawing of a cube as in Figure 1, the interior polygons A, B and C correspond to unoccluded faces of the object. The point, P, surrended by 3 unoccluded regions constitute a feature to be searched for in the input image. Once the feature point is found, other corresponding points are established between the input picture and the model. Transform between the model and the object is found from these points and final match is verified by projecting the model onto 2-D plane by the found transform. Features other than that in the above example could be used. Other

3-D OBJECT RECOGNITION AND ESTIMATION

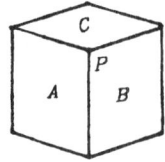

Figure 1. Line drawing of a picture of a cube.

features include a line surrended by unoccluded polygons (such as the line between polygons B and C for the cube example); a single unoccluded polygon and others. This feature matching method is effective only if the two dimensional features associated with a model is visible. Also, in the case of a complex object, it is sometimes difficult to isolate a feature in a 2-D projection which corresponds to a single unique part of the object (i.e., many different parts of an object may project to give rise to the same 2-D feature); therefore, complicating the correspondence search between the 2-D projection and the 3-D model. In the case of many models to be matched, a large set of different models may result in the same 2-D features. This approach is therefore not general for the above reasons.

Another representative approach is that of Charkravarty and Freeman [5] in which the space of all possible perspective projections of an object is factored into a set of characteristic views, where each such view defines a characteristic-view domain within each all projections are topologically identical and related by a linear transformation. The characteristic views of an object can then be hierarchically structured for efficient classification. This approach suffers in storage space requirement when the number of models to be matched is large, as there may be a large set of characteristic views even for a simple object [5]. In fact, objects are restricted to be in stable positions in [5] to limit the size of the projection set. Stable positions are the positions in which the object would come to rest if thrown on a flat, horizontal surface. This assumption is invalid in an uncontrollable environment where the supporting surface is not flat, or when there are other irregular objects between the object of interest and the supporting plane.

In both approaches discussed above, those of Roberts [13] and Chakravarty and Freeman [5], there does not exist a systematic method in extracting the necessary information for matching, given the geometric model. Roberts [13] did not describe a general method to isolate features useful for recognition given any 3-D object. Apparently, no single feature is effective for all objects in general and no single feature can be found in all objects. Chakravarty and Freeman [5] did not provide a method to exhaustively identify all possible characteristic views of an object. Certainly, in a fully automated environment, we do not expect to have a human observer to determine all the characteristic views of an object and for all objects in the recognition domain. Finally, both approaches are not susceptible to rigorous time and space complexity analysis.

In the next section, we propose and describe a systematic method of modeling 3-D objects by undirected graphs and pose the matching problem as a subgraph searching problem. Following sections discuss searching in the case of imperfect line drawing, camera transform estimation and hypothesis verification, estimation of object position and orientation, and experimental results. Finally, the last section gives concluding remarks.

GRAPH-THEORETIC MATCHING OF 3-D MODELS

In our approach, an input image is first preprocessed to extract its 2-D line drawing. Methodology for extracting line drawing from a 2-D image can be found in [16].

3-D Model Graphs

The domain of 3-D objects we are dealing with are planar-faced or curved-surface solid bodies, having vertices formed by the intersection of at most three surfaces, and edges formed by the intersection of two surfaces. We define a virtual edge to be a locus of points of tangency on a surface with the lines of projection (as shown in Figure 2). Throughout this paper, we differentiate those edges formed by the intersection of two surfaces from the virtual edges by calling the former physical edges. Quantitatively, the two types of edges can be differentiated on the basis of surface normal orientation. A physical edge is a locus of points at which there are discontinuities in surface normal orientations. A virtual edge does not show discontinuities in surface normal orientation. A virtual edge (limb) may not pass through a vertex [14]. In a 2-D projection of a scene, called an image, bodies are projected as objects, physical edges and virtual edges as lines, and vertices as junctions. We define a virtual junction to be a junction at which at least one of the lines forming the junction is a projection of a virtual edge. A face is a portion of a surface bounded by physical edges or closed on itself. A region is a connected visible part of the projection of a face; a face may correspond to zero, one, or more regions [15].

We adopt the generalized line and junction labeling scheme developed by Chakravarty [15] for planar-faced or curved-surface solid bodies as our basis in labeling junctions in a 2-D line drawing. The labeling scheme may be considered as an extension to the Hoffman-Clowes [19,20] scheme for polyhedral bodies. Six classes of generalized junctions types were identified by Shapira [14] as

3-D OBJECT RECOGNITION AND ESTIMATION

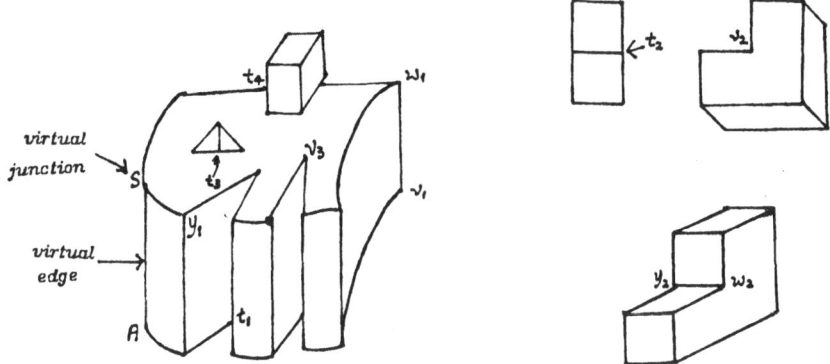

Figure 2. Labelling with generalized junction types.

shown in Figure 3. Note that class A and class S are virtual junctions classes. Class W, class Y, class T and class V junctions can be subcategorized based on the number of regions associated with the junction. We thus define 13 generalized junction types as shown in Figure 4. Figure 2 depicts a line drawing labeled with these junction types.

We model a 3-D object as an undirected graph where a node of the graph correspond to a vertex in the 3-D object and a branch connecting two nodes corresponds to an edge between two vertices in the 3-D object. The edge may be a straight physical edge or a curved physical edge. The model graph thus constructed therefore captures the topology of how vertices are connected by edges in the 3-D object. An example is as shown in Figures 5(a) and (b). Note that each vertex in the 3-D object correspond to a unique node in the graph and each edge corresponds to a unique branch in the graph. We thus have the following two lemmas.

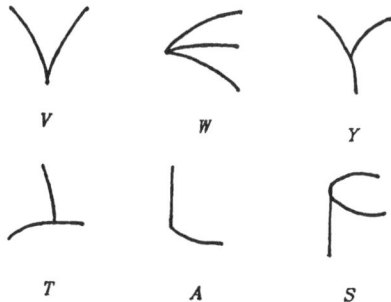

Figure 3. Generalized junction classes.

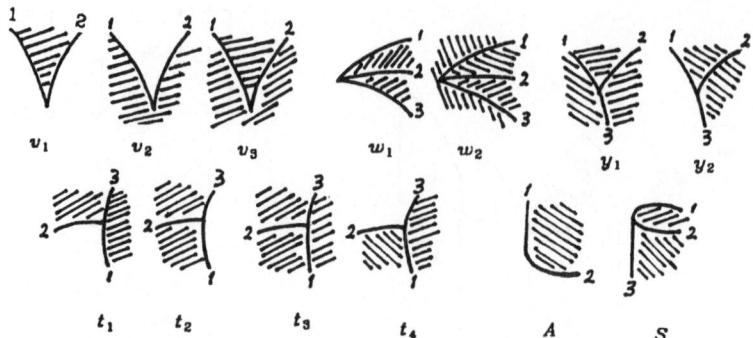

Figure 4. Generalized junction types and ordering convention for neighbor-constraint-list.

Lemma 1:

The total number of nodes in a model graph represent the total number of vertices in the 3-D objects it represents.

Lemma 2:

The total number of branches in a model graph represents the total number of edges in the 3-D objects it represents. An edge may be a straight or curved physical edge.

Since we are dealing with objects whose vertices are formed by the intersection of three surfaces, and edges formed by the intersection of two surfaces. The number of edges meeting at a vertex must be exactly three. Therefore, the following lemma holds.

Lemma 3:

The number of branches emerging from a model graph node is exactly three.

Assume that each node of a model graph is initially not connected to each other by branches and that each has three open branches attached to it. Therefore, an odd number of nodes gives rise to an odd number of total branches and an even number of nodes gives rise to an even number of total branches. Now, to form a model graph, two open branches from two different nodes must be connected to form a single branch. Therefore, the total number of branches must be initially even for no open branch to be left after the connecting process. Consequently, the total number of nodes must be initially even and we have the following two lemmas.

Lemma 4:

The total number of nodes in a model graph is even.

3-D OBJECT RECOGNITION AND ESTIMATION

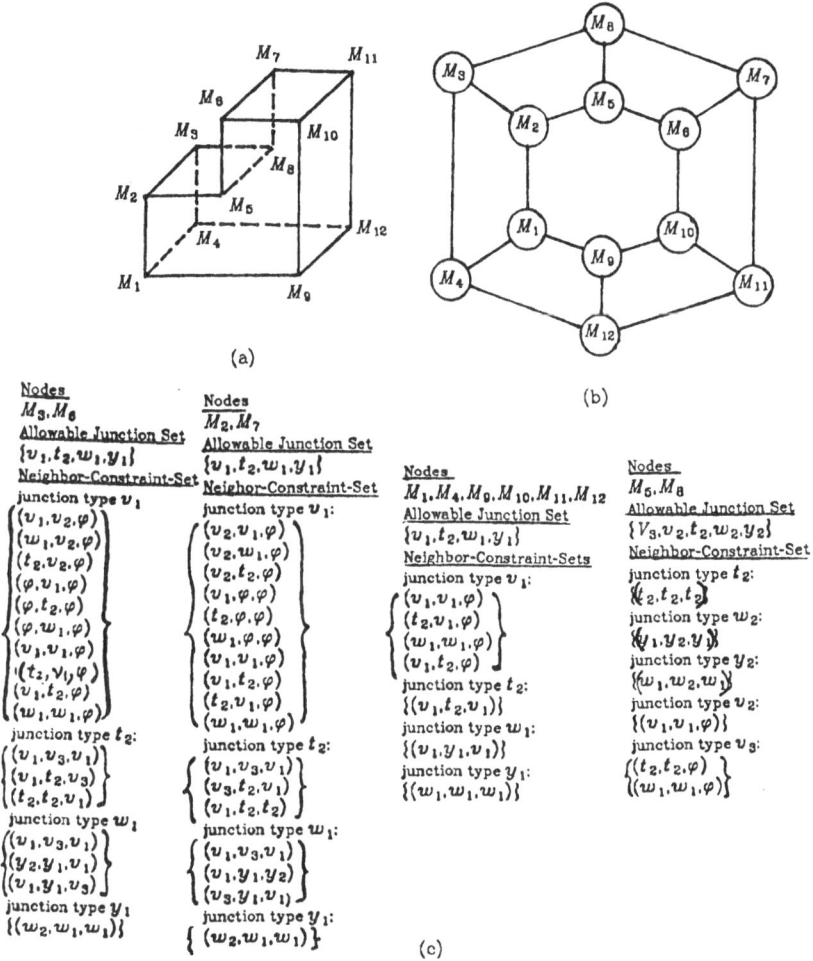

Figure 5. A 3-D object, its model graph, allowable junction sets and neighbor-constraint-sets.

Lemma 5:

The total number of branches in a model graph is $\frac{3m}{2}$ where m is the total number of nodes in the model graph.

For curved objects with convex surfaces, virtual edges and virtual junctions may occur in the 2-D projections of a 3-D object. As they do not correpond to actual physical edges and physical junctions, we do not create nodes or branches in the model graph for them. Figure 6(a) shows a curved object and Figure 6(b) is its model graph.

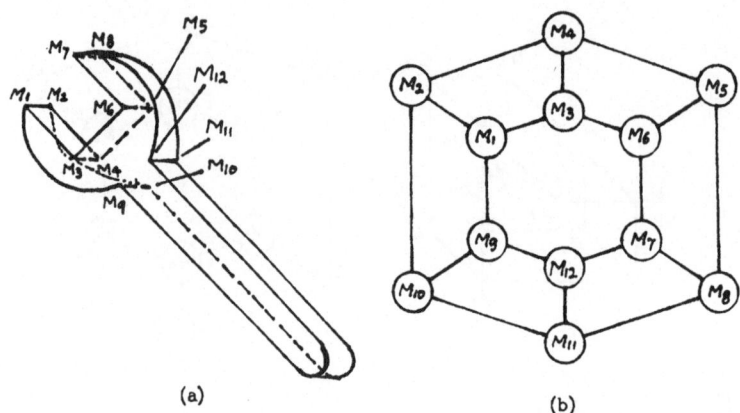

Figure 6. A convex-curved-object 'wrench' and its model graph (allowable junction types and neighbor-constraints-sets not shown).

Junction Types Constraint Propagation

In this section, constraints for the types of possible junction projections allowable between neighboring vertices of a 3-D object are expressed as the constraints between neighboring nodes of the model graph.

In the analysis that follows, let us assume that the viewpoint will be in such a position that no accidental alignment of a vertex and an edge will occur (see Figure 7). That is, when projected on a 2-D plane, all lines meeting at a junction are actually projections of physical or virtual edges meeting in 3-D space. This assumption is valid when we take stereo images of a scene, since accidental alignment at a given vertex will not occur at both positions of the stereo cameras. We define a junction projection to be a 2-D projection of a 3-D vertex. We define the neighboring vertices

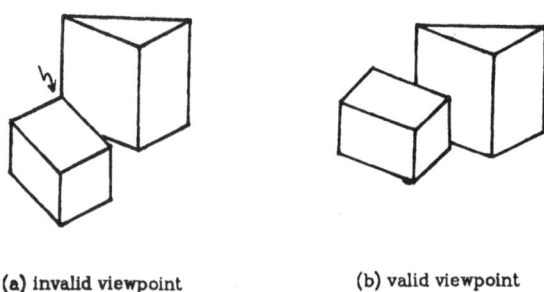

(a) invalid viewpoint   (b) valid viewpoint

Figure 7. Accidental alignment of vertex and edge.

of a given vertex to be the vertices which are connected by an edge
to the given vertex. We define the neighboring junctions of a given
junction to be the junctions which are connected to the given junction by a line in a 2-D line drawing. And we define the neighboring
nodes of a given node to be the nodes which are connected by a
branch to the given node in a model graph. We define the allowable
junction set for a vertex to be the set of all possible junction
types which can occur in the 2-D projections of the vertex. Each
member in the set is called an allowable junction type. As an
example, the set of all possible junction types for vertex M_1 in
the object of Figure 5(a) is the set $\{v_1, w_1, y_1, t_2\}$ as shown in
Figure 8. The neighboring vertices of M_1 are M_2, M_4 and M_9 are as
shown in Figure 9. Since M_2, M_4 and M_9 are connected to M_1 by
edges, each of them bears a fixed spatial relationship to M_1,
regardless of the actual spatial orientation and position of M_1.
Therefore, in a 2-D projection of the 3-D object, the types of
junction projections occur for vertices M_2, M_4 and M_9 are constrained by the type of junction projections occur for vertex M_1.

Let us denote a junction which is hidden (occluded) on a 2-D
projection by the symbol '𝔍' and include this in our set of junction types. Vertex M_1, at viewpoints VP_1, VP_2, VP_3 and VP_4 as
indicated in Figure 10(a), will have a corresponding v_1 junction
and 4 different combinations of junction types for vertices M_2, M_4
and M_9, as shown in Figure 10(b). We call each of these combinations
neighbor-constraint-list. All other viewpoints at which vertex M_1
projects into a v_1 junction type will have combinations of junction
types for vertices M_2, M_4 and M_9 to be one of these 4 lists. Note
that the neighbor-constraint-list is an ordered list of three elements, for which conventions are defined to order the elements of
the list for each junction type. For a V type junction, the ordering convention is such that if we orient the V junction to resemble
the letter 'V', then the neighboring junction connected to the left
line forming the V junction would be ordered first and the neighboring junction connected to the right line would be ordered second.
The third element of the list is a hidden junction type and will be
ordered last. Conventions we adopted for the other junction types
are as indicated in Figure 4. We define the set of all distinct
neighbor-constraint-lists associated with a particular junction type
x at a given vertex to be the neighbor-constraint-set for junction
type x at the given vertex. Therefore, for each allowable junction
type at a vertex we have a neighbor-constraint-set associated with

Figure 8. Possible junction types for vertex M_1 in Fig. 5(a).

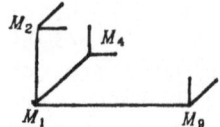

Figure 9. Neighboring vertices of M_1 for the 3-D object in Fig. 5(a).

the junction type. The allowable junction types and their neighbor-constraint-sets for vertex M_1 is as shown in Table 1. We call the allowable-junction-types at a given vertex the property of the corresponding node in the model graph, and we call the neighbor-constraint-sets the neighbor-constraints of the node. The above analysis is repeated for each vertex of a 3-D object and we generate a model graph for the object, where each node of the graph has its property of allowable junction types and neighbor-constraints. The complete model graph for the 3-D object in Figure 5(a) is shown in Figures 5(b) and (c).

Projection-Graphs

In this section, the 2-D projections of a 3-D object are modeled as graphs, and they can be shown to be subgraph isomorphisms

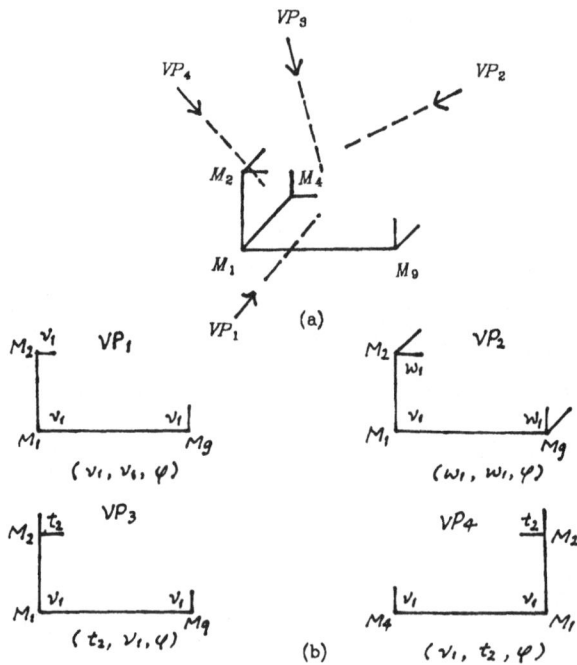

Figure 10. Junction constraint propagation.

Table 1. Allowable Junction Types and Neighbor-Constraint-Sets for Vertex M_1 in Fig. 5(a)

Allowable Junction Set
$\{v_1, t_2, w_1, y_1\}$

Neighbor-Constraint-Sets
junction type v_1:
$$\left\{\begin{array}{l}(v_1, v_1, \varphi)\\(t_2, v_1, \varphi)\\(w_1, w_1, \varphi)\\(v_1, t_2, \varphi)\end{array}\right\}$$
junction type t_2:
$\{(v_1, t_2, v_1)\}$
junction type w_1:
$\{(v_1, y_1, v_1)\}$
junction type y_1:
$\{(w_1, w_1, w_1)\}$

of the model graph. Model matching then becomes the problem of searching whether an input graph is a subgraph isomorphism of the model graph.

We define a projection graph to be a graph constructed from the 2-D line drawing of a 3-D object by the following procedure. We create a node for each junction in the 2-D line drawing which is a projection of a vertex in the 3-D object. The junction type that occurs is incorporated as a property of the node. We then connect two nodes by a branch if their corresponding junctions are connected by a line in the line drawing. Note that we do not create a node for junction types t_1, t_3 and t_4 as they are the projections of one edge occluding the other in the 3-D space and therefore is not a projection of a vertex in 3-D space. We use an open arc to represent the occluded line and do not connect it to any node as it is actually connected to a hidden junction. We do not create nodes for type A and type S virtual junctions as they do not correspond to any nodes in the model graph. We do not create a branch for a line connecting two virtual junctions as it is a virtual edge. A projection of the 3-D object in Figure 5(a) is as shown in Figure 11(a) and its projection graph is shown in Figures 11(b) and 11(c). Also, a projection of the curved object in Figure 6(a) is as shown in Figure 12(a), and its projection graph is as shown in Figures 12(b) and 12(c).

We observe that two vertices connected by an edge in 3-D space project into two junctions connected by a line in 2-D space. The projection from 3-D space to 2-D space does not change the topology of how the vertices or the junctions are connected. The projection only hides some of the lines and junction in 2-D space. These are called hidden lines and hidden junctions in computer graphics literature [17]. For example, vertices M_4 and M_8 of the 3-D object in Figure 5(a) are projected as hidden junctions invisible in the 2-D line drawing of Figure 11(a). If in fact they are visible, then we are able to reconstruct the model graph from the 2-D projection.

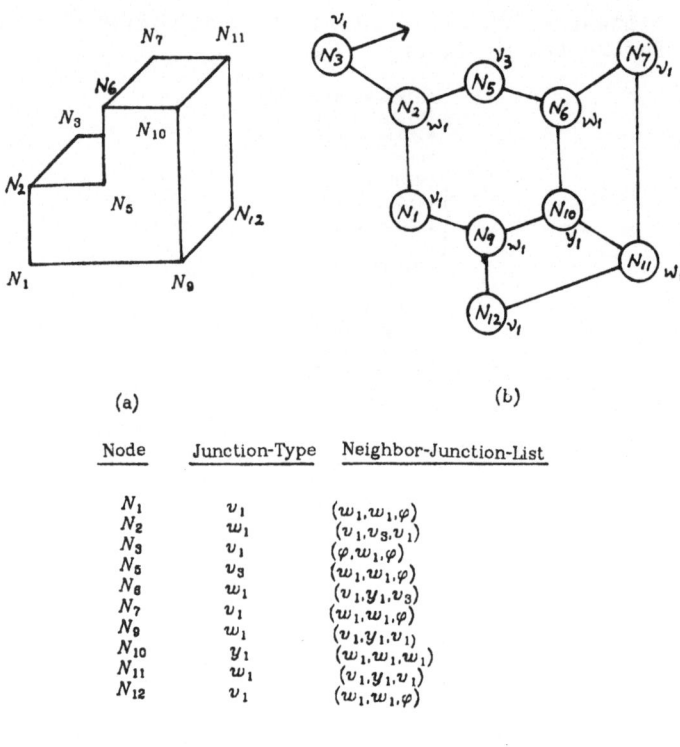

Node	Junction-Type	Neighbor-Junction-List
N_1	v_1	(w_1,w_1,φ)
N_2	w_1	(v_1,v_3,v_1)
N_3	v_1	(φ,w_1,φ)
N_5	v_3	(w_1,w_1,φ)
N_6	w_1	(v_1,y_1,v_3)
N_7	v_1	(w_1,w_1,φ)
N_9	w_1	(v_1,y_1,v_1)
N_{10}	y_1	(w_1,w_1,w_1)
N_{11}	w_1	(v_1,y_1,v_1)
N_{12}	v_1	(w_1,w_1,φ)

(c)

Figure 11. A 2-D projection, its projection graph and neighbor-junction-list.

Since every node in a projection graph corresponds to a unique node in its model graph and every branch connecting two nodes in a projection graph corresponds to a unique branch connecting the corresponding nodes in its model graph, we have the following lemma.

Lemma 6:

The projection graph constructed from a 2-D projection of a 3-D object is a subgraph isomorphism of the model graph constructed from the 3-D object.

Note that the number of branches emerging from a projection graph is at most 3 and the total number of nodes in a projection graph could not be larger than the total number of nodes in its model graph. In fact, we can prove a stronger condition on the relationships between the number of nodes in a projection graph and the number of nodes in its model graph as follows.

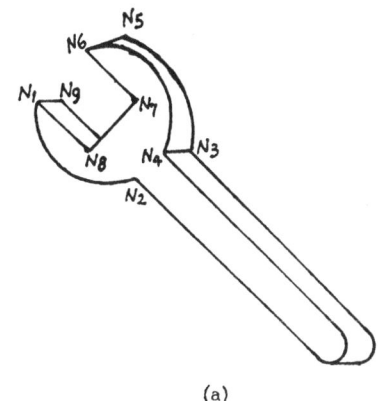

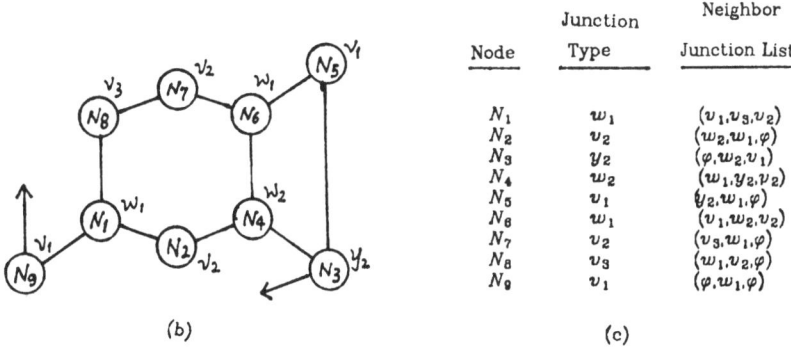

Figure 12. 2-D projection of a convex-curved object 'wrench', its projection graph, and neighbor-junction-list.

By Lemma 4, a model graph has at least the same number of nodes n in its projection graph if n is even; or the model graph has at least n+1 nodes if n is odd. The minimum number of branches in a model graph is thus $3\lceil \frac{n}{2} \rceil$ by Lemma 5. Let us denote the total number of branches (not including open arcs) in a projection graph by B. The number of hidden line segments, L_{hid}, (an occluded line is considered a hidden line segment since part of it is hidden) is then bounded below as

$$L_{hid} \geq 3\lceil \frac{n}{2} \rceil - B \qquad (1)$$

Since at most 3 hidden lines are needed to form a junction, the number of hidden junctions is bounded below by

$$J_{hid} \geq \lceil \frac{L_{hid}}{3} \rceil = \lceil \frac{n}{2} \rceil - \lfloor \frac{B}{3} \rfloor \qquad (2)$$

For example, in the projection graph of Figure 11(b),

$$L_{hid} \geq \left\lceil 3\frac{10}{2} \right\rceil - 12 = 3$$

and

$$J_{hid} \geq \left\lceil \frac{3}{3} \right\rceil = 1$$

In fact, we have 6 hidden lines and 2 hidden junctions. Since the number of visible junctions plus the minimum number of hidden junctions must be less than or equal to the number of nodes in the model graph, the following lemma holds.

Lemma 7:

The number of nodes n in a projection graph and the number of nodes m in its model graph satisfies

$$n + \left\lceil \frac{n}{2} \right\rceil - \left\lfloor \frac{B}{3} \right\rfloor \leq m \qquad (3)$$

where B is the total number of branches in the projection graph not including the open arcs.

Model Matching as Subgraph Searching

Since all possible 2-D projections of a 3-D object is a subgraph isomorphism of the model graph, it is necessary that an input projection graph be a subgraph of the model graph for the input projection to match the model. Figure 13 gives an example for a cube, where the projection graphs for all the characteristic views of the cube are shown to be subgraph isomorphisms of its model graph. Note that this is not a sufficieint condition since 2 2-D line drawings may have the same projection graph, yet representing different objects. That is, a projection graph could match more than one model graph. The projection graph conveys only information about the topology of how vertices and edges are connected in a 3-D object. We thus have the following lemma.

Lemma 8:

The subgraph isomorphism of a projection graph to a model graph is a necessary but not sufficient condition for a 2-D projection to match a 3-D object. We call this a topological match.

3-D OBJECT RECOGNITION AND ESTIMATION

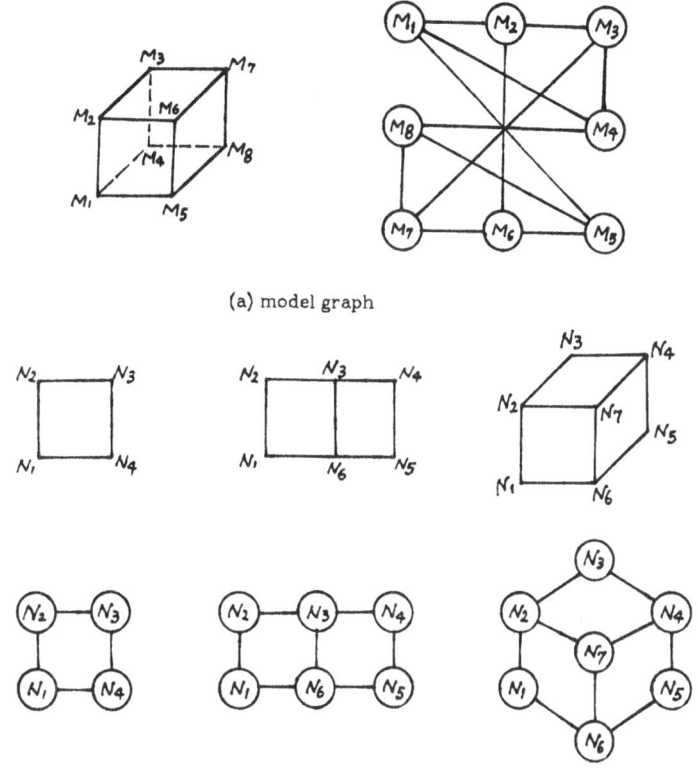

(a) model graph

(b) projection graphs

Figure 13. Subgraph isomorphism of the projection graphs for a cube.

For final verification, the 3-D geometric model must be projected onto the 2-D plane by an estimated camera transform between the input 2-D line drawing and the 3-D geometric model. The projected image is then compared with the input 2-D line drawing. We call this hypothesis verification. Model matching then becomes two sub-problems, topological matching and hypothesis-verification. Topological matching becomes a problem of searching for subgraph isomorphism between a projection graph and a model graph in our approach. Known algorithms for subgraph searching can be found in literature [18]. The time complexity can be as high as M^N in the general case where M is the number of nodes in the bigger graph and N is the number of nodes in the smaller graph. However, we are dealing with a special case of the subgraph searching problem. Since each node in the model graph has only 3 branches, and we further allow junction types propagation between neighboring nodes to serve as constraints in the matching process, the time complexity is greatly reduced.

Similar to the constructions of model graphs, we define the neighbor-junctions-list for each node in the projection graph to be an ordered list which consists of the junction types of its neighboring junctions. The ordering conventions follow those in Figure 4. A ' ' is used to denote a hidden junction. A node i in the projection graph is said to satisfy the neighbor-constraints at a node j in the model graph if the junction type x of node i is an allowable junction type of node j and the neighbor-junctions-list for node i is in the neighbor-constraint-set for junction type x at node j. In order for a node i in the projection graph to match a node j in the model graph, this constraint must be satisfied.

The algorithm for subgraph searching proceeds as follows. We label the projection graph nodes from N_1 to N_n and the model graph nodes from M_1 to M_m. We try to match node N_1 to one of nodes M_1 to M_m at which the neighbor-constraints are satisfied. Then we try to match the neighboring nodes of N_1 to the neighboring nodes of the node matched to N_1 in the model graph. And for each node N_i matched in the projection graph, we try to match its neighboring nodes to the neighboring nodes of the node matched to node N_i in the model graph. We repeat this process until all nodes in the projection graph are matched to some nodes in the model graph. If an error occurs, we backtrack until all possible matching combinations are exhausted.

We analyze the time complexity of the algorithm as follows. There are at most m choices for N_1, at most 3 choices for the first neighboring node of N_1, at most 2 choices for the second neighboring node of N_1 and 1 choice for the third neighboring node. The remaining nodes other than N_1 has at most 2 choices for its first neighboring node and 1 choice for its second neighboring node since one of the neighboring nodes was matched previously. This is as depicted in Figure 14. The time complexity is therefore

$$t = O(m2^{\lceil \frac{n}{2} \rceil}) \tag{4}$$

It is expected that a highly regular and symmetric object takes longer time in searching than an irregular object as the constraint propagations between neighboring nodes take greater effect in an irregular object.

Search Time Speed Up

Since time is a crucial factor in many practical applications, it is desirable to speed up the search process. The procedures described below do not improve the time complexity but substantially reduce the matching time in many instances.

The first procedure is by use of Lemma 7. We count the number of nodes and branches in the projection graph and the number of

3-D OBJECT RECOGNITION AND ESTIMATION

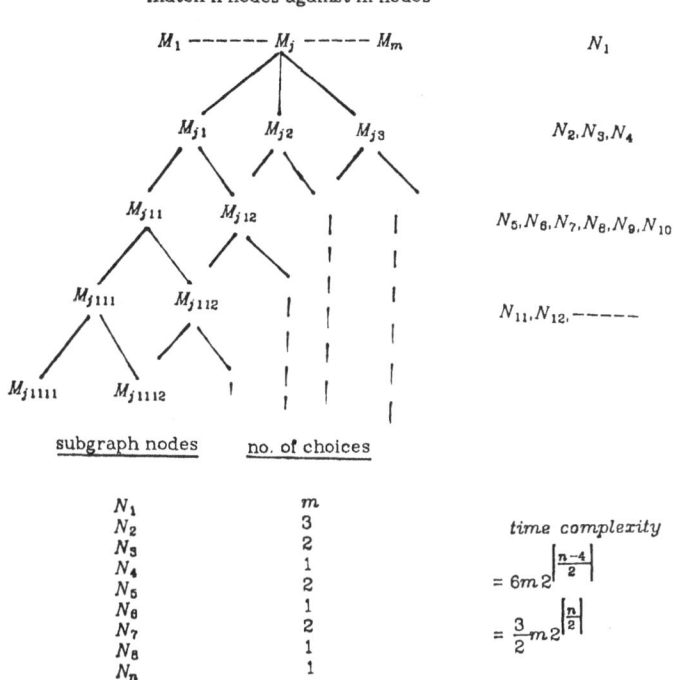

Figure 14. Time complexity of searching algorithm.

nodes in the model graph (this can be done once and the values stored). If equation (3) is not satisfied, we conclude that the projection does not match the model without going into subgraph searching.

The second procedure is to hierarchically rank the allowable junction types in the model graph. A junction type which is an allowable junction type in a less number of model graph nodes is ranked a higher precedence than a junction type which is an allowable junction type in a higher number of model graph nodes. The ranking of junction types in the model graph of Figures 5(b) and (c) is shown in Table 2. Note that junction types y_2, v_2, v_3, and w_2 are ranked the highest as they can occur only at two nodes (nodes M_5 and node M_8). In subgraph searching, the projection graph is first searched for the highest precedence junction, attempts are then made to match it with the nodes in the model graph at which these junction types could occur. This considerably reduces the search space. This method has a similarity with Roberts' method in which a most effective feature is searched first in the 2-D projection.

These two procedures can be incorporated as initial steps in a recognition scheme as in the block diagram of Figure 15. This

Table 2. Ranking of Junction Types for the Model Graph in Fig. 5(b) and (c)

Junction Types	Nodes at Which it Occurs	Number of Occurrences	Rank
w_2, y_2, v_2, v_3	M_5, M_8	2	1
y_1, v_1, w_1	$M_1, M_2, M_3, M_4, M_6, M_7,$ $M_9, M_{10}, M_{11}, M_{12}$	10	2
t_2	$M_1 \rightarrow M_{12}$	12	3

scheme has the effect of trying to optimize the model searching strategy in terms of time complexity.

As an example, we demonstrate the subgraph searching of the 2-D projection in Figure 11(a) against the model graph in Figure 5(b). We construct the projection graph for the 2-D line drawing as in Figure 11(b), and the neighbor-junction-list for each node in the projection graph as in Figure 11(c). The number of branches (excluding the open arc emerging from node N_3) in the projection graph is 12 and the number of visible junctions n is 10. The number of nodes m in the model graph is 12. Applying Lemma 7, we obtain

$$n + \left\lceil \frac{n}{2} \right\rceil - \left\lfloor \frac{B}{3} \right\rfloor \leq m$$

$$10 + \left\lceil \frac{10}{2} \right\rceil - \left\lfloor \frac{12}{3} \right\rfloor \leq 12$$

$$10 + 5 - 4 \leq 12$$

$$11 \leq 12$$

Thus, the first procedure is satisfied. The second procedure is then applied by searching for the highest ranking junction types in the projection graph. From Table 2, the highest ranking junction types are w_2, y_2, v_2, and v_3, and they can occur only at nodes M_5 and M_8 of the model graph. Of the four junction types, only junction type v_3 occurs at node N_5 in the projection graph. Thus, node N_5 is matched against nodes M_5 and M_8. Suppose node N_5 is matched against node M_5 first, there is a match since the junction type of N_5, which is v_3, is an allowable junction type for node M_5 and the neighbor-junction-list for node N_5, (w_1, w_1, φ), is in the neighbor-constraint-set for junction type v_3 at node M_5 (see Figure 5(c)).

3-D OBJECT RECOGNITION AND ESTIMATION

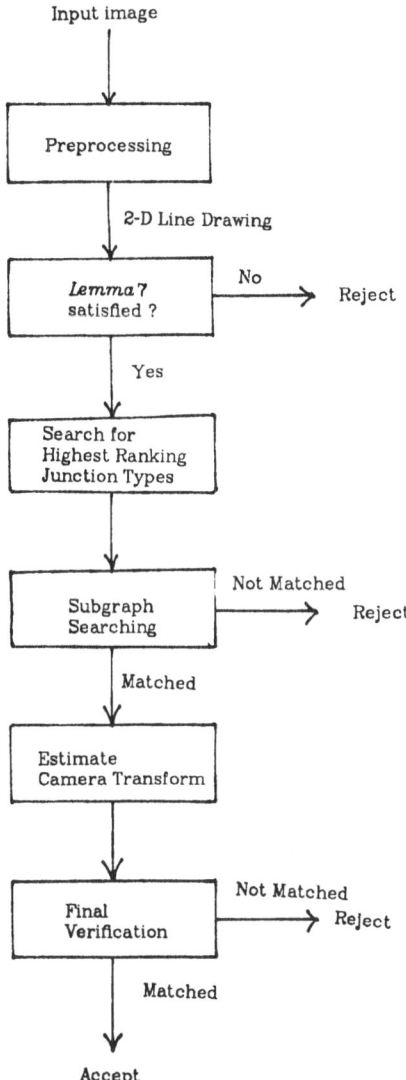

Figure 15. Block diagram of a recognition scheme.

Next, the neighboring node of node N_5, node N_2, is matched against node M_6. There is a match. Next, the other neighboring node of node N_5, node N_6, is matched against node M_2, there is a mismatch since the neighbor-junction-list for node N_6, (v_1, y_1, v_3), is not in the neighbor-constraint-set for junction type w_1 at node M_2. Therefore, the algorithm backtracks and matches node N_2 against node M_2 and node N_6 against node M_6 and there is a match. The algorithm proceeds as above until all nodes in the projection graph are matched to some nodes in the model graph.

Parallel Search Algorithm

For 2-D projections with up to 50 junctions (the projection in Figure 11(a) has 10 junctions), the time complexity is not a crucial factor since n is small. For very complex objects for which n is large, we do not need to match every node in the projection graph. We only have to match a small set of nodes (6 is all that is sufficient [16]) to estimate the transform of the object with respect to the camera. The other unmatched junctions can be verified by a projection of their corresponding 3-D object vertices in the geometric model onto the 2-D plane by the estimated camera transform.

However, it happens that there may be more than one distinct set of six nodes in the model graph with which a set of six nodes in the projection graph can be matched. This occurs in highly regular objects having many identical sub-structures. Only one of these sets of six nodes in the model graph is the correct matching set. To find the correct matching set, a camera transform is estimated and a final verification is performed with each set.

We propose a parallel algorithm for the exhaustive search of the set of six matching nodes in the model graph, given a set of six connected nodes in the projection graph. A set of six connected nodes in the projection graph is numbered from N_1 to N_6. N_1 is then matched against each of the m nodes in the model graph parallelly. And, of course, N_1 can be chosen to be the highest precedence node in the projection graph, as discussed in the last subsection. From the m parallel search processes, let k be the number of sets of six nodes matched, where

$$k \leq m \tag{5}$$

For each of these k sets of nodes matched, the camera transform is estimated and a final verification is performed. This is carried out by k parallel processes.

The time complexity for the parallel algorithm can be analyzed as follows. Since there is a set of six nodes, n = 6 in equation (4). Since the searching process is carried out parallelly at each of the m model nodes, the factor of m in equation (4) cancels out. Thus,

$$t_{topol} = O(2^{\left\lceil \frac{6}{2} \right\rceil}) = O(8) = O(\text{constant}) \tag{6}$$

Therefore, the time complexity is constant in the topological stage of model matching. For final verification, assuming that it is sufficient to verify the projections of the m vertices of the 3-D

3-D OBJECT RECOGNITION AND ESTIMATION

model onto the 2-D plane, the time complexity is linear with respect to m. Therefore,

$$t_{verify} = O(m) \tag{7}$$

Thus, the total complexity, that of topological match and final verification, is still linear in m.

$$t_{total} = t_{topol} + t_{verify} = O(m) \tag{8}$$

INCOMPLETE LINE DRAWING

Because of noise and lighting conditions, sometimes we are unable to get a perfect line drawing in preprocessing. There may be junctions or part or all of a line segment missing (see Figure 16(a)). The inherent nature of our modeling scheme allows us to model such an imperfect line drawing. We construct the projection graph of an incomplete line drawing as usual. We represent each line segment not meeting at a junction by an open arc (see Figure 16(b)). A missing line segment causes a missing branch in the projection graph. A missing junction causes a missing node in the projection graph. An open line segment corresponds to an open arc. Missing line segments or junctions do not change the topology of interconnections between junctions and line segments which are present. The obtained projection graph will still be a subgraph isomorphism of the model graph, and we can still carry out the subgraph searching algorithm described before. Therefore, we have the following lemma.

Lemma 9:

An incomplete line drawing projection of a 3-D object can be modeled as a subgraph isomorphism of the projection graph obtained from the complete line drawing at the same viewpoint.

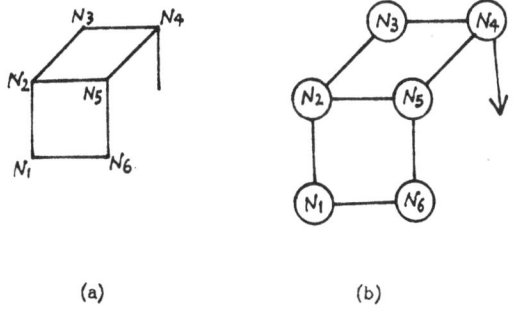

(a)   (b)

Figure 16. Imperfect line drawing.

And consequently, the projection graph obtained from an incomplete line drawing is still a subgraph isomorphism of the model graph.

As long as we have a minimum of 6 nodes in the projection graphs from an incomplete line drawing, we can still estimate the camera transform and perform final verification.

Note that in the approach where visible regions or surfaces are extracted as key matching feature [13], a missing line or a missing junction from the outline of a region may totally invalid the recognition procedure or lead to misrecognition. In the approach using characteristic views [5], a single missing junction or line could also lead to a failure of the recognition procedure. We therefore conclude that the proposed scheme is superior to other schemes in real application environment where noise and lighting condition cannot be fully controlled.

CAMERA TRANSFORM ESTIMATION AND HYPOTHESIS VERIFICATION

After the topological matching using subgraph searching method as discussed in an earlier section, the 3-D geometric model must be projected onto the 2-D plane for verification. Let

$V'' = \{v_1'', v_2'', \ldots, v_n''\}$ be the junctions in the image,

$V = \{v_1, v_2, \ldots, v_n\}$ be the matching model vertices and

$V^p = \{v_1^p, v_2^p, \ldots, v_n^p\}$ be the predicted vertices under a transformation H.

We seek to choose H to minimize the error function

$$E = \sum_{i=1}^{n} ||v_i'' - v_i^p|| \qquad (9)$$

A geometric model is said to match a given image if the computed error is less than some threshold T, that is

$$E < T \qquad (10)$$

It was mentioned in an earlier section that there may be more than one model that is topologically matched to the input image. In such cases, we choose the model which produces the minimum matching error E from all the models whose computed error satisfies Equation (10). Also, when projected onto the 2-D plane, the 2-D coordinates of the unmatched vertices must not fall outside the 2-D line drawing – that is, they must be predicted to be hidden, else, the model is an unacceptable match. This method is called hypothesis verification by picture synthesis, first used in Roberts [13] and discussed in [16].

Let us define a model point to be a 3-D coordinates in the geometric model of a 3-D object, an object point a 3-D coordinates in a 3-D object, and an image point the 2-D coordinates of the 2-D projection of a model point or an object point. In the subgraph searching procedure discussed earlier, each pair of nodes matched between the projection graph and the model graph establish a pair of correspondence points between the 2-D image and the 3-D model. Since the coordinates of both points are known in their respective spaces, the camera transformation relating the model points and the image points can be estimated. One method as used in Roberts [13] and discussed in [16] is as follows.

The model points are given in 4-dimensional homogeneous coordinates by a column vector $(x_i, y_i, z_i, w_i)^t$ and the image points are given in 3-dimensional homogeneous coordinates by column vector $(x_i'', y_i'', w_i'')^t$. The model points are given in a coordinate system attached to the model, chosen for convenience of measuring model coordinates, and the image points are given in a coordinate system with the x and y axes in the image plane. The equation relating the image points to the model points is given by

$$HV = V''D \tag{11}$$

where V is a 4-by-n matrix whose n columns are the homogeneous coordinates of the n model points, V" is a 3-by-n matrix whose n columns are the homogeneous coordinates of the n corresponding points in the image. H is a 3-by-4 matrix representing the transformation of model points to picture points and D is a n-by-n diagonal matrix which allows the scaling coordinate of HV to differ from the w_i of V". Since we have 12 unknowns in H(3X4) and n unknowns in D(nXn), a minimum of 6 pairs of correspondence points are necessary for a nondegenerate solution of Equation (11). The pseudo-inverse solution is given in [13] as

$$H = V''DV(VV^T)^{-1} \tag{12}$$

D can be solved by letting

$$Q = V''^T V''$$
$$A = V^T(VV^T)^{-1}V - I \tag{13}$$

where I is the identity matrix. If we define a new matrix such that $s_{ij} = a_{ij}q_{ji}$, then the diagonal terms of D, $\{d_1, \ldots, d_n\}$, can be formed by solving the linear equation

$$Sd = 0 \tag{14}$$

where d is a column vector formed by the diagonal elements of D. That is,

$$d = (d_1, d_2, \ldots, d_n)^t \tag{15}$$

It should be noted that this solution ignores the interdependence of the elements of matrix H. Therefore, the solution may be inaccurate. The elements of the matrix H are related to each other by the parameters of the camera -- focal length and the pan, tilt, and swing of the image plane relative to the object. More accurate but computationally expensive estimates can be obtained by using nonlinear optimization techniques as in [21-23]. Also, a set of three correspondence points between the 3-D model and the 2-D image are sufficient for estimating the solution of H if nonlinear techniques are used, rather than the six needed for the linear solution of Equation (11).

ESTIMATION OF POSITION AND ORIENTATION OF A 3-D OBJECT

In many applications, such as a robotic system with visual feedback, recognition is not sufficient for executing the desired actions on an object. In a robotic system, for example, the position and orientation of the object must be known relative to the robot. Frequently, the position and orientation of the robot are known relative to some base coordinate system. Therefore, the problem reduces to the determination of position and orientation of the object relative to the base coordinate system. Such a relationship is depicted in Figure 17 where each arrow is a homogeneous transform relationship. Since the position and orientation of the camera can be predetermined, the problem further reduces to the determination of the transform R between the camera and the object.

An ideal camera model can be described as a "pin-hole" camera as shown in Figure 18. The distance from the lens center C and the image plane I, is known as the focal length f of the camera model.

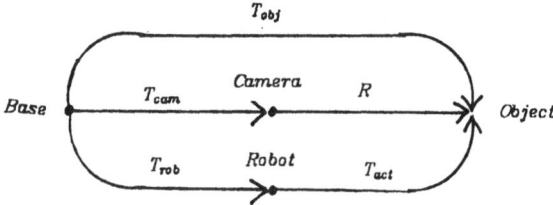

Figure 17. Transform relationship of a robotic system with visual feedback.

The image of a given point V is formed on the image plane I as point V" determined by the intersection of the ray connecting C and V with the plane I. If a Cartesian coordinate system is chosen with the origin on the image plane along the principal ray and the z axis normal to the image plane, then the point V" on the image plane and the object point V is related by the homogeneous transform P as

$$V" = PV \tag{16}$$

or

$$\begin{bmatrix} x" \\ y" \\ z" \\ w" \end{bmatrix} = \begin{bmatrix} 1 & 0 & 0 & 0 \\ 0 & 1 & 0 & 0 \\ 0 & 0 & 0 & 0 \\ 0 & 0 & 1/f & 1 \end{bmatrix} \begin{bmatrix} x \\ y \\ z \\ w \end{bmatrix}$$

The transform P is known as the perspective transformation. Note that matrix P is singular and its inverse does not exist. The transformation does not preserve any information about the distance along the z axis of an object point V since z" is always 0. Therefore, the transformation is not invertible; that is, given a picture point (x", y", z"), we cannot completely specify the corresponding object point but can only constrain it to lie along a certain straight line. Equation (16) applies only when the object and the image plane are specified in a coordinate system aligned with the camera. In general, the object model is specified in its own coordinates system. Therefore, the object coordinates must be first transformed to a coordinate system aligned with the camera before the perspective transformation can be applied. This transformation can be specified by a geometrical transformation R which is a combination of translation and sequential rotation about the three coordinate axes. These three rotation angles are often referred to as pan, tilt, and swing. Therefore,

$$R = \text{Trans Rot}_x \text{ Rot}_y \text{ Rot}_z \tag{17}$$

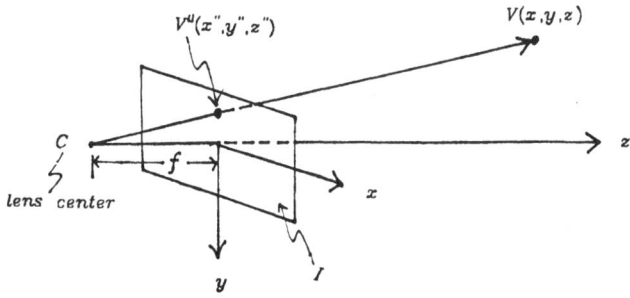

Figure 18. An ideal camera model.

where

$$\text{Trans} = \begin{bmatrix} 1 & 0 & 0 & x_0 \\ 0 & 1 & 0 & y_0 \\ 0 & 0 & 1 & z_0 \\ 0 & 0 & 0 & 1 \end{bmatrix} \quad (18)$$

$$\text{Rot}_x = \begin{bmatrix} 1 & 0 & 0 & 0 \\ 0 & \cos & -\sin & 0 \\ 0 & \sin & \cos & 0 \\ 0 & 0 & 0 & 1 \end{bmatrix} \quad (19)$$

$$\text{Rot}_y = \begin{bmatrix} \cos\psi & 0 & \sin\psi & 0 \\ 0 & 1 & 0 & 0 \\ -\sin\psi & 0 & \cos\psi & 0 \\ 0 & 0 & 0 & 1 \end{bmatrix} \quad (20)$$

and

$$\text{Rot}_z = \begin{bmatrix} \cos\theta & -\sin\theta & 0 & 0 \\ \sin\theta & \cos\theta & 0 & 0 \\ 0 & 0 & 1 & 0 \\ 0 & 0 & 0 & 1 \end{bmatrix} \quad (21)$$

correspond to translation along the x, y and z axis by x_0, y_0 and z_0 and subsequent rotations along the x, y and z axis by , ψ and θ degrees.

Frequently, we observe or process image data not directly of those on the image plane of the camera. Rather, we process data which is mapped onto some viewport such as a TV monitor. Such a mapping can be described by a matrix G where

$$G = \begin{bmatrix} 1 & 0 & 0 & x_g \\ 0 & 1 & 0 & y_g \\ 0 & 0 & 1 & 0 \\ 0 & 0 & 0 & 1/s \end{bmatrix} \quad (22)$$

and x_g and y_g correspond to translations on the viewport plane and s is a scale factor. Therefore, the transform relating object point V and image point V" can be described as

$$H' = GPR \quad (23)$$

If we correct the observed image points on the viewport plane by the amouunt $-x_g$ and $-y_g$, then x_g and y_g in Equation (22) becomes 0

3-D OBJECT RECOGNITION AND ESTIMATION

and

$$H' = GPR$$

$$= \begin{bmatrix} 1 & 0 & 0 & 0 \\ 0 & 1 & 0 & 0 \\ 0 & 0 & 1 & 0 \\ 0 & 0 & 0 & 1/s \end{bmatrix} \begin{bmatrix} 1 & 0 & 0 & 0 \\ 0 & 1 & 0 & 0 \\ 0 & 0 & 0 & 0 \\ 0 & 0 & 1/f & 1 \end{bmatrix} \begin{bmatrix} r_{11} & r_{12} & r_{13} & r_{14} \\ r_{21} & r_{22} & r_{23} & r_{24} \\ r_{31} & r_{32} & r_{33} & r_{34} \\ 0 & 0 & 0 & 1 \end{bmatrix}$$

$$= \begin{bmatrix} r_{11} & r_{12} & r_{13} & r_{14} \\ r_{21} & r_{22} & r_{23} & r_{24} \\ 0 & 0 & 0 & 0 \\ r_{31}/sf & r_{32}/sf & r_{33}/sf & r_{34}/sf+1/s \end{bmatrix}$$

$$= \begin{bmatrix} h'_{11} & h'_{12} & h'_{13} & h'_{14} \\ h'_{21} & h'_{22} & h'_{23} & h'_{24} \\ 0 & 0 & 0 & 0 \\ h'_{41} & h'_{42} & h'_{43} & h'_{44} \end{bmatrix} \quad (24)$$

Let a model point V transformed with R be represented as V'. Then,

$$V' = RV \quad (25)$$

or

$$\begin{bmatrix} x' \\ y' \\ z' \\ w' \end{bmatrix} = \begin{bmatrix} r_{11} & r_{12} & r_{13} & r_{14} \\ r_{21} & r_{22} & r_{23} & r_{24} \\ r_{31} & r_{32} & r_{33} & r_{34} \\ 0 & 0 & 0 & 1 \end{bmatrix} \begin{bmatrix} x \\ y \\ z \\ w \end{bmatrix}$$

and

$$V'' = GPV' \quad (26)$$

or

$$\begin{bmatrix} x'' \\ y'' \\ z'' \\ w'' \end{bmatrix} = \begin{bmatrix} 1 & 0 & 0 & 0 \\ 0 & 1 & 0 & 0 \\ 0 & 0 & 0 & 0 \\ 0 & 0 & \frac{1}{sf} & \frac{1}{s} \end{bmatrix} \begin{bmatrix} x' \\ y' \\ z' \\ w' \end{bmatrix}$$

Deleting row 3 from H', the resulting 3-by-4 matrix is the camera model H defined in Equation (11). Consequently,

$$H = \begin{bmatrix} h_{11} & h_{12} & h_{13} & h_{14} \\ h_{21} & h_{22} & h_{23} & h_{24} \\ h_{31} & h_{32} & h_{33} & h_{34} \end{bmatrix}$$

$$= \begin{bmatrix} r_{11} & r_{12} & r_{13} & r_{14} \\ r_{21} & r_{22} & r_{23} & r_{24} \\ \dfrac{r_{31}}{sf} & \dfrac{r_{32}}{sf} & \dfrac{r_{33}}{sf} & \dfrac{r_{34}}{sf} + \dfrac{1}{s} \end{bmatrix} \quad (27)$$

In estimating the position and orientation of the object relative to the base coordinate, we seek to estimate R and the transform of the object relative to the base coordinate is then

$$T_{obj} = T_{cam} R \quad (28)$$

where T_{cam} is known a priori. Let an object point relative to the base coordinate system be represented by V'''. Then,

$$V''' = T_{cam} V' \quad (29)$$

From Equation (27), the first two rows of R can be obtained from the first two rows of the estimated H and if we know the focal length f and scale factor s, the third row of R can be estimated from the last row of the estimated H. That is,

$$R = \begin{bmatrix} r_{11} & r_{12} & r_{13} & r_{14} \\ r_{21} & r_{22} & r_{23} & r_{24} \\ r_{31} & r_{32} & r_{33} & r_{34} \\ 0 & 0 & 0 & 1 \end{bmatrix}$$

$$= \begin{bmatrix} h_{11} & h_{12} & h_{13} & h_{14} \\ h_{21} & h_{22} & h_{23} & h_{24} \\ h_{31}sf & h_{32}sf & h_{33}sf & (h_{34}-1/s)sf \\ 0 & 0 & 0 & 1 \end{bmatrix} \quad (30)$$

This formulation has assumed that the geometric model is of a fixed known scale size. For model with variable scale size, a different formulation must be sought and it is not possible to determine the position of the object by knowing f and s.

In many situations, it can be assumed that the object lies on some known ground plane or on another object. In such cases, transform R can be estimated by knowing the orientation of the supporting plane or supporting object relative to the base coordinate system, without knowing s and f and the scale size of the model. This assumption is called ground plane assumption in [13]. As an example of application, suppose we use a zoom camera to locate

3-D OBJECT RECOGNITION AND ESTIMATION

objects on the workbench of a robot work space and assume f cannot be determined accurately since it is variable. Let the base coordinate system be set up such that the X-Y plane coincides with the X-Y plane of the workbench. Then we know that those vertices of the object on the workbench have z''' coordinates of 0. The z''' coordinates of other model vertices can be found relative to these vertices on the ground plane. By Equation (29),

$$V''' = T_{cam} V'$$

or

$$\begin{bmatrix} x''' \\ y''' \\ z''' \\ w''' \end{bmatrix} = \begin{bmatrix} t_{11} & t_{12} & t_{13} & t_{14} \\ t_{21} & t_{22} & t_{23} & t_{24} \\ t_{31} & t_{32} & t_{33} & t_{34} \\ 0 & 0 & 0 & 1 \end{bmatrix} \begin{bmatrix} x' \\ y' \\ z' \\ w' \end{bmatrix}$$

Therefore,

$$z''' = t_{31} x' + t_{32} y' + t_{33} z' + t_{34} w' \tag{31}$$

or

$$z' = \frac{z''' - t_{31} x' - t_{32} y' - t_{34} w'}{t_{33}} \tag{32}$$

x' and y' can be determined from Equations (11), (25) and (26) as

$$\begin{aligned} w' &= w \\ w'' &= h_{31} x + h_{32} y + h_{33} z + h_{34} w \\ x' &= x^{vp} w'' \\ y' &= y^{vp} w'' \end{aligned} \tag{33}$$

where

$$x^{vp} = \frac{x''}{w''} \tag{34}$$

and

$$y^{vp} = \frac{y''}{w''} \tag{35}$$

are the image coordinates observed on the viewing image plane corrected with offset $-x_g$ and $-y_g$. Then z' can be solved by Equation (32). We found the x', y' and z' coordinates for four V' points, then

$$[V'_1 \ V'_2 \ V'_3 \ V'_4] = R [V_1 \ V_2 \ V_3 \ V_4] \tag{36}$$

or

$$R = [V_1' \; V_2' \; V_3' \; V_4'] \; [V_1 \; V_2 \; V_3 \; V_4]^{-1} \tag{37}$$

After R is found, T_{obj} can be found by Equation (28).

EXPERIMENTAL RESULTS

Recognition

The digitized image of a 3-D object 'bracket' is as shown in Figure 19. Local edges of the digitized image is detected using a first order gradient operator, and thresholding is done to produce a binary image as in Figure 20. A thinning and line completion algorithm is then used to thin lines that are too thick and to complete lines that are slightly broken due to preprocessing errors. The image after the thinning and line completion operation is as shown in Figure 21. A junction detection algorithm is then applied to detect the junctions and their coordinates in the image plane. The projection graph is then constructed from the junctions and lines detected. This is shown in Figure 22. The subgraph searching algorithm is used to match the projection graph against the model graph and a topological match is found. Six of the nodes matched is used to estimate the camera transform using Equation (12), and the result

Figure 19. Digitized image of a 3-D object 'bracket'.

3-D OBJECT RECOGNITION AND ESTIMATION

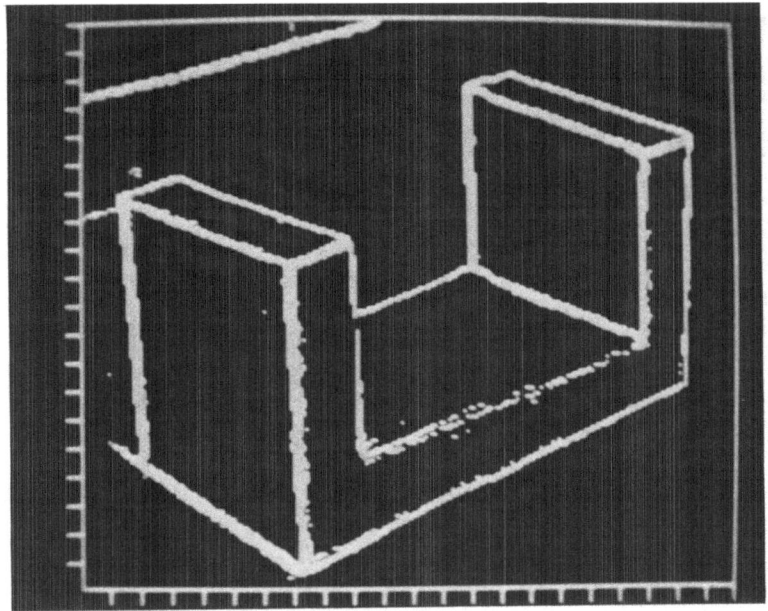

Figure 20. Digitized image after gradient operation and thresholding.

is shown in Table 3. The wire-frame model is then projected onto the image plane using this estimated transform, and the result is shown in Figure 23. This is seen to be in close resemblance to Figure 21.

Determination of Position and Orientation

Simulation experiments are conducted for the determination of object position and orientation. A base coordinate system is placed at the front left corner of a flat workbench as shown in Figure 24. The 3-D object 'bracket' is placed on the workbench with a transform of T_{obj} relative to the base coordinate system (see Table 4). A TV camera is placed at a position and orientation described by transform T_{cam} relative to the base coordinate system (as shown in Table 5). The simulated input image line drawing is shown in Figure 25. Table 6 shows the coordinates (after corrections for viewport offset of $-x_g = -50.0$ and $-y_g = -175.0$) for the junctions in the line drawing of Figure 25. Assume that we know the value of T_{cam}, we are to estimate T_{obj} from the input image.

The projection graph for the line drawing in Figure 25 is shown in Figure 26. Correspondence points are found using subgraph searching and the estimated camera transform between the object model and the image is shown in Table 7. We used a nonlinear

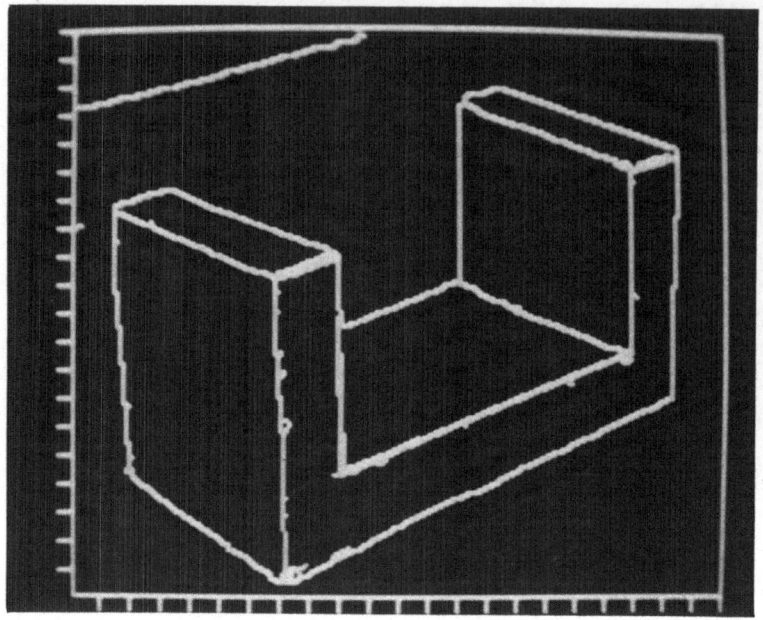

Figure 21. Line drawing after thinning and line completion operation.

technique [21-23] for the estimation of H, rather than using Equation (12), to provide a more accurate estimate of H. Thus, the performance of the position and orientation determination procedure can be evaluated independently from the large errors introduced in estimating H by Equation (12). The focal length f of the camera used was 30.0 and the scale factor s was 90.0. By Equation (30), the estimated value of R is shown in Table 8. Using Equation (28), the estimated T_{obj} is given in Table 9. This is compared to the actual value in Table 4.

Assuming we do not know the focal length and the scale factor, we use the ground plane assumption in estimating the object transform T_{obj}. Since the object is assumed to lie flat on the workbench, junctions N_8, N_{11}, N_{14} must have z''' coordinates of 0 and therefore junction N_2 must have z''' coordinates of 88 which is the height of the 'bracket'. Using Equation (32) and Equation (33), the x', y' and z' coordinates for junctions N_2, N_8, N_{11}, and N_{14} are also determined. Using Equation (37), the estimated transform R is given in Table 10. Using Equation (28), the estimated object transform T_{obj} is given in Table 11. This again is compared to the actual value of T_{obj} in Table 4.

3-D OBJECT RECOGNITION AND ESTIMATION

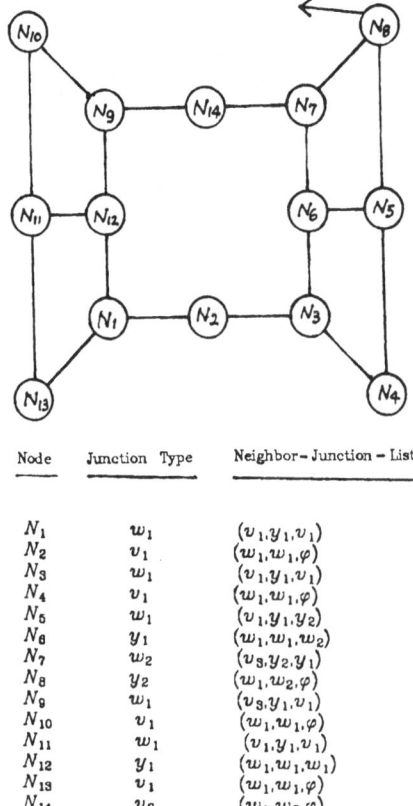

Node	Junction Type	Neighbor-Junction-List
N_1	w_1	(v_1, y_1, v_1)
N_2	v_1	(w_1, w_1, φ)
N_3	w_1	(v_1, y_1, v_1)
N_4	v_1	(w_1, w_1, φ)
N_5	w_1	(v_1, y_1, y_2)
N_6	y_1	(w_1, w_1, w_2)
N_7	w_2	(v_3, y_2, y_1)
N_8	y_2	(w_1, w_2, φ)
N_9	w_1	(v_3, y_1, v_1)
N_{10}	v_1	(w_1, w_1, φ)
N_{11}	w_1	(v_1, y_1, v_1)
N_{12}	y_1	(w_1, w_1, w_1)
N_{13}	v_1	(w_1, w_1, φ)
N_{14}	v_3	(w_1, w_2, φ)

Figure 22. Projection graph constructed from line drawing in Fig. 21.

CONCLUDING REMARKS

We have presented a graph-theoretic modeling scheme for 3-D planar and curved objects. The scheme is useful in the recognition as well as determination of position and orientation of 3-D objects. We have shown that junction semantics of a 2-D line drawing and junction constraints propagation between neighboring junctions are useful in the recognition of 3-D objects. Determination of correspondence points between the input image and the 3-D geometrical model is crucial in estimating the camera transform and in

Table 3. Estimated Camera Transform for the Line Drawing of Fig. 21

0.72009	-0.65208	-0.04062	77.52567
-0.29116	-0.29906	-1.01573	181.96817
0.00000	-0.00000	0.00000	1.00000

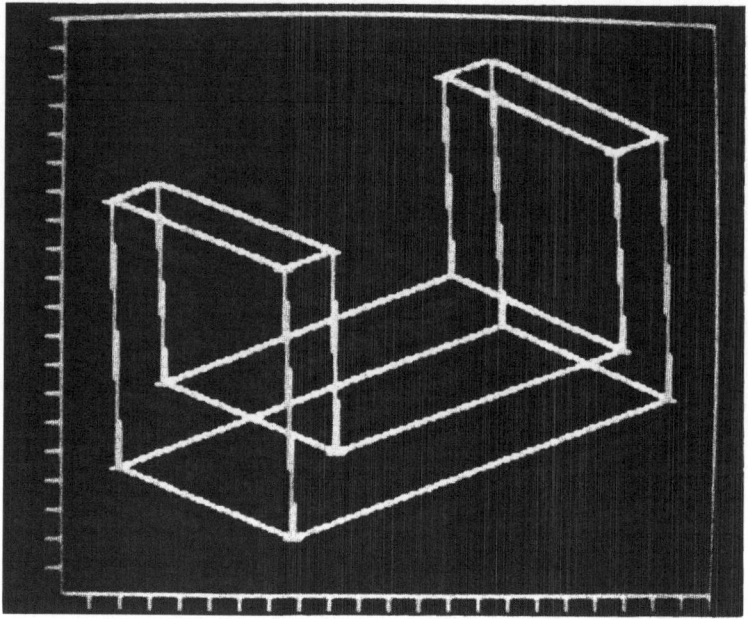

Figure 23. Projection of wire frame model onto image plane using estimated camera transform.

determining the position and orientation of a 3-D object. The intrinsic nature of our modelling scheme allows us to pose this problem as node-matching procedure between the projection graph and the model graph. The proposed scheme is efficient in memory requirement, as it requires one model graph for each model. A parallel search algorithm has been devised based on the proposed scheme and was shown to have time complexity of $O(1)$ in the topological match stage and time complexity of $O(m)$ in the final verification. Finally, the projection graph from an imperfect line drawing can also be shown to be subgraph isomorphism of the model graph and therefore the scheme is effective in noisy and poorly lighted environment where perfect line drawing could not be extracted.

ACKNOWLEDGMENT

This work was supported by the NSF Grant ECS 81-19886.

REFERENCES

[1] K. Fukunaga, Introduction to Statistical Pattern Recognition, Academic Press, New York, 1972.

3-D OBJECT RECOGNITION AND ESTIMATION

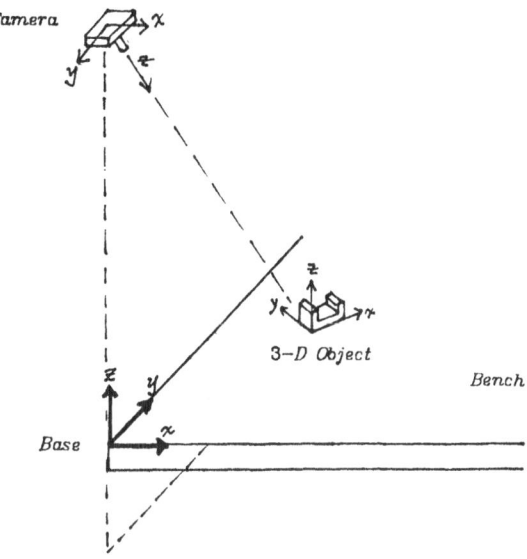

Figure 24. Simulated experimental setup.

[2] K. S. Fu, Syntactic Pattern Recognition and Applications, Prentice Hall, New Jersey, 1982.
[3] W. A. Perkins, "A model-based vision system for industrial parts," IEEE Trans. on Computers, C-27, (2), pp. 126-143, February 1978.
[4] S. W. Holland, L. Rossol, and W. R. Ward, "CONSIGHT-1: A Vision Controlled Robot System for Transferring Parts from Belt Conveyors," Computer Vision and Sensor Based Robots, edited by G. C. Dodd and L. Rossol, Plenum Press, New York, 1979.
[5] I. Chakravarty and H. Freeman, "Characteristic views as a basis for three-dimensional object recognition," SPIE Vol. 336 Robot Vision, pp. 37-45, 1982.
[6] N. Chen, J. Birk, and R. Kelly, "Estimating workpiece pose using the feature point method," IEEE Trans. on Automatic Control, AC-25, (6), December 1980.
[7] R. A. Brooks, "Symbolic reasoning among 3-D models," Artificial Intelligence, Vol. 17, pp. 285-348, August 1981.

Table 4. Actual Value for the Transform of the Object Relative to the Base

0.86603	-0.50000	0.00000	528.00000
0.50000	0.86603	0.00000	792.00000
0.00000	0.00000	1.00000	0.00000
0.00000	0.00000	0.00000	1.00000

Table 5. Transform for the Camera Relative to the Base

1.00000	0.00000	0.00000	528.00000
0.00000	-0.86603	0.50000	-792.00000
0.00000	-0.50000	-0.86603	2743.56860
0.00000	0.00000	0.00000	1.00000

[8] D. Marr and H. K. Nishihara, "Representation and recognition of the spatial organization of three-dimensional shapes," MIT A.I. Laboratory Memo 377, August 1976.
[9] L. G. Shapiro, R. M. Haralick, J. D. Moriarty, and P. G. Mulgaonkar, "Matching three-dimensional models," Proc. IEEE Computer Society Conference on PRIP, Dallas, TX, pp. 534-541, August 1981.
[10] H. G. Barrow and J. M. Tennenbaum, "Interpreting line drawing as three-dimensional surfaces," Artificial Intelligence, Vol. 17, pp. 75-116, August 1981.
[11] S. E. Yam, "Three-dimensional pattern analysis for industrial robotic systems," Ph.D. thesis, School of Electrical Engineering, Purdue University, August 1976.
[12] A. A. G. Requicha, "Representation for rigid solids: Theory, methods, and systems," Computing Surveys, Vol. 12, No. 4, December 1980.

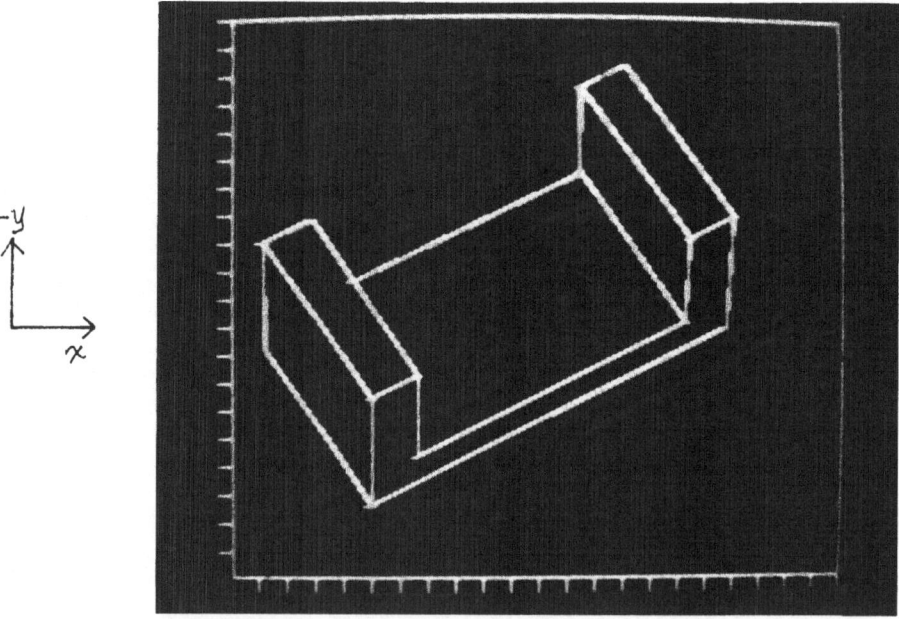

Figure 25. Simulated input line drawing.

3-D OBJECT RECOGNITION AND ESTIMATION

Table 6. Coordinate Values for the Line Drawing in Fig. 25 (after correction for offset $-x_g = -50.0$ and $y_g = -175.0$)

Nodes	X	Y	W
N1	-21	-101	1
N2	16	-46	1
N3	16	-17	1
N4	74	-119	1
N5	75	-149	1
N6	113	-94	1
N7	111	-65	1
N8	-36	-55	1
N9	-37	-93	1
N10	0	-38	1
N11	0	0	1
N12	91	-156	1
N13	129	-102	1
N14	126	-63	1

[13] L. G. Roberts, "Machine Perception of Three-Dimensional Solids," in J. Tipett (ed.), Optical and Electro-Optical Information Processing, Cambridge, MA: MIT Press, pp. 159-197, 1965.

[14] R. Shapira, "Computers reconstruction of bodies bounded by quadric surfaces from a set of imperfect projections," Tech. Rep. CRL-48, Rensselaer Polytechnic Inst., Troy, NY, 1976.

[15] I. Chakravarty, "A generalized line and junction labelling scheme with application to scene analysis," IEEE Transactions on Pattern Analysis and Machine Intelligence, PAMI-1, (2), April 1979.

[16] R. Nevatia, Machine Perception, Prentice-Hall, Inc., New Jersey, 1982.

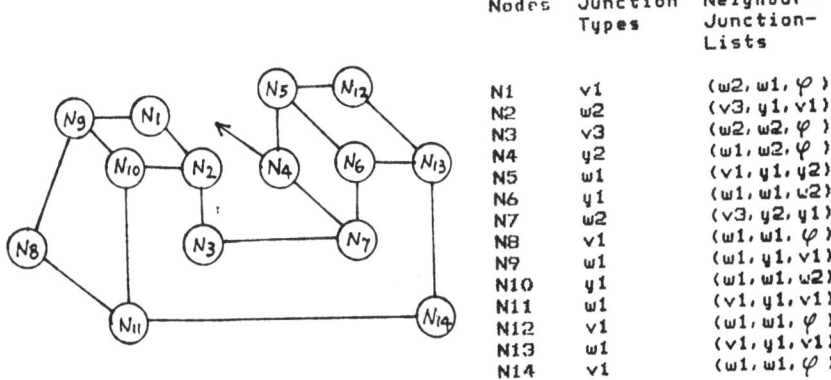

Nodes	Junction Types	Neighbor-Junction-Lists
N1	v1	(w2, w1, φ)
N2	w2	(v3, y1, v1)
N3	v3	(w2, w2, φ)
N4	y2	(w1, w2, φ)
N5	w1	(v1, y1, y2)
N6	y1	(w1, w1, w2)
N7	w2	(v3, y2, y1)
N8	v1	(w1, w1, φ)
N9	w1	(w1, y1, v1)
N10	y1	(w1, w1, w2)
N11	w1	(v1, y1, v1)
N12	v1	(w1, w1, φ)
N13	w1	(v1, y1, v1)
N14	v1	(w1, w1, φ)

Figure 26. Projection graph constructed for the line drawing in Fig. 25.

Table 7. Estimated Camera Transform for the Simulated Line Drawing

0.86603	-0.50000	0.00000	528.00000
0.50000	0.86603	0.00000	792.00012
-0.00000	-0.00000	1.00000	0.00000
0.00000	0.00000	0.00000	1.00000

Table 8. Estimated Geometrical Transform Between Object and Camera by Knowing Focal Length and Scale Factor

0.86603	-0.50000	0.00000	0.00000
-0.43301	-0.75000	-0.50000	0.00000
0.25000	0.43301	-0.86603	3168.00024
0.00000	0.00000	0.00000	1.00000

Table 9. Estimated Object Transform by Knowing Focal Length and Scale Factor

0.86603	-0.50000	0.00000	0.00000
-0.43301	-0.75000	-0.50000	0.00000
0.00009	0.00016	-0.00032	1.18444

Table 10. Estimated Geometrical Transform Between Object and the Camera Using Ground Plane Assumption

0.86603	-0.50000	0.00000	0.00002
-0.43301	-0.75000	-0.50000	-0.00003
0.25000	0.43301	-0.86603	3167.34546
0.00000	0.00000	0.00000	1.00000

Table 11. Estimated Object Transform Using Ground Plane Assumption

0.86603	-0.50000	-0.00000	528.00024
0.50000	0.86603	-0.00000	791.67236
0.00000	0.00000	1.00000	0.00002
0.00000	0.00000	0.00000	1.00000

[17] W. K. Giloi, Interactive Computer Graphics, Prentice-Hall, New Jersey, 1978.
[18] J. A. Bondy and U. S. R. Murty, Graph Theory with Applications, North Holland, Inc., 1981.
[19] M. B. Clowes, "On seeing things," Artificial Intelligence, Vol. 1, pp. 79-116, 1970.
[20] D. A. Huffman, "Impossible Objects as Nonsense Sentences," in Machine Intelligence, Vol. 6, B. Meltzer and D. Michie, Ed., American Elsevier, New York, pp. 295-323, 1971.
[21] I. Sobel, "On calibrating computer controlled cameras for perceiving 3-D scenes," Artificial Intelligence, Vol. 5, No. 2, 1974.
[22] D. B. Gennery, "Modelling the environment of an exploring vehicle by means of stereo vision," Stanford Artificial Intelligence Laboratory Memo AIM-339 (Ph.D. thesis), June 1980.
[23] M. A. Fischler and R. C. Bolles, "Random samples consensus: A paradigm for model filtering with applications to image analysis and automated cartography," Communications of the ACM, Vol. 24, No. 6, pp. 381-396, June 1981.

INDUSTRIAL COMPUTER VISION

R. C. Gonzalez

Electrical Engineering Department
University of Tennessee
Knoxville, TN 37996   U.S.A.

ABSTRACT

This chapter is an overview of the principal concepts and techniques used in the design and implementation of state-of-the-art industrial vision systems. Attention is focused on six major areas: visual sensing, preprocessing, segmentation, description, recognition, and interpretation.

INTRODUCTION

Computer vision may be defined as the process of extracting, characterizing, and interpreting information from images of a three-dimensional world. This process, also commonly referred to as machine vision, may be subdivided into six principal areas: (1) sensing, (2) preprocessing, (3) segmentation, (4) description, (5) recognition, and (6) interpretation. Sensing is the process that yields a visual image. Preprocessing deals with techniques such as noise reduction and enhancement of details. Segmentation is the process that partitions an image into objects of interest. Description deals with the computation of features (e.g., size, shape) suitable for differentiating one type of object from another. Recognition is the process that identifies these objects (e.g., wrench, bolt, engine block). Finally, interpretation assigns meaning to an ensemble of recognized objects.

It is convenient to group these various areas according to the sophistication involved in their implementation. We consider three levels of processing: low-, medium-, and high-level vision. While there are no clear-cut boundaries between these subdivisions, they do provide a useful framework for categorizing the various processes that are inherent components of a machine vision system. For instance, we associate with low-level vision those processes that are primitive in the sense that they may be considered "automatic reactions" requiring no intelligence on the part of the vision system. In our discussion, we shall treat sensing and preprocessing as low-level vision functions. This will take us from the image formation process itself to compensations such as noise reduction, and finally to the extraction of primitive image features such as intensity discontinuities. This range of processes may be compared to the sensing and adaptation process a human goes through in trying to find a seat in a dark theater immediately after walking in during a bright afternoon. The intelligent process of finding an unoccupied space cannot begin until a suitable image is available.

We will associate with medium-level vision those processes that extract, characterize, and label components in an image resulting from low-level vision. In terms of our six subdivisions, we will treat segmentation, description, and recognition of individual object as medium-level vision functions. High-level vision refers to processes that attempt to emulate cognition. While algorithms for low- and medium-level vision encompass a reasonably well-defined spectrum of activities, our knowledge and understanding of high-level vision processes is considerably more vague and speculative. These limitations lead to the formulation of constraints and idealizations intended to reduce the complexity of this task.

The categories and subdivisions discussed above are suggested to a large extent by the way machine vision systems are generally implemented. It is not implied that these subdivisions represent a model of human vision nor that they are carried out independently of each other. We know, for example, that recognition and interpretation are highly interrelated functions in a human. These relationships, however, are not yet understood to the point where they can be modeled analytically. Thus, the subdivision of functions addressed in this discussion may be viewed as a practical approach for implementing state-of-the-art machine vision systems, given our level of understanding and the analytical tools currently available in this field.

VISUAL SENSING

Imaging Devices

Visual information is converted to electrical signals by the use of visual sensors. The most commonly used visual sensors are vidicon cameras and solid state diode arrays. Vidicons are the usual vacuum tube cameras used as TV imaging devices. Their widespread use is due to their availability and reasonable cost, even though undesirable characteristics such as signal drift, noise, and distortion frequently necessitate readjustment of these devices. An input video signal is digitized and transferred to a computer as an image of size ranging typically from 64 x 64 to 512 x 512 pixels, depending on the resolution requirements of a given application.

Sensors based on semiconductor technology consist of arrays of photosensitive elements and are available in a wide variety of shapes and sizes. These devices are rugged and inexpensive, with rapid and variable scan rates and long operating life. Spatial distortion, blooming and drift are generally quite acceptable in these sensors; consequently, solid-state imaging devices are more suitable for rugged robotics applications and industrial environments.

Solid-state arrays are available as linear and area arrays. If the scene to be imaged is in continuous, uniform motion (as in belt conveyors), a linear array can be used to scan a line across the conveyor, and the motion of an object in the direction perpendicular to scan produces the desired two-dimensional image. The motion of the conveyor, however, has to be as accurate as the smallest resolution element. In a one-dimensional array, each element is read every N time intervals whereas in a two-dimensional array each element is read every N^2 time intervals; therefore, the two-dimensional array maintains a higher output data rate while allowing a long integration time for noise reduction. Linear arrays with resolution of up to 2048 elements are available.

Imaging Methods

The system configuration used in imaging plays a central role in the complexity of decision-making and recognition algorithms. The intelligent design of imaging methods may solve problems in automatic inspection or robotics that otherwise would have been very hard or impossible to solve. In practice, every application has peculiar considerations which must be taken into account early in the design of the vision system.

The vision system may use a single camera or a pair of cameras. Even though most computer vision systems today use a single camera, a pair of cameras provide much more information about the scene. The images obtained from the two cameras can be used independently, each one providing a portion of the required information; or they can be used in conjunction with each other, providing a pair of stereo pictures. An example of the former approach is discussed by Saraga and Skoyles [48] where one of the cameras is mounted vertically above the work table to yield a plan view of the object on the table. The attitude of the object is determined from images produced by the horizontal camera, and the object is located through the use of vertical camera images. In stereo vision systems, pictures from two cameras are analyzed to extract three-dimensional information about the scene. Although there a number of processing difficulties with stereo computer vision, such as finding corresponding points in two images, the application of stereo imaging is becoming more and more attractive [60].

Illumination

Illumination of a scene in an inspection or robot application is an important factor affecting the complexity of vision algorithms. Arbitrary lighting of the environment is often not acceptable because it often results in low-constrast images, specular reflections, shadows, and extraneous details. Some approaches to illumination design for inspection and robotics are introduced in the following discussion.

A well-designed lighting system illuminates the scene so that the complexity of the resulting image is minimized, while the information required for inspection or manipulation is enhanced. Four basic schemes (Figure 1) are used for illuminating parts to be inspected or manipulated: diffuse lighting, backlighting, structured lighting, and directional lighting. The diffuse-lighting approach may be employed when objects are characterized by smooth, regular surfaces. Backlighting is ideally suited for applications

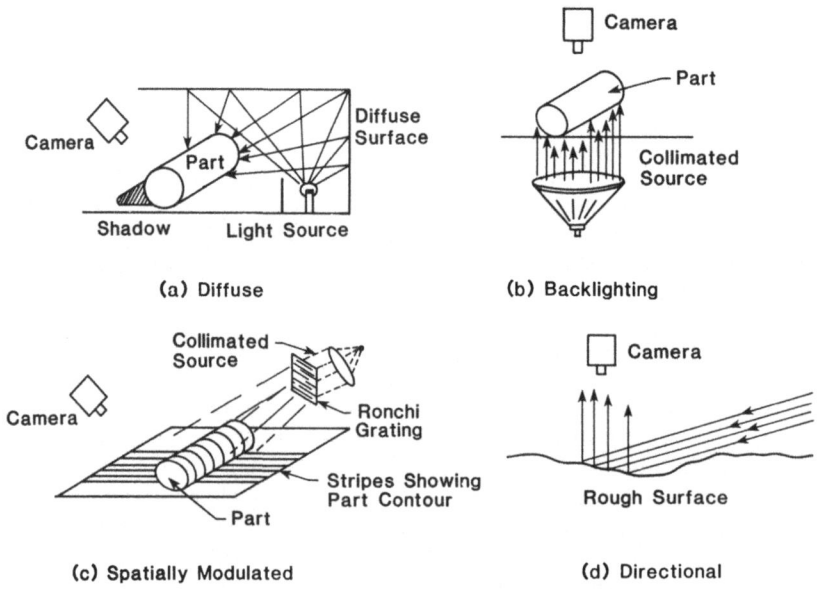

Figure 1. Basic lighting schemes (from Mundy [32]).

in which the silhouettes of objects are sufficient for recognition and other relevant measurements. The structured lighting method consists of projecting points, stripes, or grids on an object. The curvature of the object distorts the light pattern. The distortion is detected in the image and is used to find the object curvature. This method of lighting may result in missing some edges of the object. It is, however, possible to include a point source of light as well, and using the same camera at the same position obtain an image of the object which includes edge information. The directional lighting approach is ideally suited for the inspection of rough surfaces. Flaws on the surface can be detected with careful choice of projected light. For flaw-free surfaces, little light is scattered upwards to the camera but when a flaw is present, the scattered light is considerable and the defect can be detected.

Camera Model

A camera model is a geometric relationship between 3-dimensional world points and their corresponding imaged points. This relationship is a transformation from a 3-dimensional coordinate system to the image coordinate system.

The situation is depicted in Figure 2, which shows a world coordinate system (X,Y,Z) used to locate both the camera and 3D points (denoted by $\underline{w}$). This figure also shows the camera coordinate system (x,y,z) and image points (denoted by $\underline{c}$). It is assumed that the camera is mounted on a gimbal which allows pan through an angle α and tilt through angle θ. In this discussion, pan is defined as the angle between the x and X axes, and tilt as the angle between the z and Z axes. The offset of the center of the gimbal from the origin of the world coordinate system is denoted by vector $\underline{w}_o$, and the offset of the center of the imaging plane with respect to the gimbal center is denoted by a vector $\underline{r}$, with components (r_1,r_2,r_3). The approach we will follow is to bring the camera and world coordinate systems into alignment by applying a set of transformations. After this has been accomplished, we simply apply a perspective transformation to obtain the image - plane coordinates of any given world point.

Suppose that, initially, the camera was in normal position, in the sense that the gimbal center and origin of the image plane were at the origin of the world coordinate system, and all axes were aligned. Starting from normal position, the geometrical arrangement of Figure 2 can be achieved in a number of ways. We assume the following sequence of steps: (1) displacement of the gimbal center from the origin, (2) pan of the x-axis, (3) tilt of the z-axis and (4) displacement of the image plane with respect to the gimbal center.

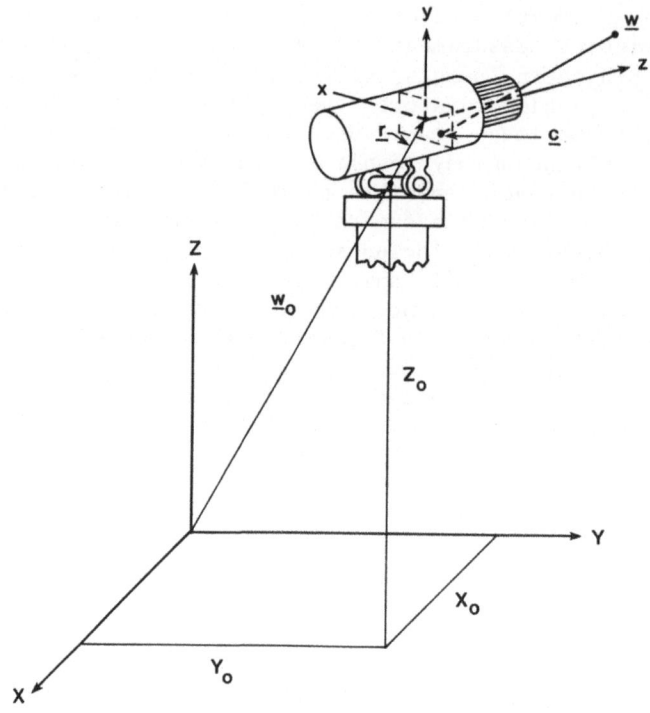

Figure 2. Imaging geometry with two coordinate systems.

The sequence of mechanical steps just discussed obviously does not affect the world points since the set of points seen by the camera after it was moved from normal position is quite different. However, we can achieve normal position again simply by applying exactly the same sequence of steps to all world points. Since a camera in normal position satisfies the arrangement for application of the perspective transformation, our problem is thus reduced to applying to every world point a set of transformations which correspond to the steps given above.

Translation of the origin of the world coordinate system to the location of the gimbal center is accomplished by using the following transformation matrix:

$$\underline{G} = \begin{bmatrix} 1 & 0 & 0 & -X_o \\ 0 & 1 & 0 & -Y_o \\ 0 & 0 & 1 & -Z_o \\ 0 & 0 & 0 & 1 \end{bmatrix} \qquad (1)$$

In other words, a homogeneous world point $\underline{w}_h$ that was at coordinates (X_o, Y_o, Z_o) is at the origin of the new coordinate system after the transformation $\underline{G}\,\underline{w}_h$.

As indicated earlier, the pan angle is measured between the x and X axes. In normal position these two axes are aligned. In order to pan the x-axis through the desired angle, we simply rotate it by θ. The rotation is about the z-axis and is accomplished by using the matrix

$$\underline{R}_\theta = \begin{bmatrix} \cos\theta & \sin\theta & 0 & 0 \\ -\sin\theta & \cos\theta & 0 & 0 \\ 0 & 0 & 1 & 0 \\ 0 & 0 & 0 & 1 \end{bmatrix} \qquad (2)$$

At this point in the development the z and Z axes are still aligned. We tilt the z-axis simply by rotating it by an angle α about the x-axis. This is accomplished by using the transformation matrix

$$\underline{R}_\alpha = \begin{bmatrix} 1 & 0 & 0 & 0 \\ 0 & \cos\alpha & \sin\alpha & 0 \\ 0 & -\sin\alpha & \cos\alpha & 0 \\ 0 & 0 & 0 & 1 \end{bmatrix} \qquad (3)$$

These two matrices can be combined into a single transformation matrix: $\underline{R} = \underline{R}_\alpha \underline{R}_\theta$. Finally, displacement of the origin of the image plane by vector $\underline{r}$ is achieved by the transformation matrix.

$$\underline{C} = \begin{bmatrix} 1 & 0 & 0 & -r_1 \\ 0 & 1 & 0 & -r_2 \\ 0 & 0 & 1 & -r_3 \\ 0 & 0 & 0 & 1 \end{bmatrix} \qquad (4)$$

Thus, by applying to $\underline{w}_h$ the series of transformations $\underline{C}\,\underline{R}\,\underline{G}\,\underline{w}_h$ we have brought the world and camera coordinate systems into coincidence. The image-plane coordinates of a point $\underline{w}_h$ are finally obtained by using the well-known perspective transformation matrix

$$\underline{P} = \begin{bmatrix} 1 & 0 & 0 & 0 \\ 0 & 1 & 0 & 0 \\ 0 & 0 & 1 & 0 \\ 0 & 0 & \frac{-1}{\lambda} & 1 \end{bmatrix} \quad (5)$$

where λ is the focal length of the camera [11]. In other words, a homogeneous world point which is being viewed by a camera satisfying the geometrical arrangement shown in Figure 2 has the following homogeneous representation in the camera coordinate system:

$$\underline{c}_h = \underline{P}\,\underline{C}\,\underline{R}\,\underline{G}\,\underline{w}_h \quad (6)$$

We obtain the Cartesian coordinates (x,y) of the imaged point by dividing the first and second components of $\underline{c}_h$ by the fourth. Expanding Equation (6) and converting to Cartesian coordinates yields

$$x = \lambda \frac{(X-X_o)\cos\theta + (Y-Y_o)\sin\theta - r_1}{-(X-X_o)\sin\theta\sin\alpha + (Y-Y_o)\cos\theta\sin\alpha - (Z-Z_o)\cos\alpha + r_3 + \lambda} \quad (7)$$

and

$$Y = \lambda \frac{-(X-X_o)\sin\theta\cos\alpha + (Y-Y_o)\cos\theta\cos\alpha + (Z-Z_o)\sin\alpha - r_2}{-(X-X_o)\sin\theta\sin\alpha + (Y-Y_o)\cos\theta\sin\alpha - (Z-Z_o)\cos\alpha + r_3 + \lambda} \quad (8)$$

which are the image coordinates of a point $\underline{w}$ whose world coordinates are (X,Y,Z).

PREPROCESSING

Preprocessing deals with functions such as enhancement, restoration, and geometric corrections. In this section we focus attention on fundamental approaches to this problem. The reader is referred to Gonzalez and Wintz [16] and to Rosenfeld and Kak [45] for details on specific algorithms based on these concepts.

Spatial-Domain Methods

The term spatial-domain refers to the aggregate of pixels composing an image, and spatial-domain methods are procedures that operate directly on these pixels. Preprocessing functions in the spatial domain may be expressed as

$$g(x,y) = h[f(x,y)] \quad (9)$$

where f(x,y) is the input image, g(x,y) is the resulting (preprocessed) image, and h is an operator on f, defined over some neighborhood of (x,y). It is also possible to let h operate on a set of input images, such as performing the pixel-by-pixel sum of K images for noise reduction.

The principal approach used in defining a neighborhood about (x,y) is to use a square or rectangular subimage area centered at (x,y), as shown in Figure 3. The center of the subimage is moved from pixel to pixel starting, say, at the top left corner, and applying the operator at each location (x,y) to yield g(x,y). Although other neighborhood shapes, such as a circle, are sometimes used, square arrays are by far the most predominant because of their ease of implementation.

The simplest form of h is when the neighborhood is 1 x 1 and, therefore, g depends only on the value of f at (x,y). In this case h becomes an intensity mapping or transformation, T, of the form

$$s = T(r) \qquad (10)$$

where, for simplicity, we have used s and r as variables denoting, respectively, the intensity of f(x,y) and g(x,y) at any point (x,y).

One of the spatial-domain techniques used most frequently is based on the use of so-called convolution masks (also referred to as templates, windows, or filters). Basically, a mask is a small (e.g., 3 x 3) two-dimensional array, such as the one shown in

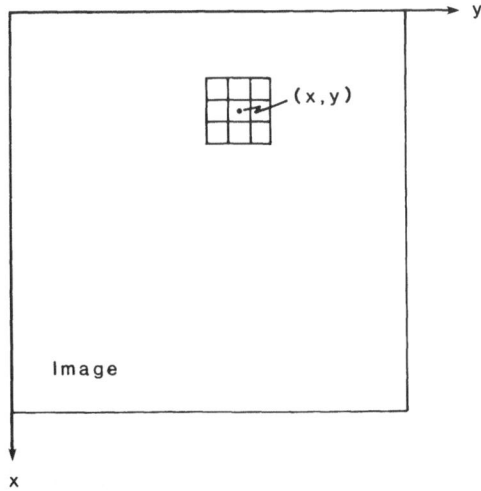

Figure 3. A 3x3 neighborhood about a point (x,y) in an image.

Figure 3, whose coefficients are chosen to detect a given property in an image. As an introduction to this concept, suppose that we have an image of constant intensity which contains widely isolated points whose intensities are different from the background. These points can be detected by using the mask shown in Figure 4. The procedure is as follows: The center of the mask (labeled 8) is moved around the image, as indicated above. At each pixel position in the image, we multiply every pixel that is contained within the mask area by the corresponding mask coefficient; that is, the pixel in the center of the mask is multiplied by 8, while its 8-neighbors are multiplied by -1. The results of these nine multiplications are then summed. If all the pixels within the mask area have the same value (constant background), the sum will be zero. If, on the other hand, the center of the mask is located at one of the isolated points, the sum will be different from zero. If the isolated point is in an off-center position the sum will also be different from zero, but the magnitude of the response will be weaker. These weaker responses can be eliminated by comparing the sum against a threshold.

As shown in Figure 5, if we let w_1, w_2, ..., w_9 represent mask coefficients and consider the 8-neighbors of (x,y), we may generalize the preceding discussion as performing the following operation

$$h[f(x,y)] = w_1 f(x-1,y+1) + w_2 f(x-1,y) + w_3 f(x-1,y-1)$$
$$+ w_4 f(x,y-1) + w_5 f(x,y) + w_6 f(x,y+1)$$
$$+ w_7 f(x+1,y-1) + w_8 f(x+1,y) + w_9 f(x+1,y+1) \quad (11)$$

on a 3 x 3 neighborhood of (x,y). This type of processing can be considerably more powerful than the simple example of point detection discussed above. For instance, neighborhood operations can be used for noise reduction, to obtain variable image thresholds, to compute measures of texture, and to obtain the skeleton of an object.

-1	-1	-1
-1	8	-1
-1	-1	-1

Figure 4. A mask for detecting isolated points different from a constant background.

w_3 (x-1,y-1)	w_2 (x-1,y)	w_1 (x-1,y+1)
w_4 (x,y-1)	w_5 (x,y)	w_6 (x,y+1)
w_7 (x+1,y-1)	w_8 (x+1,y)	w_9 (x+1,y+1)

Figure 5. A general 3x3 mask showing coefficients and corresponding image pixel locations.

Frequency-Domain Methods

The term frequency-domain refers to an aggregate of complex pixels resulting from taking the Fourier transform of an image. The concept of "frequency" is often used in interpreting the Fourier transform and arises from the fact that this particular transform is composed of complex sinusoids. Due to extensive processing requirements, frequency-domain methods are not nearly as widely used in vision as spatial-domain techniques. However, the Fourier transform does play an important role in areas such as the analysis of object motion and object description. In addition, many spatial techniques for enhancement and restoration are founded on concepts whose origins can be traced to a Fourier transform formulation. The material in this section will serve as an introduction to these concepts. A more extensive treatment of the Fourier transform and its properties may be found in Gonzalez and Wintz [16].

We begin the discussion by considering functions of one variable, $f(x)$, for $x = 0, 1, 2, \ldots, N-1$. The forward Fourier transform of $f(x)$ is defined as

$$F(u) = \frac{1}{N} \sum_{x=0}^{N-1} f(x) e^{-j2\pi ux/N} \qquad (12)$$

for $u = 0, 1, 2, \ldots, N-1$. In this equation $j = \sqrt{-1}$ and u is the so-called frequency variable. The inverse Fourier transform of $F(u)$ yields $f(x)$ back and is defined as

$$f(x) = \sum_{u=0}^{N-1} F(u)e^{j2\pi ux/N} \qquad (13)$$

for $x = 0, 1, 2, \ldots, N-1$. The validity of these expressions, called the Fourier transform pair, is easily verified by substituting Equation (12) for $F(u)$ in Equation (13), or vice versa. In either case we would get an identity.

A direct implementation of Equation (12) for $u = 0, 1, 2, \ldots, N-1$ would require on the order of N^2 additions and multiplications. Use of a fast Fourier transform (FFT) algorithm significantly reduces this number to $N\log_2 N$. Similar comments apply to Equation (13) for $x = 0, 1, 2, \ldots, N-1$. A number of FFT algorithms are readily available in a variety of computer languages.

The two-dimensional Fourier transform pair of an $N \times N$ image is defined as

$$F(u,v) = \frac{1}{N} \sum_{x=0}^{N-1} \sum_{y=0}^{N-1} f(x,y)e^{-j2\pi(ux+vy)/N} \qquad (14)$$

for $u,v = 0, 1, 2, \ldots, N-1$ and

$$f(x,y) = \frac{1}{N} \sum_{u=0}^{N-1} \sum_{v=0}^{N-1} F(u,v)e^{+j2\pi(ux+vy)/N} \qquad (15)$$

for $x,y = 0, 1, 2, \ldots, N-1$. It is possible to show through some manipulation that each of these equations can be expressed as separate one-dimensional summations of the form shown in Equation (12). This leads to a straightforward procedure for computing the two-dimensional Fourier transform using only a one-dimensional FFT algorithm: We first compute and save the transform of each row of $f(x,y)$, thus producing a two-dimensional array of intermediate results. These results are multiplied by N and the one-dimensional transform of each column is computed. The final result is $F(u,v)$. Similar comments apply for computing $f(x,y)$ given $F(u,v)$. The order of computation from a row-column approach can be reversed to a column-row format without affecting the final result.

The Fourier transform can be used in a number of ways by a vision system. For example, by treating the boundary of an object as a one-dimensional array of points and computing their Fourier transform, selected values of $F(u)$ can be used as descriptors of boundary shape. The one-dimensional Fourier transform has also been used as a powerful tool for detecting object motion. Applications of the discrete two-dimensional Fourier transform in image reconstruction, enhancement, and restoration are abundant although, as mentioned earlier, the usefulness of this approach in industrial machine vision is still quite restricted due to the extensive computational requirements needed to implement this transform. We point out before leaving this section, however, that the two-dimensional, continuous Fourier transform can be computed (at the speed of light) by optical means. This approach, which requires the use of precisely-aligned optical equipment, is used in industrial environments for tasks such as the inspection of finished metal surfaces. Although further discussion of this topic is outside the scope of our present discussion, the interested reader is referred to the book by Goodman [17] for an excellent introduction to Fourier optics.

SEGMENTATION

Segmentation is the process that breaks up a sensed scene into its constituent parts or objects. Virtually hundreds of segmentation algorithms have been proposed in the literature over the past 15 years (see, for example, Fu and Mui [10], Gonzalez and Wintz [16], and Rosenfeld [43]). Not surprisingly, this is still an active area of research because of its important as the first processing step in any practical computer vision application. Although image segmentation has proved to be a difficult task in unconstrained situations such as automatic target detection, the problems encountered in industrial applications can be considerably simplified by special lighting techniques such as those discussed earlier.

Segmentation algorithms are generally based on one of two basic principles: discontinuity and similarity. The principal approach in the first category is edge detection. The principal approaches in the second category are thresholding and region growing. Most edge detection techniques for industrial applications are based on the use of spatial convolution masks in order to reduce processing time. The idea is to move a mask over the entire image area, one pixel location at a time and, at each location, to compute a measure proportional to discontinuity (e.g., the gradient) in the image area directly under the mask, as discussed in the previous section. Thresholding is by far the most widely used approach for segmentation in industrial applications of computer vision. There are two reasons for this. First, thresholding techniques (in their simpler forms) are fast and, in addition, they are quite straightforward to implement in hardware. Second, the lighting environment is usually a controllable factor in industrial application; this results in images that often readily lend themselves to a thresholding approach for object extraction. Region growing techniques are applicable in situations where objects cannot be differentiated from each other or the background by thresholding or edge detection. Although region growing has been used extensively in scene analysis, it has not found wide applicability in industrial applications because (1) this method is usually impractical from a computational and/or hardware implementation point of view, and (2) many of the problems which would require region growing for segmentation can usually be handled by special lighting or other enhancement techniques. A good overview of region growing algorithms may be found in a paper by Zucker [65]. A considerable amount of work has also been reported in the literature dealing with techniques that attempt to incorporate contextual information in the segmentation process. This includes the use of relaxation [44], plan-guided analysis [21,27,59], and the use of semantic information [54].

For the most part, the preceding discussion deals with segmentation of two-dimensional data. Since true vision is a 3D problem, it is widely accepted that a key to the development of versatile vision systems capable of operating in unconstrained environments lies in being able to process three-dimensional scene information. Although research in this area spans more than a ten-year history, we point out that factors such as cost, speed, and complexity have inhibited the use of three-dimensional vision techniques in industrial applications.

Three-dimensional information about a scene may be obtained in three principal forms. If range sensing is used, we obtain the (x,y,z) coordinates of points on the surface of objects. The use of stereo imaging devices yields 3D coordinates, as well as intensity information about each point. In this case, we represent each point in the form $f(x,y,z)$, where the value of f and (x,y,z) gives the intensity of that point (the term voxel is often used to denote a 3D point and its intensity). Finally, we may infer 3D relationships from a single two-dimensional image of a scene. In other words, it is often possible to deduce relationships between objects such as "above," "behind," "in front of," etc. Since the exact 3D location of scene points generally cannot be computed from a single view, the relationships obtained from this type of analysis is often referred to as 2 1/2-D information. The following material is representative of current work in 3D segmentation.

Fitting Planar Patches to Range Data

One of the simplest approaches for segmenting and describing a three-dimensional structure given in terms of range data points, (x,y,z), is to first subdivide it into small planar "patches" and then combine these patches into larger surface elements according to some criterion. This approach is particularly attractive for polyhedral objects whose surfaces are smooth with respect to the resolution of the sensed scene.

We illustrate the basic concepts underlying this approach by means of the example shown in Figure 6. Part (a) of this figure

INDUSTRIAL COMPUTER VISION   359

shows a simple scene and Figure 6(b) shows a set of corresponding
3-D points. These points can be assembled into small surface elements by, for example, subdividing the 3-D space into cells and
grouping points according to the cell which contains them. Then,
we fit a plane to the group of points in each cell and calculate a
unit vector which is normal to the plane and passes through the
centroid of the group of points in that cell. A planar patch is
established by the intersection of the plane and the walls of the
cell, with the direction of the patch being given by the unit
normal, as illustrated in Figure 6(c). All patches whose directions are similar within a specified threshold are grouped into
elementary regions (R), as shown in Figure 6(d). These regions are
then classified as planar (P), curved (C), or undefined (U) by
using the directions of the patches within each region (for example, the patches in a planar surface will all point in essentially
the same direction). This type of region classification is illustrated in Figure 6(e). Finally (and this is the hardest step), the
classified regions are assembled into global surfaces by grouping
adjacent regions of the same classification, as shown in Figure 6(f).
It is noted that, at the end of this procedure, the scene has been
segmented into distinct surfaces, and that each surface has been
assigned a descriptor (e.g., curved or planar).

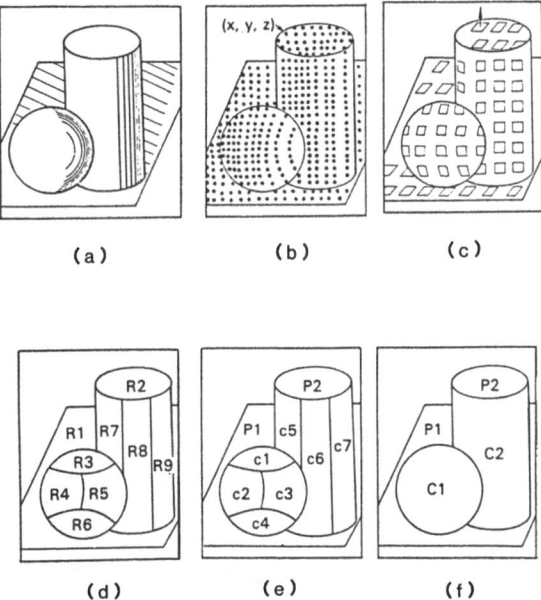

Figure 6. Three-dimensional surface description based on
          planar patches (from Shirai [49]).

Use of the Gradient

When a scene is given in terms of voxels, the 3D gradient can be used to obtain patch representations (similar to those discussed in the previous section) which can then be combined to form surface descriptors. It is well-known that the gradient vector is normal to the direction of maximum rate of change of a function, and the magnitude of this vector is proportional to the strength of that change. These concepts can be used to segment 3D structures in a manner analogous to that used for two-dimensional data.

Given a function $f(x,y,z)$, its gradient vector at coordinates (x,y,z) is given by

$$\underline{G}[f(x,y,z)] = \begin{bmatrix} G_x \\ G_y \\ G_z \end{bmatrix} = \begin{bmatrix} \frac{\partial f}{\partial x} \\ \frac{\partial f}{\partial y} \\ \frac{\partial f}{\partial z} \end{bmatrix} \quad (16)$$

The magnitude of $\underline{G}$ is given by

$$G[f(x,y,z)] = (G_x^2 + G_y^2 + G_z^2)^{1/2} \quad (17)$$

This equation is often approximated by absolute values to simplify computation:

$$G[f(x,y,z)] \approx |G_x| + |G_y| + |G_z| \quad (18)$$

The implementation of the 3D gradient can be carried out using operators analogous in form to those used in two dimensions [11]. Figure 7 shows a 3 x 3 x 3 operator proposed [64] for computing G_x. The same operator oriented along the y-axis is used to compute G_y, and oriented along the z-axis to compute G_z. A key property of these operators is that they yield the best (in a least-square-error sense) planar edge between two regions of different intensities in a 3D neighborhood.

The center of each operator is moved from voxel to voxel and applied in exactly the same manner as their two-dimensional counterparts, as discussed earlier. That is, the response of these operators at any point (x,y,z) yields G_x, G_y, and G_z, which are then substituted into Equation (16) to obtain the gradient vector at (x,y,z) and into Equation (17) or (18) to obtain the magnitude. It is of interest to note that the operator shown in Figure 7 yields a zero output in a 3 x 3 x 3 region of constant intensity.

INDUSTRIAL COMPUTER VISION   361

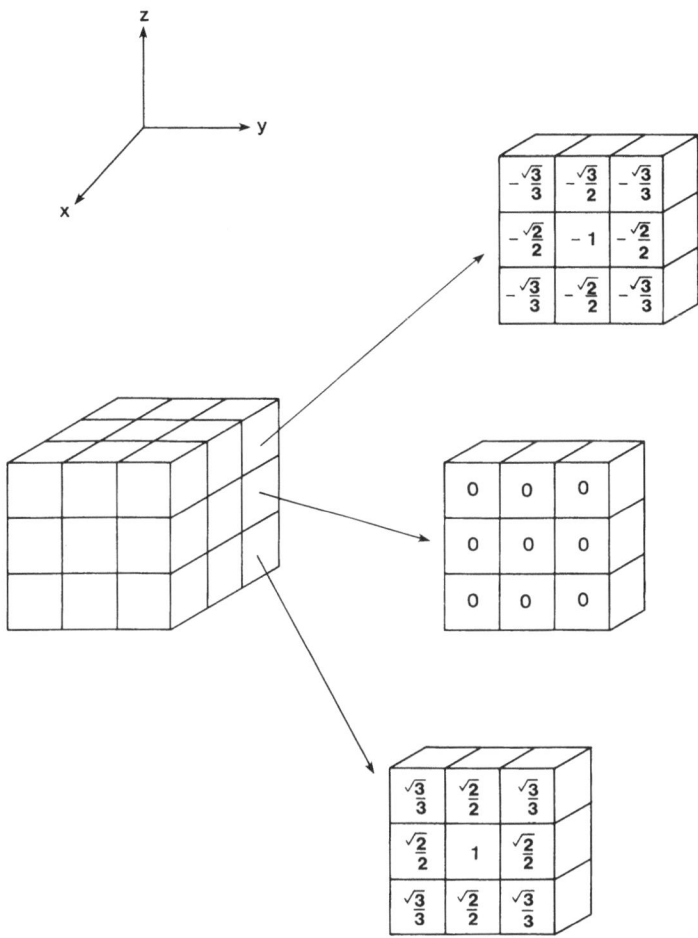

Figure 7. A 3x3x3 operator for computing the gradient component G_x (adapted from Zucker and Hummel [64]).

It is a straightforward procedure to utilize the gradient approach for segmenting a scene into planar patches analogous to those discussed in the previous section. It is not difficult to show that the gradient vector of a plane $ax + by + cz = 0$ has components $G_x = a$, $G_y = b$, and $G_z = c$. Since the operators discussed above yield an optimum planar fit in a 3 x 3 x 3 neighborhood, it follows that the components of the vector $\underline{G}$ establish the direction of a planar patch in each neighborhood, while the magnitude of $\underline{G}$ gives an indication of abrupt changes of intensity within the patch; that is, it indicates the presence of an intensity edge within the patch. An example of such a patch representation using the gradient operators is shown in Figure 8. Since each planar patch surface passes through the center of a voxel, the borders of these patches

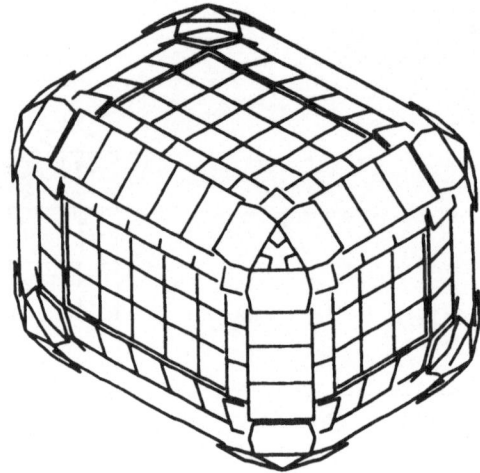

Figure 8. Planar patch approximation of a cube using the gradient (from Zucker and Hummel [64]).

may not always coincide. Patches that coincide are shown as larger uniform regions in Figure 8. Once patches have been obtained, they can be grouped and described in the form of global surfaces as discussed in the previous section. Note, however, that additional information in the form of intensity and intensity discontinuities is now available to aid the merging and description process.

DESCRIPTION

The description problem in computer vision is one of extracting features from an object for the purpose of recognition. Ideally, these features should be independent of object location and orientation and should contain enough discriminatory information to uniquely differentiate one object from another. Descriptors for industrial computer vision are based primarily on shape and amplitude (e.g., intensity) information. Shape descriptors attempt to capture invariant geometrical properties of an object. This has been an illusive goal which has led to an impressive number of proposed techniques [34]. Approaches for shape analysis and description are generally either global (region) oriented or boundary oriented. Global techniques include principal axes analysis [16], texture [20], [58], [57], two- and three-dimensional moment invariants [25], [47], geometrical descriptors such as perimeter squared/area and the extrema of a region [37,38], [51], topological properties such as the Euler number [16], and decomposition into primary convex subsets [35]. Boundary-oriented techniques include Fourier descriptors [63,40,42,56], chain codes [8], and graph representations, of which strings and trees are special cases, [35,15,5]. Boundary feature extraction is often preceded by

INDUSTRIAL COMPUTER VISION

linking procedures which fit straight line segments or polynomials to the edge points resulting from segmentation.

Techniques for describing two-dimensional structures are well-documented in the literature, as indicated by the preceding references. In this section we focus attention on procedures for describing 3D structures. Although, as indicated in the previous section, present industrial vision systems are based for the most part on two-dimensional techniques, the capability to process 3D information is generally accepted as one of the principal factors which will enhance the flexibility of these systems.

Line and Junction Labeling

With reference to the discussion in the previous section, edges in a 3D scene are determined by discontinuities in range and/or intensity data. Given a set of surfaces and the edges between them, a finer description of a scene may be obtained by labeling the lines corresponding to these edges and the junctions which they form.

As illustrated in Figure 9, we consider three basic types of lines. A convex line (labeled +) is formed by the intersection of two surfaces which are part of a convex solid (e.g., the line formed by the intersection of two sides of a cube). A concave line (labeled -) is formed by the intersection of two surfaces belonging to two different solids (e.g., the intersection of one side of a cube with the floor). An occluding line (labeled with an arrow) is the edge of a surface which obscures a surface. The occluding matter is to the right of the line looking in the direction of the arrow, and the occluded surface is to the left.

After the lines in a scene have been labeled, their junctions provide clues as to the nature of the 3D solids in the scene.

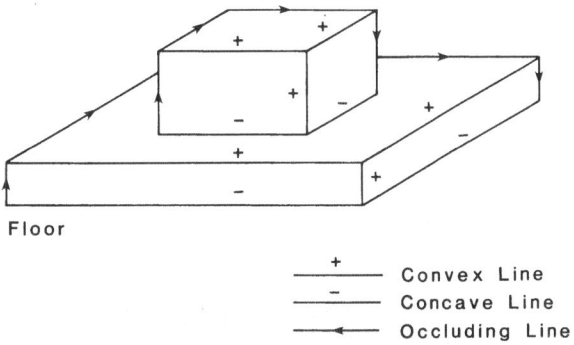

Figure 9. Three basic line labels.

Physical constraints allow only a few possible combinations of line labels at a junction. For example, in a polyhedral scene, no line can change its label between vertices. Violation of this rule leads to impossible physical objects, as illustrated in Figure 10.

The key to using junction analysis is to form a dictionary of allowed junction types. For example, it is easily shown that the junction dictionary shown in Figure 11 contains all valid labeled vertices of trihedral solids (i.e., solids in which exactly three plane surfaces come together at each vertex).

Once the junctions in a scene have been classified according to their match in the dictionary, the objective is to group the various surfaces into objects. This is typically accomplished via a set of heuristic rules designed to interpret the labeled lines and sequences of neighboring junctions. The basic concept underlying this approach can be illustrated with the aid of Figure 12. We note in Figure 12(b) that the blob is composed entirely of an occluding boundary, with the exception of a short concave line, indicating where it touches the base. Thus, there is nothing in front of it, and it can be extracted from the scene. There is a vertex of type (10). This is strong evidence (if we know we are dealing with trihedral objects) that the three surfaces involved in that vertex form a cube. Similar comments apply to the base after the cube surfaces are removed. Removing the base leaves the single object in the background, which completes the decomposition of the scene.

Generalized Cones

A generalized cone (or cylinder) is the volume described by a planar cross section as it is translated along an arbitrary space curve (the spine), held at a constant angle to the curve, and transformed according to a sweeping rule. In machine vision, generalized cones provide viewpoint-independent representations of

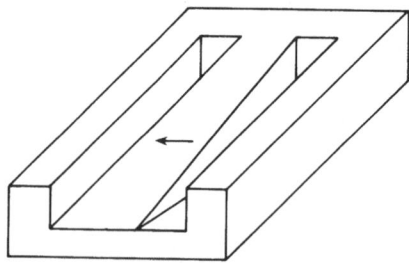

Figure 10. An impossible physical object. Note that the line marked by an arrow changes label between the upper and lower vertices.

INDUSTRIAL COMPUTER VISION 365

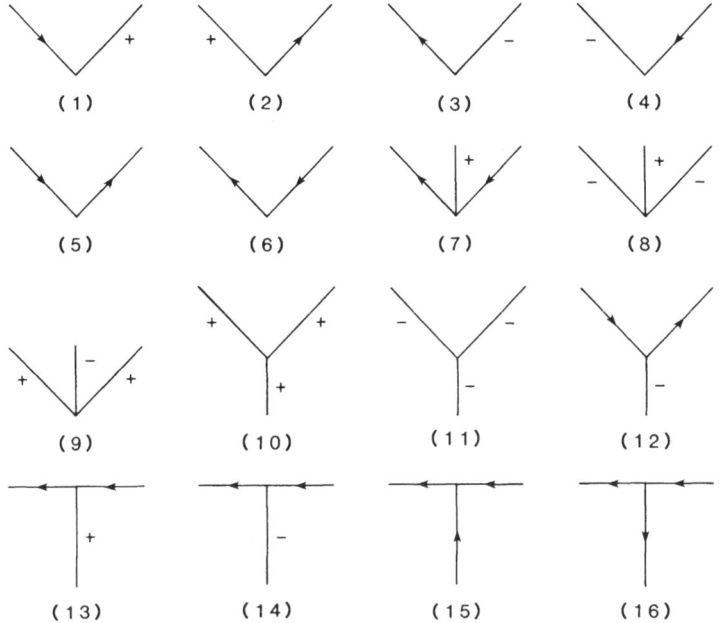

Figure 11. Junction dictionary for trihedral solids.

three-dimensional structures which are useful for description and model-based matching purposes.

Figure 13 illustrates the procedure for generating generalized cones. In Figure 13(a) the cross section is a ring, the spine is a straight line and the sweeping rule is: translate the cross section normal to the spine while keeping its diameter constant. The result is a hollow cylinder. In Figure 13(b) we have essentially the same situation, with the exception that the sweeping rule holds the diameter of the cross section constant and then allows it to increase linearly past the midpoint in the spine.

When matching a set of 3D points against a set of known generalized cones, we first determine the center axis of the points and then find the closest set of cross sections that will fit the data as we travel along the spine. In general, considerable trial and error is required, particularly when one is dealing with incomplete data.

RECOGNITION

Recognition is a labeling process; that is, the function of recognition algorithms is to identify each segmented object in a scene and to assign a label (e.g., wrench, seal, bolt) to that

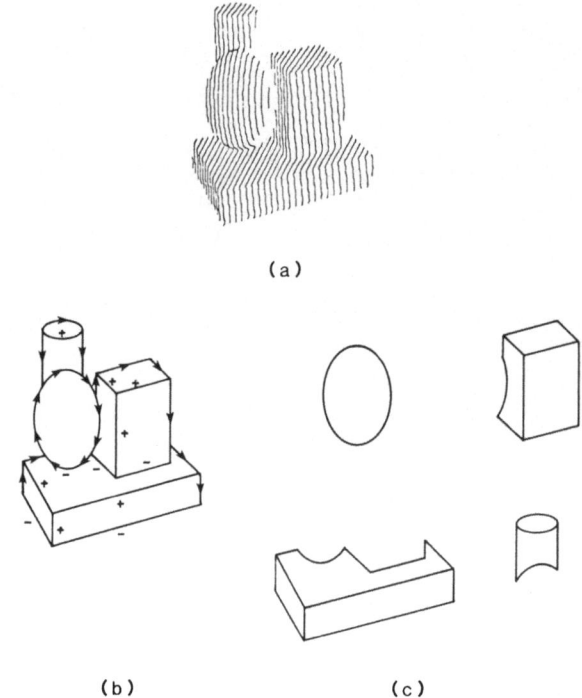

Figure 12. (a) Scene. (b) Labeled lines. (c) Decomposition via line and junction analysis (adapted from Shirai [49]).

object. For the most part, the recognition stages of present industrial vision systems operate on the assumption that objects in a scene have been segmented as individual units. Another common constraint is that images be acquired in an known viewing geometry (usually perpendicular to the workspace). This decreases variability in shape characteristics and simplifies segmentation and description by reducing the possibility of occlusion. Variability in object orientation is handled by choosing rotation-invariant descriptors or by using the principal axis of an object to orient it in a predefined direction.

Recognition approaches in use today can be divided into two principal categories: decision-theoretic and structural. As will be seen in the following discussion, decision-theoretic methods are based on quantitative descriptions (e.g., statistical texture) while structural methods rely on symbolic descriptions and their relationships (e.g., sequences of directions in a chain-coded boundary).

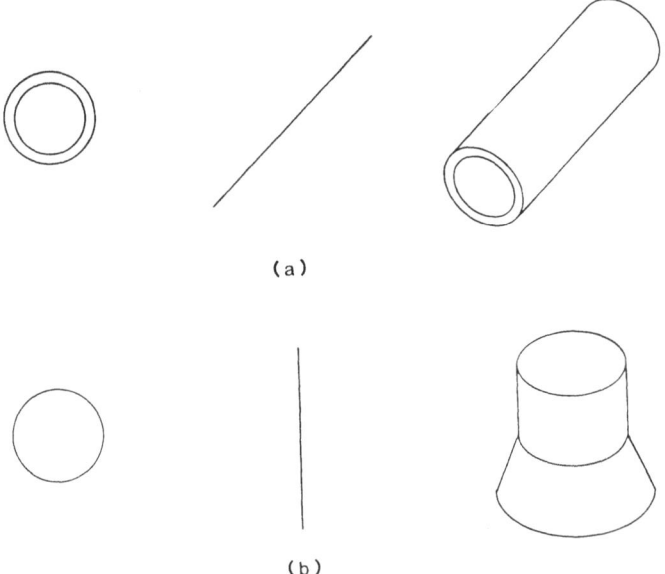

Figure 13. Cross sections, spines, and their corresponding generalized cones. In (a) the cross section remained constant during the sweep, while in (b) its diameter increased linearly past the midpoint in the spine.

Decision-Theoretic Methods

Decision-theoretic pattern recognition is based on the use of decision (discriminant) functions. Let $\underline{x} = (x_1, x_2, \ldots, x_n)^T$ represent a column pattern vector with real components, where x_i is the ith descriptor of a given object (e.g., area, average intensity, perimeter length). Given M object classes, denoted by $\omega_1, \omega_2, \ldots, \omega_M$, the basic problem in decision-theoretic pattern recognition is to identify M decision functions, $d_1(\underline{x}), d_2(\underline{x}), \ldots, d_M(\underline{x})$, with the property that, for any pattern vector $\underline{x}^*$ belonging to class ω_i,

$$d_i(\underline{x}^*) > d_j(\underline{x}^*) \tag{19}$$

$$j = 1, 2, \ldots, M$$

$$j \neq i$$

In other words, an unknown object represented by vector $\underline{x}^*$ is recognized as belonging to the ith object class if, upon substitution of $\underline{x}^*$ into all decision function, $d_i(\underline{x}^*)$ yields the largest value.

The predominant use of decision functions in industrial vision systems is for matching. Suppose that we represent each object class by a prototype (or average) vector:

$$\underline{m}_i = \frac{1}{N} \sum_{k=1}^{N} \underline{x}_k \qquad (20)$$

$$i = 1, 2, \ldots, M$$

where the $\underline{x}_k$ are the sample vectors known to belong to class ω_i. Given an unknown $\underline{x}^*$, one way to determine its class membership is to assign it to the class of its closest prototype. If we use the Euclidean distance to determine closeness, the problem reduces to computing the following distance measures:

$$D_j(\underline{x}^*) = \| \underline{x} - \underline{m}_j \| \qquad (21)$$

$$j = 1, 2, \ldots, M$$

where $\| \underline{a} \| = (\underline{a}^T \underline{a})^{1/2}$ is the Euclidean norm. We then assign $\underline{x}^*$ to class ω_i if $D_i(\underline{x}^*)$ is the smallest distance. It is not difficult to show that this is equivalent to evaluating the functions

$$d_j(\underline{x}^*) = (\underline{x}^*)^T \underline{m}_j - \frac{1}{2} \underline{m}_j^T \underline{m}_j \qquad (22)$$

$$j = 1, 2, \ldots, M$$

and selecting the largest value. This formulation agrees with the concept of a decision function, as defined in Equation (19).

Another application of matching is in searching for an instance of a subimage $w(x,y)$ in a larger image $f(x,y)$. At each location (x,y) of $f(x,y)$ we define the correlation coefficient as

$$\gamma(x,y) = \frac{\sum_s \sum_t [w(s,t) - m_w][f(s,t) - m_f]}{\left[\sum_s \sum_t [w(s,t) - m_w]^2 \sum_s \sum_t [f(s,t) - m_f]^2 \right]^{1/2}} \qquad (23)$$

where it is assumed that $w(s,t)$ is centered at coordinates (x,y). The summations are taken over the image coordinates common to both regions, m_w is the average intensity of w, and m_f is the average intensity of f in the region coincident with w. It is noted that, in general, $\gamma(x,y)$ will vary from one location to the next and that its values are in the range $[-1,1]$, with a value of 1 corresponding to a perfect match. The procedure, then, is to compute $\gamma(x,y)$ at each location (x,y) and to select its largest value to determine the best match of w in f (the procedure of moving $w(x,y)$ throughout f(x,y) is analogous to Figure 3). The quality of the match can be controlled by accepting a correlation coefficient only if it exceeds a preset value (e.g., 0.9). Since this method consists of directly comparing two regions, it is clearly sensitive to variations in object size and orientation. Variations in intensity are normalized by the denominator in Equation (23).

Structural Methods

The techniques discussed in the previous section deal with patterns on a quantitative basis, ignoring any geometrical relationships which are inherent in the shape of an object. Structural methods, on the other hand, attempt to achieve object discrimination by capitalizing on these relationships.

Central to the structural recognition approach is the decomposition of an object into pattern primitives. This idea is easily explained with the aid of Figure 14. Part (a) of this figure shows a simple object boundary, and Figure 14(b) shows a set of primitive elements of specified length and direction. By starting at the top left, tracking the boundary in a clockwise direction, and identifying instances of these primitives, we obtain the coded boundary shown in Figure 14(c). Basically, what we have done is represent the boundary by the string aaabcbbbcdddcd. The known length and direction of these primitives, together with the order in which they occur, establishes the structure of the object in terms of this particular representation. The objective of this section is to introduce the reader to techniques suitable for handling this and other types of structural pattern descriptions.

String matching. Suppose that two object contours C_1 and C_2 are coded into strings $a_1 a_2 \ldots a_n$ and $b_1 b_2 \ldots b_m$, respectively. Let A represent the number of matches between the two strings, where we say that a match has occurred in the jth position if $a_j = b_j$. The number of symbols that do not match up is given by

$$B = \max(|C_1|, |C_2|) - A \qquad (24)$$

where $|C|$ is the length (number of symbols) of string C. It can be shown that $B = 0$ if and only if C_1 and C_2 are identical.

A simple measure of similarity between strings C_1 and C_2 is defined as the ratio

$$R = A/B$$
$$= A/[\max(|C_1|, |C_2|) - A] \qquad (25)$$

Based on the above comment regarding B, R is infinite for a perfect match and zero when none of the symbols in C_1 and C_2 match (i.e.,

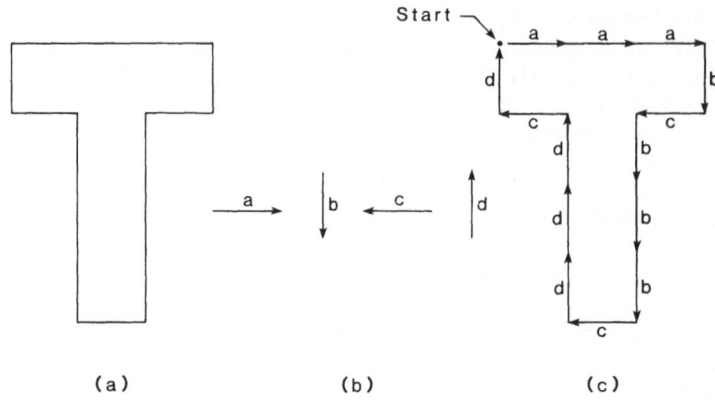

Figure 14. (a) Object boundary. (b) Primitives. (c) Boundary coded in terms of primitives, resulting in the string aaabcbbbcdddcd.

A = 0 in this case). Since the matching is done on a symbol-by-symbol basis, the starting point on each boundary when creating the string representation is important. Alternatively, we can start at arbitrary points on each boundary, shift one string (with wrap-around), and compute Equation (25) for each shift. The number of shifts required to perform all necessary comparisons is max $(|C_1|, |C_2|)$.

Figures 15(a) and (b) show a sample boundary from each of two classes of objects. The boundaries were approximated by a polygonal fit (Figures 15(c) and (d)) and then strings were formed by computing the interior angle between the polygon segments as the polygon was traversed in a clockwise direction. Angles were coded into one of eight possible symbols corresponding to 45° increments, $s_1: 0 < \theta \leq 45°$, $s_2: 45° < \theta \leq 90°$, ..., $s_8: 315° < \theta \leq 0°$.

The results of computing the measure R for five samples of object 1 against themselves are shown in Figure 15(c) where the entries correspond to values of R = A/B and, for example, the notation 1.c refers to the third string for object class 1. Figure 15(d) shows the results for the strings of the second object class. Finally, Figure 15(e) is a tabulation of R values obtained by comparing strings of one class against the other. The important thing to note is that all values of R in this last table are considerably smaller than any entry in the preceding two tables, indicating that the R measure achieved a high degree of discimination between the two classes of objects. For instance, if string 1.a had been an unknown, the smallest value in comparing it with the other strings of class 1 would have been 4.67. By contrast, the largest value in a comparison against class 2 would have been 1.24. Thus, classification of this string into class 1 based on

INDUSTRIAL COMPUTER VISION

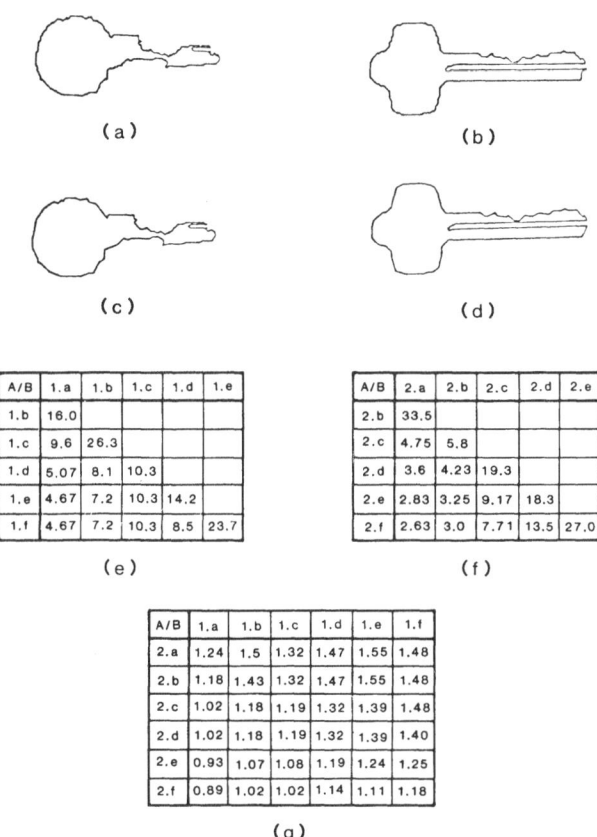

Figure 15. (a) and (b) Sample boundaries of two different object classes, (c) and (d) their corresponding polygonal approximations, and (e) through (f) tabulations of R = A/B (from Sze and Yang [52]).

the maximum value of R would have been a simple, unambiguous matter.

Syntactic methods. Syntactic techniques are by far the most prevalent concepts used for handling structural recognition problems. Basically, the idea behind syntactic pattern recognition is the specification of structural pattern primitives and a set of rules (in the form of grammar) which govern their interconnection. We consider first string grammars and then extend these ideas to higher-dimensional grammars.

String Grammars

Suppose that we have two classes of objects, ω_1 and ω_2, which are represented as strings of primitives, as outlined earlier. We may interpret each primitive as being a symbol permissible in some grammar, where a grammar is a set of rules of syntax (hence the name syntactic pattern recognition) for the generation of sentences formed from the given symbols. In the context of the present discussion, these sentences are strings of symbols which in turn represent patterns. It is further possible to envision two grammars, G_1 and G_2, whose rules are such that G_1 only allows the generation of sentences which correspond to objects of class ω_1 while G_2 only allows generation of sentences corresponding to objects of class ω_2. The set of sentences generated by a grammar G is called its language and denoted by L(G).

Once the two grammars G_1 and G_2 have been established, the syntactic pattern recognition process is, in principle, straightforward. Given a sentence representing an input, unknown pattern, the problem is one of deciding in which language the pattern represents a valid sentence. If the sentence belongs to $L(G_1)$, we say that the pattern belongs to object class ω_1. Similarly, we say that the object comes from class ω_2 if the sentence is in $L(G_2)$. A unique decision cannot be made if the sentence belongs to both $L(G_1)$ and $L(G_2)$. If the sentence is found to be invalid over both languages it is rejected.

When there are more than two pattern classes, the syntactic classification approach is the same as described above, except that more grammars (at least one per class) are involved in the process. In this case the pattern is assigned to class ω_i if it is a sentence of only $L(G_i)$. A unique decision cannot be made if the sentence belongs to more than one language, and (as above) a pattern is rejected if it does not belong to any of the languages under consideration.

When dealing with strings, we define a grammar as the four-tuple

$$G = (N, \Sigma, P, S) \tag{26}$$

where

  N is a finite set of nonterminals or variables,
  Σ is a finite set of terminals or constants,
  P is a finite set of productions or rewriting rules, and
  S in N is the starting symbol.

It is required that N and Σ be disjoint sets. In the following discussion nonterminals will be denoted by capital letters:

A,B,...,S,... Lowercase letters at the beginning of the alphabet
will be used for terminals: a,b,c,... Strings of terminals will
be denoted by lowercase letters toward the end of the alphabet:
v,w,x,y,z. Strings of mixed terminals and nonterminals will be
denoted by lowercase Greek letters: $\alpha,\beta,\theta,\ldots$ The empty sentence
(the sentence with no symbols) will be denoted by λ. Finally,
given a set V of symbols, we will use the notation V* to denote the
set of all sentences composed of elements from V.

String grammars are characterized primarily by the form of
their productions. Of particular interest in syntactic pattern
recognition are regular grammars, whose productions are always of
the form A→a or A→aB with A and B in N, and a in Σ, and context-
free grammars, with productions of the form A→α with A in N, and α
in the set $(N \cup \Sigma)^* - \lambda$, that is, α can be any string composed of
terminals and nonterminals, except the empty string.

The preceding concepts are best clarified by an example.
Suppose that the object shown in Figure 16(a) is represented by its
skeleton, and that we define the primitives shown in Figure 16(b)
to describe the structure of this and similar skeletons. Consider
the grammar G = (N,Σ,P,S,) with N = {A,B,S,}, Σ = {a,b,c}, and
production rules

P:

1) S→aA
2) A→bA
3) A→bB
4) B→c

where the terminals a, b, and c are as shown in Figure 16(b). As
indicated earlier, S is the starting symbol from which we generate
all strings in L(G). If, for instance, we apply production (1)
followed by two applications of production (2), we obtain: S=>aA=>
abA=>abbA, where "=>" indicates a string derivation starting from S
and using production rules from P. It is noted that we interpret
the production S→aA and A→bA as "S can be rewritten as aA" and "A
can be rewritten as bA." Since we have a nonterminal in the string
abbA and a rule which allows us to rewrite it, we can continue the
derivation. For example, if we apply production (2) two more
times, followed by production (3) and then production (4) we obtain
the string abbbbbc which corresponds to the structure shown in
Figure 16(c). It is important to note that no nonterminals are
left after application of production (4) so the derivation terminates after this production is used. A little thought will reveal
that the grammar given above has the language L(G) = $\{a\ b^n c | n \geq 1\}$,
where b^n indicates n repetitions of the symbol b. In other words,
G is capable of generating the skeletons of wrench-like structures

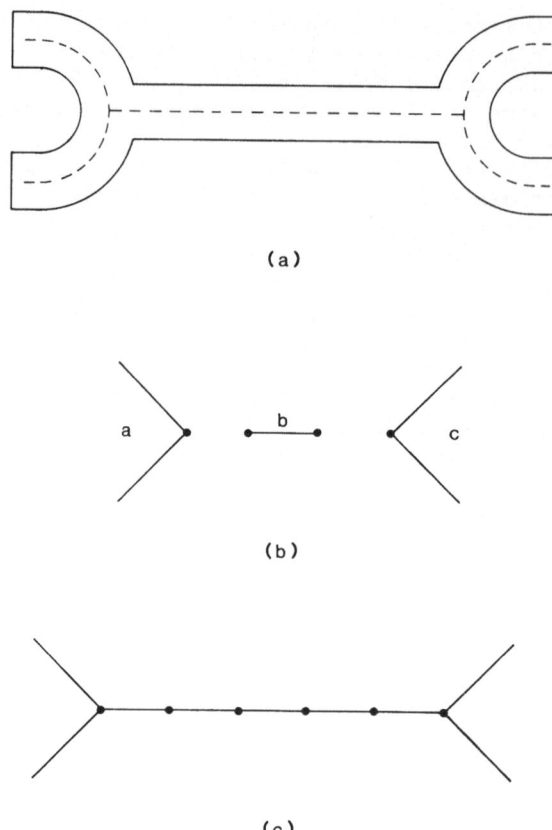

Figure 16. (a) Object represented by its skeleton. (b) Primitives. (c) Structure generated using a regular string grammar.

with bodies of arbitrary length within the resolution established by the length of primitive b.

Use of Semantics

In the above example we have implicitly assumed that the interconnection between primitives takes place only at the dots shown in Figure 16(b). In more complicated situations the rules of connectivity, as well as other information regarding factors such as primitive length and direction, and the number of times a production can be applied, must be made explicit. This is usually accomplished via the use of semantics. Basically, syntax establishes the structure of an object or expression, while semantics deal with its meaning. For example, the FORTRAN statement A = B/C is syntactically correct, but it is semantically correct only if $C \neq 0$.

In order to fix these ideas, suppose that we attach semantic information to the wrench grammar just discussed. Typically, this information is attached to the production as follows:

Production	Semantic Information
S→aA	Connections to "a" are made only at the dot. The direction of "a", denoted by r, is given by the direction of the perpendicular bisector of the line joining the endpoints of the two undotted segments. These line segments are 3 cm each.
A→ba	Connections to "b" are made only at the dots. No multiple connetions are allowed. The direction of "b" must be the same as that of "a" and the length of "b" is 0.25 cm. This production cannot be applied more than 10 times.
A→bB	The direction of "a" and "b" must be the same. Connections must be simple and only at the dots.
B→c	The direction of "c" and "a" must be the same. Connections must be simple and only at the dots.

It is noted that, by using semantic information, we are able to use a few rules of syntax to describe a broad (yet, limited as desired) class of patterns. For instance, by specifying the direction θ, we avoid having to specify primitives for each possible object orientation. Similarly, by requiring that all primitives be oriented in the same direction, we eliminate from consideration nonsensical wrench-like structures.

Recognition

Thus far, we have seen that grammars are generators of patterns. In the following discussion we consider the problem of recognizing if a given pattern string belongs to the language L(G) generated by a grammar G. The basic concepts underlying syntactic recognition can be illustrated by the development of mathematical models of computing machines, called automata, which, given an input pattern string, have the capability of recognizing whether or not the pattern belongs to a specified language or class. We will focus attention only on finite automata, which are the recognizers of languages generated by regular grammars.

A finite automaton is defined as a five-tuple

$$A = (Q, \Sigma, \delta, q_0, F) \qquad (27)$$

where Q is a finite, nonempty set of states, Σ is a finite input alphabet, δ is a mapping from $Q \times \Sigma$ (the set of ordered pairs formed from elements of Q and Σ) into the collection of all subsets of Q, q_0 is the starting state, and F (a subset of Q) is a set of final or accepting states. The terminology and notation associated with Equation (26) are best illustrated by a simple example. Consider an automaton given by Equation (27) where $Q = \{q_0, q_1, q_2\}$, $\Sigma = \{a, b\}$, $F = \{q_0\}$, and the mappings are given by $\delta(q_0, a) = \{q_2\}$, $\delta(q_0, b) = \{q_1\}$, $\delta(q_1, a) = \{q_2\}$, $\delta(q_1, b) = \{q_0\}$, $\delta(q_2, a) = \{q_0\}$, $\delta(q_2, b) = \{q_1\}$. If, for example, the automaton is in state q_0 and an "a" is input, its state changes to q_2. Similarly, if a "b" is input next, the automaton moves to state q_1, and so forth. It is noted that, in this case, the initial and final states are the same.

A state diagram for this automaton is shown in Figure 17. The state diagram consists of a node for each state and directed arcs showing the possible transitions between states. The final state is shown as a double circle and each arc is labeled with the symbol that causes that transition. A string w of terminal symbols is said to be accepted or recognized by the automaton if, starting in state q_0, the sequence of symbols in w causes the automaton to be in a final state after the last symbol in w has been input. For example, the automaton in Figure 17 recognizes the string w = abbabb, but rejects the string w = aabab.

Higher-Dimensional Grammars

The grammars discussed above are best suited for applications where the connectivity of primitives can be conveniently expressed in a string-like manner. In the following discussion we consider two examples of grammars capable of handling more general interconnection between primitives and subpatterns.

INDUSTRIAL COMPUTER VISION 377

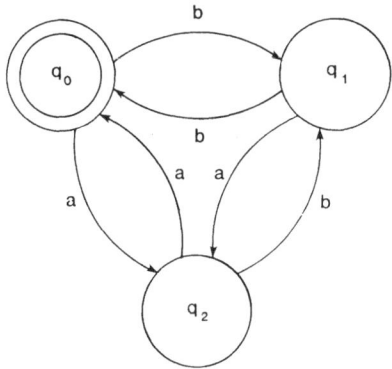

Figure 17. A finite automaton.

A tree grammar is defined as the five-tuple

$$G = (N, \Sigma, P, r, S) \qquad (28)$$

where N and Σ are, as before, sets of nonterminals and terminals, respectively; S is the start symbol which can, in general, be a tree; P is a set of productions of the form $T_i \rightarrow T_j$, where T_i and T_j are trees; and r is a ranking function which denotes the number of direct descendants of a node whose label is a terminal in the grammar. An expansive tree grammar has productions of the form

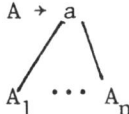

where $A, A_1, \ldots, A_n$ are nonterminals, and "a" is a terminal.

As an illustration, the skeleton of the structure shown in Figure 18(a) can be generated by means of a tree grammar with productions

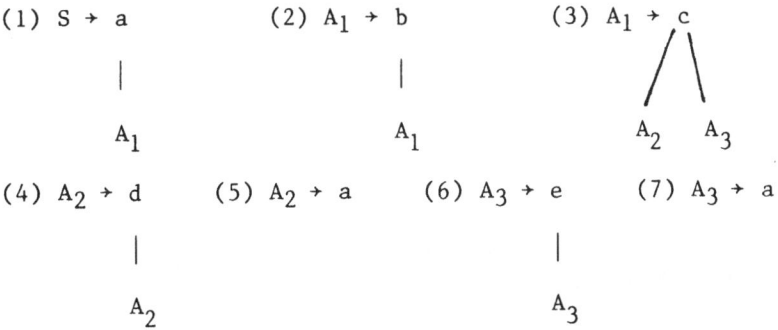

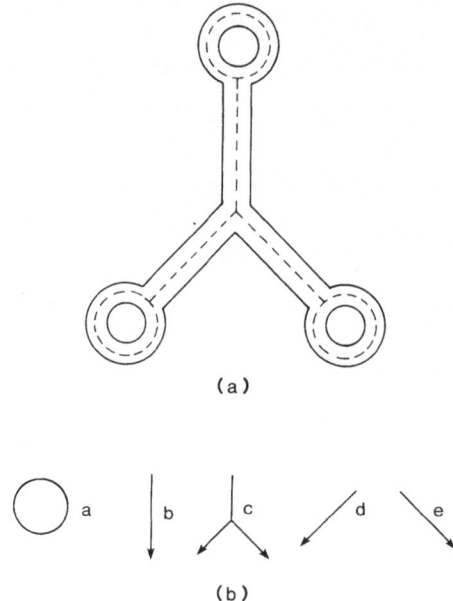

Figure 18. (a) An object and (b) Primitives used for representing the skeleton by means of a tree grammar.

where connectivity between linear primitives is head-to-tail, and connections to the circle primitive can be made anywhere on its circumference. The ranking functions in this case are $r(a) = \{0,1\}$, $r(b) = r(d) = r(e) = \{1\}$, $r(c) = \{2\}$. It is noted that restricting the use of productions (2), (4) and (6) to be applied to the same number of times generates a structure in which all three legs are of the same length. Similarly, requiring that productions (4) and (6) be applied the same number of times produces a structure that is symmetrical about the vertical axis in Figure 18(a).

INTERPRETATION

In this discussion, we view interpretation as the process which endows a vision system with a higher level of cognition about its environment than that offered by any of the concepts discussed thus far. When viewed in this way, interpretation clearly encompasses all these methods as an integral part of understanding a visual scene. Although this is one of the most active research topics in machine vision, the reader is reminded of the comments made earlier regarding the fact that our understanding of this area is really in its infancy. In this section we touch briefly upon a number of topics which are representative of current efforts toward advancing the state of the art in machine vision.

The power of a machine vision system is determined by its ability to extract meaningful information from a scene under a broad range of viewing conditions and using minimal knowledge about the objects being viewed. There are a number of factors which make this type of processing a difficult task, including variations in illumination, occluding bodies, and viewing geometry. Earlier we discussed techniques designed to reduce variability in illumination and thus provide a relatively constant input to a vision system. The back- and structured-lighting approaches discussed in that section are indicative of the extreme levels of specialization employed by current industrial systems to reduce the difficulties associated with arbitrary lighting of the workspace. Among these difficulties we find shadowing effects which complicate edge finding, and the introduction of nonuniformities on smooth surfaces which often results in their being detected as distinct bodies. Clearly, many of these problems result from the fact that relatively little is known about modeling the illumination-reflectance properties of 3D scenes. The line and junction labeling techniques represent an attempt in this direction, but fall short of explaining the interaction of illumination and reflectivity in quantitative terms. A more promising approach is based on mathematical models which attempt to infer intrinsic relationships between illumination, reflectance, and surface characteristics such as orientation [24,29,26].

Occlusion problems come into play when we are dealing with a multiplicity of objects in an unconstrained working environment. Consider, for example, the scene shown in Figure 19. A human observer would have little difficulty in, for example, determining the presence of two wrenches behind the sockets. For a machine, however, interpretation of this scene is a totally different story. Even if the system were able to perform a perfect segmentation of object clusters from the background, all the two-dimensional procedures discussed thus far for description and recognition would perform poorly on most of the occluded objects. Three-dimensional descriptors would have a better chance, but even they would yield incomplete information. For instance, several of the sockets would appear as partial cylindrical surfaces, and the middle wrench would appear as two separate objects.

Processing scenes, such as the one shown in Figure 19, requires the capability to obtain descriptions which inherently carry shape and volumetric information, and procedures for establishing relationships between these descriptions, even when they are incomplete. Ultimately, these issues will be resolved only through the development of methods capable of handling 3D information obtained either by means of direct measurements or via geometric reasoning

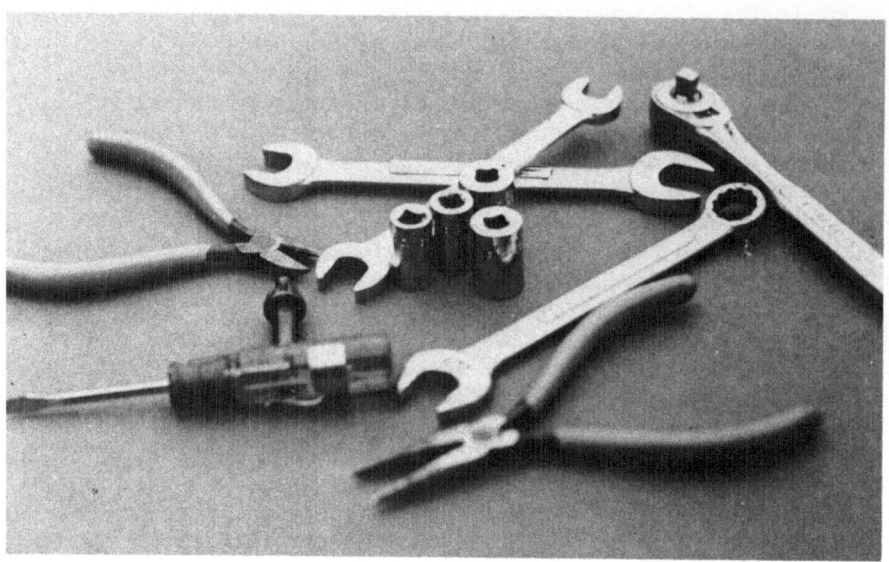

Figure 19. Oblique view of a three-dimensional scene.

techniques capable of inferring (but not necessarily quantifying) 3D relationships from intensity imagery.

As an example of this type of reasoning, the reader would have little difficulty in arriving at a detailed interpretation of the objects in Figure 19, with the exception of the object occluded by the screwdriver. The capability to know when interpretation of a scene or part of a scene is not an achievable task is just as important as correctly analyzing the scene. The decision to look at the scene from a different viewpoint (Figure 20) to resolve the issue would be a natural reaction in an intelligent observer.

One of the most promising approaches in this direction is research in model-driven vision [6]. The basic idea behind this approach is to base the interpretation of a scene on discovering instances of matches between image data and 3D models of volumetric primitives or entire objects of interest. Vision based on 3D models has another important advantage: it provides an approach for handling variances in viewing geometry. Variability in the appearance of an object when viewed from different positions is one of the most serious difficulties in machine vision. Even in two-dimensional situations where the viewing geometry is fixed, object orientation can strongly influence recognition performance if not handled properly. One of the advantages of a model-driven approach is that, depending on a known viewing geometry, it is possible to project the 3D model onto an imaging plane in that orientation and

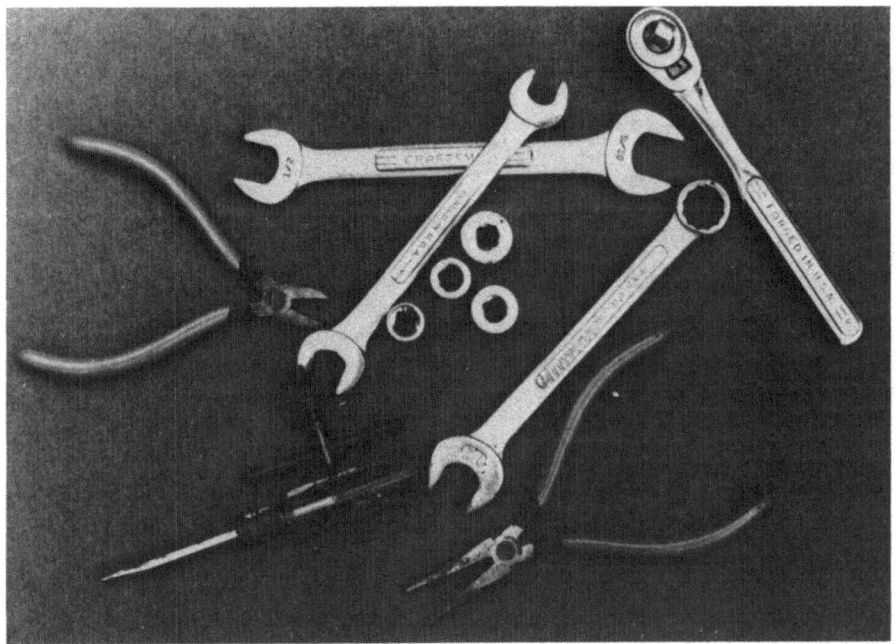

Figure 20. Another view of the scene shown in Figure 19.

thus simplify the match between an unknown object and what the system would expect to see from a given viewpoint.

CONCLUDING REMARKS

For the purpose of subdividing computer vision techniques into reasonable categories according to their function, we considered the vision process as consisting of sensing, preprocessing, segmentation, description, recognition, and interpretation. These areas were explained in some detail and their use in the context of industrial applications was discussed. Computer vision is a dynamic field which has been extensively investigated and reported in the literature. The references provided at the end of this chapter complement our discussion and give the reader a starting point for a broader study of this area, as well as related topics.

REFERENCES

[1] H. G. Barrow and J. M. Tenenbaum, "Computational vision," Proc. IEEE, vol. 69, pp. 572-595, 1981.

[2] J. Birk, R. Kelley, N. Chen, and L. Wilson, "Image feature extraction using diameter-limited gradient direction histograms," IEEE Trans. PAMI, vol. 1, no. 2, pp. 228-235, 1979.

[3] J. M. Brady (ed.), Computer Vision, North-Holland Publishing Company, Amsterdam, 1981.

[4] E. Bribiesca, "Arithmetic operations among shapes using shape numbers," Pat. Recog., vol. 13, no. 2, pp. 123-137, 1981.

[5] E. Bribiesca and A. Guzman, "How to describe pure form and how to measure differences in shape using shape numbers," Pat. Recog., vol. 12, no. 2, pp. 101-112, 1980.

[6] R. A. Brooks, "Symbolic reasoning among 3-D models and 2-D images," Artificial Intelligence, vol. 17, pp. 285-348, 1981.

[7] R. O. Duda and P. E. Hart, "Use of the Hough transformation to detect lines and curves in pictures," Comm. ACM, vol. 15, pp. 11-15, 1972.

[8] H. Freeman, "Shape description via the use of critical points," Proc. Conf. Pat. Recog. and Image Processing, IEEE Publ. 77CH1208-9 C, pp. 169-174, 1977.

[9] K. S. Fu, Syntactic Methods in Pattern Recognition, Academic Press, New York, 1974.

[10] K. S. Fu and J. K. Mui, "A survey of image segmentation," Pat. Recog., vol. 13, pp. 3-16, 1982.

[11] K. S. Fu, R. C. Gonzalez, and C. S. G. Lee, Introduction to Robotics, McGraw-Hill, New York, (to appear), 1985.

[12] R. C. Gonzalez, "How vision systems see," Machine Design, vol. 55, no. 10, pp. 91-96, 1983.

[13] R. C. Gonzalez and R. Safabakhsh, "Computer vision techniques for industrial applications," Computer, vol. 15, no. 12, pp. 17-32, 1982.

[14] R. C. Gonzalez, J. J. Edwards, and M. G. Thomason, "An algorithm for the inference of tree grammars," Int'l. Joint Comput. Inform. Sci., vol. 5, no. 2, pp. 145-164, 1976.

[15] R. C. Gonzalez and M. G. Thomason, Syntactic Pattern Recognition: An Introduction, Addison-Wesley, Reading, MA, 1978.

[16] R. C. Gonzalez and P. Wintz, Digital Image Processing, Addison-Wesley, Reading, MA, 1977.

[17] J. W. Goodman, Introduction to Fourier Optics, McGraw-Hill Book Co., New York, 1968.

[18] A. K. Griffith, "Edge detection in simple scenes using a-priori information," IEEE Trans. Comput., vol. 22, pp. 371-381, 1973.

[19] E. L. Hall, Computer Image Processing and Recognition, Academic Press, New York, 1979.

[20] R. M. Haralick, K. Shanmugan, and I. Dinstein, "Textural features for image classification," IEEE Trans. SMC, vol. SMC-3, no. 6, pp. 610-621, 1973.

[21] C. A. Harlow and S. A. Eisenbeis, "The analysis of radiographic images," IEEE Trans. Comput., vol. C-22, pp. 678-689, 1973.

[22] S. W. Holland, L. Rossol, and M. R. Ward, "CONSIGHT-1: A vision-controlled robot system for transferring parts from belt conveyors," in Computer Vision and Sensor-Based Robots (G. G. Dodd and L. Rossol, eds.), Plenum Press, New York, pp. 81-97, 1979.

[23] R. Horaud and J. P. Charras, "Automatic inspection and orientation of external screws," Proc. 5th Int'l Joint Conf. Pat. Recog., pp. 264-268, 1980.

[24] B. K. P. Horn, "Understanding image intensities," Artificial Intelligence, vol. 8, pp. 201-231, 1977.

[25] M. K. Hu, "Visual pattern recognition by moment invariants," IEEE Trans. Info. Theory, vol. 8, pp. 179-187, 1962.

[26] I. Katsushi and B. K. P. Horn, "Numerical shape from shading and occluding boundaries," Artificial Intelligence, vol. 17, pp. 141-184, 1981.

[27] R. B. Kelley, J. R. Birk, H. A. S. Martins, and R. Tella, "A robot system which acquires cylindrical workpieces from bins," IEEE Trans. SMC, vol. SMC-12, no. 2, pp. 204-213, 1982.

[28] C. S. G. Lee, J. Campbell, and G. Saridis, "Voice-controlled trainable manipulators with visual feedback," Proc. IEEE Conf. Decision & Control, pp. 1423-1428, 1979.

[29] D. Marr, "Visual Inf. Processing: The structure and creation of visual representations," Proc. Int'l Joint Conf. Artif. Intell., Tokyo, Japan, pp. 1108-1126, 1979.

[30] L. Mero and T. Vamos, "Medium level vision," in Progress in Pattern Recognition (L. Kanal and A. Rosenfeld, eds.), North-Holland Publishing Co., New York, 1981.

[31] U. Montanari, "On the optimal detection of curves in noisy pictures," Comm. ACM, vol. 14, pp. 335-345, 1971.

[32] J. L. Mundy, "Automatic visual inspection," Proc. 1977 Conf. Decision and Control, pp. 705-710, 1977.

[33] T. Pavlidis, "Algorithms for shape analysis of contours and waveforms," IEEE Trans. PAMI, vol. PAMI-2, no. 4, pp. 301-312, 1981.

[34] T. Pavlidis, "A review of algorithms for shape analysis," CGIP, vol. 7, pp. 243-258, 1978.

[35] T. Pavlidis, Structural Pattern Recognition, Springer-Verlag, New York, 1977.

[36] T. Pavlidis and F. Ali, "A hierarchical syntactic shape analyzer," IEEE Trans. PAMI, vol. PAMI-1, pp. 2-9, 1979.

[37] W. A. Perkins, "Area segmentation of images using edge points," IEEE Trans. PAMI, vol. PAMI-2, no. 1, pp. 8-15, 1980.

[38] W. A. Perkins, "Simplified model-based part locator," Proc. 5th Int'l Joint Conf. Pat. Recog., pp. 260-263, 1980.

[39] W. A. Perkins, "Model-based vision system for scenes containing multiple parts," Proc. Int'l Joint Conf. Artificial Intelligence, pp. 678-684, 1977.

[40] E. Persoon and K. S. Fu, "Shape discrimination using Fourier descriptors," IEEE Trans. SMC, vol. SMC-7, no. 3, pp. 170-179, 1977.

[41] U. Ramer, "Extraction of line structures from photographs of curved objects," CGIP, vol. 4, pp. 81-103, 1975.

[42] C. W. Richard, Jr. and H. Hemani, "Identification of three-dimensional objects using Fourier descriptors of the boundary curve," IEEE Trans. SMC, vol. SMC-4, no. 4, pp. 371-378, 1974.

[43] A. Rosenfeld, "Picture processing," CGIP, vols. 1-7, 1972-78.

[44] A. Rosenfeld, R. A. Hummel, and S. W. Zucker, "Scene labeling by relaxation operations," IEEE Trans. SMC, vol. SMC-6, pp. 420-433, 1976.

[45] A. Rosenfeld and A. C. Kak, Digital Picture Processing, Academic Press, New York, 1982.

[46] A. Rosenfeld and D. L. Milgram, "Web Automata and Web Grammars," in Machine Intelligence-7 (B. Miltzer and D. Michie, eds.), Wiley, New York, 1972.

[47] F. A. Sadjadi and E. L. Hall, "Three-dimensional moment invariants," IEEE Trans. PAMI, vol. PAMI-2, no. 2, pp. 127-136, 1980.

[48] P. Saraga, and D. R. Skoyles, "An experimental visually-controlled pick and place machine for industry," Proc. 3rd Int'l Joint Conf. Pat. Recog., pp. 17-21, 1976.

[49] Y. Shirai, "Three-dimensional computer vision," in Computer Vision and Sensor-Based Robots (G. G. Dodd and L. Rossol, eds.), Plenum Press, New York, 1979.

[50] Y. Shirai and H. Inoue, "Guiding a robot by visual feedback in assembling tasks," Pat. Recog., vol. 5, pp. 99-108, 1973.

[51] W. E. Snyder and D. A. Tang, "Finding the extrema of a region," IEEE Trans. PAMI, vol. PAMI-2, no. 3, pp. 266-269, 1980.

[52] T. W. Sze and Y. H. Yang, "A simple contour matching algorithm," IEEE Trans. Pattern Anal. Mach. Intell., vol. PAMI-3, no. 6, pp. 676-678, 1981.

[53] G. Y. Tang and T. S. Huang, "A syntactic-semantic approach to image understanding and creation," IEEE Trans. PAMI, vol. PAMI-1, pp. 135-144, 1979.

[54] J. M. Tenenbaum and H. G. Barrow, "IGS: A paradigm for integrating image segmentation and interpretation," Proc. 3rd Int'l Joint Conf. Pat. Recog., pp. 504-513, 1976.

[55] J. T. Tou and R. C. Gonzalez, Pattern Recognition Principles, Addison-Wesley, Reading, MA, 1974.

[56] T. P. Wallace and O. R. Mitchell, "Analysis of three-dimensional movement using Fourier descriptors," IEEE Trans. PAMI, vol. PAMI-2, no. 6, pp. 583-588, 1980.

[57] H. Wechsler and T. Citron, "Feature extraction for texture classification," Pat. Recog., vol. 12, no. 5, pp. 301-311, 1980.

[58] J. S. Weszka and A. Rosenfeld, "An application of texture analysis to material inspection," Pat. Recog., vol. 8, pp. 195-200, 1976.
[59] M. Yachida, M. Ikeda, and S. Tsuji, "A plan-guided analysis of cineangiograms for measurement of dynamic behavior of heart wall," IEEE Trans. PAMI, vol. PAMI-2, pp. 537-543, 1980.
[60] Y. Yakimovsky and R. Cunningham, "A system for extracting three-dimensional measurements from a stereo pair of TV cameras," CGIP, vol. 7, pp. 195-210, 1978.
[61] K. C. You and K. S. Fu, "Distorted shape recognition using attributed grammars and error-correcting techniques," CGIP, vol. 13, pp. 1-16, 1980.
[62] K. C. You and K. S. Fu, "A syntactic approach to shape recognition using attributed grammars," IEEE Trans. SMC, vol. MC-9, pp. 334-345, 1979.
[63] C. T. Zahn and R. Z. Roskies, "Fourier descriptors for plane closed curves," IEEE Trans. Comput., vol. 12, pp. 269-281, 1972.
[64] S. W. Zucker and R. A. Hummel, "A three-dimensional edge operator," IEEE Trans. Pattern Anal. Mach. Intell., vol. PAMI-3, no. 3, pp. 324-331, 1981.
[65] S. W. Zucker, "Region growing: Childhood and adolescence," CGIP, vol. 5, pp. 382-399, 1976.

INTEGRATING VISION AND TOUCH FOR GRASPING OF AN OBJECT

Ruzena Bajcsy

Computer and Information Science Department
University of Pennsylvania
Philadelphia, PA   19104   U.S.A.

ABSTRACT

The aim of this paper is to present considerations that go into the design of a system that will be able to grasp a variety of objects. Grasping depends on the assumption of two properties: The shape and size of the three-dimensional object and on the task that follows after grasping; hence, our interest in the integrated approach to 3-D shape recognition. We shall present an overview of currently available and being developed algorithms for representation and recognition of three-dimensional shape of an object as perceived by visual and/or tactile sensors. For the visual sensor we assume that we have available stereo cameras or their equivalent. As the tactile sensor we use an articulated multifingered hand equipped with tactile sensory arrays as the data acquisition device. We shall present available configurations of these devices. Then we shall investigate the sensory processing, in particular, what representation schemas should be considered. We shall argue that the object-centered representation as opposed to observer-centered representation is more important for grasping. Finally, a rule based schema for the control strategies will be outlined. As examples, first some artificial geometric objects then some real laboratory objects from the blocks world will be analyzed.

INTRODUCTION AND MOTIVATION

During the last decade advances have been made towards a realization of robots. The news media is full of reports on expansion of the use of robots in industry and justification of this use in terms of the increase in productivity, quality control, etc. The robots currently available are by and large unsophisticated and simple minded. They work in a very constrained and well-controlled environment with tasks that are restricted and deterministic. By this we mean control of illumination, a small number of different objects and/or parts with well specified shapes, materials and other physical properties. In addition, exact sequencing of operations on the assembly line and the procedures for manipulation of these objects is usually known. Under these restrictions it is relatively easy to design an automation (deterministic by definition) which will perform satisfactorily, and one does not need a robot. Indeed, frequently this is the case in today's factories. The difference between the automated machines and today's robots is the computer, which allows a robot to be reprogrammable and hence more flexible than a fixed automated machine. It is the flexibility which is so impressive about humans as opposed to machines which we are striving towards in our quest for future robots. This flexibility comes in part from the fact that people use more than one sensory input during their recognition procedure; hence, our interest in the integration of multisensors.

In this paper we shall concentrate on a subset of capabilities of a flexible future robot; that is, what is required to be able to grasp an object. There are many ways to grasp, hence one needs to identify constraints that limit the large number of possibilities to a few. Assuming that there is a free space between the hand and the object, these constraints will be determined from the 3-D shape of the object and the task that will follow after the object is grasped. Hence, our motivation to study the 3-D shape in an integrated fashion, i.e., via vision and touch. We shall assume that each sensor is processed by an individual processing module and then the results of the two processes are integrated during the recognition process. What will follow will be a short review of current vision and tactile modules, what each can do independently and then their integration in a recognition task.

VISION MODULE

The vision module is composed of three basic parts:

The data acquision part
The feature extraction part
The recognition part

Before we launch into the details of each of the parts, perhaps it will be useful to formulate the overall task of the vision module in a more informal way. Firstly, one has to recognize that the peculiarity about the visual data acquisition is that it takes the three-dimensional information and projects it into two-dimensional information. During this projection there is a loss of information.

Hence, the most difficult task of the vision module is how to reconstruct the three-dimensional information from these two-dimensional projections. An example of such a reconstruction method is stereo computation, for review see [1]. In our laboratory [12] we are also developing a stereo system of which some examples are shown in Figures 1a through 1e. The figures 1a and 1b represent the stereo pair. The figures 1c and 1d show the samples of matches determined by the algorithm. Finally, the figure 1e displays the disparity map. In addition to the loss of information due to projection there is a loss of information due to illumination, reflectance properties of the object and the inadequacy of the imaging device. The feature extraction process is determined by the different representations and primitives that are being considered. Finally, the recognition process will deal with problems of how much a priori knowledge will be needed, given the data, in order to recognize different objects.

THREE-DIMENSIONAL OBJECT REPRESENTATION AND FEATURE EXTRACTION

Given that we have scanned the 3-D object completely (from all angles), the question is what are the suitable representation, or parametrizations, or data reduction mechanisms for purposes such as storing, recognition and description. Geometrically, a three-dimensional object can be described by its surface characteristics, especially its symmetry surfaces and symmetry axis. From the point of view of topology, additional properties come into focus, such as: connectedness, number of holes, handles, etc. Although the data is perceived in the observer-centered coordinate system, as shown for example in Figures 2a, b, c, d, and e. For grasping purposes, however, one needs the object-centered coordinate system. This latter assumes that we have the complete scan of the data (not only half of the hemisphere), but perhaps with coarser resolution. In other words, for this task it is less important to know every detail of the surface but rather to recognize the overall shape of the object, such as whether it has joints, holes, is round, elongated or flat. Such classification is possible as we have shown in [4]. For example, in Figure 3 the cube is represented first by a large sphere (indicative of the cube symmetry) and then the 8 small spheres corresponding to the 8 corners.

In addition to this description the important shape information for grasping is surface patches with zero or small curvature so that the fingers can make as much contact as possible. This, of course, can be measured as shown in Figures 1e and 2e, and surfaces can be identified (see also [5]).

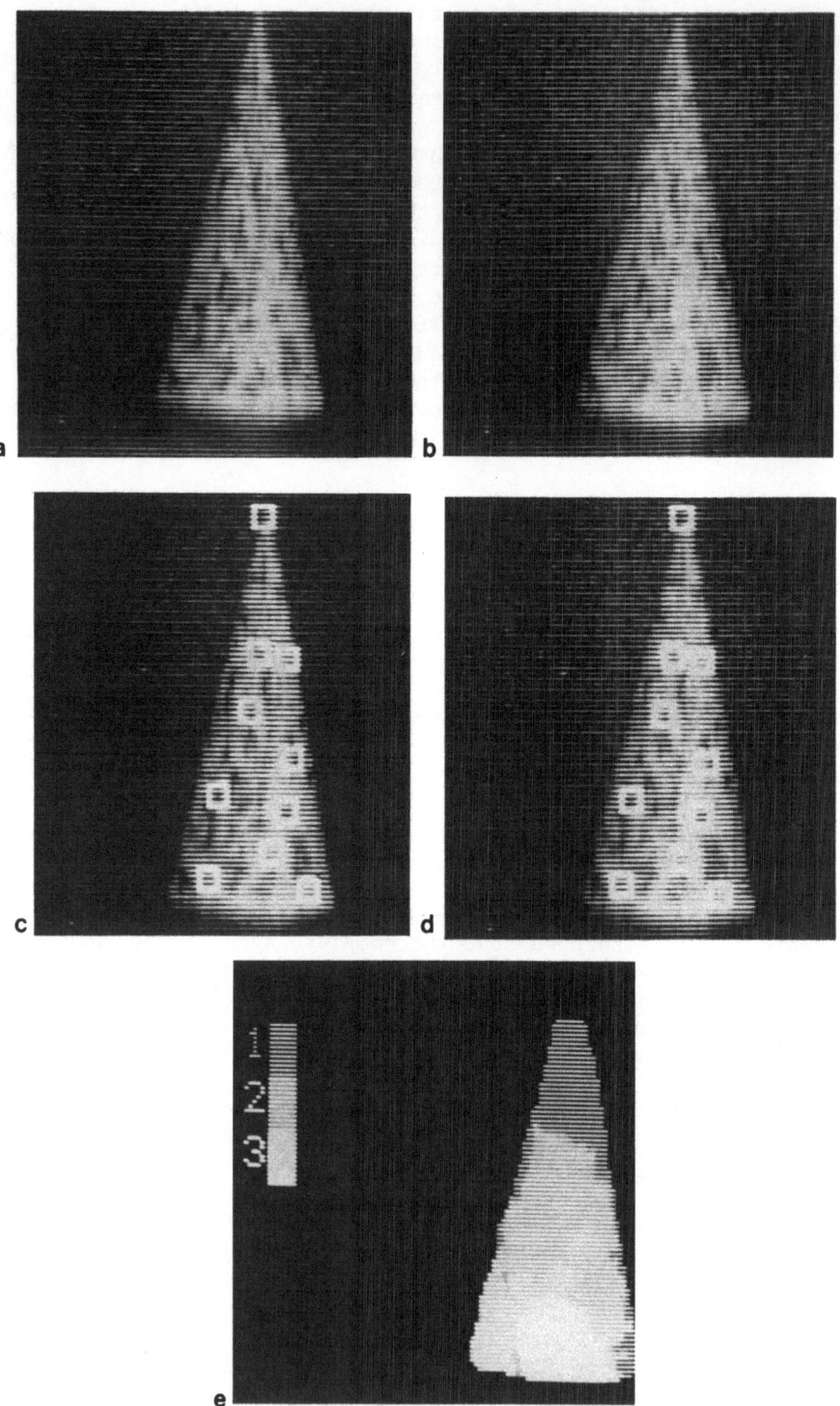

Figure 1

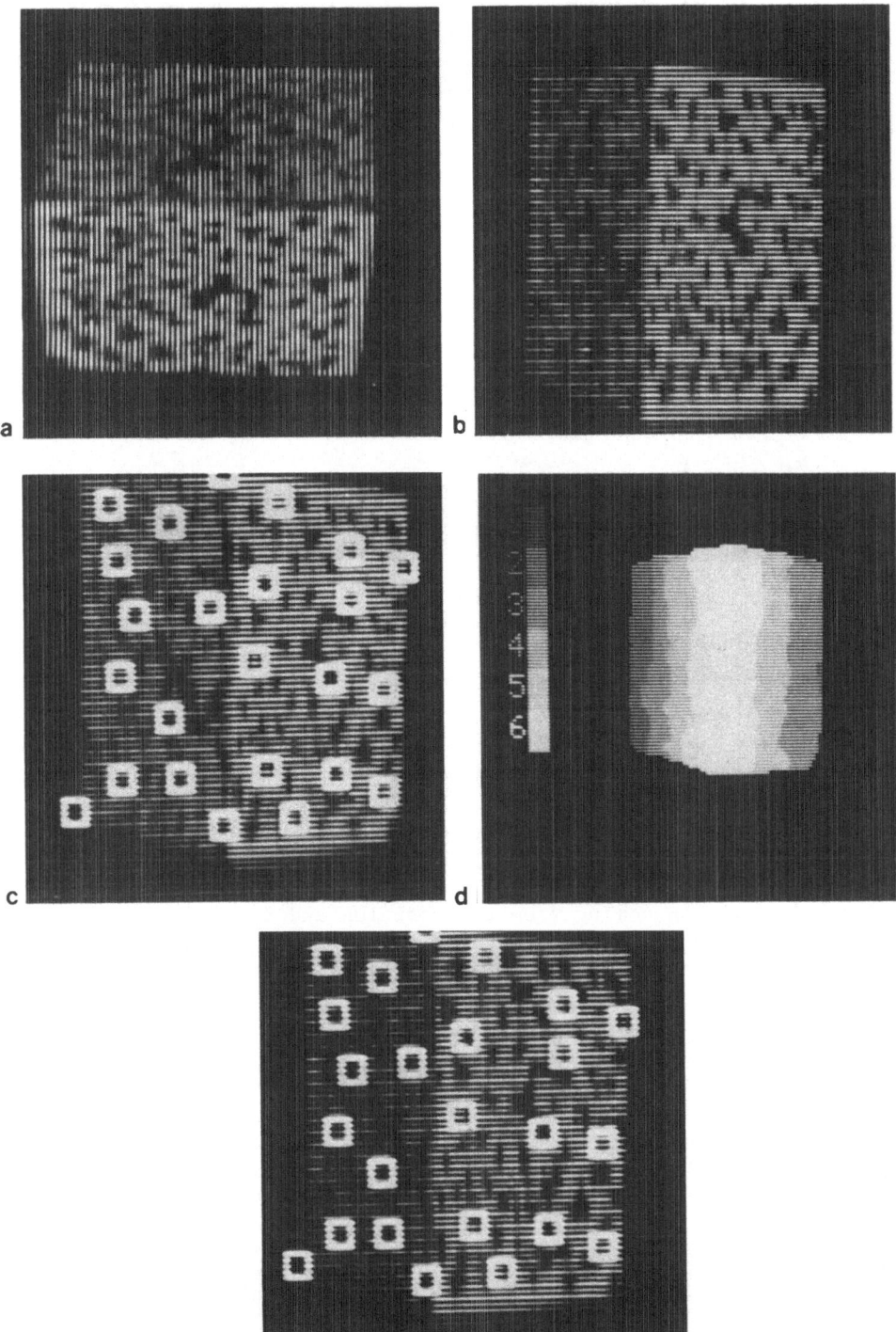

Figure 2

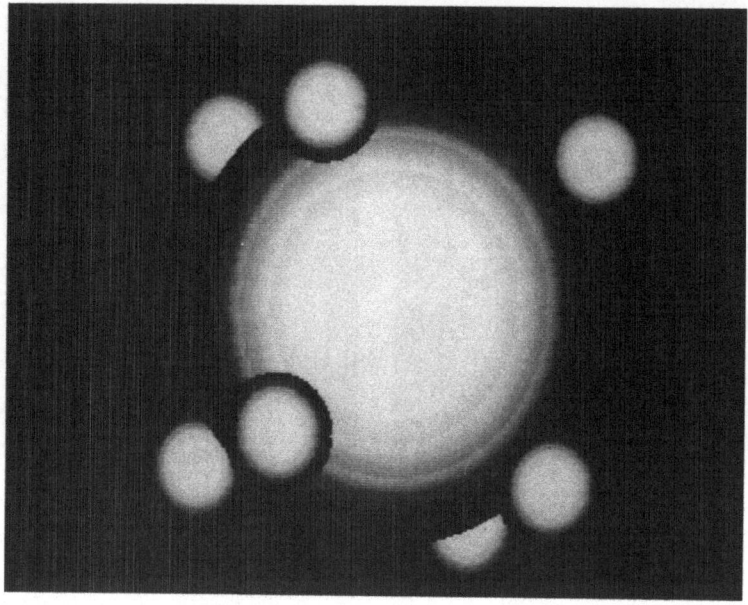

Figure 3.

THE TACTILE MODULE

Just like the two eyes equipped with all the processing power deliver the visual information about the world, the hand with its joints and skin can be viewed as a data acquisition system of tactile and kinematic information. Under the tactile module we include both the tactile sensation due to the skin and the kinematic information coming from the position and force sensation based on the joints of the fingers as well as on the wrist. Similarly, as in the previous section, there are three parts to this module:

The data acquisition part
The representation of the tactile and kinematic information
The recognition part

In the beginning of this section we must point out that research in tactile sensation is far less developed both in psychology and robotics literature. The psychological literature maintains that vision is the superior modality, although this is now beginning to be questioned. In any case, for a flexible robot there is need to have more than just one sensory input, especially if one considers situations when illumination may not exist, or during grasping situations.

INTEGRATING VISION AND TOUCH

THE DATA ACQUISITION SYSTEM

The data acquisition system is composed of two different systems: one is the hand, which is composed of multijointed fingers with position and force sensors in the joints, and the other is the skin, which is an array of pressure sensitive sensors distributed over pads which are placed between the joints. We shall not review all the articulated hands currently available in various laboratories; for that see [10]. Here we just wish to state that we assume availability of such devices equipped with tactile sensors. In our laboratory, for example, we have experimented with two such devices: a rigid finger with 133 tactile sensors obtained from LAAS in Toulouse, France and the other, a Lord Corp. Pad. As for the hand, recently we have developed a three-fingered hand with 7 degrees of freedom, as shown in Figure 4 (see also [17]). Currently, we are working on an array of tactile sensors which will be attached to each finger of the above hand. We feel, however, that the most important task is in the processing of the tactile information and the control of the articulated hand.

INFORMATION PROCESSING OF TACTILE SENSORY DATA

Recently, we and others [13,14] have started to focus on the different kinds of information processing algorithms that are

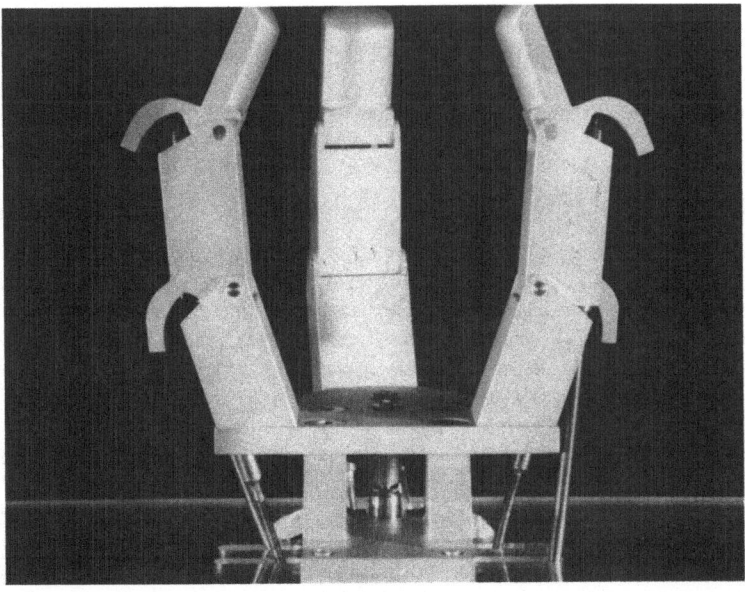

Figure 4.

desirable for tactile data. They can be put into the following categories:

1. Hardness measurements which allow one to classify the material according to its resiliency,

2. surface texture measurements,

3. surface normal measurements, boundary outlines and surface curvature measurements,

4. reconstruction of shapes from several tactile measurements, called also haptic information.

In our previous reports [6,3] we have shown that all the above properties can be measured with a certain accuracy depending on the properties of the tactile sensor itself.

Recently, we have completed an algorithm [2,15] which, using the finger, follows surfaces. This algorithm uses the following two rules:

1. Rule of continuity: based on the previous path, first follow the same direction as before.

2. Role of discontinuity: if you reach an edge move first north, east and then west in order to find the next surface.

INTEGRATION OF THE VISUAL AND TACTILE INFORMATION

Before one can think of integration of two or more modalities one needs firstly to understand the capabilities and the weaknesses of each independently. That was the reason we covered vision and touch separately.

There are, however, additional needs that follow from the integration process. Those are: the form of representation of the information that is collected from vision and touch and the control structure that will be employed using the two modalities for the recognition purposes. Hence, we view the integration process as a specially designed representation and control structure that uses two modalities: vision and touch. One can ask what is different about the representation of the visual and tactile information if treated independently versus if they are going to be used in an integrated system.

Comparison between vision and touch suggests that while the visual and the tactile sensors both have a 2-D receptive field, they do measure different physical properties of the world, i.e., reflectance and pressure. The eye and hand movement both provide 3-D information, but the eye has a higher spatial resolution than the tactile sensor. These similarities and differences of course imply different requirements for the knowledge representation and the control structure needed for recognition if the two modalities are available.

REPRESENTATION NEEDED FOR INTEGRATION

It is in this part where the difference as contrasted to the situation when one has only one kind of a sensor will be evident; that is to say, all the knowledge needed to supplement the inadequacy of one sensor only can be now assumed to be provided by the other sensor. Hence, the fixed a priori knowledge necessary for successful recognition will be much less since more and greater variety of measurements can be made. What we will have to provide is a mechanism (slots) for gathering this information and extracting the appropriate features. What kind of knowledge do we need?

1. Knowledge about the individual sensor. That is, for the camera we need the parameters about the focus, zoom, pan tilt, spatial resolution, aperture and light sensitivity, etc.

2. For the hand-touch acquisition device, we need the geometric, kinematic and dynamic model of the device. We have developed an internal model of the hand, as shown in Figures 5a and 5b, in order to model the control, the work spaces of the hand. For details, again see [16].

3. Signal processing procedures which will extract low level properties: edges, regions, texture descriptions, feature descriptions like corners, points of extreme, surface normals, elementary surface descriptions, hardness of material, etc.

4. Geometric models of three-dimensional objects, i.e., the classes of objects to be recognized.

5. Models of physics that are relevant for the object and its environment. That involves illumination and photogrammetry rules, law of gravity, law of statics and force distribution, law of elasticity, law of viscosity, etc.

THE CONTROL STRATEGIES

Here we shall consider the control strategies just for one purpose and that is to reconstruct the three-dimensional object from the partial visual and tactile measurements. There are three

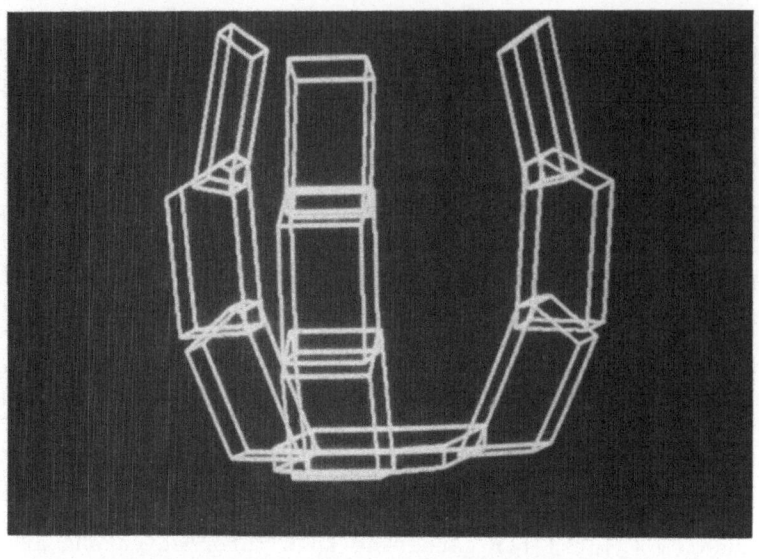

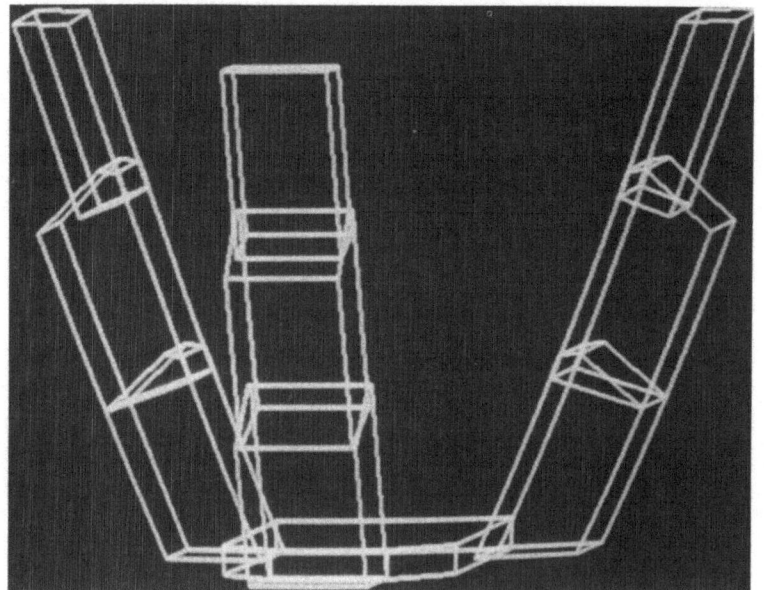

Figure 5.

stages during the data acquisition and recognition process. They are: investigation, measurement, and reconstruction.

During the investigation stage the eye and the hand are searching for the object. Once they have found the object they perform certain measurements and finally the reconstruction takes place. In general, the vision sensor is used with higher priority than the touch. It provides a more global view of the scene and it is used in the beginning to make a hypothesis. The touch is then used to verify this hypothesis. In the case when the visual recognition leads to multiple and/or ambiguous interpretations the touch will be used for resolving this ambiguity. We have performed such an experiment (see [11]). In the case of poor illumination the hand will be used first for making a hypothesis or just simply taking actions. For example, if one walks into a dark room, the first thing one does is look by touch for the switch to turn on the light. Or another example is if one looks for an object in a bag one first feels the object, then pulls it out for visual examination.

There is very little concrete work done in this area. Recently, we have started such work by implementing a simple scenario, where we used the visual information to predict the planar surfaces of a prism and then used the finger (as opposed to a hand) for verifying the visible surfaces as well as detecting the nonvisible surfaces [29]. Clearly this is just a mere beginning!

CONCLUSION

The above review represents the state of the art in computer vision tactile information processing as it is in the research and academic communities. Commercially available vision modules are by and large more primitive than those described in this paper. The vision modules usually use one camera (no binocular vision) with orthogonal projection (no perspective distortion) and only black and white as opposed to gray valued pictures. The software available can locate a small variety of parts if they are isolated, and recently even if they are touching or partially occluded.

The tactile sensors are even less developed than the vision modules. There is only one currently available and that is of the Lord Corp., which has an array of 8 by 8 pressure sensitive sensors on a pad. There is no software available to go with the sensor. Hence, there is need for a great deal of work.

The integration effort is just beginning. The advantage of having such an integrated system is in providing more complete measurements about the object and hence relying less on restrictions of the domain, set of predetermined shapes, in short, having more flexibility.

ACKNOWLEDGMENT

This work was in part supported by: NSF grant MCS-82-07294, Air Force grant AFOSR 82-NM-299, NIH grant 5-R01-HL-29985-02, NIH grant NS-10939-11 as part of Cerebro Vascular Research Center by DEC Corp., IBM Corp. and Lord Corp.

REFERENCES

[1] S. Barnard and M. Fischler, "Computational stereo," ACM Comp. Surveys, Vol. 14, No. 4, December 1982.
[2] L. Blumenthal, "A surface follower and edge detector for the finger," Computer and Information Science Dept., Univ. of Pennsylvania, Senior Project, May 1983.
[3] V. Buono, "The lord sensor," Computer and Information Science Dept., Univ. of Pennsylvania, Senior Project, May 1983.
[4] R. Mohr and R. Bajcsy, "Packing volumes by spheres," IEEE Trans. on PAMI, Vol. 6, No. 1, January 1983.
[5] C. A. Dane, "An object-centered three-dimensional model builder," Ph.D. dissertation, Computer Science Dept., Univ. of Pennsylvania, Philadelphia, 1982.
[6] R. Bajcsy, "What can one learn from one finger experiments?," Proceedings of the International Symposium of Robotics Research, 1983, Bretton Woods, NH, August 1983.
[7] G. I. Kinoshita, S. Aida, and M. Mori, "A pattern classification by dynamic tactile sense information processing," Pattern Recognition, Vol. 7, pp. 243-251, 1975.
[8] T. Okada and S. Tsuchiya, "Object recognition by grasping," Pattern Recognition, Vol. 9, pp. 111-119, 1977.
[9] H. Ozaki, S. Waku, A. Mohri, and M. Takata, "Pattern recognition of a grasped object by unit-vector distribution," IEEE Trans. on Systems, Man and Cybernetics, Vol. SMC-12, No. 3, pp. 315-324, May/June 1982.
[10] R. Bajcsy, "Shape From Touch," in Advances in Automation and Robotics, (G. N. Saridis, ed.) JAI Press, 1984.
[11] P. Allen, "Visually driven tactile recognition and acquisition," Image Processing and Robotics, Washington, DC, June 1982.
[12] L. DeRisi, "3-D data acquisition system," M.S. thesis, Univ. of Pennsylvania, Philadelphia, 1982.
[13] L. Harmon, "Automated tactile sensing," Int. J. Robotics Research, 1(2), pp. 3-33, 1982.
[14] W. D. Hillis, "A high resolution image touch sensor," Int. J. Robotics Research, 1(2) pp. 33-44, 1982.
[15] R. Bajcsy and K. I. Goldberg, "A new approach to robotic tactile perception," submitted to J. Cognition and Brain Theory, 1983.
[16] R. Bajcsy, M. J. McCarthy, and J. C. Trinkle, "Feeling by grasping," Proc. IEEE Int. Conf. on Robotics, Atlanta, 1984.
[17] J. D. Abramowitz, J. W. Goodnow, and B. Paul, "Pennsylvania Articulated Mechanical Hand," Proc. Int. Conf. on Computers and Engineering, ASME, Chicago, August 1983.

SPHERICAL SHADING CORRECTION OF EYE FUNDUS

IMAGE BY PARABOLA FUNCTION

Kozo Okazaki* and Shinichi Tamura†

*Department of Electrical Engineering
 Faculty of Engineering
 Tottori University
 Tottori, Tottori 680, Japan
†Department of Information and Computer Sciences
 Faculty of Engineering Science
 Osaka University
 Toyonaka, Osaka 560, Japan

ABSTRACT

In this paper we are presenting four methods of correcting spherical shading for eye fundus images by using a parabola function. The estimation of the correcting function is difficult since the characteristics of the shading are unknown and we must estimate the parameters from a limited field photograph.

(1) The hill climbing method; all the results obtained don't converge into the true values, but they depend rather on the initial values.

(2) The pattern search method; this is a simplification of the hill climbing method. The results obtained do not always converge into the true values.

(3) The razor search method; to improve the convergence of the pattern search method, we apply the razor search method additionally to the pattern search method.

(4) The slice method; when we binarize a spherical shaded image of eye fundus photograph by thresolding, a circular area is obtained. We can estimate the shading correcting function from this area. By this algorithm, we obtained the good corrected images that depict the blood vessels.

INTRODUCTION

In recent years, a great variety of biomedical images have been dealt with by digital processing techniques. Ocular fundus imaging has also been dealt with by some researchers.[1-6] We have studied the assembly of eye fundus photographs[7] and the leakage analysis for fluorescein ocular fundus angiography.[8] In these cases, we observed the cocentric shadings occasionally because of the inequality of illumination, or disagreement between the optical axes of the eye and camera. Dynamic thresholding, 2nd order curved surface approximation methods and the FFT method have been proposed for correction of shaded images. However, the spherical shading correction of eye fundus image is very difficult. Since there is limitation of photographic field size, the parameter estimation of obtaining the correcting function becomes nonlinear estimation.

We present in this paper four methods of correcting spherical shading for eye fundus images, by using a parabola function.

HILL CLIMBING METHOD

Assume we have a cost function

$$J(\underline{C}) = \sum_{\underline{X}} Q(\underline{X},\underline{C}) \tag{1}$$

for a scalar $Q(\underline{X},\underline{C})$, which has a gradient $\nabla_{\underline{C}} Q(\underline{X},\underline{C})$, where the summation is taken all over the pixels with limited size and

$$\underline{X} = (x,y) \text{ ; pixel} \tag{2}$$

$$\underline{C} = (A, X_0, Y_0, Z_0) \text{ ; parameter of correcting function} \tag{3}$$

$$Q(\underline{X},\underline{C}) = [g(\underline{X}) - f(\underline{X},\underline{C})]^2 \tag{4}$$

$$f(\underline{X},\underline{C}) = A(x-X_0)^2 + A(y-Y_0)^2 + Z_0 \text{ ; correcting function} \tag{5}$$

$g(\underline{X})$; image function

The hill climbing method is a method of obtaining $\underline{C}^*$ which minimizes $J(\underline{C})$ by iteratively modifying parameters as follows:

$$\underline{C}^{(k+1)} = \underline{C}^{(k)} + \Delta\underline{C} \tag{6}$$

$$\Delta\underline{C} = \gamma \frac{\partial J}{\partial \underline{C}} \bigg|_{\underline{C}=\underline{C}^{(k)}} \tag{7}$$

SPHERICAL SHADING CORRECTION

$$\underline{Y} = \begin{bmatrix} \gamma_1 & 0 & 0 & 0 \\ 0 & \gamma_2 & 0 & 0 \\ 0 & 0 & \gamma_3 & 0 \\ 0 & 0 & 0 & \gamma_4 \end{bmatrix} \qquad (8)$$

To obtain $\underline{C}^*$, we must determine a step size $\underline{\gamma}$ and initial values $\underline{C}^{(0)} = (A^{(\overline{0})}, X_0^{(0)}, Y_0^{(0)}, Z_0^{(0)})^T$ that give rapid convergence.

In Table 1, results of computer simulations are shown for the artificial images (70x70 pixels) which are generated by a computer. In the table the figures are converged values of $\underline{C}^{(k)}$ for some γ's. The mark ** means the oscillated case and the values show the maximum and minimum ones. All the results obtained don't converge into the true values, but they rather depend on the initial value $\underline{C}^{(0)}$ and the step size $\underline{\gamma}$. The reason may be the nonlinearity in estimating the parameter $\underline{C}$. It is due to the product of the parameter A and X_0, A and Y_0, and the limitation of the visual field of eye fundus photograph.

To determine the initial values, an all-checking method is presented in the following: the ranges of each parameter A, X_0, Y_0 and Z_0 are restricted in practice. Therefore, we may take, for example, five points for each parameter. Hence, we can calculate directly the cost function for five cases. We can determine the initial parameters which give the minimum cost function of all 5^4 cases.

PATTERN SEARCH

The pattern search method is a simplification of the hill climbing method. It need not use differentials of the cost function, but directly calculate it and search the minimum point. We examine the two pattern search methods.

All Vicinity Pattern Search

Let

$$\underline{C}^{(k)} = [C_1^{(k)}, C_2^{(k)}, C_3^{(k)}, C_4^{(k)}]^T \qquad (9)$$

be the parameter which is determined after k-th iteration, where $C_1^{(0)} = A^{(0)}$, $C_2^{(0)} = X_0^{(0)}$, $C_3^{(0)} = Y_0^{(0)}$, $C_4^{(0)} = Z_0^{(0)}$. Then, $\underline{C}^{(k+1)}$ is such parameter that attains the minimum of the cost function $J(\underline{C})$ of all combination points (3^4 cases) of

$$\begin{bmatrix} c_1^{(k)}+\Delta_1^1, & c_2^{(k)}+\Delta_2^1, & c_3^{(k)}+\Delta_3^1, & c_4^{(k)}+\Delta_4^1 \\ c_1^{(k)}, & c_2^{(k)}, & c_3^{(k)}, & c_4^{(k)} \\ c_1^{(k)}-\Delta_1^1, & c_2^{(k)}-\Delta_2^1, & c_3^{(k)}-\Delta_3^1, & c_4^{(k)}-\Delta_4^1 \end{bmatrix}^T \quad (10)$$

The converged condition is decided by

$$\| \underline{c}^{(k+1)} - \underline{c}^{(k)} \| < \varepsilon_1 \quad (11)$$

and if this condition is satisfied, then set the initial parameter $[c_1^{(0)}, c_2^{(0)}, c_3^{(0)}, c_4^{(0)}]^T = \underline{c}^{(k+1)}$, replace $\varepsilon_2 \leftarrow \varepsilon_1$, $\Delta_1^2 \leftarrow \Delta_1^1$, $\Delta_2^2 \leftarrow \Delta_2^1$, $\Delta_3^2 \leftarrow \Delta_3^1$, $\Delta_4^2 \leftarrow \Delta_4^1$, and repeat (10) and (11). The values for ε_1 and $\underline{\Delta}^1$ are determined in advance, and $\Delta^{m+1} = \beta\Delta^m$; $m = 1, 2, \ldots$, where $0<\beta<1$, and $\Delta^m = (\Delta_1^m, \Delta_2^m, \Delta_3^m, \Delta_4^m)^T$.

Pattern Search in Serial Order Perturbation[9]

When we must estimate many parameters like in the design of antennas, calculation minimizing the cost function is very difficult. In these cases, the pattern search in serial order perturbation is often used. Let (9) be the parameter which is determined after the k-th iteration. Parameter $\underline{c}^{(k+1)}$ is determined as follows: First, the variable $c_1^{(k)}$ is changed by $\pm\Delta_1^{(k)}$. The cost functions are calculated for three cases of

$$\left. \begin{array}{l} [c_1^{(k)}+\Delta_1^{(k)}, c_2^{(k)}, c_3^{(k)}, c_4^{(k)}]^T \\ [c_1^{(k)}, c_2^{(k)}, c_3^{(k)}, c_4^{(k)}]^T \\ [c_1^{(k)}-\Delta_1^{(k)}, c_2^{(k)}, c_3^{(k)}, c_4^{(k)}]^T \end{array} \right\} \quad (12)$$

and its minimum point (parameter (13) attaining the minimum of the cost function) is denoted by $\underline{c}^{(k)}{}_1$, i.e.,

$$\underline{c}^{(k)}{}_1 = [c_1^{(k)*}, c_2^{(k)}, c_3^{(k)}, c_4^{(k)}]^T \quad (13)$$

where $c_1^{(k)*} \varepsilon \{c_1^{(k)}+\Delta_1^{(k)}, c_1^{(k)}, c_1^{(k)}-\Delta_1^{(k)}\}$. Next, the same operation is done concerning $c_2^{(k)}$ also among three points

$$\left.\begin{array}{c}[c_1^{(k)*},\ c_2^{(k)}+\Delta_2^{(k)},\ c_3^{(k)},\ c_4^{(k)}]^T \\[6pt] [c_1^{(k)*},\ c_2^{(k)},\ c_3^{(k)},\ c_4^{(k)}]^T \\[6pt] [c_1^{(k)*},\ c_2^{(k)}-\Delta_2^{(k)},\ c_3^{(k)},\ c_4^{(k)}]^T\end{array}\right\} \quad (14)$$

Among (14), let a parameter which makes the cost function minimum be $\underline{C}^{(k)}{}_2$, where

$$\underline{C}^{(k)}{}_2 = [c_1^{(k)*},\ c_2^{(k)*},\ c_3^{(k)},\ c_4^{(k)}]^T \quad (15)$$

After four iterative operations, the point $\underline{C}^{(k)}{}_4$ is denoted as $\underline{B}^{(k+1)}$. Then, $\underline{C}^{(k+1)}$ is determined as follows:

if $J[\underline{B}^{(k)} + \alpha(\underline{B}^{(k+1)} - \underline{B}^{(k)})] < J(\underline{B}^{(k+1)})$,

Then $\underline{C}^{(k+1)} = \underline{B}^{(k)} + \alpha(\underline{B}^{(k+1)} - \underline{B}^{(k)})$,

else $\underline{C}^{(k+1)} = \underline{B}^{(k+1)}$, and $\underline{\Delta}^{(k+1)} \leftarrow \underline{\Delta}^{(k)}$

where α is predetermined such that $\alpha > 1$, $\underline{B}^{(0)} = \underline{C}^{(0)}$, and $\underline{\Delta}^{(k)} = (\Delta_1^{(k)}, \Delta_2^{(k)}, \Delta_3^{(k)}, \Delta_4^{(k)})^T$, and $0 < \beta < 1$. The convergence condition is the same as (11).

Figure 1 shows the results of the computer simulation of convergence of the method (i) all vicinity pattern search and (ii) pattern search in serial order perturbation by using the same artificial images as Table 1. Though the method (i) needs much calculation, it shows more stable convergence than that of (ii). The step size closely relates the convergence. If $\underline{\Delta}$ is too small, the amount of the calculation increases. If $\underline{\Delta}$ is too large, the convergence process becomes oscillating.

RAZOR SEARCH[10]

The results obtained above do not always converge to true values. This is a disadvantage. To compensate for it, we further apply the razor search method to the converged results of the pattern search.

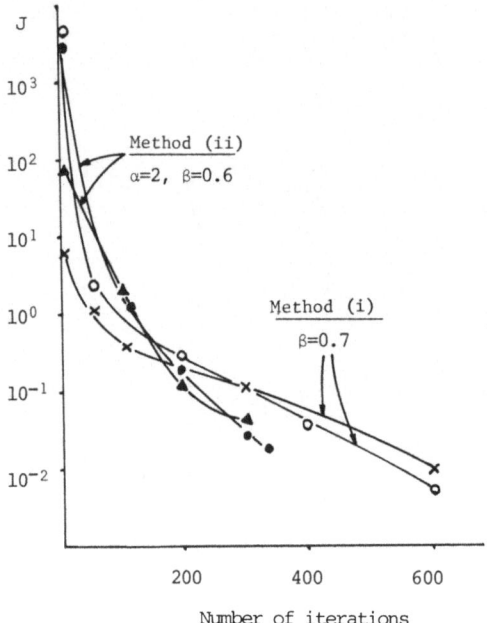

Figure 1. Convergence curve of performance function $J(\underline{C})$ vs. number of iterations.

We obtain the first converged values $^1\underline{C}$ of the parameter by the pattern search. Next, we determine the new initial values of parameters by (17)

$$^2\underline{C}(0) = {}^1\underline{C} \, (\underline{I} + \rho\underline{R}) \tag{17}$$

$$\underline{R} = \begin{bmatrix} R_1 & 0 & 0 & 0 \\ 0 & R_2 & 0 & 0 \\ 0 & 0 & R_3 & 0 \\ 0 & 0 & 0 & R_4 \end{bmatrix} \tag{18}$$

where ρ is a scale factor and is given in advance (the order of 0.1), R_1, R_2, R_3, R_4 is the random number such R_1, R_2, R_3, $R_4 \in [0,1]$. This new value $^2\underline{C}(0)$ is used as the initial value of the 2nd pattern search. We try these iterations 10 times, and we determine the final estimation of the parameter by the minimum among them. The random number matrix $\underline{R}$ will give chances of escaping from local minimums to the parameter $\underline{C}$.

Table 1. Converged Values of the Parameter $\underline{C}^{(k)}$ in the Simulation
Real values; $(A, X_0, Y_0, Z_0) = (150, 130, 110, 140)$
Initial values; $(A^{(0)}, X_0^{(0)}, Y_0^{(0)}, Z_0^{(0)}) = (15, 40, 60, 20)$
$\gamma_2 = \gamma_3 = 200$; const.,
Number of iterations = 12

	A	X_0	Y_0	Z_0
$\gamma_1 = 100$ $\gamma_4 = 5$	** 92 103	128	108	** 157 137
$\gamma_1 = 200$ $\gamma_4 = 4$	** 53 57	128	108	** 154 140
$\gamma_1 = 300$ $\gamma_4 = 3$	** 63 66	128	108	** 162 140

SLICING METHOD

This method reduces the calculation time than that of pattern search. If we binarize a spherically shaded image such as an eye fundus photograph by thresholding, a circular area is obtained. We can directly estimate the correcting parameters from these circular areas. By this algorithm, we can obtain good spherical shading correction. The steps of this algorithm are as follows:

(i) Using N different levels of threshold, we obtain N binarized images.

(ii) Assuming the binarized images have circular areas, we obtain equations of the circle by the hill climbing method. By this, we can obtain the radius r_i and the center (x_{0i}, y_{0i}) easily.

(iii) We obtain X_0 and Y_0 from the centers of the circles

$$X_0 = \frac{1}{N} \sum_i^N x_{0i}, \qquad Y_0 = \frac{1}{N} \sum_i^N y_{0i}$$

(iv) We obtain the coefficient of the shading correcting function of (5) using X_0, Y_0 and r_i by the least-squared method. Thus, we can determine the correcting function $f(\underline{X},\underline{C})$.

In Figure 2, an example of an almost normal fluorescein eye fundus photograph is shown; gray level is 0-127 and composed of 140x140 pixels. We obtained (19) as the correcting function for Figure 2.

$$f(\underline{X},\underline{C}) = -0.0025(x-35)^2 - 0.0025(y-100)^2 + 78 \qquad (19)$$

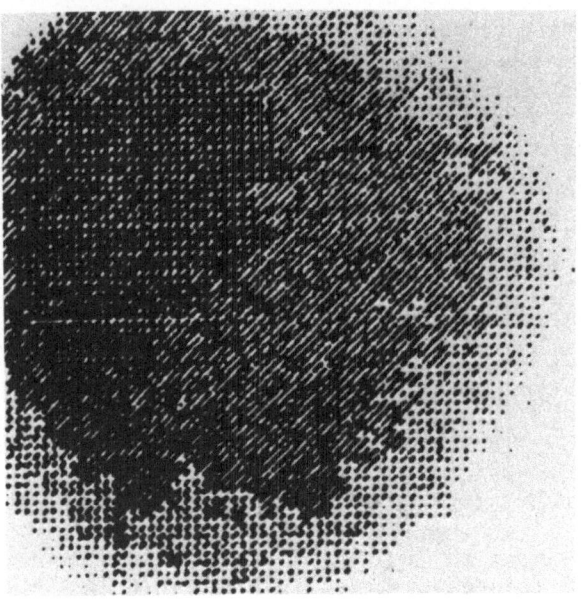

Figure 2. Original fluorescein eye fundus photograph; (almost normal), (140x140 pixels).

Figure 3 and Figure 4 show the effect of the shading correction by the slicing method. The practical correcting function $g*(\underline{X})$ contains bias term and gray level expansion as follows:

$$g*(\underline{X}) = 2.5[g(\underline{X})-f(\underline{X},\underline{C})+23] \qquad (20)$$

We assume in this paper the correcting function to be a parabola of revolution. Inspecting the corrected images on the basis of a blood vessel, the shaded images are improved fairly well. However, the practically derived shading correcting function may not always be the parabola of revolution. This is the remaining problem.

REFERENCES

[1] B. H. McCormic, J. S. Read, R. T. Borovec, B. C. Amendola, and M. H. Goldbaum, "Image Processing in Television Ophthalmoscopy," Digital Processing of Biomedical Images, K. Preston, Jr. and M. Onoe, eds., pp. 399-424, University of Tokyo Press, Tokyo, 1976.

[2] S. Yamamoto and H. Yokouchi, "Automatic Recognition of Color Fundus Photographs," Digital Processing of Biomedical Images, K. Preston, Jr. and M. Onoe, eds., pp. 385-398, University of Tokyo Press, Tokyo, 1976.

SPHERICAL SHADING CORRECTION

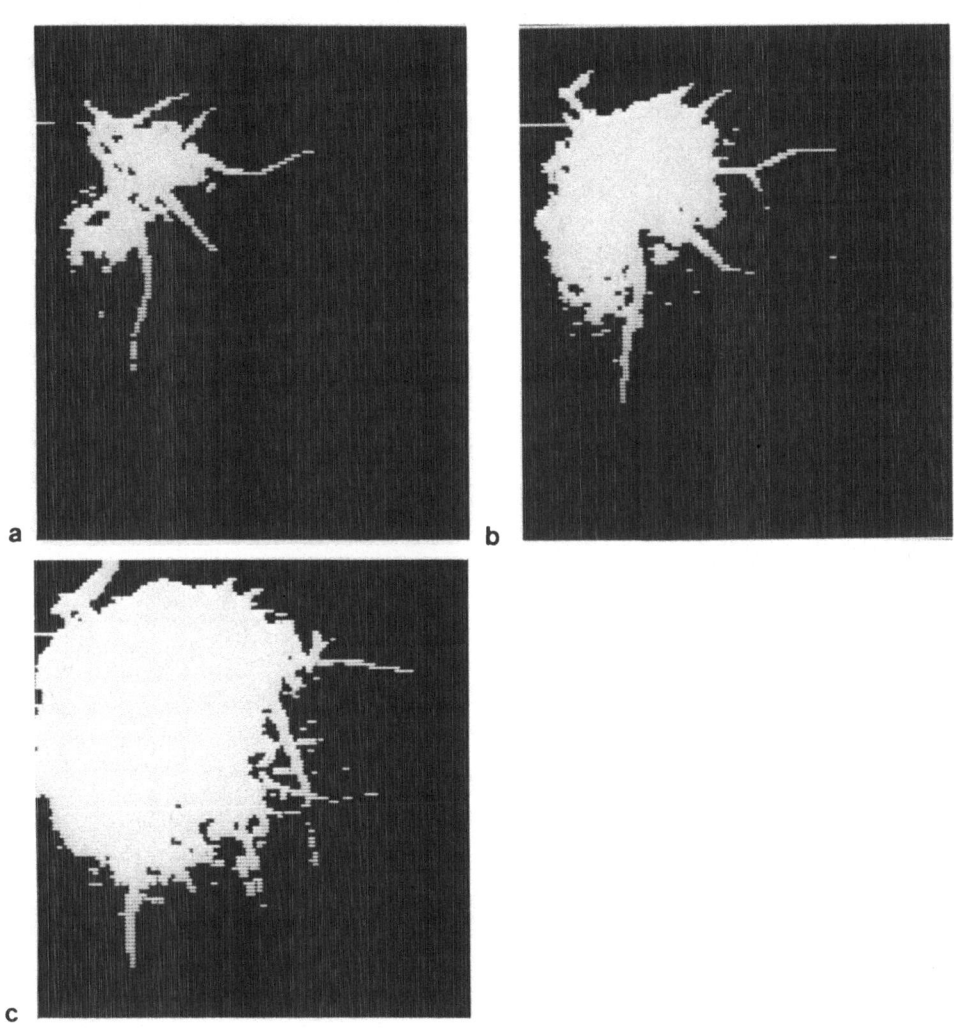

Figure 3. Binarized images of Fig. 2 (Before correction)
    (a) slice level = 78
    (b) slice level = 74
    (c) slice level = 67

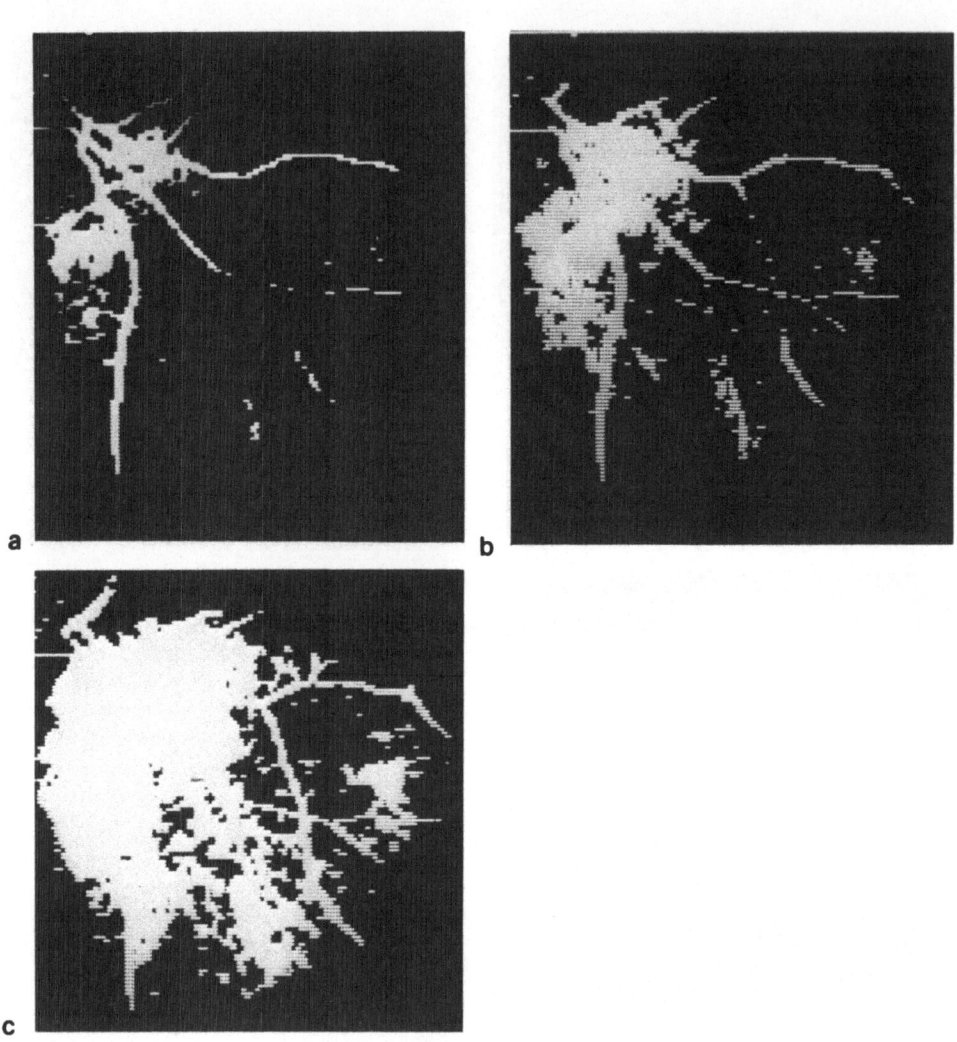

Figure 4. Binarized images after the shading correction
    (a) slice level = 57
    (b) slice level = 52
    (c) slice level = 43

[3] K. Akita and H. Kuga, "Towards understanding color ocular fundus images," Proceedings of the International Joint Conference on Artificial Intelligence, Tokyo, pp. 7-12, 1979.

[4] R. S. Ledley, "Computerized Pattern Recognition in Health Care Delivery," Health Hand Book, G. K. Chacko, ed., pp. 533-567, North-Holland, Amsterdam, 1979.

[5] M. Zingirian, C. Braccini, S. Gaglio, G. Marino, and V. Tagliasco, "A processing system oriented to the management of images acquired by different ophthalmological diagnosis methods," Proceedings of Computers in Ophthalmology, IEEE, pp. 111-122, 1979.

[6] J. S. Read, A. C. Petersen, B. H. McCormic, and M. F. Goldberg, "The television ophthalmoscope image processor: Methods and applications," Proceedings of Computers in Ophthalmology, IEEE, pp. 123-132, 1979.

[7] M. Tanaka, S. Tamura, and K. Tanaka, "A technique for the automatic assembly of eye fundus photographs using blood vessel structure," Proceedings of the International Conference on Cybernetics and Society, Tokyo and Kyoto, pp. 266-270, 1978.

[8] S. Tamura, K. Tanaka, S. Ohmori, K. Okazaki, A. Okada, and M. Hoshi, "Semiautomatic leakage analyzing system for time series fluorescein ocular fundus angiography," Pattern Recognition, 16(2), pp. 149-162, 1983.

[9] R. Hooke and T. A. Joeves, "Direct search solution of numerical and statistical problems," J. ACM, 8(2), p. 212, 1961.

[10] J. W. Bandler and P. A. MacDonald, "Optimization of microwave networks by razor search," IEEE Trans. Microwave Theory & Tech., MTT-17(8), p. 552, 1969.

AUTOMATIC WELDING: INFRARED SENSORS FOR PROCESS CONTROL

Bryan A. Chin and Nels H. Madsen

Department of Mechanical Engineering
Auburn University
Auburn University, AL 36849

ABSTRACT

A feasibility study was performed to determine if infrared thermography could be used to detect perturbations in the arc welding process which result in defects. Data were gathered using an infrared camera with a resolution of .2C which was trained on the molten metal pool during welding. Several defects were then intentionally induced and the resulting thermal images were preserved on film. These images revealed that different types of weld defects induce different characteristic changes in the thermal image by detectably altering the temperature field around the weld. These perturbations in the temperature field can be used to identify and locate defects such as arc misalignment, plate gap, puddle impurities etc. Macrostructural examinations permitted investigations into the relationships between weld puddle penetration depth and the temperature field. Using computer aided processing of these thermal images, it is expected that the welding process can be controlled to a higher degree than is presently possible.

INTRODUCTION

Automatic welding is commonplace in industry today. However, most, if not all, involves preprogramming for performance of a repetitive task. These systems are incapable of correcting for perturbations which arise during the welding process. Control of the welding process requires the identification and monitoring of perturbations in a wide variety of parameters. The varied nature of these parameters and the large number of variables involved have thwarted previous attempts at closed loop control of the welding process. The sensor(s) employed by a control system need to meet rather strenuous requirements. First and foremost of these is that any such sensor must be able to identify and discriminate between perturbations that might affect the quality of the weld, including such geometrical perturbations as a change of direction in the seam between the parts being welded, gaps in the seam or a misalignment of the parts being welded. The sensor should also be able to locate and identify contaminants and impurities in the weld puddle and ahead of it while monitoring the extent of the heat-affected zone and the cooling rate of the metal. The sensor(s) should, additionally, be able to perform these tasks in a time frame that is consistent with automatic control of the welding process. This paper presents results that could aid in the assembly of such a controlled welding system.

The high temperatures associated with metal arc welding and the appropriate thermophysical properties (primarily thermal diffusivity) of the parts to be welded cause very strong spatial temperature gradients (of over 10^5 C/m) in the vicinity of the weld pool surface. Modern infrared thermography equipment permits rapid sensing of these gradients with a high degree of accuracy. During an ideal weld these gradients should show repeatable and regular patterns. However, imperfections should cause a discernable change in the thermal profiles. Based on this hypothesis, experiments were carried out to determine the quality and magnitude of the thermal changes as a function of specific induced weld flaws, arc or plate misalignment, penetration, and contamination.

Since World War II, engineers have been striving to develop fully automatic welding machines. Today, machines have progressed to a state where preprogrammed parameters can be set as a function of time or position. A few companies have begun to market systems with limited in-process feedback sensing for applications such as V-notch seam tracking. Welding automation research has concentrated principally on systems for seam or joint tracking [1-6], magnetic control of arcs for positioning [7-11], and penetration control through puddle characterization [12-15]. Attempts to adapt intelligent vision systems to seam tracking and weld puddle control have also been made [16-20]. Ultrasonics [21-23], acoustic emission [24-26], and infrared temperature [27-31] sensing systems have also been used as sensors, primarily as after-the-fact inspection devices. Recently, infrared thermography has been used to monitor cooling rates in welds as a possible means of on-line control of heat input [32]. Although the above research has led to increased knowledge of arc welding physics, an acceptable sensing system for in-process welding and quality control has not been developed. The development of improved sensors is required to achieve wide scale automation of the welding process.

EXPERIMENTAL PROCEDURE

To demonstrate the feasibility of using infrared thermography to detect and identify various impending weld problems, a set of experiments was performed. The results of these experiments are detailed in this section. The experiments were conducted using infrared imaging equipment made by Inframetrics Corporation and by A.G.A., Inc. The objectives of the experiments were to determine the sensitivity of the thermal field and infrared detector to geometric variations, arc position, contaminants, penetration depth, and weld machine parameters and to determine the time required to identify changes in weld status.

Both infrared cameras were equipped with color monitors and Polaroid cameras. Both instruments have a manufacturer specified resolution of 0.2C. The Inframetrics camera has a maximum range from -20C to 2500C, while the A.G.A. camera has a range of 1000C. The Inframetrics instrument also came equipped with a color enhancement system with isothermal line scan and a 3x telescope system.

All experiments were conducted using 12" x 12" plain carbon steel plates (AISI 1040) ranging in thickness from 0.275" to 0.325". The edges to be joined were milled for a precise fit. A Miller model 330A/BP-AC/DC inert gas welder with water-cooled torch and a 3/16" tungsten electrode was used to produce the welds. The torch was mounted to a precision positioning table allowing movement in three orthogonal directions (see Figure 1). Movements were controllable to within $\pm$ 0.01 inches. After the initial stationary arc experiments, penetration was maintained between 80% and 95% to

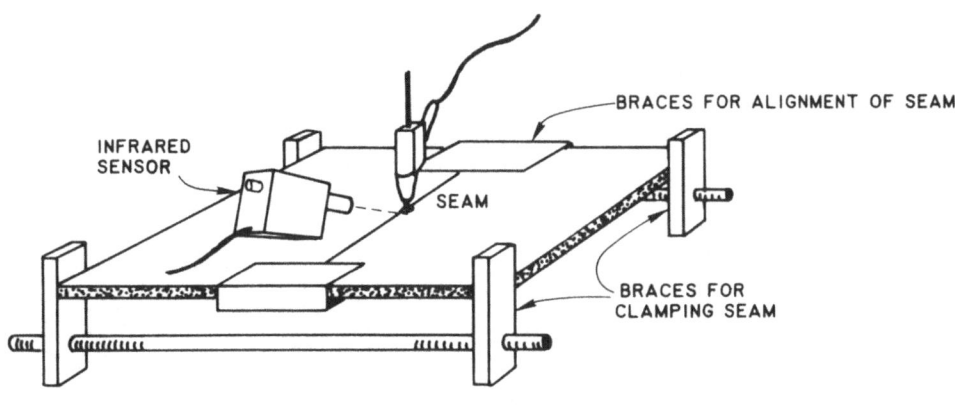

EXPERIMENTAL SETUP

Fig. 1. Schematic showing restraint of plates and placing of cameras.

assure that no burnthrough occurred which might have damaged monitoring equipment placed on the opposite side of the plates.

EXPERIMENTAL RESULTS

Initial measurements were made using a stationary arc positioned on the seam of two plates. The plates were clamped tightly together as shown in Figure 1 to restrain motion due to thermal expansion. Backside measurements were investigated first to avoid molten metal splatter, frontside positioning and arc reflection problems. Photograph 1a is the resulting two dimensional thermal scan with the arc positioned within $\pm$ 0.01" of the seam center. The color bar at the bottom of the photograph shows the color assignment to regions of temperature ranging from blue (coolest temperature) to white (hottest temperature). The white circle separating the green and blue regions represents an isothermal profile on the surface.

Photograph 1b is a thermal line scan of Photograph 1a in which temperature is plotted versus distance across the center of Photograph 1a. The distribution is symmetrical indicating positioning of the arc over the seam center. From this profile, one can also determine the average diameter of the molten metal pool. The inflections around the peak represent the isothermal band of the molten metal-solid metal interface (and/or the emissivity variations associated with the phase transformation).

Photographs 2a and 2b show the temperature distribution which results when the arc is positioned 0.1" to the left of the seam center. The two dimensional thermal scan of Photograph 2a is composed of half-moon shapes or portions of circles. This asymmetrical temperature distribution is caused by contact resistance at the seam. The degree of arc off-center appears to be proportional to the difference in the radii of the half-moons of the isotherms. The temperature distance profile shown in Photograph 2b shows that the temperature distribution is highly nonsymmetrical about the peak temperature. A marked difference in the thermal distribution symmetry for an arc placed on the seam versus a placement at a distance of 0.1" off the seam is seen by comparing Photographs 1 and 2. The results obtained tend to indicate that seam tracking may be based on the difference in radii of isothermal curves to the left and right of the seam. A simple linear control could be used to move the arc until both radii were equal.

In addition to seam tracking, the infrared sensors should be able to identify geometrical variations encountered in the welding process such as joint gaps and mismatches. Photograph 3 shows theresulting two dimensional thermal distribution for seam gaps. The gaps in the seam cause an indentation in the constant

AUTOMATIC WELDING 415

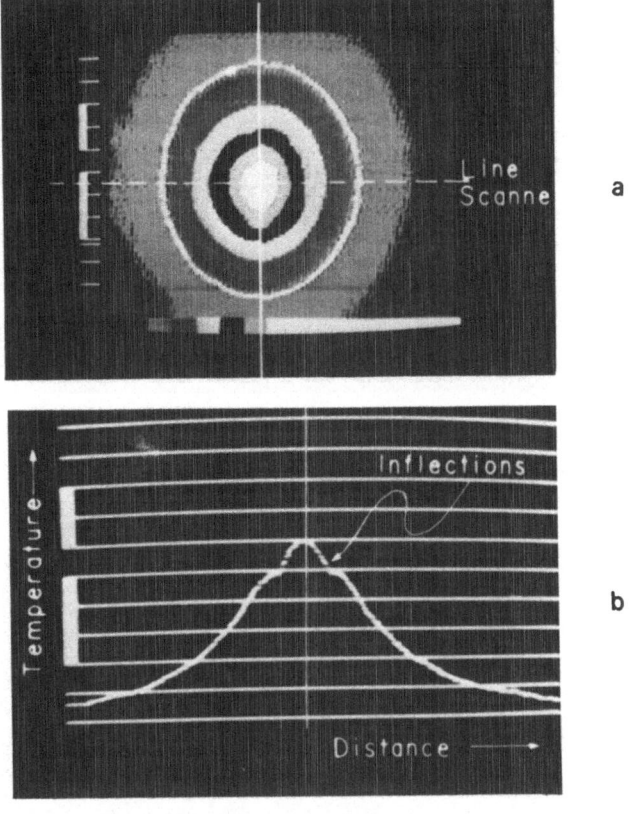

Photograph 1 a) Two dimensional thermal scan of plate surfaces
showing a symmetrical temperature distribution.
Arc placed within ± 0.01" of the seam.

b) Temperature versus distance scan through the
center of the configuration shown in part 1a.
Inflections pinpoint the weld puddle boundary.

temperature lines corresponding to a decrease from the peak temperatures of the metal surrounding the gaps. Photograph 3a shows these indentations which yield a 'butterfly pattern' thermal surface. Photograph 3b shows this drop in the temperature-distance profile in a region toward the bottom of Photograph 3a. The sensitivity of the infrared sensor in identifying arc position is also shown in these photographs. The halfmoons would be of the same size if the arc were positioned directly on the seam center. Also the peak temperature would be of the same height on both sides of the gap in Photograph 3b.

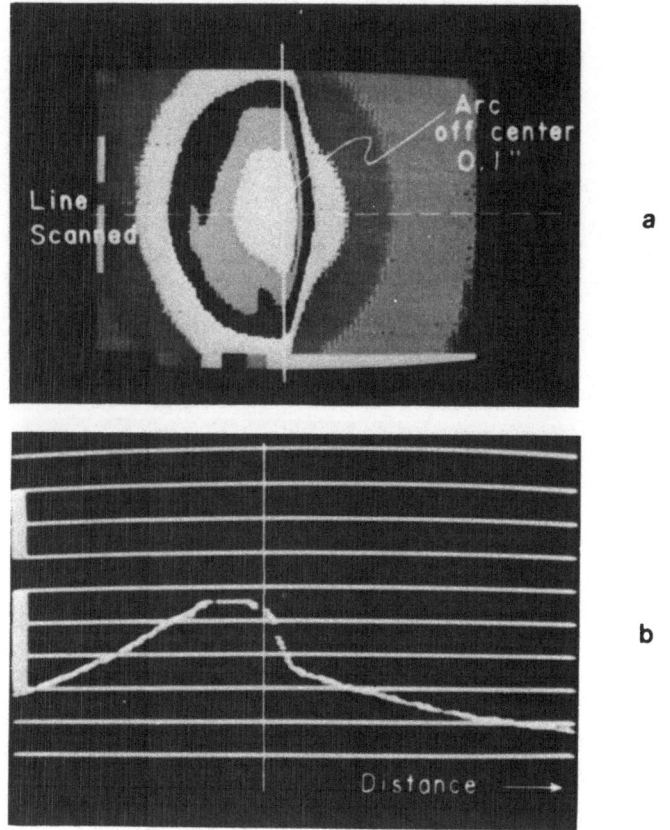

Photograph 2 a) Two dimensional surface scan with the arc placed 0.1" to the left of the seam. Note the significantly asymmetric temperature distribution.

b) Temperature versus distance scan through center of weld puddle emphasizes the asymmetry of the temperature distribution caused by an off seam arc.

MOVING ARC MEASUREMENTS

The next group of experiments that were conducted involved the measurements of temperature distributions produced by a moving arc. The camera was positioned for frontside measurements as in Figure 1 and the arc moved at a speed of 2.0 inches per minute down the seam of two matched and fully restrained plates.

Monitoring thermal distributions at a distance in front of the molten metal pool allows one to 'preview' upcoming plate geometry variations, contaminants and seam position. Using this technique of previewing, Photographs 4 and 5 show the ability to identify seam position and contaminants. Sequential Photographs 4a and 4b

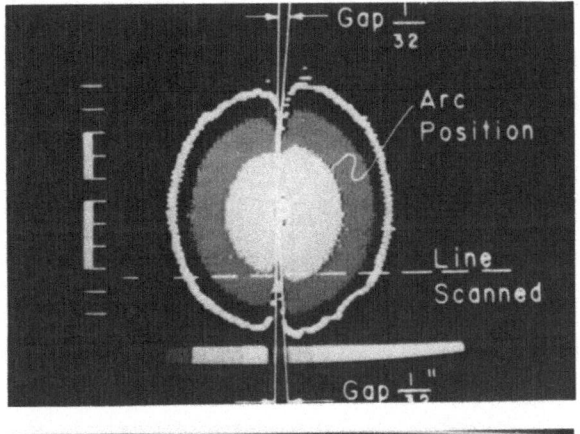

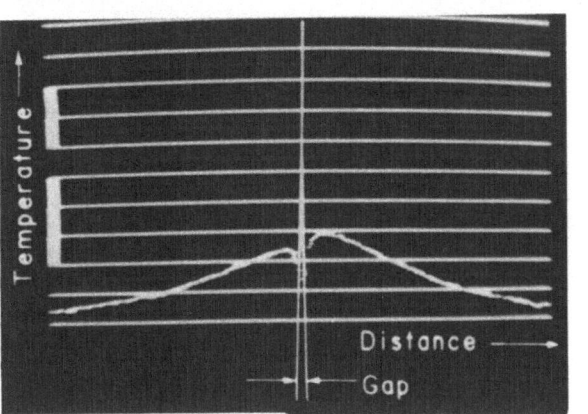

Photograph 3 a) A seam gap causes the isotherms to develop a notch, pointing to the center, where they cross the seam (white line on periphery).

b) Temperature versus distance scan across the bottom of photograph 3a. Two peaks with a notch between them can be seen at the points where the seam gap occurs. The peaks are not symmetrical indicating that the arc was displaced slightly to the right.

show the effect of a zig-zag seam on the thermal distribution. As the torch approaches the seam shift, the isotherm (dark blue) is offset toward the upcoming seam. Similar results were obtained for other seam configurations such as 30° and 45° angles. This notch in the isotherm can be identified from the change in isotherm slope. Its path could then be used as input in an arc-positioning control algorithm. Contaminants can also be detected for moving arc cases. In Photograph 5, the impurity is seen as a closed 'cold spot'. This temperature distribution resulted from a 0.0625" diameter Al_2O_3 particle which was placed along a seam being traversed by the arc at a speed of 2" per minute. The particle becomes clearly

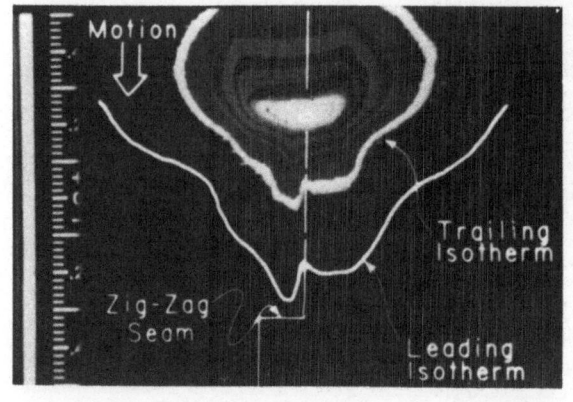

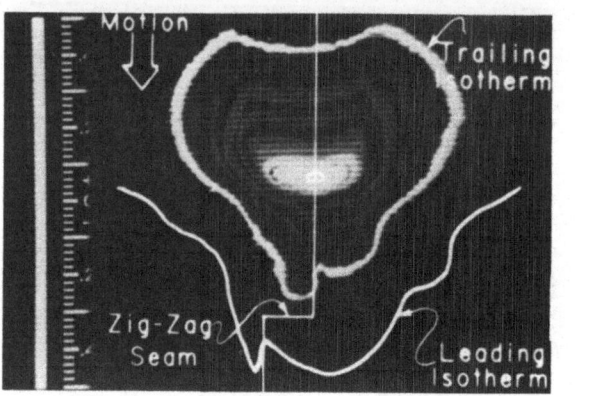

Photograph 4 a) Effect of a change in seam direction on the shape of the isotherms. The notches, observed in the isotherms, are aligned indicating a straight seam.

b) As the seam shifts so does the notch in the isotherm.

visible as much as 1.5" ahead of the arc center. This ability to identify impending changes in weld status is critical to the development of successful control schemes.

DEPTH OF PENETRATION

The integrity and strength of a weld depends a great deal upon the depth of penetration of the molten pool into the metal. Therefore a special effort was made to determine what characteristics of the thermal field, if any, were related to the depth of penetration and in what way. Data concerning the depth of penetration were gathered from microstructural examinations of the welds. This consisted of cutting the plates through the weld and then polishing,

AUTOMATIC WELDING 419

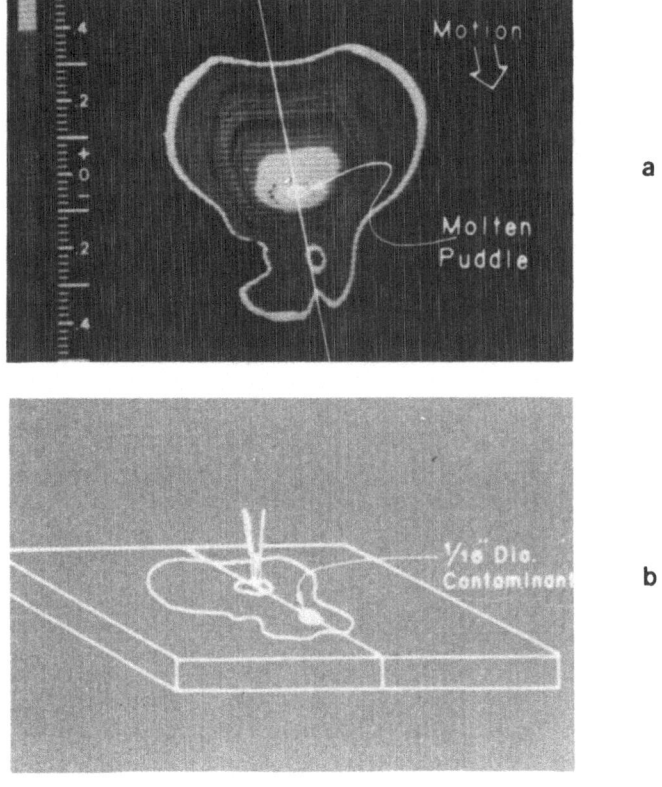

Photograph 5 a) An impurity placed in the path of the arc is
visible some distance ahead of the arc.

etching and photographing the exposed surface. Photograph 6
contains three such photographs showing various stages of puddle
penetration. Note that the temperature dependent microstructural
changes allow the identification of surfaces that reached
particular temperatures during the welding process.

Figure 2 shows the theoretical temperature distribution through
a cross-section of a plate, of arbitrary thickness t_1 and thermal
conductivity k_1 in contact with a backing material, of thermal
conductivity k_2, on one side and under the influence of an arc on
the other side. Malmuth, Hall, Davis, and Rosen [33] have shown
that for a given thickness of material this temperature distribution is a function only of the ratio k_1/k_2. Figure 2 shows the two
extreme cases for this thermal field ($k_1/k_2 = 0$ and $k_1/k_2 = \infty$) and
one intermediate case ($k_1/k_2 < 1$). Two important conclusions may
be drawn from the work done by Malmuth et al. They are (a) the
thermal distribution across the surface determines the thermal
field throughout the material, and (b) the shape of the isotherms

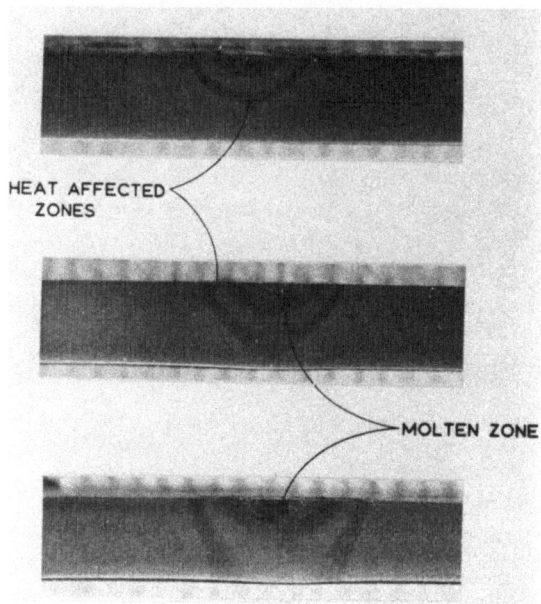

Photograph 6 Photographs showing the heat affected zones for various depths of penetration.

inside the weld metal is determined solely by the surface position of that isotherm (the thickness and k_1/k_2 being fixed). These two facts allow us to develop a method for predicting the depth of penetration of an isotherm given only its shape and position on the surface as well as the thickness, t, and the ratio k_1/k_2 for the material.

Figure 3a depicts the relationship that would allow us to affect control over the depth of pentration. The quantities A & B are the lengths of the surface principal axes and the vertical

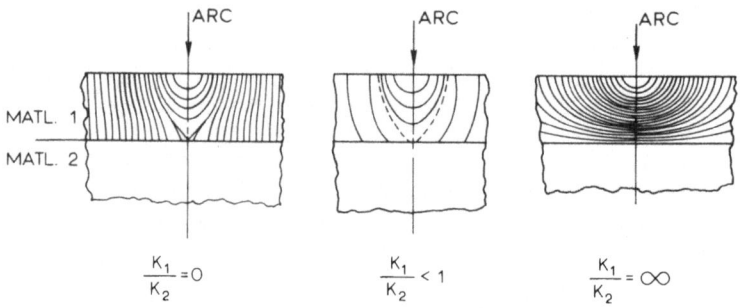

Fig. 2. Schematic showing the effect of material conductivity on the shape of cross-sectional isotherms.

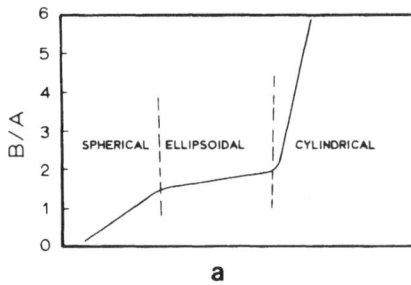 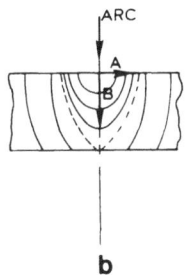

Fig. 3 a) Curve showing the general relationship between cross-sectional isothermal radii or depth of penetration 'B' and surface isothermal radii 'A'.

b) Schematic of plate cross-section showing general shape of isotherms and the quantities 'A' and 'B'.

principal axes of the isotherms, if they are modeled as ellipses, as illustrated in Fig. 3b. The exact shape of the curve in Fig. 3a is a function only of the ratio k_1/k_2 and to a small extent the thickness t (it is independent of the temperature). Both the work done here and that done by Malmuth et al. support this contention. Thus when the temperature at a point on the surface of the plate reaches the melting point of the metal full penetration is achieved. The location of this point depends only upon the thermal conductivity of the metal and its thickness. Therefore control algorithms can be developed as soon as the functional dependence of the transition radius upon plate thickness has been quantified. Preliminary work suggests a roughly linear relationship.

Figure 4 adds some of the existing data field to the characteristic curve to allow a direct comparison to be made between the theoretical and experimental measurements. Here the inner and outer isotherms correspond to the inner and outer zones that can be seen in Photograph 6. The inner isotherm corresponds to the boundary of the molten metal puddle. Another investigation of the relationship between surface thermal field and penetration depth was undertaken. The volume of information contained in a thermal image may become difficult to manage. Thus in this set of experiments attention was focused on line scans. Weld speed was varied to modify energy input, and a line normal to the seam was scanned. Figure 5 portrays the relationship between the peak value of the line scan and penetration depth. These results help confirm the hypothesis that easily observable and measurable characteristics of the thermal field will reliably predict penetation depth.

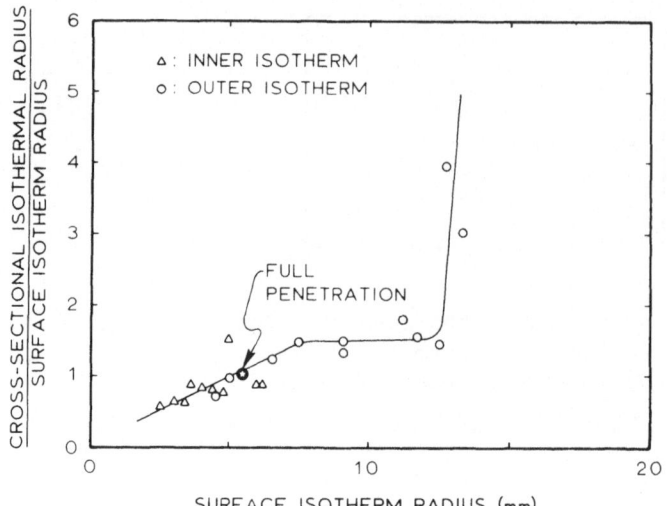

Fig. 4. Actual experimental relationship between cross-sectional isothermal radii and surface isothermal radii. Note that the inner isotherm points fall along the same curve as the outer isotherm.

DISCUSSION OF RESULTS

The experimental results may be partially summarized as shown in Table 1. As can be seen from the table, each weld defect creates its own distinctive and discernable pattern of isotherms on the plate surface. Further, the characteristics of a stationary arc are similar to those of a moving arc except that the isotherms trail out behind the moving arc. It is also apparent from the results that previewing thermal distributions ahead of the molten metal pool is an important characteristic which will allow longer times for identification and correction of weld process perturbations. Simple corrective actions for seam tracking are suggested by the symmetry characteristics of the thermal distribution. Similarly, measurement of the distance from arc center to a fixed temperature isotherm in front of the pool (opposite sides of the joint) could be used for seam tracking. From the results presented, there appears to be a direct relation between the degree of arc misalignment and isothermal radii.

The depth of penetration of the molten puddle into the metal can be controlled, once the characteristic curve of the metal has been determined, by simply monitoring the radius of the weld puddle.

The entire thermal distribution is not required for all aspects of weld process control. For instance, monitoring the temperature

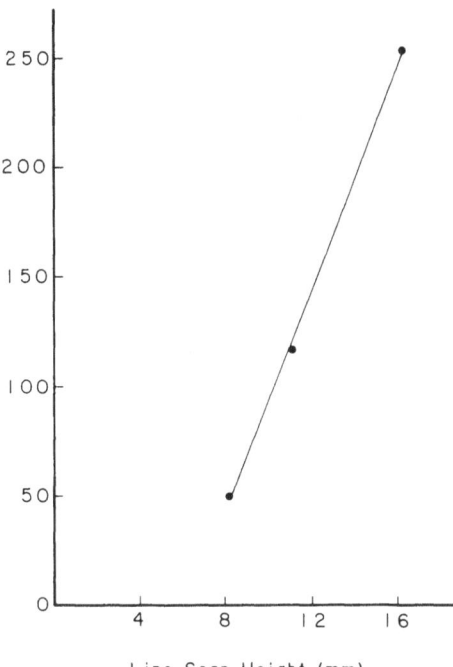

Fig. 5. Peak surface temperature as a prediction of penetration depth.

at two fixed positions in front of the arc may be all that is required for seam tracking of symmetrical geometry parts. Additionally because of the distinct temperature distributions achieved, it is anticipated that integration of all weld perturbation sensors into one large control sequence may not be required. Non-coupled logic routines for specific corrective actions are envisioned. For instance, arc positioning involves identifying plate temperature symmetry and moving the arc in a horizontal plane. Such action would be independent of weld puddle defect control which would involve identifying molten pool characteristics and for instance

Table 1. Summary of Experimental Results

	Weld Defect	Photograph	Description of Thermal Field
Stationary arc	none	1	symmetrical isotherms
	arc misalignment	2	asymmetrical half-moons
	seam gap	3	symmetrical half-moons with distinct indentions at gap
Moving arc	zig-zag seam	4	leading isotherm shifts sequentially
	impurities	5	cold spot in molten metal pool

current pulsing for puddle agitation and redistribution of contaminants.

There were two areas of interest that were not investigated in depth in these preliminary experiments. They are both germane to infrared sensed welding and are discussed here briefly.

Measurement of absolute temperatures by infrared thermography requires that the emissivity of the radiating surface be known. In the experiments which have been described, the principal interest was on relating patterns of the temperature distribution field to various geometric perturbations such as joint gap, mismatch and arc position. These experiments have relied on an analysis of symmetry or the lack of symmetry about the plate seam. As such a detailed knowledge of the surface emissivity was not required.

However, to obtain the greatest amount of information about the welding process from the infrared sensors, it would be advantageous to know both absolute temperature and emissivity as a function of position. This information could be used to monitor temperature history of the plates for microstructure control and to identify surface contaminants. Unfortunately most metals are poor emitters of infrared radiation, their reflectivity being high at high temperatures. The emissivity of the metal surface is also dependent on its condition (rough, oxidized, etc.). Coatings can be used to provide the specimen surface with a high uniform emittance. However differences in coating thickness and bonding can result in erroneous measurements.

Techniques have been developed which reduce the problems associated with variable surface emissivities. These techniques have been successful in non-destructive testing of nuclear fuel pin welds [30]. These techniques known as Emittance Independent Infrared Analysis (EIIA), involve the simultaneous scan of dual infrared detectors. Although changes in temperature for a surface of constant emissivity, result in both a change in intensity and distribution, emittance variations affect principally the intensity and not the distribution (assuming gray-body behavior). If the intensities at two wavelengths are simultaneously measured over the same area of the target, their ratio becomes a function of only the temperature of the target surface. Hence this ratio is independent of emissivity. By monitoring both the absolute values of intensities and the ratios of intensities, changes in emissivity can be detected. These changes in emissivity can be used to identify conditions such as oxidation or surface contamination.

The ultimate applicability of infrared thermography to welding process control will depend upon how rapidly weld problems can be detected, identified and appropriate corrective action taken. Detectable changes in thermal distribution are produced very rapidly by changes in weld conditions. The thermal image changes immediately (to within experimental accuracy) when the arc is moved from a stationary position. Further work will be required to determine the minimum time needed to detect meaningful changes in the temperature distribution.

The time required to determine the temperature field over the area of concern is also important. The scanning time of the infrared detector used in the experiments is 17 milliseconds (a full 250 x 192 matrix scan of the field of view) and it is conservatively estimated to require less than 10 milliseconds to transmit the data for processing. It is difficult to estimate the required processing time since this will be a function of the specific logic which must be performed to identify and initiate corrective sequences. Additional experimentation is needed to determine the diagnostics required.

The time available to generate the desired correction can be estimated from the experimental data of the previous section. Consider the case of a plate gap as shown in Photograph 7. The arc traverse speed was 2" per minute. If one conservatively assumes that an isotherm located 0.25" in front of the arc were being monitored (corresponding to the green band in Photograph 7), there would be 7.5 seconds to identify and respond to the problem. It therefore appears that adequate response time exists for plate geometry diagnostics. (Similar arguments applied to Photographs 4 and 5 indicate even longer times available to respond to seam variations and impurities.) Detailed experimentation is required to develop diagnostics. However, the initial results (millisecond detection times and dramatic effects) appear very favorable.

These experiments have demonstrated that infrared thermography can identify the weld problems delineated at the start of this section. The photographs show that geometric variations, arc position, contaminants, penetration depth, and weld machine variables can all be detected within a time frame consistent with automatic control of the welding process. It is expected that the use of a dual scanning system will permit control over surface contaminants and material microstructure. These characteristic temperature distributions may be the missing link required for integration with computer-aided processing to permit closed-loop control of the weld process.

CONCLUSIONS

Specific weld defects were intentionally induced into an arc welding process and the resulting surface isotherms were observed using a scanning infrared camera for both stationary and moving arcs. Each weld defect produced a distinctively different surface

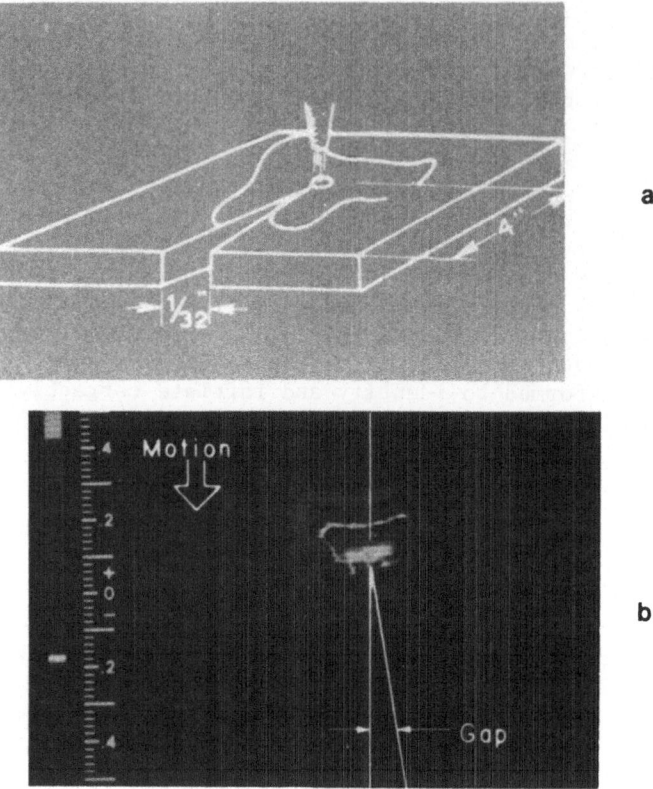

Photograph 7 Geometry defects-plate gap.

temperature distribution. The infrared camera is capable of monitoring arc position relative to the seam and can be used to identify plate geometry defects such as plate gaps and offset. Changes in the patterns of surface isotherms in front of a moving arc are directly related to the presence of impurities and obstructions. Depth of penetration can be directly related to the radius of the molten metal puddle and can therefore be predicted and controlled. Defects as small as 1/32" in diameter can be observed in the molten metal pool prior to solidification. It appears therefore, that infrared thermography can be used as a sensor for incorporation into a closed loop feedback system for continuous process and quality controlled welding.

ACKNOWLEDGEMENTS

The authors would like to acknowledge the support and use of equipment provided by Inframetrics Incorporated, Eklund AGA Infrared and ESAB North America Incorporated. Support of students working on this project was made possible by a grant-in-aid from the Welding Research Council.

REFERENCES

[1] A. J. Moorehead and R. W. Reed, "Internal bore welding of 2 1/4 Cr-1Mo steel tube to tubesheet joints," Welding Journal, Jan. 1980.
[2] T. Nozaki and Y. Higo, "The development of an arc welding robot for shipbuilding," Joint Automatic Control Conference, FA7-7, 1980.
[3] G. E. Cook, "Feedback control of process variables in arc welding," Joint Automatic Control Conference, FA7-B, 1980.
[4] D. P. Edmonds, G. M. Goodwin, and G. M. Slaughter, "Development of automated pipe and tube welding techniques for aluminum," Welding Journal, Feb. 1977.
[5] P. T. Houldcraft, "Automation and automatic welding," British Welding Journal, 13(4), p. 212, 1966.
[6] J. G. Balliger and H. L. Harrison, "Automated welding using spatial seam tracing," Welding Journal, p. 787, Nov. 1971.
[7] J. A. Bachelis and I. V. Varlamov, "Movement of the electric arc in a magnetic field," Automatic Weld, 19(4), p. 43, 1966.
[8] S. L. Mandelberg, B. G. Sidorenko, and V. E. Lopata, "Control of arc welding with the help of a traveling magnetic field," Automatic Weld, 29(9), p. 1, 1976.
[9] G. K. Hicken, N. D. Stucki, and H. W. Randall, "Application of magnetically controlled welding arcs," Welding Journal, p. 264, April 1976.
[10] T. N. Jayarajan and C. E. Jackson, "Magnetic control of gas tungsten-arc welding process," Welding Journal, p. 377S, August 1972.
[11] A. E. Guile, "Magnetic fields in arc welding," IEEE Conference on Gas Discharges, No. 70, p. 489, 1970.
[12] M. S. L'vov and A. P. Igoshin, "A system for the automation of arc welding with self adjustment by depth of penetration," Automatic Weld, 24(5), p. 42, 1971.
[13] A. M. Naidenov, "Mechanical control of the transfer of electrode metal," Automatic Weld, 22(12), p. 35, 1969.
[14] E. P. Vilkas, "Automation of gas tungsten arc welding process," Welding Journal, Vol. 45, No. 5, p. 410, May 1966.
[15] W. M. McCampbell, G. E. Cook, L. E. Nordholt, and G. J. Merrick, "Development of weld intelligence system," Welding Journal, Vol. 45, No. 3, p. 139S, March 1966.
[16] G. J. Vanderbrug, J. S. Albus, and E. Barkmeyer, "A vision system for real time control of robots," Proceedings 9th International Symposium on Industrial Robots, Washington, DC, March 1979.

[17] R. Eskenazi and J. M. Wilf, "Low level processing for real-time image analysis," JPL Report 79-79 NSAS Contract NAS7-100.
[18] Ya. S. Vaisband, A. B. Voitsekhovskii, and A. P. Zhurishkin, "A television system for automatically guiding electrodes along butt welds," Automatic Weld, 24(7), p. 47, 1971.
[19] N. Yokoshima and H. Takagi, "Online adaptive control of narrow gap CO_2 welding with industrial television camera," Joint Automatic Control Conference, FA7-F, 1980.
[20] B. E. Paton, N. V. Podola, V. G. Kvachev, and A. A. Ursat'ev, "The mathematical modelling of welding processes to create systems for forecasting the quality of joints and for optimum control," Automatic Weld, 24(7), p. 1, 1971.
[21] S. J. Mech and T. E. Michaels, "Development of ultrasonic examination methods for austenitic stainless steel weld inspection," Materials Evaluation, p. 61, July 1977.
[22] L. Alder, K. V. Cook, B. R. Dewey, and R. T. King, "The relationship between ultrasonic rayleight waves and surface residual stress," Materials Evaluation, p. 93, July 1977.
[23] L. Alder, K. V. Cook, H. L. Whaley, and R. W. McClung, "Flaw size measurement in a weld sample by ultrasonic frequency analysis," Materials Evaluation, p. 44, March 1977.
[24] W. D. Jolly, "The application of acoustic emission to in-process inspection of welds," Materials Evaluation, p. 135, June 1970.
[25] D. M. Romrell, "Acoustic emission weld monitoring of nuclear components," Welding Journal, p. 81S.
[26] F. N. Kiselevskii, N. R. Shvydkii, and Yu. I. Streletskii, "The automatic analysing of radiographic pictures of welded joints," Automatic Weld, 24(7), p. 10, 1971.
[27] P. W. Ramsey, J. J. Chyle, J. N. Kuhr, P. S. Myers, M. Weiss, and W. Groth, "Infrared temperature sensing systems for automatic fusion welding," Welding Journal, p. 337-S, August 1963.
[28] D. R. Green and J. A. Hassberger, "Infrared electro-thermal method for nondestructively testing welds in stainless steel pipes," Materials Evaluation, Vol. 37, No. 11, p. 54, 1979.
[29] D. R. Green, "Experimental electro-thermal method for nondestructively testing welds in stainless steel pipes," Materials Evaluation, Vol. 37, No. 11, p. 54, 1979.
[30] D. R. Green, "Principles and applications of emittance-independent infrared nondestructive testing," Applied Optics, Vol. 7, No. 9, p. 1779, Sept. 1968.
[31] L. D. McCullough and D. R. Green, "Electrothermal nondestructive testing of metal structures," Materials Evaluation, Vol. xxx, No. 4, p. 87, April 1972.
[32] W. E. Lukens and R. A. Morris, "Infrared temperature sensing of cooling rates for arc welding control," Welding Journal, Vol. 61, No. 1, p. 27, Jan. 1982.

[33] N. D. Malmuth, W. F. Hall, B. I. Davis, and C. D. Rosen, "Transient thermal phenomena and weld geometry in GTAW," Welding Journal, Vol. 53, No. 9, p. 388-S, Sept. 1974.
[34] B. A. Chin, N. H. Madsen, and J. S. Goodling, "Infrared thermography for sensing the arc welding process," Welding Journal, Vol. 62, No. 9, p. 227-5, Sept. 1983.
[35] B. A. Chin, J. S. Goodling, and N. H. Madsen, "Infrared thermography shows promise for sensors in robotic welding," Robotics Today, Vol. 5, No. 1, p. 85, Feb. 1983.

Computer-Aided Manufacturing

PRESENT STATE AND FUTURE TRENDS IN THE DEVELOPMENT OF PROGRAMMING LANGUAGES FOR MANUFACTURING

U. Rembold and W. K. Epple

Department of Computer Science III
University of Karlsruhe
Federal Republic of Germany

ABSTRACT

For the development of programming languages, or in a broader sense, of software tools and languages, a milestone has been reached in two distinct areas, namely factory-information processing and factory automation. Many excellent but specialized tools and languages have been developed to date. The emphasis has been to custom design software for many activities in manufacturing such as production control, factor data collection, material handling and process control.

This paper discusses software for numerical control of machine tools, robots, process control and for conventional data processing purposes. Until to date, the languages and computers for these applications have been developed independently. For this reason it is very difficult to design a homogeneous computer integrated manufacturing system. The future modernization of factories will strongly be influenced by the development of general purpose programming languages and compatible computing systems. Furthermore, the controls of processing equipment have to be made computer compatible. For the factory of the ninetieth, expert systems will be available which perform the majority of all routine tasks.

INTRODUCTION

Manufacturing creates between 60-80% of the real wealth of the major industrialized countries. For this reason most of these countries have numerous industry and government supported programs to increase the manufacturing productivity.

Within the last two decades, many important innovations were made in the development of new design aids, manufacturing processes, engineering materials and manufacturing systems. The digital computer has contributed to the most significant advances. This so far almost untapped resource has the greatest potential to improve the manufacturing productivity, compared with any other invention which contributed to the industrial revolution. With conventional production know-how it becomes increasingly difficult and expensive to improve manufacturing processes. The computer offers possibilities to enhance the manufacturing technology in many areas. It can directly control production and quality control equipment and adapt these quickly to changing customer orders and new products. The computer makes possible to instantly evaluate data, to assess the flow of information in the plant and to immediately initiate corrective actions to optimize the manufacturing process. Probably the most important asset of the computer is its capability to integrate the entire manufacturing system. Its ability to make decisions will contribute to the conception and design of flexible manufacturing systems (FMS) which can be reconfigured to changing marketing requirements and new manufacturing processes.

Unfortunately, the use of the computer has a serious drawback. For the average manufacturing engineer it is very difficult and for the factory worker virtually impossible to communicate with it.

About 25 years ago, there were essentially only assembly languages available. During the course of time, many higher programming languages were developed. They are so numerous that even an expert has difficulties to know them all. The same problem exists for manufacturing. There are not only one but several programming languages which have been developed for different manufacturing applications. They can be divided into the following categories:

(1) Numerical control of machine tools,
(2) robot control,
(3) process control, and
(4) commercial data processing.

In this paper, a description of programming languages and programming-support-tools for each of these categories is given.

The development of programs for numerical control will be simplified by means of easy-to-use independent work stations. In the near future, it will be possible to generate control programs with the aid of a graphic representation of a workpiece and by animating the machining process on a display.

In the case of robot control the essential features of the most important programming languages will be discussed. The future of robot programming is characterized by what is called "implicit programming" and the use of expert-systems.

For process control, there exist different real-time languages (e.g. ADA, PEARL, CORAL, RTL/2) which take into consideration parallelism and the problem of interfacing the technical process with the computer. In this area, programming support environments like APSE or UNIX and the use of descriptive languages together with automated tools will simplify program development.

Finally, in the commercial data processing area of a manufacturing organization, there is a wide use of COBOL as the principal programming language. Several methods and tools for program development have been proposed and are in use. In this field, programming will be strongly influenced by new languages like PROLOG and the results of research work on artificial intelligence.

The computers of the 5th generation will process knowledge at a high level and will render solutions at this level. The dialog with the computer will be done by natural means of communication. Such a system, for example, could be used to lead a factory. Its expert system contains the status of all orders, information on the manufacturing process and other resources, knowledge about alternative manufacturing methods and the boundaries of the manufacturing environment. This system is capable to learn from past and present operations and can generate with the aid of knowledge and deduction rules production plans for new products. The communication to describe these products is done via graphical input. Orders for the factory are entered via speech communication. With the aid of the expert system, the computer is capable to plan production runs and dates and to control the production process.

Although there are many programmng languages and program development aids available, the software as well as the planned expert systems used for the control of manufacturing processes still are very complex, and for this reason also very expensive. It will be the combined task of the manufacturing engineer and the computer scientist to conceive user-friendly and easy-to-apply software tools.

PROGRAMMING OF MACHINE TOOLS

The APT Language

APT (Automatically Programmed Tools) is the first high order language to program NC machine tools. It can be recognized as the origin of many other languages and dialects. The first attempt to conceive this language dates back to the late fifties. An organized development effort started in 1961, when the APT long range program was created and contracted to the Illinois Institute of Technology Research Institute (IITRI). Over 130 companies supported through joint funding the creation of this language.

APT uses English-like instructions to describe the geometry of a part. The input statements are of four different types:

- Geometric statements are used to define the part configuration. With these the programmer is able to describe geometric elements such as points, lines, circles, ellipses, planes, cylinders, cones and general conics and quadrics with different surfaces.

- Motion commands which control the path of the cutter along the surface of a workpiece. The repertoire includes start-up and point-to-point instructions, modifiers to change cutter movement direction and methods to describe the cutter.

- Postprocessor commands which control different machine functions such as spindle speed, feedrate, acceleration, deceleration and coolant supply.

- Special control instructions which generate translations, rotations and output listings. There are also possibilities to program loops, jump instructions and subroutines.

The contour of the part is described by the programmer with the aid of a sequence of instructions. Figure 1 shows how the tool is directed to generate the contour of the workpiece with the help of control surfaces. The tool travels along the part and drive surfaces until it encounters the check surface. Here its path is changed and the tool follows a newly defined part surface. This process is repeated until the entire part contour has been generated. A so-called processor is used to transform the part program into a tool path description, Figure 2. The translator compiles the program language into computer executable instructions which are processed by the arithmetic unit. The mathematical calculations

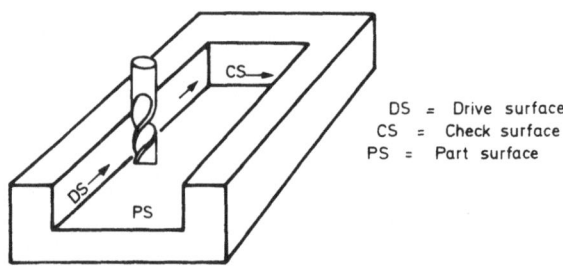

Figure 1. APT principle.

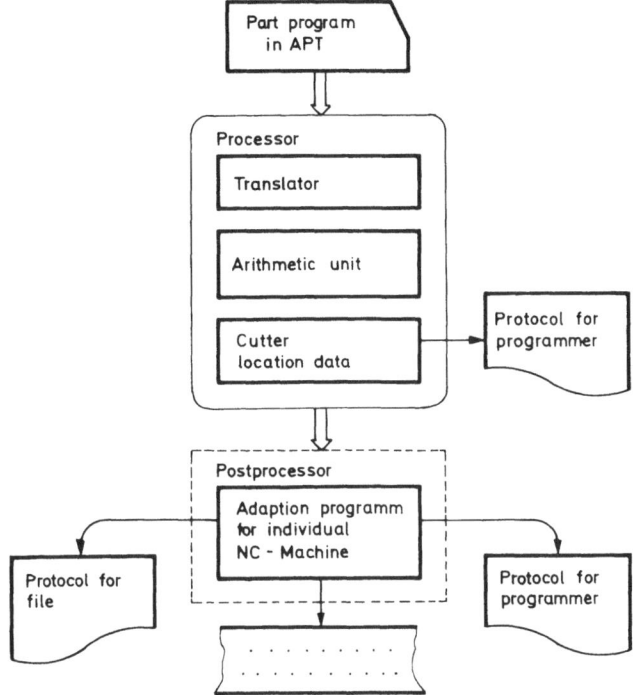

Figure 2. Translation of an APT program.

needed to generate cutter location coordinates, are done by this unit. It allows the inspection of the actual path the cutter will follow. The output of the processor is a machine tool independent program.

Since the control unit of a machine tool usually is machine dependent, the cutter location data are adapted by a postprocessor to machine specific code. The program to generate the part contour consists of dimensional and technological data which are necessary to operate the machine tool.

APT was conceived as a generalized program development tool to handle many different machining processes. For this reason the processor is very large and needs a considerable amount of memory space. This fact, however, limits the use of APT to manufacturers who have access to large computers. During the sixties and seventies, attempts were made to simplify the programming system so that it could be handled by smaller computers. This has led to the conception of APT like languages and subsets of APT. As a matter of fact, several minicomputer manufacturers now offer APT program development systems for their equipment. In the following section, the German development EXAPT (EXtended APT) will be discussed.

The EXAPT Programming System

Syntactically, EXAPT is identical to APT. In addition to the geometrical description of the workpiece technological data of the machine tool and of the workpiece can be considered. The EXAPT system is of modular design and can be operated on smaller computers. The universities of Aachen, Berlin and Stuttgart developed this system during the late sixties. Now it can be purchased from the EXAPT Verein located at Aachen, Federal Republic of Germany. The fundamental concept of the EXAPT system is shown in Figure 3 [1]. With this system the most important 2 and 3 axes machining operations can be handled. The concept of EXAPT is as follows:

BASIC EXAPT. This module is the basis of all other modules. It can be used similarly as APT to program a tool path to generate a workpiece surface. In addition feeds, speeds and the geometry of

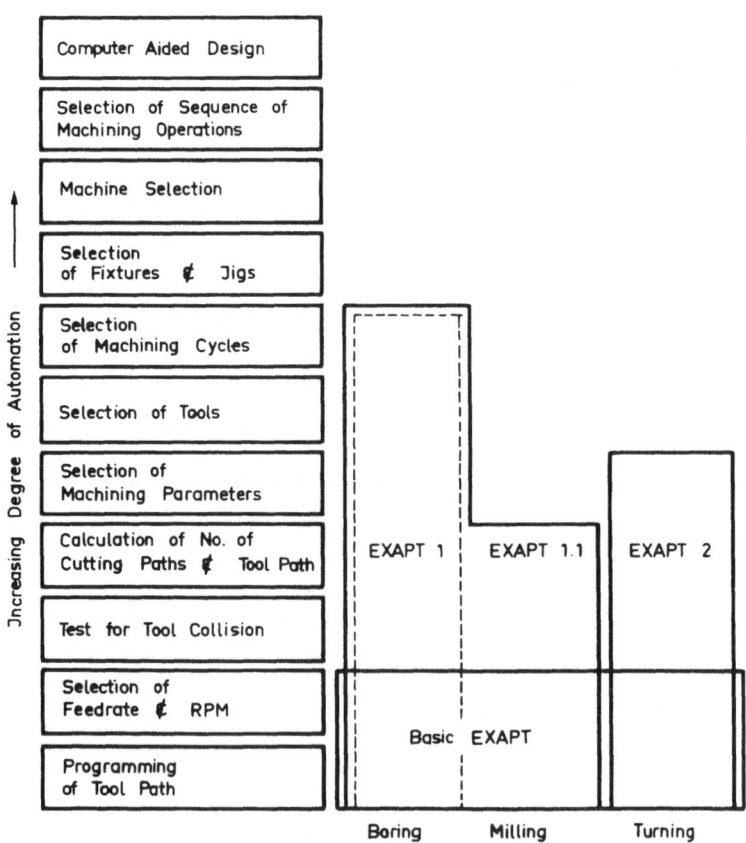

Figure 3. Degree of automation of EXAPT.

tools can be described. The degree of automation which can be obtained is limited.

EXAPT 1. This programming module was devised for boring operations. The following capabilities were added to the BASIC EXAPT features:

- test for tool collision,
- automatic calculation of number of required cutting paths,
- automatic selection of machining parameters,
- selection of tools, and
- selection of machining cycles.

EXAPT 1.1. With EXAPT 1.1 boring and milling operations can be programmed. For boring all features of EXAPT 1 are available. For milling the additional capabilities are test for tool collision and calculation of automatic cutter path segmentation.

EXAPT 2. This module was designed to handle turning and boring operations on lathes. It is possible to describe the contour of the raw workpiece and that of the finished workpiece. Additions to EXAPT 2 are:

- test for tool collision,
- automatic cutting path segmentation, and
- selection of machining parameters.

Figure 3 shows how with increasing degree of automation the EXAPT system is being integrated into design. At the present time a strong development effort is made to close the gap between CAD and CAM and to add the following features, for example, for boring:

- selection of machining sequences,
- selection of machine tools, and
- selection of fixtures and jigs.

The automation of the programming process for milling and turning is not that easy. However, in the long term, there will also be solutions found to tie CAD together with CAM for these machining operations.

The process of preparing the NC tape with an EXAPT module is shown in Figure 4 [1]. The part program is entered into the computer and interpreted by the processor. Analogous to APT in the first phase the geometrical data are processed. In the second phase the technological data are incorporated into the program with the aid of information obtained from the tool, material data and machining files. The output of the processor is a machine independent part program. The following postprocessor serves the same function as that used with APT.

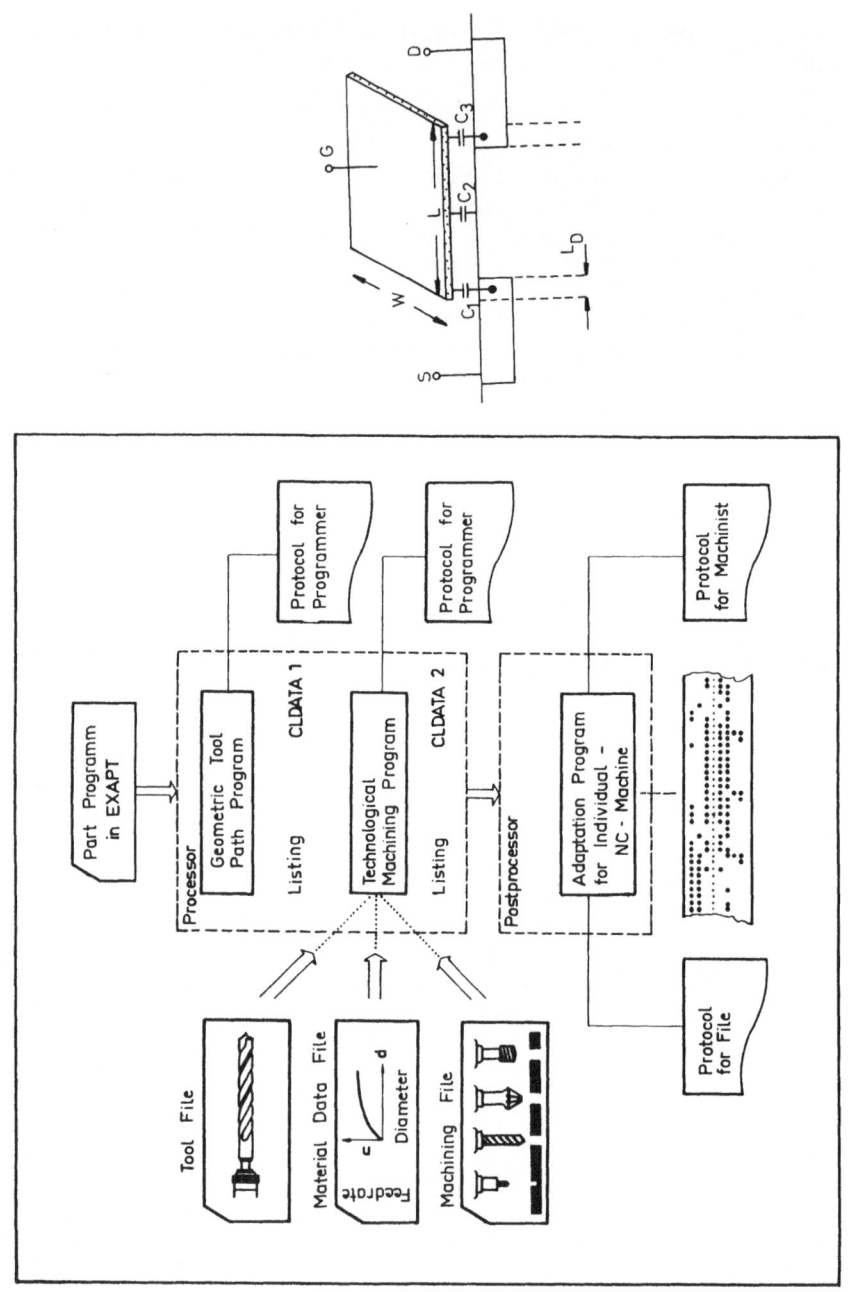

Figure 4. Sequence of programming a N/C machine tool with EXAPT.

A workpiece to be machined is represented in Figure 5a. The corresponding part program for EXAPT is shown in Figure 5b [2]. The program consists of the following parts:

- header data defining the part, the machine tool and the coordinate system,
- geometrical data describing the workpiece,
- technological data describing the machining operations,
- executive instructions to direct the machining operation, and the
- end instruction to complete the machining cycle.

Interactive Symbolic Programming

When workpieces are complex, program development by the APT and EXAPT principle may be quite time consuming and cumbersome. For this reason, several machine tool manufacturers have developed symbolic programming facilities. In the simplest case, programming is done via a keyboard directly at the machine tool. Each key contains an instruction, for example, "Drill, Countersink, Grid pattern, Keyway, etc.". With the help of the keyboard the programmer enters step by step the part program into the computer which converts the instructions into machine code. Usually this programming method is designed for specific applications such as turning.

More advanced symbolic programming systems use a keyboard in connection with an interactive terminal. The FAPT TURN system, developed by the Siemens Corporation, will be discussed in detail. The program development system consists of a computer, a CRT, a keyboard, a bubble memory and a tape reader/punch combination. The rotational part shown in Figure 6 [3] will be taken as an example to show how programming works. The different steps which are greatly simplified are explained below.

(1) Initiation of the programming system.

(2) Selection of the raw material; there are 17 different choices.

(3) Selection of the surface finish.

(4) Positioning of the coordinate system, Figure 7.

(5) Raw part selection, cylinder, hollow cylinder or a special shape, Figure 6.

(6) Generation of the finished part contour, Figures 6 and 7. This is done with the help of symbol keys. The sequence of keys to be depressed is shown in the lower part of Figure 6. Basically, lines and circles can be

entered. The system will ask more information when a line or a circle is indicated. For example, dimensions, diameters, etc.

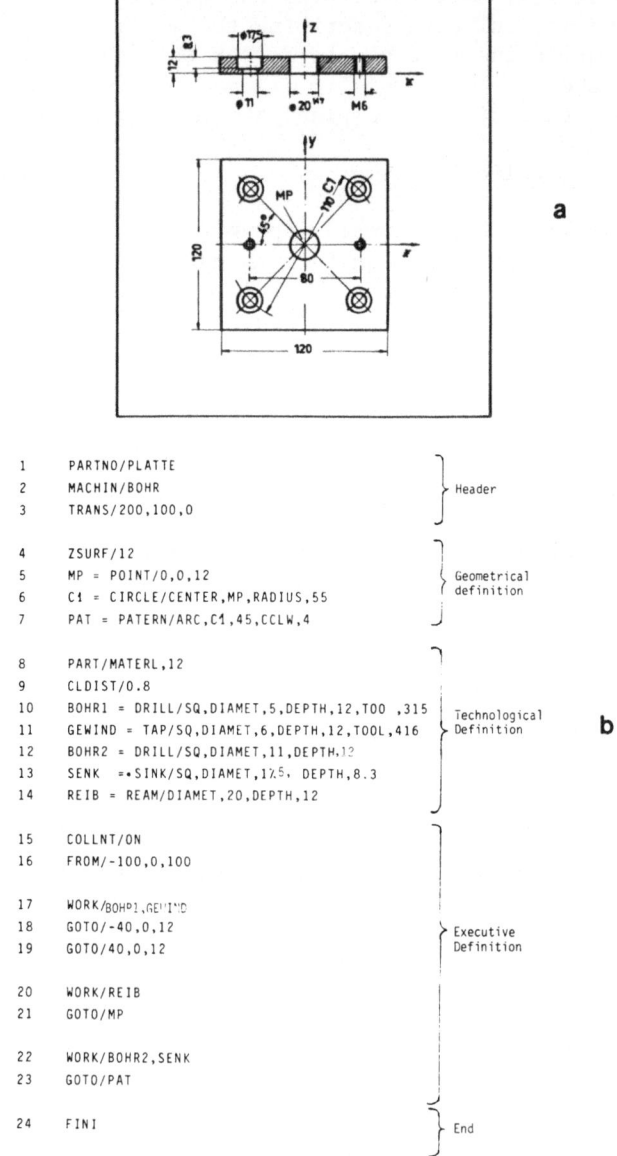

```
1    PARTNO/PLATTE                                          ⎫
2    MACHIN/BOHR                                            ⎬ Header
3    TRANS/200,100,0                                        ⎭

4    ZSURF/12                                               ⎫
5    MP = POINT/0,0,12                                      ⎬ Geometrical
6    C1 = CIRCLE/CENTER,MP,RADIUS,55                        ⎮ definition
7    PAT = PATERN/ARC,C1,45,CCLW,4                          ⎭

8    PART/MATERL,12                                         ⎫
9    CLDIST/0.8                                             ⎮
10   BOHR1 = DRILL/SQ,DIAMET,5,DEPTH,12,TOO ,315            ⎬ Technological
11   GEWIND = TAP/SQ,DIAMET,6,DEPTH,12,TOOL,416             ⎮ Definition
12   BOHR2 = DRILL/SQ,DIAMET,11,DEPTH,12                    ⎮
13   SENK  = SINK/SQ,DIAMET,1X.5, DEPTH,8.3                 ⎮
14   REIB = REAM/DIAMET,20,DEPTH,12                         ⎭

15   COLLNT/ON                                              ⎫
16   FROM/-100,0,100                                        ⎮

17   WORK/BOHR1,GEWIND                                      ⎮
18   GOTO/-40,0,12                                          ⎬ Executive
19   GOTO/40,0,12                                           ⎮ Definition

20   WORK/REIB                                              ⎮
21   GOTO/MP                                                ⎮

22   WORK/BOHR2,SENK                                        ⎮
23   GOTO/PAT                                               ⎭

24   FINI                                                   } End
```

Figure 5a. A workpiece to be machined.

Figure 5b. EXAPT 1 program for the workpiece of Fig. 5a.

DEVELOPMENT OF PROGRAMMING LANGUAGES

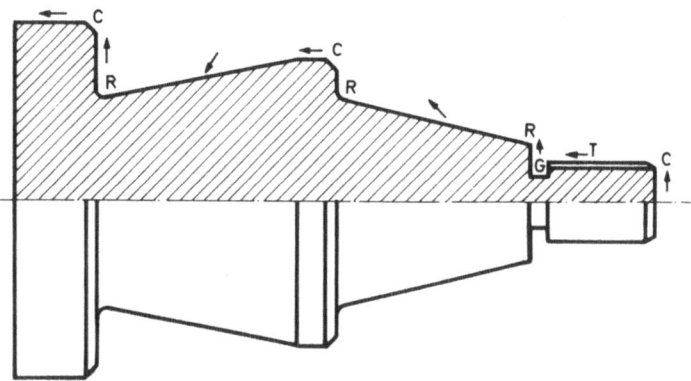

Symbol keys to describe a workpiece contour

Rotational workpiece to be machined

Keys to be depressed to program workpiece contour

Figure 6. Programming of a rotational workpiece with the FAPT system.

(7) Generation of the thread by indicating length, pitch and cutting direction.

(8) Groove cutting and dimensions.

(9) Chamfer and dimensions.

(10) Programming of mathematical functions such as sin, cos, square root, etc.

(11) Specification of machine tool reference point, Figure 7.

(12) Tool position selection, Figure 7.

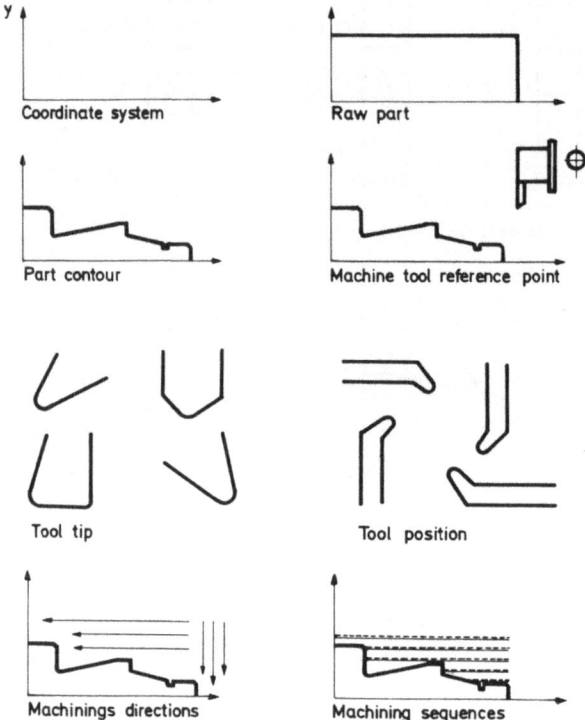

Figure 7. Interactive CRT display of different programming phases.

(13) Determination of tool holder.

(14) Determination of cutting parameters, feed and speed.

(15) Selection of machining directions.

(16) Contour segmentation.

For machining of certain segments it may be necessary to redefine the surface finish. This can be done with the aid of parameters when the surfaces are described.

Special Purpose Languages

Many factories produce special purpose parts which have common features. For example, the workpiece shown in Figure 8 [4] has a typical design found in sand mold castings. Since it is very difficult to program such a part by hand, a computer is normally used. In this particular case, a manufacturer of precision patterns (Anderson Industries, Muskegon, Mich., USA) introduced a programming system

DEVELOPMENT OF PROGRAMMING LANGUAGES 445

which can easily handle this type of parts. The objectives of the programming system are:

- short programming instructions,
- simple syntax,
- no hand computation,
- the ability to handle many different geometries and machining operations, and
- the ability to program a wide spectrum of machine tools, e.g., from point-to-point machines to 5-axis continuous path equipment.

The lower half of Figure 8 shows a section of the program which describes the center cavity of the part. There is a 10 field record into which only numbers are entered. The points are needed to describe the contour are 1, 2, 3, 5, 6, 7, 8. The contour points of the radii and parabolic fillers as well as the coordinates of point 4 are calculated by the programming system.

The known points are encircled in the figure and are designated as codepoints. They are enumerated in field 7. They may be defined absolutely or incrementally in reference to the previously entered codepoint. For example, point 8 is defined incrementally from codepoint 7. The number 500 in field 7 separates the geometrical description of the cavity from the milling instructions. First the compiler calculates the coordinates of point 4, see entry 19 in field 7. It is located at a distance of 6.48 from codepoint 6 at a positive angle of 61° from the x-axis. The -4 at the end of line 19 is an instruction to calculate point 4. The 50 in the next line turns on the spindle and coolant. It commands the tool from the home position to its work position at a travel clearance of 1" above the part. The tool is lowered by 1 inch to contact point 5. The next line has a 2 in field one and a 1 in field two. These numbers define the line at which the parabola is tangent to point 2. Likewise the 1 in field four and the 3 in field 5 define the second line to which the parabola is tangent to codepoint 3. The cutter is moved from point 5 to point 2 and then along the parabola to point 3. The next line entry 23 instructs the cutter to move from point 3 to the beginning of the circular element, defined by points 1 and 2 and the radius originating at point 7. The -1 in this line indicates that the radius of the filler circle is 1" and that the cutter approaches the circular element from its inside. The next line indicates that the cutter travels from the circular elements whose center is point 7, to the element whose center is point 8. Line 24 is the instruction to bring the cutter from the circular element about point 8 to point 4. Line 28 instructs the cutter to move back to point 6, and the instruction 70 brings the cutter back to its home position.

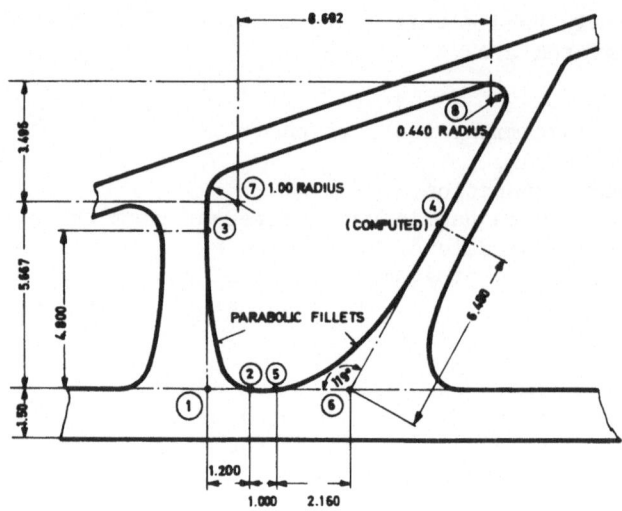

Figure 8. Example of a special purpose NC programming language.

This coding method might look complex to a reader who is not familiar with it. However, it is of great benefit and easy to use when there are always similar parts to be coded.

Generative Programming by the Machine Tool Control

With the EXAPT language it is possible to calculate the number of cutting paths needed to machine a part. For this purpose feed and speed parameters must be available for different material and tool combinations. For future programming it will be possible to optimize the cutting operation with the aid of intelligent controllers built into the machine tool. Therefore, it is necessary to equip the machine tool with sensors, Figure 9. The part program which describes the raw part and contour of the workpiece is sent to the control computer of the lathe. In addition, preliminary feed and speed conditions are given. The part is chucked and

DEVELOPMENT OF PROGRAMMING LANGUAGES

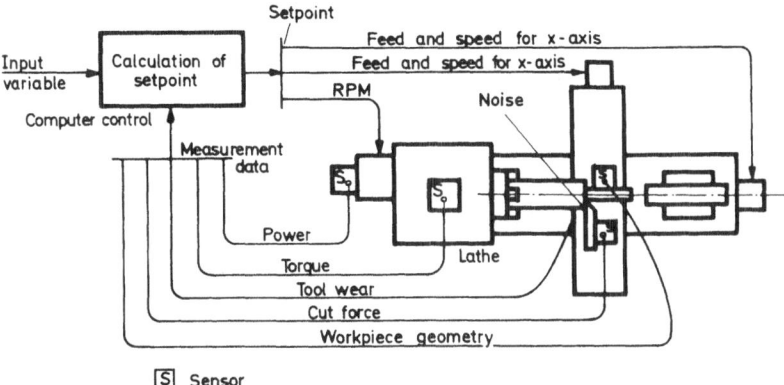

Figure 9. Principle of an adaptive control system for a lathe.

machining is initiated, Figure 10 [5]. First the tool rapidly approaches the raw part at its smallest diameter until a proximity sensor detects the workpiece. At this instance, the computer directs the tool drive to assume the normal cut speed. The tool now engages the workpiece and tries to start machining. A sensor tells the computer that the cut force is too high. The tool is retracted and a new cut attempt is made at a further out face diameter. This procedure is repeated until the permissible cut force is obtained. From here on, the feedback control commences its machining operation and partitions its own cuts. The operation may be done under the supervision of an optimization model. Optimization criteria may be minimum cost or maximum throughput.

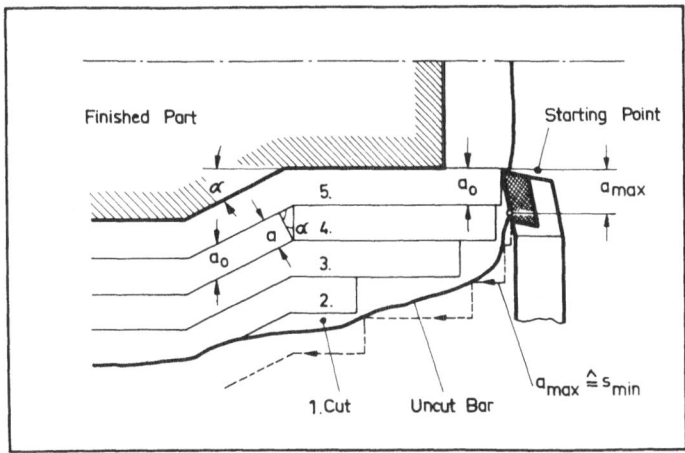

Figure 10. Finished part oriented cut selection with contour parallel cuts.

PROGRAMMING LANGUAGES FOR ROBOTS

Within recent years the industrial robot has matured to a universal tool which can handle efficiently many manufacturing operations. However, its greatest potential lies in the automation of assembly. It first was used for simple material handling work and for welding. For this purpose it must be equipped with sensors and with a comfortable programming systems. In order to describe the work of a robot, the language must have conventional constructs and also such which are robot specific. With these requirements a programming language for robots usually becomes very complex. Researchers first started to develop simple assembly languages. Valuable experience was gathered from their use and lead to various concepts of explicit high order languages. Presently, there are languages being developed for implicit programming. With these it will be possible to solve assembly tasks with the help of expert systems.

Design Considerations to Program Robots

The programming methods are similar to those used by data processing. Operations and data are described with the help of character strings. The user enters the sequence of instructions which describe the movement of the robot into the computer. The information is translated by the compiler into machine code. This can be done off-line independently of the robot. However, it is not possible to program positions and orientations. These parameters have to be entered on-line by a teach-in method. For this reason no pure textual programming is possible.

Robots are programmed by the teach-in method or by textual languages. Both methods do not only allow the description of spacial movements, but they also permit branching, looping and the use of subroutines. Presently, the most frequently used programming method by industry is teach-in. However, modern robots are increasingly being employed to handle complex tasks such as assembly work. Here, teach-in may become very cumbersome and there is an ongoing trend towards the use of higher order languages. Several robot manufacturers offer a combination of textual programming and teach-in. A spacial point can be entered by defining it and by assigning a name to it. The name is used by the program. It is also possible to lead the effector to the spacial point and to enter in the teach-in mode with the aid of a function key an instruction into the control computer of the robot. This procedure is supported by an editor. The system enters the text of a motion instruction into the program. It is parametrized with a defined speed and the position of the point. With this method it is possible to program a complete sequence of movements and to obtain readable software. The program can be easily changed.

In the future, however, the textual method will predominate for programming of assembly work in the factory. It offers the following advantages:

- The program is readable to programmers and other users,

- the program can easily be changed and expanded also by other programmers.

- the program may have variables, during programming only a data type is assigned to it, not its value,

- the program can be stored in readable form which is important for documentation, and

- the program can be written off-line without the availability of a robot, no definition of the operating points is necessary.

In principle, the instruction set of a programming language for robots is similar to that of the teach-in method. Thus, for each basic symbol of the language, a function key may be provided and installed in a keyboard. For simple assembly tasks this programming method may render adequate service. However, textual programming in combination with teach-in method is needed when the following conditions exists:

- complex assembly work is done which require frequent program branching and subroutine calls,

- when sensor signals of the robot have to be processed,

- when external data have to be accessed, e.g., from a data file of a manufacturing system, and

- when an internal world model is used and its data are manipulated.

In addition, textual programming permits the implementation of special strategies which assure an orderly and safe assembly. The following operations may be included:

- Measurement of a position during program execution: For example, if the exact height of a workpiece is unknown, the effector can be lowered slowly until it touches the object surface. A force sensor records this event and the position of the workpiece is then entered into the computer with the help of an instruction.

- Supervision of an operation: The success of an assembly is monitored by sensors to watch for misalignment, breakage or other missing parts. For example, the insertion of a pin into a hole can be monitored with a force torque sensor. During insertion the effector may move slightly in direction of the axes perpendicular to the centerline of the pin. A force signal in either direction will indicate that the pin is surrounded by its mating hole, otherwise no hole is present. In the latter case, the robot will be alerted and a corrective move can be done.

- Search strategies: During bolting of mating operations, the workpiece may not be in its exact position. In this case the robot may try to locate the center of a hole by a search operation. For example, it may try to target the object by a spirally shaped search procedure.

For the following reasons, the number of such special programming instructions should be kept to a minimum

- to reduce the assembly time of the object,
- to save memory space,
- to reduce compilation time, and
- to reduce the number of programming errors.

A Survey of Existing Programming Languages

There are numerous programming languages under development. A ranking of different programming languages is shown in Figure 11 [6,7]. The majority of these are explicit assembly or compiler languages. Here, every movement has to be described explicitly by the programmer. The number of statements to program an assembly task will be quite numerous, even for a simple problem. Presently, there is only one implicit programming language known and that is AUTOPASS. It has limited capabilities to describe simple assembly primitives. A typical instruction is "Pick up a Bolt, Insert it into a Hole." There are three languages which are based on the NC-language concept. They may be used in connection with loading and unloading of NC-machine tools. In this case the programmer only has to be familiar with the NC-language concept. The NC type languages, however, become very complex when variables have to be defined and when sensor data derived from moving objects have to be processed. Several of the listed languages are conceived for universal applications, others are designed for a specific type of robot.

Figure 12 [7] shows desired features of programming languages for robots. In addition to the constructs of conventional languages, there should be several ones specific to robots. For example, typical data types are vector, frame, rotation and translation. It

DEVELOPMENT OF PROGRAMMING LANGUAGES

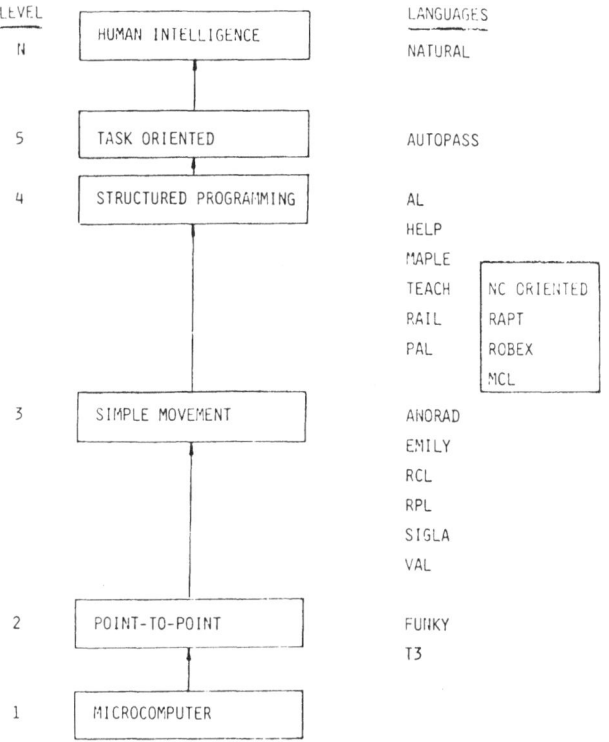

Figure 11.

also should be possible to describe to the robot an effector trajectory and how to handle the synchronization of the work of several arms. The robot must be able to operate the effector and the work tools under program control. In addition, there must be language constructs available which can handle sensor signals to which the robot is capable to react.

The many languages presently available suggest that the output of a compiler is a standard intermediate code, Figure 13. In this case the robot manufacturer has to lay out his control system in such a manner that the interface of the controller accepts the intermediate code. Thus, it is possible to use for different robots different languages via a standardized interface.

Concepts for New Programming Languages

Presently, there is a considerable amount of research work done to develop new programming languages and systems for robots. Two different development trends can be observed:

```
Teach-in programming
Control structure
Subroutines
Nested loops
Data types
Comments
Trajectory calculation
Effector commands
Tool commands
Parallel operation
Process peripherals
Force-torque sensors
Touch sensors
Approach sensors
Vision systems
```

Figure 12.

(1) The individual robot is made autonomous. It obtains the capability to adapt itself to the work environment, to make own decisions and to take actions to solve unforeseeable situations. In other words, the robot will be provided with an amount of limited intelligence which is needed to perform its assembly task.

(2) The entire manufacturing process will be completely automated to eliminate the necessity for human intervention. The robot becomes an integral part of a manufacturing facility which is supervised by a hierarchy of computers. The description of the workpieces and that of the assembly is automatically generated from the design process and transferred from the CAD-system to the control computer of the robot. There is no need to program the robot directly and to describe to it the geometrical shape of the object, its surface description, gripping position, etc. This information is all available from a central CAD data base.

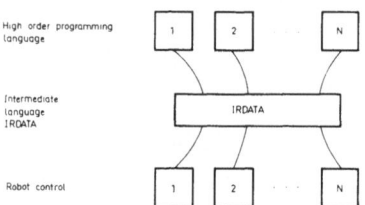

Figure 13.

Both of these developments will find their application in different industries:

- The autonomous and flexible robot, equipped with sensors and a runtime system, will be used by small and medium size companies. This system can be adapted and reprogrammed quickly to new production runs. It may make its own decision with the aid of an expert system, Figure 14. The drawback of this device is its complexity and the unavoidable high development effort.

- Systems with several robots, where each unit performs a specific task will find their entrance in mass production. They will be the domain of larger companies. Such an integrated manufacturing system will include machine tools, robots, material handling, peripherals, conveyors, etc., Figure 15 [8]. The investment cost will be quite high. With increasing product diversity the intelligent robot may also be installed in flexible manufacturing systems which can handle many product variants.

Independently of these two developments several high order programming languages have been developed to solve assembly tasks. Most of them use the frame concept. Their use, however, requires that the programmer has the capability to view the assembly object in a three dimensional space. For example, he must be able to visualize the permissible path of an effector to move from one point in the assembly space to another. Typical features of a high order programming system will be explained for the SRL language developed by the University of Karlsruhe.

Based on an extensive study of the VAL [9] and AL [10] systems and the result of a comparison of several existing languages for industrial robots, a new language SRL was developed. The frame concept with its geometric data types and geometric operators was taken from AL. Experience obtained from an AL-implementation lead to the conception of the following new features:

(1) Generalization of the geometric data types "vector, rotation and frame" with the help of the structured data types of PASCAL.

(2) A time compound statement to distinguish between a compound statement for syntactical reasons, a block and a sequence of statements to be executed sequentially.

(3) A general structure for parallel, cyclic or delayed execution of parts of the program.

(4) Input-output to digital or analog ports and sensors.

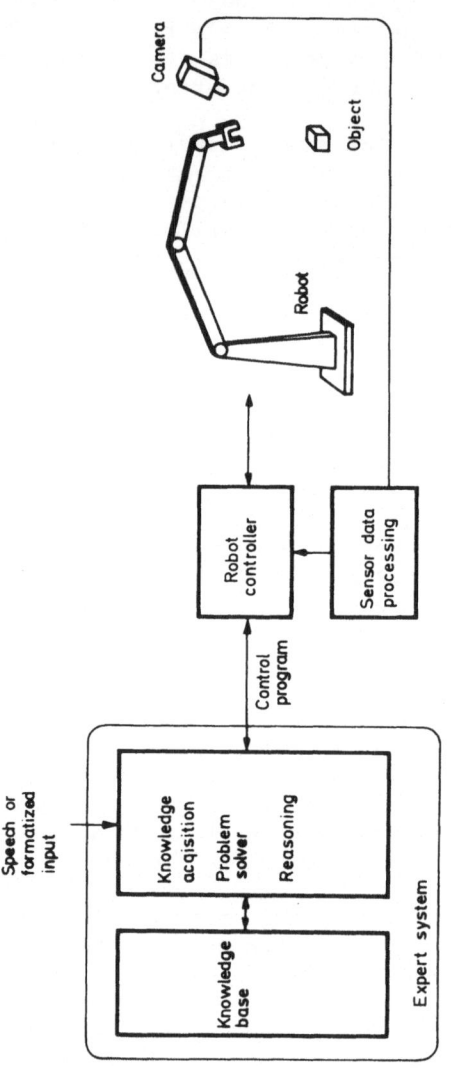

Figure 14. Future autonomous robot.

DEVELOPMENT OF PROGRAMMING LANGUAGES

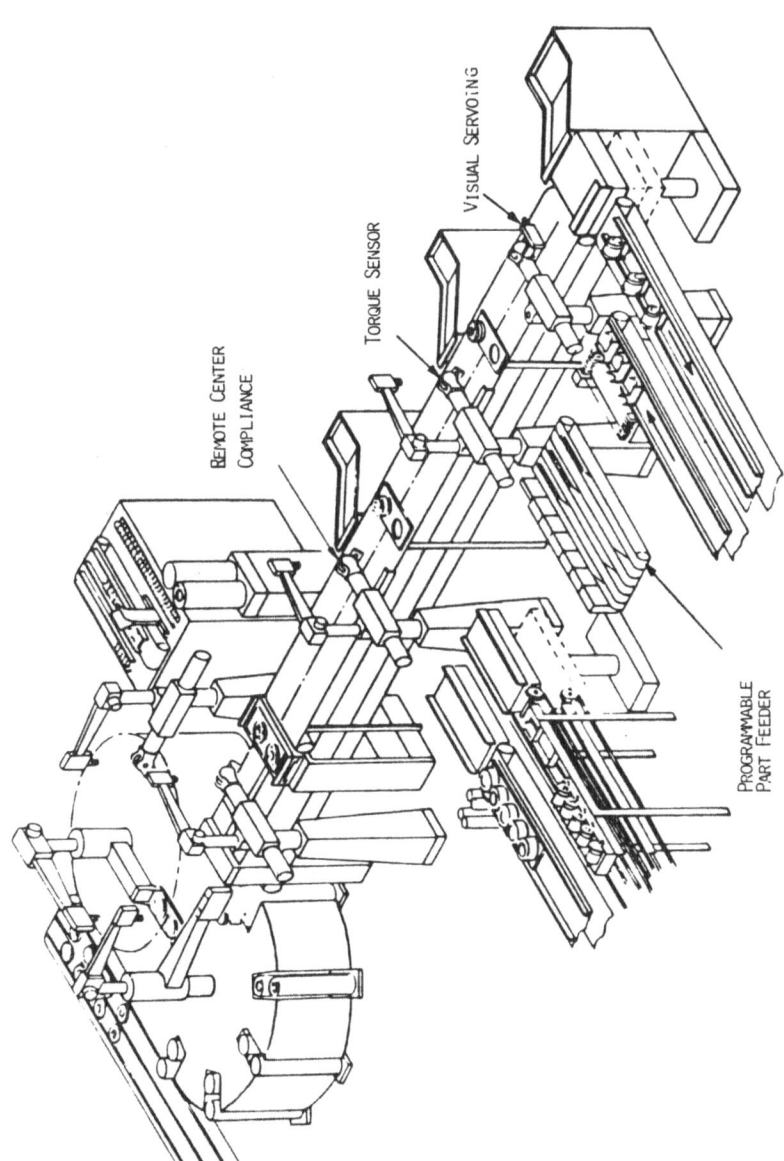

Figure 15. Adaptable programmable assembly system.

(5) Specification of system components like robots, sensors or interrupts.

(6) General sensor interface.

(7) Several move statements for different kinds of interpolation.

Most of the points mentioned above were derived from the fact that AL was designed hardware dependent. Therefore, AL uses only for the MOVE statement measuring data from a force and torque sensor.

To overcome hardware dependence and to support structured and self-documenting programming SRL includes the above stated language constructs. As a new facility, SRL has an interface to a general world model at program runtime. The world model can contain data about objects and their attributes, like workpieces, fixtures, robots, frames and trajectories.

Another fundament feature of SRL is the language PASCAL. The data concept and file management are taken from PASCAL because this language gives the user a very flexible and problem-oriented data structure.

The goal of the development of SRL is the design of a language which can easily be learned and adapted for further developments and applications. It will also provide an interface between future planning modules and the "traditional" programming system. A planning module will be used to generate SRL statements from a task (goal) oriented specification. This will replace explicit programming for every action, Figure 16. Therefore, SRL has to be well-structured and universal in nature, and it has to include all features of robot programming and process control. The standard data types and the RECORDs of structured data types, defined by the programmer are from PASCAL. There are new data types added to improve handling of synchronization between the program and external events. Predefined records can be used for geometrical computation needed for robot moves. The standard data types of SRL are:

```
INTEGER    ⎫
REAL       ⎪
BOOLEAN    ⎬ from PASCAL
CHAR       ⎭
VECTOR     ⎫
ROTATION   ⎬ from AL
FRAME      ⎭
SEMAPHOR   ⎫
SYSFLAG    ⎬ for synchronization
```

DEVELOPMENT OF PROGRAMMING LANGUAGES

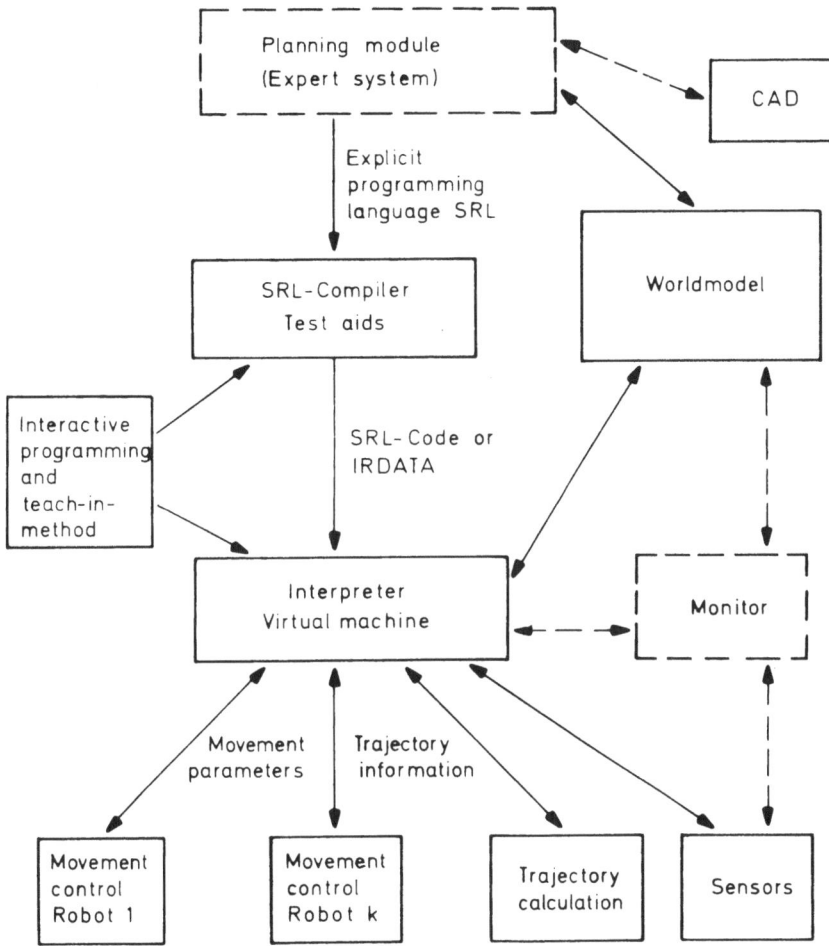

Figure 16.

The programmer describes textually with a frame in SRL notation the position and orientation of the robot effector. It consists of a position vector and a rotation. The rotation can be defined by one or several rotations, and it is stored as a 3 by 3 rotation matrix. The user has access to the matrix.

A semaphor is used for synchronization and queueing of tasks within a program. The system flag SYSFLAG is introduced to synchronize programs. The programmer has no direct access to the data type SEMAPHOR or SYSFLAG, but he can use the statements SIGNAL and WAIT to handle them. SRL includes the structured data types ARRAY, RECORD and FILE of PASCAL. Furthermore, the programmer can define his own problem-oriented data types as in PASCAL by enumeration and subrange. There are also pointers included in SRL. With respect to data types, the programmer can write records and its components in any expression as it is done in PASCAL.

Example:

> save-distance:=posvector.x + tolerance;

Programming with a Natural Language

Natural language programming systems are being developed to aid untrained personnel to teach the desired movements of a task to a robot, Figure 17. Because of the complexity of a natural language, not the entire vocabulary can be used. Usually simple syntax and semantic are selected which allow to describe the task of the robot by a quasi natural language. The programming system performs the syntactical and semantical analysis of the speech input, it extracts the pertinent information and makes a plausibility check. Thereafter, a corresponding formal robot program is executed. These systems need large memory and require long execution times.

Implicit Programming Languages

Systems to automatically program robots are under development since the seventieth. Here the programmer does not need to formulate explicitly every instruction of a task, e.g., "MOVE ARM TO POS. 1". However, he gives task-oriented instructions, e.g., "FASTEN FLANGE WITH 4 BOLTS". The system tries to interpret this instruction and plans its execution. The system searches in its library for different operators which will perform the required robot actions. Starting with the initialization state, each succeeding state is planned until the final state has been reached. The result of this search is a sequence of operators and states which can be visualized as an operation plan. This plan is equivalent to a program obtained from a programming language. A system capable to set up automatically such a plan is called a problem solver. The individual steps to assemble the plan may be as follows:

> First the flange has to be recognized. Then the effector picks up the flange and places it on the mating part. Now, a check is made with a sensor for proper alignment. The next steps are to fasten a screwdriver to the effector, to locate a bolt, to pick it up and to insert it into a bolt hole of the flange. In order to assure proper tightness, a torque sensor supervises the fastening operation. Thereafter, the other 3 bolts are inserted. In the last step the presence of all 4 bolts is verified by a vision system.

DEVELOPMENT OF PROGRAMMING LANGUAGES

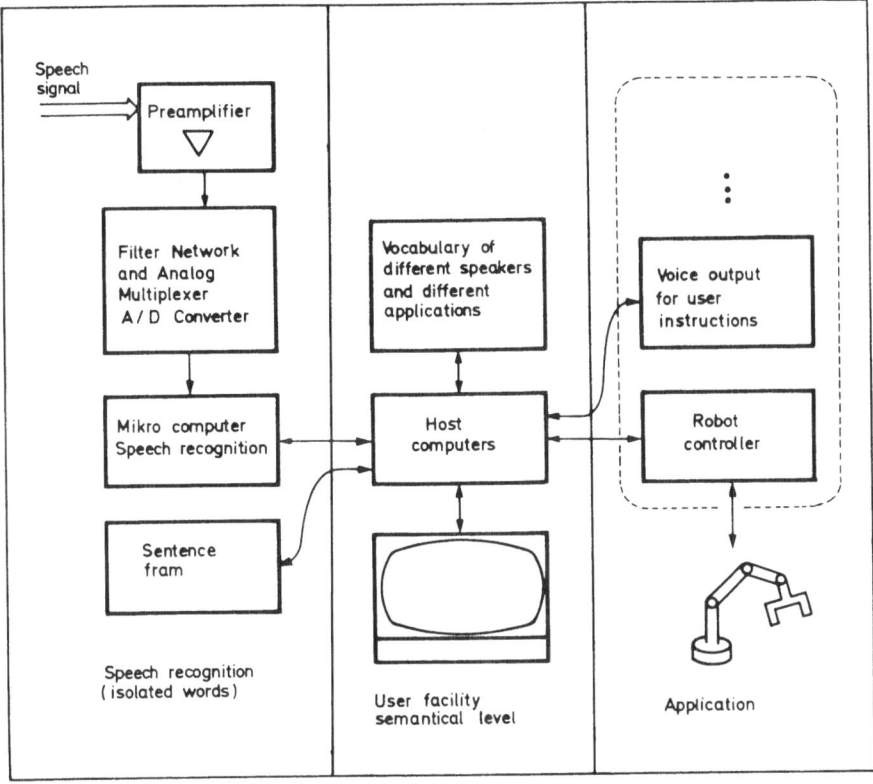

Figure 17. Acoustic programming.

Programming Aids

In addition to the language there must be a powerful programming system available, consisting of several software packages and of a low cost program development computer. Figure 18 shows a comprehensive programming system for assembly robots.

The user describes to the robot the object and the workpiece with the help of an application oriented language. This information is processed by a geometry procesor and entered into a world model. Likewise the movement of the robot is functionally described by implicit instructions and a syntactical analysis is performed. This program is combined with information of the world model. The result is sent to the SRL-compiler via a generating model. It is also possible to communicate interactively with the SRL-compiler to enter or edit instructions. The output of the SRL-compiler in form of interpretative code is loaded down to the control computer of the robot. Sensor signals from the robot can be brought back to the sensor data processing module. In case an object or a workpiece

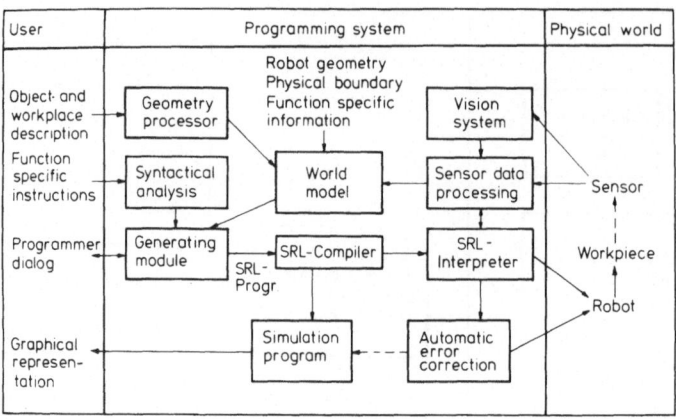

Figure 18.

had changed its position, this module will send instructions to the world model to update it. The same information is needed by the SRL-interpreter to correct the movements of the effector. There is also a simulation program available which allows the programmer to display graphically the work environment of the robot, to check its movements and to detect possible collisions.

The graphical emulation system is part of the programming system or of the real-time controller, Figure 19 [11]. In the early stage of the robot design, its kinematic attributes (joints, links, end-effectors) the assembly cell as well as its environment can be described on a graphic display. Trajectory planning, the interpolation in cartesian coordinates and the corresponding coordinate transformation can be tested and optimized. By adding a program for the simulation of the robot's dynamics the response of the axis motor drives and their control can be traced to evaluate the dynamics of the robot. For debugging of assembly programs, the emulated robot is interfaced to the programming system which defines multiple moving tasks. With an off-line program test facility, the workpiece and the robot components can be emulated without the risk of collision. When it is certain that all assembly sequences are performed without conflict, the program can be transferred to the robot control computer for execution in realtime to move the mechanical manipulator. Verification of the assembly can be performed in this stage. Figures 20 and 21 show how a robot is constructed from basic components with the aid of a simulator.

PROCESS CONTROL

Languages for process control - also named real-time programming languages - are characterized as follows:

DEVELOPMENT OF PROGRAMMING LANGUAGES

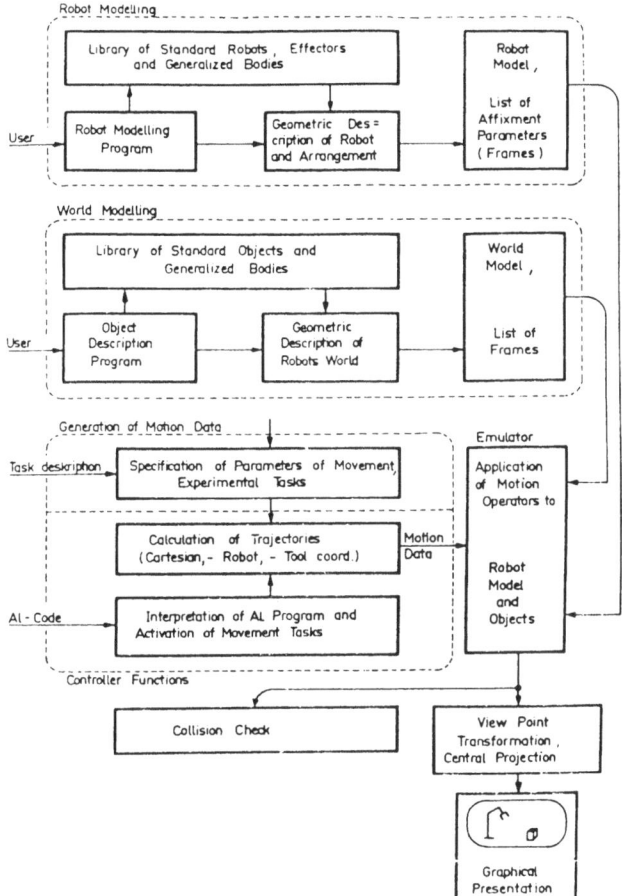

Figure 19.

(1) They permit the description of parallelism. Depending on the structure of technical processes, there are usually several activities, which have to be controlled concurrently. Process control languages contain therefore concepts for the definition of tasks.

(2) They allow the formulation of time-expressions and the interconnection of time and activities.

(3) They enable the synchronization and communication between different tasks. The programmer can define which task uses which data of another task.

(4) Some of the existing languages make possible the description of the interface between the technical process and

the computer. Usually they possess more data types than commercial data processing languages.

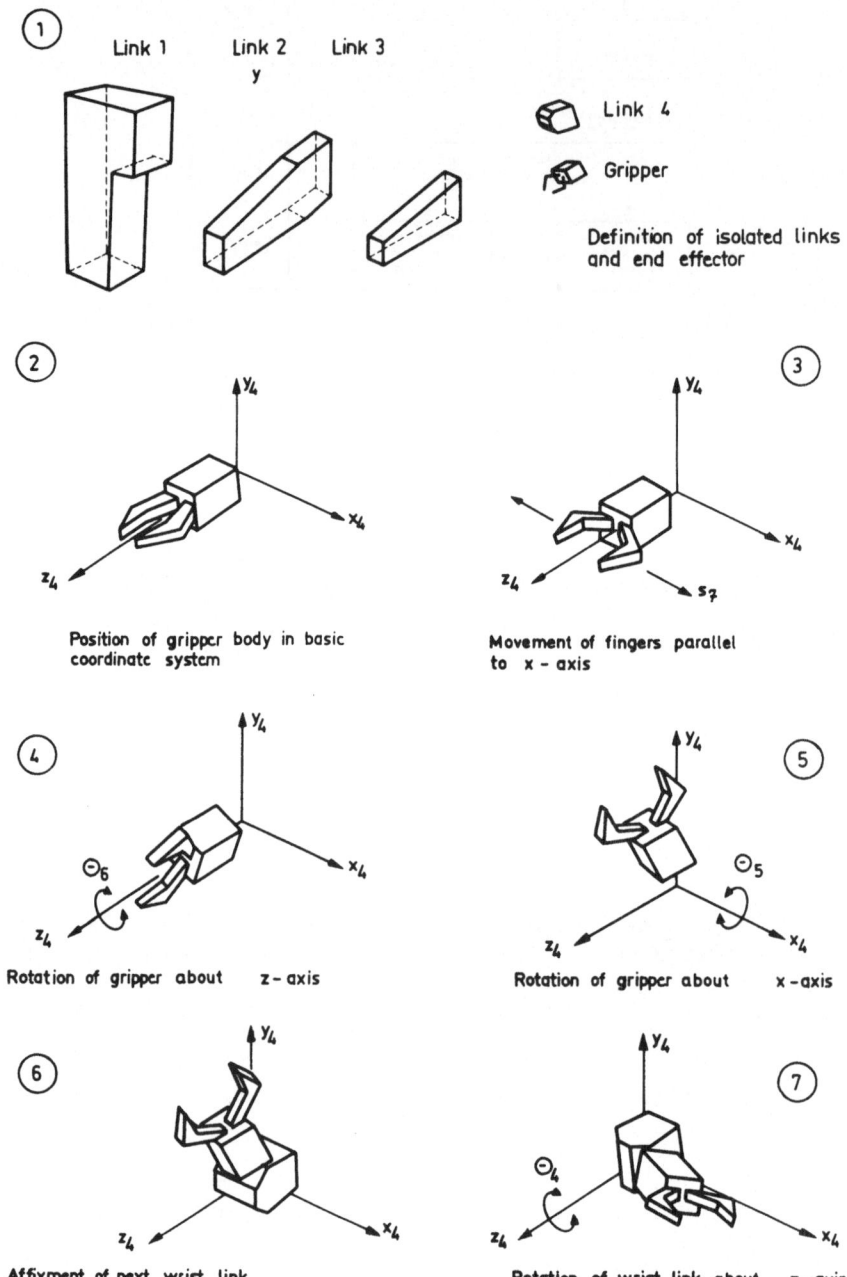

Figure 20. Modelling and composition of basic arm elements.

DEVELOPMENT OF PROGRAMMING LANGUAGES   463

(5) They support the reaction to spontaneous events of the technical process.

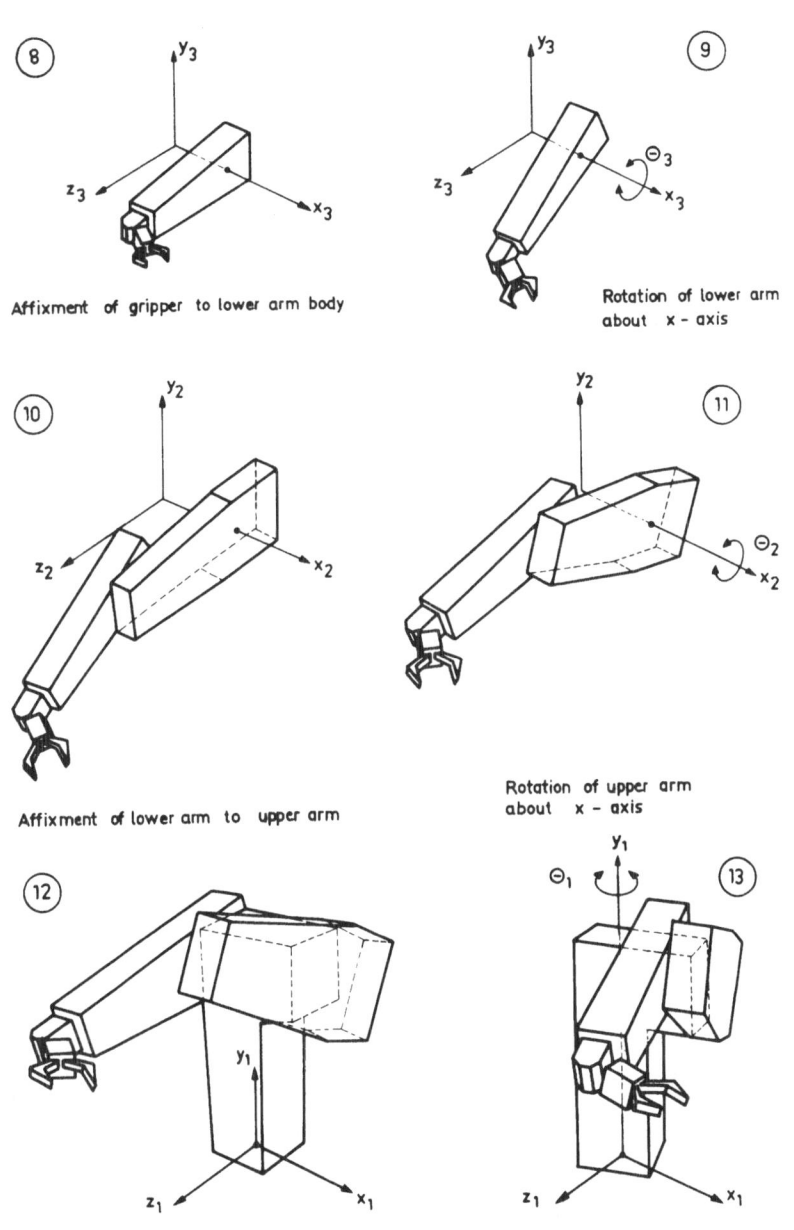

Figure 21. Successive composition of the PUMA 600.

(6) Some of them assist structured programming. Thus, they facilitate the implementation of reusable process control software.

Current languages for process control were developed either as extension of existing scientific programming languages or as completely new languages. Process-FORTRAN, Process-BASIC are of the first type, PEARL (Process and Experiment Automation Realtime Language), RTL/2 and ADA represent the second type.

Extension of Existing Programming Languages

Process-FORTRAN is an example of the extension of a language. It emerged from FORTRAN IV. Special features were included by adding new procedures. They can be activated by

CALL<Procedure Name>(<Parameters>).

These procedures enable:

- the manipulation of single bits or a group of bits,
- the scheduling of different tasks, and
- the input and output of measurement and control information to and from the technical process.

The bit-manipulation is very useful in process control because it is often necessary to process a binary coded state information. Usually a group of such state information is represented in a single word, and bit-manipulation-procedures are necessary to extract the desired state information, e.g., the state of a (binary) switch.

Procedures for task scheduling include the definition of different priorities for different tasks and the specification of which tasks depend on which events (of the technical process). It is possible to activate a task after a predefined time period or to bring it into an inactive state.

The procedures for input and output of measurement and control information permit the transfer of data to and from interfaces. They may contain digital/analog or analog/digital convertors.

Pearl - A Process and Experiment Automation Realtime Language

PEARL is a language which was developed as a system implementation language for process control application. A PEARL-program consists of several modules which can be compiled separately. Thus, it supports the modular decomposition of complex programs. Each module consists of a system part and a problem part.

The system part contains the description of the hardware-configuration on which the program has to run. All connections between the process control computer and the peripheral devices must be described in this part. These devices are either standard peripherals or disks, displays or special process control equipment. Furthermore, it is possible to assign names to these peripheral devices. They are used later on in the problem part for input and output operations.

The assigned names represent the logical connection between system part and problem part. Therefore, it is possible to adapt a process control system implemented in PEARL easily to a new environment. Only the system part has to be modified to be compatible with the new environment.

The problem part contains the algorithms which solve the desired process function. Procedures, tasks and data are components of this part. Tasks enable the independent reaction to different events of the technical process. Each task can be activated either by an external or a time event. Furthermore, it is possible to change the state of a task through the functions ACTIVATE, TERMINATE, CONTINUE and SUSPEND. In addition to the data types used by scientific languages, PEARL is equipped with data types for:

- time expressions (clock and duration),
- string handling, and
- input and output.

The available control structures are similar to those of other higher programming languages.

PEARL offers various facilities to solve the synchronization problem. The logical synchronization comprises synchronization of independent tasks and the activation of actions by events. Both operations can be described by means of semaphor-variables. These variables are manipulated by REQUEST and RELEASE operations. The time-synchronization can be described by new data-types (clock and duration) and new functions. The following statements give an example:

```
AT   10:30:00         SUSPEND TASK 1;
AT   10:00:00         EVERY 10 SEC UNTIL 11:00 ACTIVATE TASK 2;
AT   12:00:00         ACTIVATE TASK 3;
AFTER 10 MIN          RESUME;
```

ADA

ADA is a programming language which includes many facilities offered by classical high-level languages such as PASCAL. It also has facilities often found in specialized languages. ADA is suitable for programming embedded computer systems. This language has the usual control structures of high-level languages.

It offers following features:

- possibility to define types and subprograms,
- support of the module concept,
- separate compilation,
- real-time programming with facilities to model parallel tasks including exception handling.

Furthermore, systems programming is possible.

An ADA program consists of several units. A unit is either a subprogram (procedure or function), a package, a task, a generic subprogram or a generic package. A unit consists of a specification and a body. The specification contains the information which is visible to other units. It represents the interface of a unit. The body comprises the implementation details. All parts described within a body do not have to be visible to other units. The body usually is composed of a declarative part which defines the logical entities to be used in the unit and a sequence of statements which specifies the execution of the unit.

A subprogram is either a procedure or a function. Its specification part contains the name of a procedure or a function and a description of parameters. Every parameter must be characterized by its name and type. Additionally, it has to be described whether it is an input-, an output- or an input and output parameter.

A package is the basic unit for defining a collection of logically related entities. Examples of such entities are data, types, a collection of related subprograms, a set of type declarations and associated operations. The specification of a package contains all components which are visible to all other units, unless they are specified as private.

A task unit is the basic unit for defining a task whose sequence of actions may be executed in parallel with those of other tasks. The specification of a task is composed of the name of the task and optionally a list of entries. An entry can be called by other tasks. The calling of an entry resembles the invocation of a procedure or a function. If the specification part of a task contains an entry, then the body must have a corresponding (the name must be the same) accept statement.

A generic unit (generic subprogram or generic package) is a template which may be parameterized and from which corresponding subprograms or packages can be obtained. A generic unit cannot be called, it can only be instantiated. The result of an instantiation

is an instance which can be called. The generic feature enhances a
modular approach because the same algorithm can be used for different types of data. In ADA the rendezvous-technique is used to
describe the synchronization between independent tasks. A rendezvous is realized by an entry call inside the calling task and the
corresponding accept statement in another task. The following
example represents a rendezvous between the tasks tl and t2:

```
task tl is
  entry ALERT (MESSAGE:  in INFO);
end tl

task body tl is
begin
  loop
    :
    accept ALERT (MESSAGE:  in INFO) do
      :
    end ALERT;
    :
  end loop;
end tl;

task body t2 is
MESS:   INFO
begin
  .
  .
  tl.ALERT (MESS);
  .
  .
end t2;
```

where tl and t2 are independent tasks. The rendezvous mechanism is
as follows: the first task to initiate the rendezvous (represented
by accept ALERT in tl and tl.ALERT in t2) waits for its partner.
Then the calling task t2 is delayed until tl has finished the
exchange part (end ALERT). When the exchange is completed, each
task continues independently its execution. The rendezvous technique is the only mechanism in ADA to describe the logical synchronization. It is also used to couple an event in the technical
process with an action of the process control system.

For time synchronization, ADA offers two predefined datatypes, the duration and the time. It also has a delay statement
which inhibits further execution of the task for a specified time.

In ADA there exist no statements which facilitate the information transfer to and from the technical process. The problem can be solved by implementing new I/O-packages. However, this requires a thorough knowledge of ADA.

ADA as well as PEARL (and other not mentioned languages) enable the implementation of real-time systems. PEARL offers a wider range of real-time function, e.g., process I/O-and task-scheduling, whereas the strengths of ADA are its strong typing and its package concept which encapsulates data and the routines that modify these data.

Tools for the Development of Process Control Systems

The necessity of using tools for software development, especially for requirements specification, is no longer disputed. One of these tools, a high-level software development environment called SARS, will be described [12].

The software development tools of the SARS system presented here make it possible to capture requirements in a form the user can understand, and to analyze them for completeness and consistency. In contrast to similar computer-assisted systems, SARS offers the advantage of graphical input. Since it is well-known that human beings can assimilate complex information much more easily when it is presented in a graphical form; this is a significant step to assist the user.

Specifying a system is not a straightforward and well structured process. Roughly spoken, it is an iteration of three activities:

(1) The collection of information about the system to be automated and its environment.

(2) The structuring and storing of the collected information.

(3) The analysis and improvement of the stored information.

SARS is a high-level software development environment for process control systems. It stands for System for Application Oriented Requirements Specification. It consists of various media and a series of tools. These are the following principal components (Figure 22):

- Information System,
- MARS (Method for Application Oriented Requirements Specification),
- LARS (Language for Application Oriented Requirements Specification),
- Textual and Graphical Editor, and
- Dialog System.

DEVELOPMENT OF PROGRAMMING LANGUAGES 469

<u>SARS Information System</u>. The first step of a specification is a collection of ideas and of informal requirements. The specification by means of SARS will start with the transformation of these informal descriptions step by step into a formal notation (LARS). The basis of informal requirements are ideas and instructions of the costumers. The SARS specification also contains reports of project members (e.g., minutes, reports to the costumer and notes from the system analysts).

The SARS Information System stores information about the project. Among them are:

- datebooks,
- deadlines,
- project management information,
- personal data,
- first solutions,
- known methods and aids,
- references to information, and
- data sheets.

Due to the iterative nature of a formal specification process there are usually several solutions (versions). Therefore, alternate solutions are stored in a central data base which is available to the user.

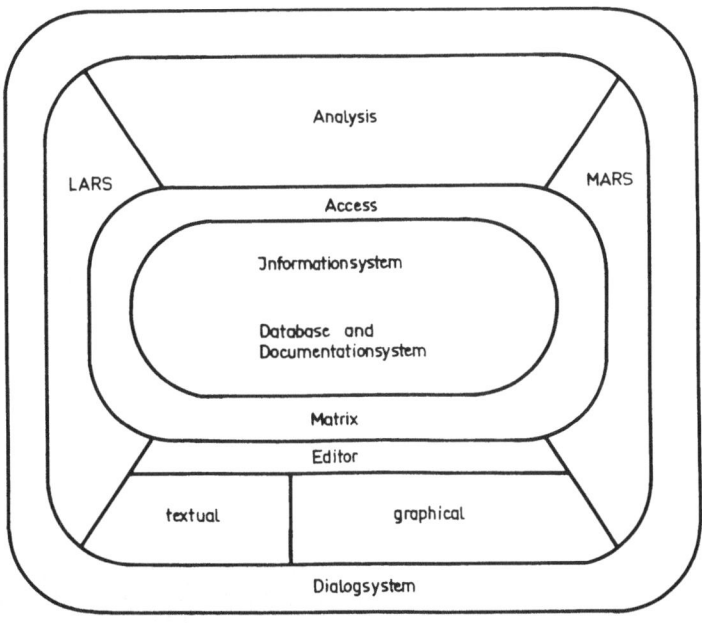

Figure 22. Components of SARS.

The nucleus of SARS is an information system that manages all data of the specification work station. The descriptive media of SARS are:

- formal language texts (LARS) and
- graphics.

There is a common data management system, which can be accessed by all software components, and which controls all elements of the data base. Underlying this data base is a relational data base, in which the specification is stored. The data base is accessed through an access matrix (Figure 23). It represents the user's logical view of the data base. One column of this matrix is reserved for the description of the entire process that is to be automated. The other columns are generated by the project leader. Their number depends on the number of subprocesses and are named according to the subprocesses.

The rows of the access matrix contain the basic components of the LARS language. This part of the access matrix cannot be changed by the user. To allow the user to store additional information like reports, notes and so on, further lines can be created which represent "private" classifications.

Pictures and tables represent the formal language. The graphic work station consists of a graphic display and a digitizer. Graphical representation aids fast information transfer to the user. The interrelation between LARS components can be seen immediately. Formal graphic input is syntactically analyzed and translated into the internal language representation. The actual transformation between LARS text and LARS graphics will be done by the system.

The common data management for all descriptive media makes selected retrieval possible which is necessary to produce reports for different user groups.

The MARS Method. This method was the conceptual point of departure in the creation of SARS. The following two considerations characterize the method:

(1) Every automation system to be developed - or, more concretely - the software for that system, reflects the environment associated with the system. Thus, in the office domain, for instance, the existing or the future organizational structure is as much a binding aspect of a requirements analysis as it is in process control the existing or planned structure of a technical process.

(2) Every action to be performed by a behavioral unit like a computer can be described with the Stimulus-Response

model. Either it is evident that a causal event (stimulus) will have an effect in reaction to that event (response), or one attempts through an analysis to trace a certain effect (response) back to an identifiable cause (stimulus). A simple example: On the birthday of the operator (stimulus: a particular date becomes valid) the

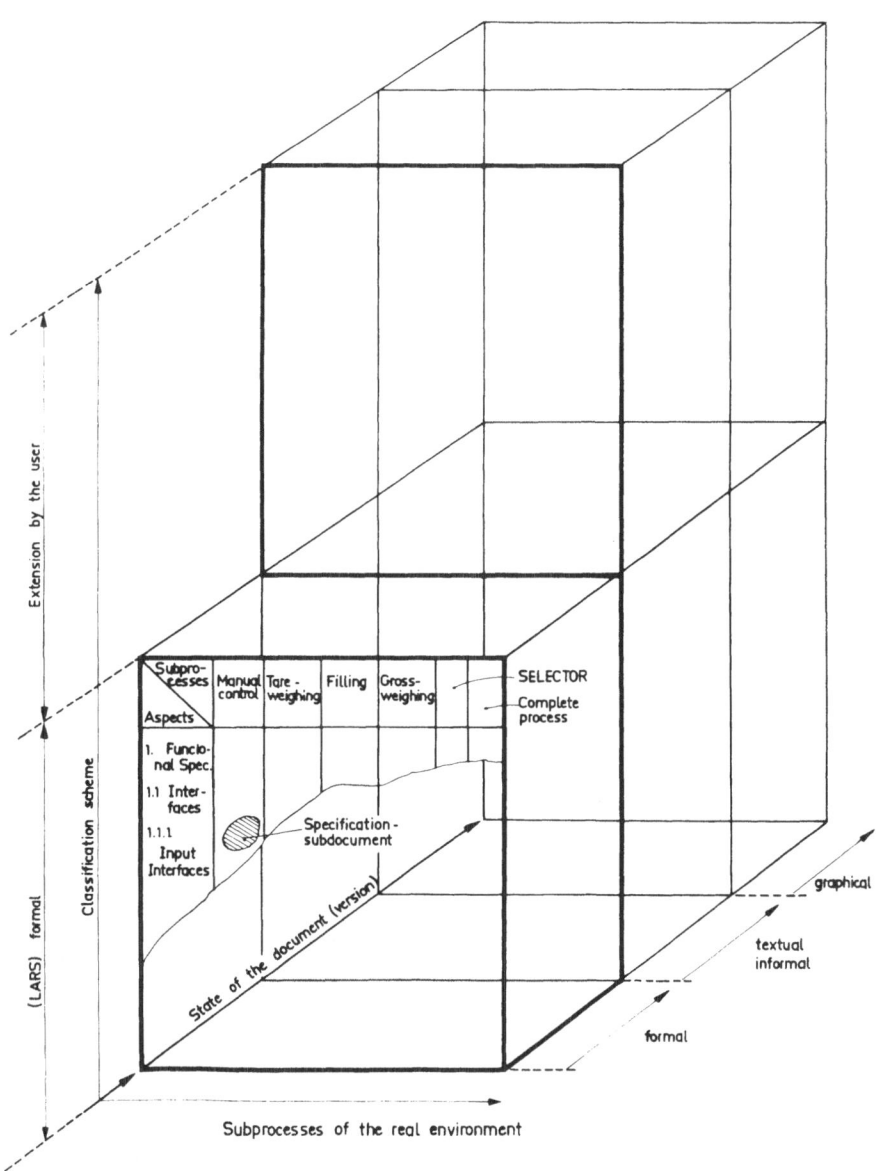

Figure 23.

computer is to print out a Marilyn Monroe poster (response: the printout).

Both principles determine the methodology of the requirement specification. The "requirement specification" defines all of the necessary stimulus-response (SR) functions according to the subprocesses. Although these SR-objects are still independent and neutral with respect to the later implementation, they can nevertheless be understood as autonomous computing processes. The SR funtion blocks are refined in further steps to nets. The syntax of these (Stimulus-Response-Memory) nets corresponds approximately to R-nets [13].

The LARS Language. The primary components of LARS are Interfaces, SRM-Nets and Guards (Figure 24). Through the Interfaces the

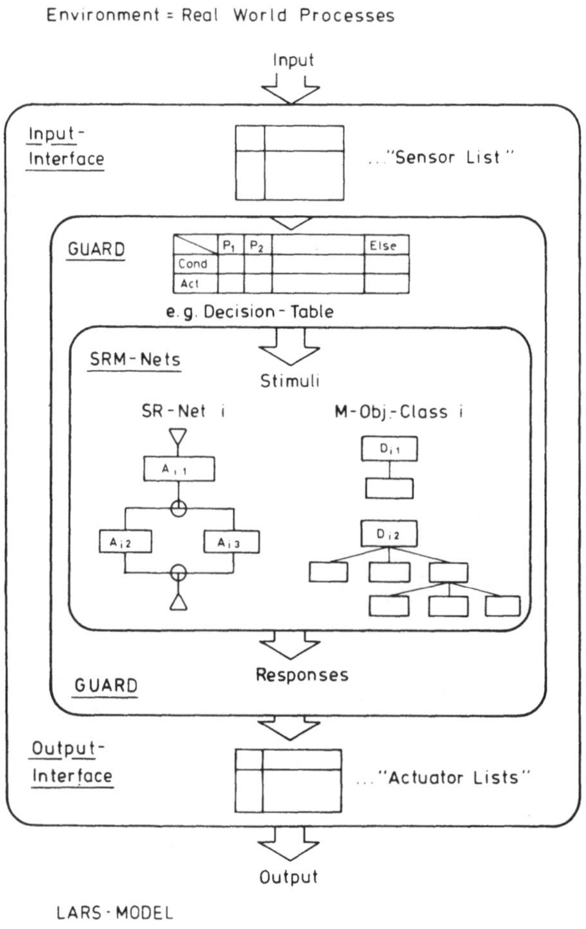

Figure 24.

user can model input and output signals in a way a non-computer scientist is accustomed to. SRM-Nets are stimulus-response-memory nets which can be regarded as basis functional components. Data objects are data and data structures which are used by the nets. The Guard was introduced as supervising constituent which controls the flow of input and output signals and data. It determines which data objects are transferred to which SRM-Nets and which nets must be activated or have to be synchronized.

The LARS language allows:

- A formal description of SRM-Nets and the data used by them (Net LARS).

- The definition of the interfaces between the control system and the external world (Interface LARS).

- The description of the logical connections between the interfaces and the SRM-Nets (Guard LARS).

There exists in SARS a compiler for LARS, with the help of which it is possible to analyze requirements specifications formulated in LARS for completeness and consistency. SARS contains two editors. One manipulates the textual representation, and the other manipulates the graphical representation.

Special characteristics of the SARS system are the graphical representation of information, the formal definition of the specification language and the harmonious integration of the various tools by means of a dialog system.

COMMERCIAL DATA PROCESSING

Currently there are many different programming languages available which are used for commercial data processing, such as Algol, Basic, Cobol, Fortran, PL/1, Pascal, and RPG. All these languages provide concepts to describe the flow of control within a program. Usually they include conditional statements (if-then-else) and loops (a repeated sequence of statements). The language most frequently used for commercial data processing is Cobol. The feature of this language will be described briefly:

Cobol

One reason for the widespread use of Cobol is its orientation towards the English language. All instructions are coded using English words. A Cobol programs consists of four divisions namely

- identification division,

- environment division,
- data division, and
- procedure division.

The identification division serves to identify the program. It contains the program name and some organizational information like author, installation, date and remarks.

The environment division assigns the input and output files to specific devices. It is divided into a configuration section and an input-output section. The configuration section supplies data concerning the computer on which the Cobol program will be compiled and executed. The input-output section comprises information concerning the specific devices used in the program. This is the only section which will change significantly if the program is to be run on different computers.

The data division consists of a file section and a working storage section. It describes in detail the field designations of the input and output file. Each field has a corresponding picture clause denoting the size and type of data. If the program requires constants or work areas, they are also contained in this section.

The procedure division contains all the instructions to be executed by the computer. It is detailed into paragraphs. Each paragraph defines an independent routine. The instruction set is comparable with that of an assembly language. But in contrast to assembly language, the word-symbols are easier to understand because they were chosen in accordance with natural English.

From the scientific point of view Cobol is a poor language. It supports neither the module concept nor structured programming. The flow of control is realized by means of the harmful go to statement. There are no means to implement abstract data types. Nevertheless, in practice there is a widespread use of Cobol.

In the near future Cobol will be replaced by languages like Pascal and ADA. It is very likely that languages currently used primarily in scientific and advanced industrial environments, e.g., LISP and PROLOG will have a broader application. PROLOG will be the language of the Japanese fifth generation computers.

FUTURE TRENDS

Currently automation of the factory (material, tools, machines) and of factory information processing is widely kept independent of each other. In the future both of these activities must be developed in parallel. Otherwise, it will be very difficult, if not impossible, to link together different computers running under different operating systems and programmed with different languages. It also must be possible to interface the manufacturing units with the computer. The need to improve productivity and quality is well recognized. But to be really successful in the modernization of manufacturing, or in a broader sense in the modernization of an entire factory, the following requirements have to be fulfilled:

- The mechanical, electronic and computer software worlds must be brought together. It is necessary to improve the communication between mechanical, electronic and software people.

- Before an existing manufacturing process is automated a thorough analysis has to be done. There should be not only a single subprocess automated, but the process should be studied. If necessary it should be modified in order to arrive at a global and homogeneous solution.

An important part of such a solution will be realized by software. A programming language like ADA, which can be adapted to different applications by the definition of suitable packages, can essentially contribute to a homogeneous solution.

At the beginning of the nineteenth expert systems will be used to perform many routine tasks of a manufacturing organization. The communication with the computer will be done with graphical, voice or fill-in-the-blanks programming aids. The computer will have decision rules to plan and control the manufacturing process. It will be able to learn from past and present operations and will plan with the entrance of an order all manufacturing activities. A typical integrated manufacturing system is shown in Figure 25. The order may be processed as follows:

- Upon entry of an order the expert system determines if the item is a current product, a variant or new. The order from a current product passes design and enters directly manufacturing planning. For a new product or a variant engineering must furnish design data and manufacturing documents. There will be expert systems available which contain design rules, mathematical models and analysis tools. Depending on the degree of automation the design is done either completely automatically or with some manual assistance.

- In manufacturing planning the material, processes and process sequences are selected. The planning module will have a know-how about the entire plant resources, manufacturing methods and fabrication alternatives. There will be a cost analysis made as well as a make or buy decision.

- The product order and the manufacturing plan activates production scheduling. There are long and short term schedules which are used to meet the manufacturing due dates. The new order will be queued into these schedules. A knowledge data base contains information on all orders available inventory and the status of the manufacturing equipment of the plant. Equipment scheduling will be done with the aid of group technology and Operations Research tools. In case of problems with due dates or manufacturing equipment, an alternative manufacturing schedule will be conceived.

- In the next stage tooling is activated and procurement issues orders for parts and raw material from outside vendors. When the order is released for manufacturing, the control computers

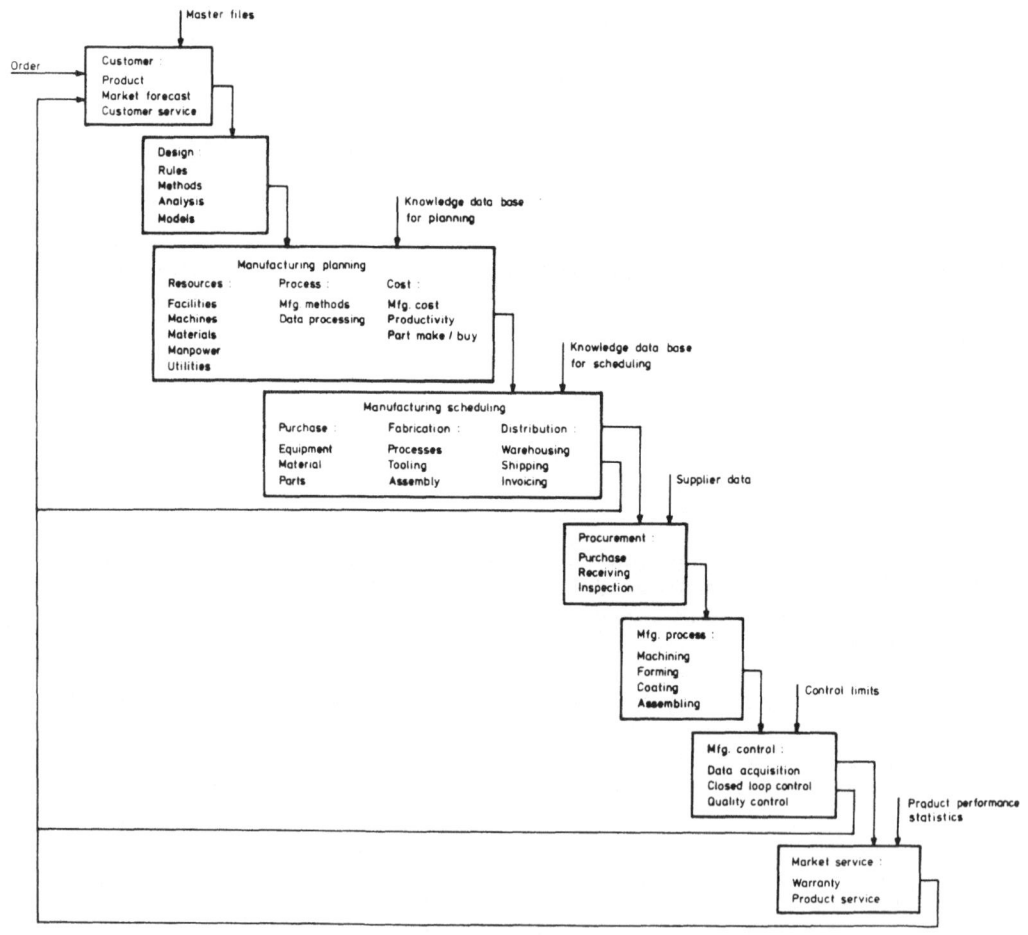

Figure 25. The manufacturing activities.

and data collection devices will supervise the fabrication process. There will be mechanisms to evaluate the capability of processes, to automatically set up test limits and to detect faulty equipment.

- The last activities are quality control testing and distribution of the product. Quality control data from the process and warranty information from the customer will be important feedback information for the expert system to maintain an efficient and a high quality product.

An expert system which can autonomously solve many of these manufacturing functions may be conceived as follows:

- It should understand the problem description and the requirement specification.
- It must be able to synthesize processing procedures.
- It must be capable to optimize between the machine system functions and the processing procedures.
- It must be able to synthesize response based on the outputs from the machine systems.
- It must understand verbal and graphic instructions via natural means of communication.

An expert system with these features must have following capabilities:

- It must understand the language used at the man-machine interface.
- It must know solutions to the problems to be solved.
- It must know the machine system.

Figure 26 [14] shows the concept of the 5th generation computer. It consists of application software, system software and computer hardware. The core of this computer is the system and application software. With the help of these, man is capable to instruct the hardware, for example, to optimize a manufacturing plant for a given product spectrum.

REFERENCES

[1] EXAPT-NC-Programmiersystem, EXAPT-Verein, Aachen, Germany.
[2] R. Langebartels, "Programmierung von numerisch gesteuerten Werkzeugmaschinen mit den Fertigungstechnisch orientierten Programmiersprachen APT und EXAPT," VDI-Z, Band III, Nr. 12, 1969.
[3] Programmierplätze für NC-Bearbeitungsprogramme: System P-D / P-F, SYMBOLIC FAPT TURN, Siemens Corporation, Munich, Germany.

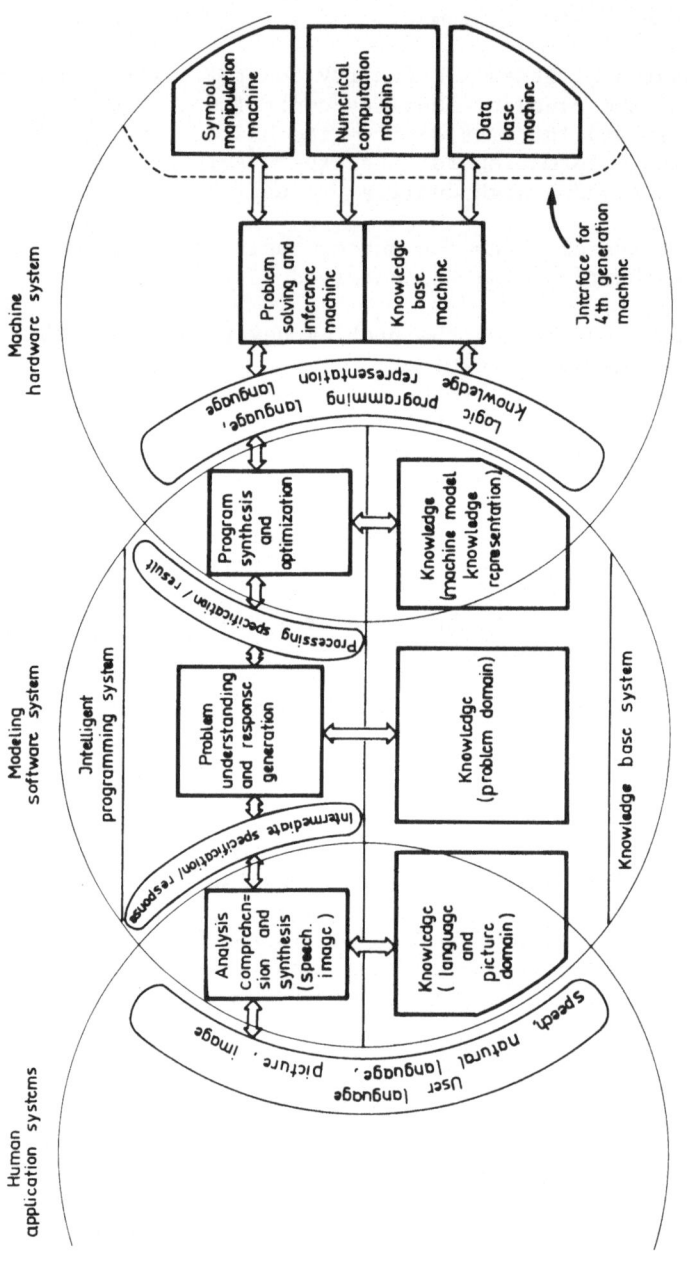

Figure 26. The 5th generation computer concept.

[4] "NC Machining by the Numbers," Manufacturing Engineers, May 1979.

[5] F. Leonards, W. Müller, F. Otto, and H. Sinning, "Prozeßlenksysteme für die Drehbearbeitung," PDV Berichte, Gesellschaft fur Kernforschung mbH Karlsruhe, August 1976.

[6] U. Rembold, C. Blume, R. Dillmann, and G. Mörtel, "Technische Anforderungen an zukunftige Montageroboter," Part 4, VDI-Zeitschrift, Nr. 21, Nov. 1. 81.

[7] S. Bonner and G. Kang, "A comparative study of robot languages," Computer, Dec. 1982.

[8] "The Advent of Adaptable Programmable Assembly Systems," Manufacturing Engineering," April 1979.

[9] C. Blume, "VAL - A robot control system of unimation," University of Karlsruhe, not published, 1980.

[10] S. Mujtaba and R. Goldmann, "AL Users' Manual," Stanford University, 1979.

[11] R. Dillmann, "A graphical emulation system for robot design and program testing," Conference Proceedings of 13th ISIR/Robots 7, Chicago, IL, July 1983.

[12] W. Epple, M. Hagemann, M. Klump, and G. Koch, "SARS-System zur anwendungsorientierten Anforderungsspezifikation," Internal Report, University of Karlsruhe, 1983.

[13] M. W. Alford, "A requirements engineering methodology for real time processing requirements," IEEE Trans. on Software Engineering, Vol. SE-3, pp. 69-69, 1980.

[14] I. Motorola, et al., "Challenge for knowledge information processing systems," Proceedings of the International Conference on Fifth Generation Computer Systems, Japan, Oct. 1981.

ON-LINE IDENTIFICATION AND SUPPRESSION OF TIME VARYING MACHINING CHATTER IN TURNING VIA DYNAMIC DATA SYSTEM (DDS) METHODOLOGY

Shing-Yuan Tsai and Shien-Ming Wu

MIRL, ITRI, Taiwan, R.O.C.
University of Wisconsin
Madison, Wisconsin, U.S.A.

ABSTRACT

The time varying stability of the machining process necessitates a technique of on-line chatter identification and control. The Dynamic Data System (DDS) methodology which has been applied in the field of manufacturing and proved itself as a powerful tool to identify the machining process under working conditions is implemented for this purpose.

Mathematical models of the machining process with an inherent stochastic nature were developed as discrete ARMA (n,n-1) models. Based on off-line analysis, the peak of power spectral density corresponding to the dynamic mode of workpiece fundamental natural frequency served as a simple and reliable index of stability for on-line chatter identification. Using this criterion, the influence of speed and feed on stability were studied and the strategy of changing speed and feed incrementally to find stable cutting conditions without sacrificing productivity was proposed. Due to the current capability of microcomputer, a simple and fast adaptive modeling technique was adopted for the on-line machining process identification. A forecasting control of chatter scheme was implemented to predict the generation of chatter. In addition, the vibration signal was purified from the results of the dynamic analysis of the chuck-workpiece-tailstock system, and a self-learning scheme was established to determine the threshold level of stability in each cutting process.

A chatter suppression controller was designed and interfaced to the PT15 CNC lathe. Cutting tests demonstrated the successful use of the theoretical and technical development by the DDS approach.

INTRODUCTION

The stability of the machining process forms a key factor in the determination of metal removal rate and is the cause of poor surface finish and annoying noise. With recent rapid advances in computer numerical control and computer aided manufacturing, the chatter problem is further confounded. Unexpected chatter is likely to cause undesirable damages, resulting in expensive down time.

The analysis and suppression of chatter have received great attention during the last two decades. Based on fundamental studies of the structural and cutting dynamics [1,2,3], the machining process has become more efficient. These techniques, however, do not eliminate the necessity for an operator to take corrective actions whenever needed. Thus developing an active on-line chatter control scheme becomes essential. Various attempts were made by designing active chatter control schemes [4,5,6]. The influence of the various operating parameters and the varying properties of the machining process could not be accounted for in most attempts.

Since the stability of the machining process is the result of the interaction of the dynamics of the machine tool structure and the cutting process, and their dynamic interaction, therefore, effective control of chatter requires analytical models and methods to facilitate on-line identification of the process allowing a continuous monitoring of its time varying characteristics. The Dynamic Data System (DDS) methodology emerged as a new tool for the identification of the combined dynamics of the machine tool structure and cutting process [7,8]. By means of this methodology, a scheme for computer control of machining chatter was proposed to predict the damping ratio identified on-line, and the predicted values of the damping ratio were used to control cutting speed and feed to suppress the onset of chatter. The theoretical and technical development for the on-line prediction and suppression of the generation of chatter using DDS approach have been further developed and successfully applied in the MIRL PT15 CNC lathe [12,13] and reported in this paper.

MATHEMATICAL BACKGROUND

The fundamentals of the DDS modeling approach relevant to this work are briefly presented and further details could be obtained from available literature [14,15].

When the dynamic data from a system, such as the machining process, can be considered as a stationary stochastic series then it is possible to develop an Autoregressive Moving Average Model

(ARMA) to represent the data. Such a form in continuous time is represented by the differential equation:

$$(D^n + \alpha_{n-1}D^{n-1} + \ldots + \alpha_0)X(t) = (1 + b_1 D + \ldots + b_m D^m)Z(t)$$

$$E[Z(t)] = 0, \quad E[Z(t)Z(t+k)] = \sigma_z^2(k) \tag{1}$$

where $X(t)$ = system response; $Z(t)$ = white noise; $\delta(k)$ = Dirac Delta function; E = expectation operator; D = differential operator; $\alpha_0 \alpha_1, \ldots, \alpha_{n-1}$ = autoregressive parameters; $b_1, b_2, \ldots, b_m$ = moving average parameters; σ_z^2 = variance of $Z(t)$.

The above model is denoted by AM (n,m). For computer analysis data are obtained by sampling the continuous system at uniform interval Δ. The discrete representation of the uniformly sampled data is given by

$$(1 - \phi_1 B - \phi_2 B^2 - \ldots - \phi_n B^n)x_t = (1 - \theta_1 B - \theta_2 B^2 - \ldots - \theta_m B^m)a_t$$

$$E[a_t] = 0; \quad E[a_t a_{t-k}] = \delta_k \sigma_a^2 \tag{2}$$

where a_t = discrete white noise; δ_k = Kronecker Delta function; $\phi_1, \phi_2, \ldots, \phi_n$ = autoregressive parameters; $\theta_1, \theta_2, \ldots, \theta_m$ = moving average parameters; σ_a^2 = variance of a_t.

It is a general ARMA (n,m) model, and a definite parametric relationship exists between the eq. (1) and eq. (2). The procedure of modeling is a decomposition of the energy of the stochastic system over the smallest number of dynamic modes. When the order of the model is increased, a new mode is introduced. The various dynamic modes of the system are given by the characteristic roots (or the eigen values) of the characteristic equation of eq. (2) given by

$$\lambda^n - \sum_{j=1}^{n} \phi_j \lambda^{n-j} = 0 \tag{3}$$

A dynamic mode is normally associated with a single degree of freedom and is given by a pair of complex conjugate roots of eq. (3) for underdamped or a pair of real roots for overdamped modes. For this work, only underdamped modes are relevant and if the roots are denoted by λ and λ^*, the natural frequency and damping ratio of the mode is given by

$$\omega_n = \frac{1}{\Delta} \sqrt{\frac{[\ln(\lambda \lambda^*)]^2}{4} + [\cos^{-1}(\frac{\lambda + \lambda^*}{2\sqrt{\lambda \lambda^*}})]^2} \tag{4}$$

$$\zeta = \sqrt{\frac{[\ln(\lambda\lambda^*)]^2}{[\ln(\lambda\lambda^*)]^2 + 4[\cos^{-1}(\frac{\lambda+\lambda^*}{2\sqrt{\lambda\lambda^*}})]^2}} \tag{5}$$

One of the basic characteristics of the model defined by eq. (3) is its impulse response, or Green's function. The Green's function of the discrete stochastic process yields its orthogonal decomposition in terms of a_t, which allows the representations of eq. (2) in the form of

$$x_t = \sum_{k=0}^{\infty} \sum_{j=1}^{n} g_j \lambda_j^k a_{t-k} \tag{6}$$

or

$$x_t = \sum_{k=0}^{\infty} G_k a_{t-k} \tag{7}$$

$$g_j = \frac{\prod_{k=1}^{m} (1 - \nu_k \lambda_j^{-1})}{\prod_{\substack{k=1 \\ k \neq j}}^{n} (1 - \lambda_k \lambda_j^{-1})} \tag{8}$$

Where G_k is Green's function, λ_j are the roots of the autoregressive polynomial eq. (3), and ν_k are the roots of the moving average polynomial

$$\nu^m - \sum_{j=1}^{m} \theta_j \nu^{m-j} = 0 \tag{9}$$

Furthermore, the contribution of each mode to the total power of the signal can also be analyzed by the variance, γ_0, as follows

$$\gamma_0 = \sum_{\ell,m=1}^{n} g_\ell g_m \sigma_a^2 \frac{1}{1 - \lambda_\ell \lambda_m} \tag{10}$$

Each term in eq. (10) represents the contribution due to the modes $(\lambda_\ell, \lambda_m)$ to the total power of the system.

The commonly used characteristic of a dynamic system, i.e., its power spectrum can be readily obtained once the adequate model of the process is available. The normalized power spectral density is defined as

$$S(f) = \frac{2\sigma_a^2}{\gamma_0} \frac{\left|1 - \sum_{k=1}^{m} \theta_k e^{-j2k\pi f\Delta}\right|^2}{\left|1 - \sum_{k=1}^{n} \phi_k e^{-j2k\pi f\Delta}\right|^2} \tag{11}$$

where f is the frequency in H_z and $j = \sqrt{-1}$.

Once the parameters of the model are available, the future value of the series can be obtained by the process of forecasting. It can be shown that the best linear forecast of ℓ step ahead $x_{t+\ell}$, at time t, denoted $\hat{x}_t(\ell)$

$$\hat{x}_t(\ell) = G_\ell a_t + G_{\ell+1} a_{t-1} + G_{\ell+2} a_{t-2} + \ldots \tag{12}$$

The 95% probability limits on the forecast are given by

$$\hat{x}_t(\ell) \pm 1.96\sigma_a (1 + G_1^2 + G_2^2 + \ldots + G_{\ell-1}^2)^{1/2} \tag{13}$$

and the forecasting error is given by

$$e_t(\ell) = x_{t+\ell} - \hat{x}_t(\ell) \tag{14}$$

IDENTIFICATION OF THE MACHINING PROCESS TO OBTAIN AN CRITERION FOR ON-LINE IDENTIFICATION OF CHATTER

Actual cutting tests were conducted on a Gisholt turret lathe and a MIRL PT15 CNC lathe to obtain a reliable criterion as the index of stability of the machining process. The block diagram representation of the cutting tests and data acquisition setup is shown in Figure 1.

For the purpose of the present analysis, a set of data taken from the cutting process exhibited all characteristics of stable and unstable processes will be presented. These data will serve as an example in determining a criterion which can represent the response of the generation of chatter.

The signal when chatter occurred and was then suppressed by a reduction in speed is shown in Figure 2. The following initial conditions were used:

Workpiece	SAE 1045, 900 mm x 64 mm
Depth of cut	3 mm
Feed	0.26 mm/rev
Spindle speed	475 rpm
Coolant	none
Cutting tool	Holder - KTGPR-146C INSTP-43
	Insert - TPG433 KC810

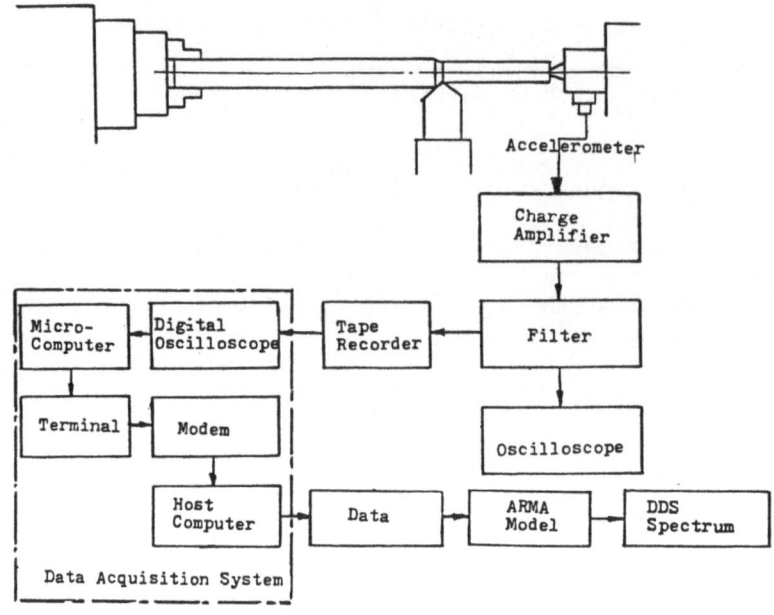

Figure 1. Experimental setup and data acquisition system.

This signal, which contains 4096 points, was passed through a 60~200 HZ band pass filter using a 1 ms sampling interval. The 4096 data points were separated into eight sets of data to fit the ARMA models. The dispersion analysis of the adequate ARMA models and the DDS spectra are shown in Table 1 and Figure 3, respectively.

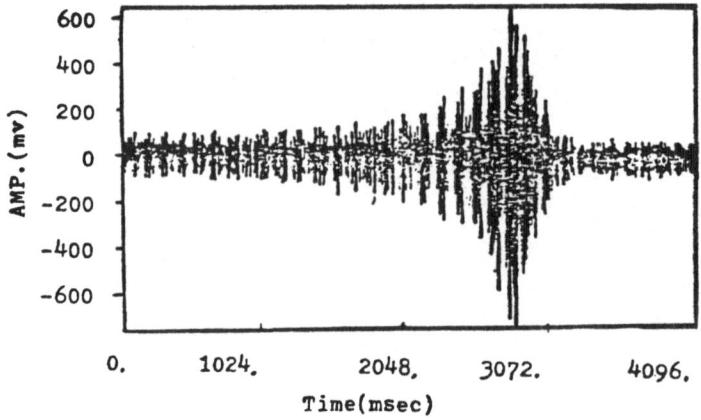

Figure 2. Vibration signal.

Table 1. Dispersion Analysis of the Adequate ARMA Model for 4096 Continuous Data Points Divided into 8 Segments for a Workpiece 900 mm x 64 mm

Segment	Var.	Mode 1	Mode 2	Segment	Var.	Mode 1	Mode 2
1	λ $\|\lambda\|$ f_n ζ %	0.575±j0.750 0.945 145 0.061 17%	0.425±j0.903 0.998 180 0.00146 83%	5	λ $\|\lambda\|$ f_n ζ %	0.628±j0.774 0.997 141 0.003 88%	0.422±j0.855 0.954 177 0.043 12%
2	λ $\|\lambda\|$ f_n ζ %	0.613±j0.752 0.746 141 0.034 23%	0.425±j0.902 0.997 180 0.0026 77%	6	λ $\|\lambda\|$ f_n ζ %	0.636±j0.760 0.991 139 0.01 97%	0.194±j0.949 0.967 282 0.018 2%
3	λ $\|\lambda\|$ f_n ζ %	0.622±j0.767 0.987 141 0.014 36%	0.425±j0.900 0.995 180 0.004 64%	7	λ $\|\lambda\|$ f_n ζ %	0.526±j0.708 0.88 148 0.134 100%	
4	λ $\|\lambda\|$ f_n ζ %	0.622±j0.775 0.993 142 0.007 56%	0.424±j0.902 0.996 180 0.003 44%	8	λ $\|\lambda\|$ f_n ζ %	0.586±j0.705 0.917 139 0.098 1%	0.426±j0.905 0.999 180 0.00015 99%

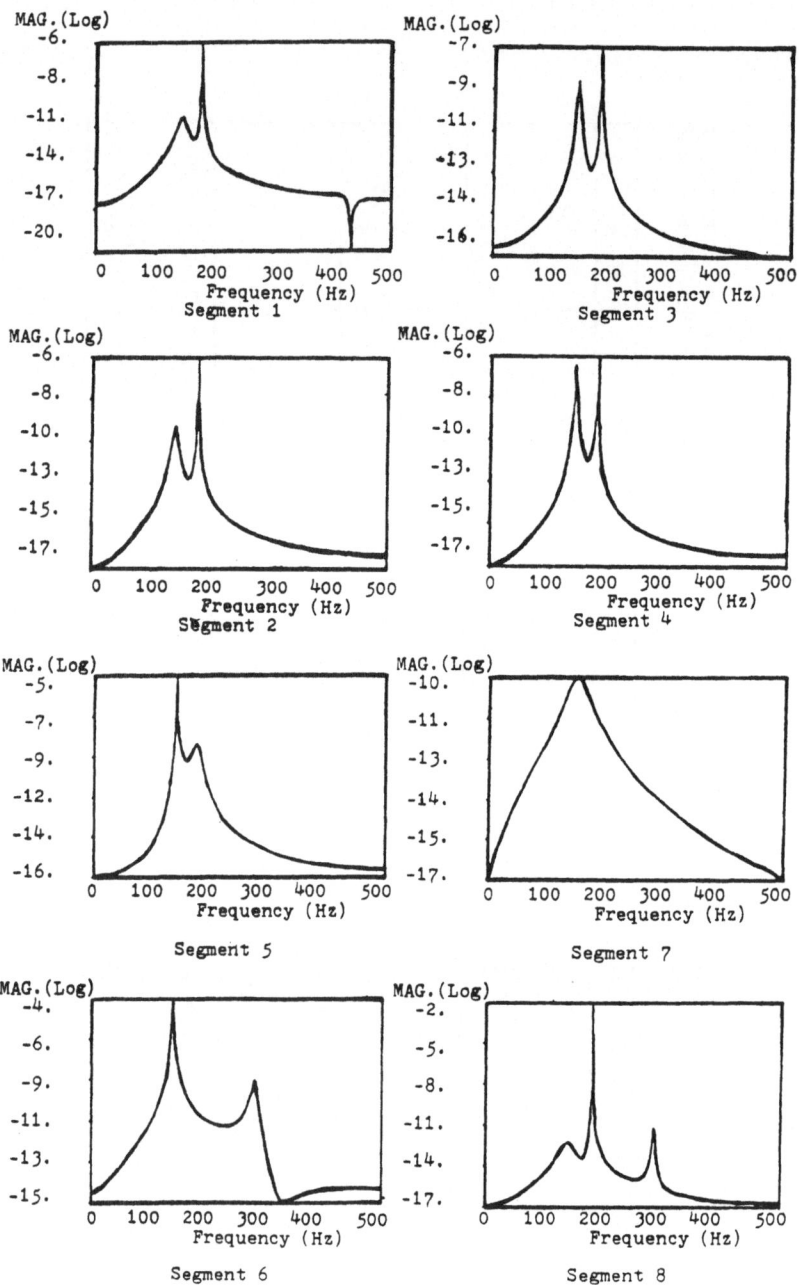

Figure 3. The DDS spectra for 4096 data points divided into eight segments.

The chatter starts to develop from the beginning to 3000 ms as shown in Figure 2, while the damping ratio corresponding to the dynamic mode of the workpiece fundamental natural frequency decreases gradually from 0.061 to 0.003 as shown in Table 1. At the same time, the peak of PSD corresponding to the dynamic mode of the workpiece fundamental natural frequency increased gradually, as shown in Figure 3. After chatter is suppressed, the damping ratio becomes 0.134 in segment 7 and 0.098 in segment 8 (Table 1). The corresponding peak of PSD decreases at the same time. Therefore, the damping ratio, or the peak of PSD corresponding to the dynamic mode of the workpiece fundamental natural frequency, can be used as an index of stability for the machining process. To simplify the calculation for on-line purpose [13], the peak of PSD have been implemented as the criterion.

DYNAMIC ANALYSIS OF THE CHUCK-WORKPIECE-TAILSTOCK SYSTEM

In the previous section, the approximate range of the fundamental mode workpiece natural frequency for given lathe has shown a very important factor for on-line identification of the machining process. In this section, the range of the workpiece fundamental natural frequency will be obtained by experimental method.

The workpiece, clamped in the chuck, was excited by an electromagnetic exciter. A block diagram of the experimental setup is shown in Figure 4. The random excitation force, generated by the noise generator, was applied on the workpiece. The vibration signal taken from the workpiece and tailstock were passed through charge amplifier and a 70~500 HZ band pass filter, then sampled at 1 ms intervals. Five hundred and twelve data points were used to fit an ARMA model. The results are shown in Table 2.

From the experimental results, a filter with a 70~170 HZ band pass could be taken to account for the range of the workpiece fundamental natural frequency and purify the analog vibration signal.

ON-LINE DETERMINATION OF THE DYNAMIC STABILITY LIMIT

A self-learning scheme is developed which can be implemented for the on-line determination of the threshold level of stability in each cutting process. Since this stability limit is not deterministic and will be automatically generated from the beginning of each cutting process, it is termed the "Dynamic Stability Limit" (DSL).

It can be seen that regardless of cutting conditions, workpiece, and type of chatter, once the machining process starts to shift from stable to unstable, the vibration signal will change from stationary to non-stationary. Thus, it is feasible to define the stability

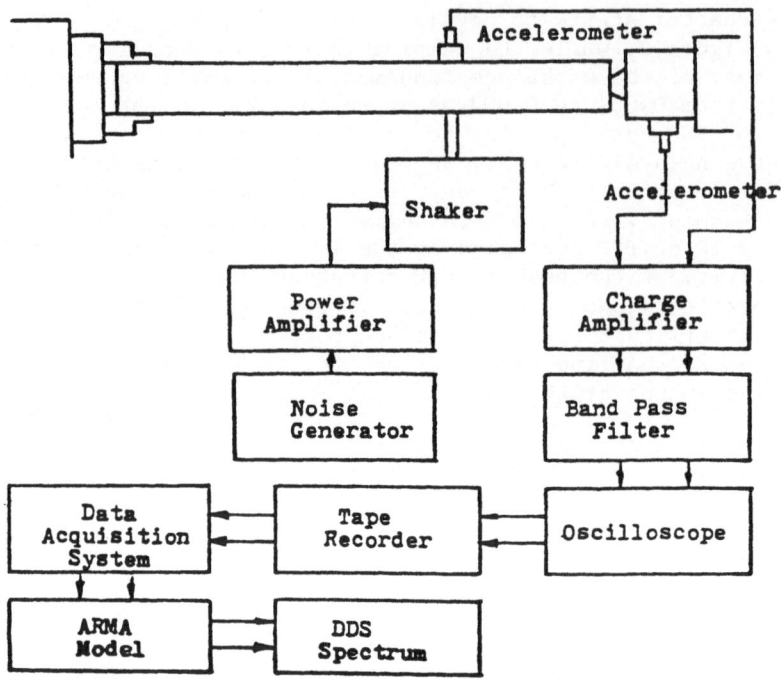

Figure 4. Experimental setup for random excitation.

limit from the relative variation of the signal at the transitional state.

From the work done in [13], the concept of a control chart used for statistical quality control was adopted to measure the value of the DSL. The confidence region of a control chart is constructed from the mean (M) of the control variable, plus and minus, a given number of standard deviations of the data -- say three -- covering up to 99.9% of the population. Based on the off-line analysis, the value of M+3σ is used as the dynamic stability limit.

ON-LINE IDENTIFICATION AND PREDICTION OF THE MACHINING PROCESS

It has been described before that the machining process can be represented in discrete time by an ARMA model. This model can also be approximated by a pure autoregressive model (AR) of the form

$$X_t - \sum_{i=1}^{n} \phi_i X_{t-1} = a_t \qquad (15)$$

Table 2. Workpiece Fundamental Natural Frequency by Random Excitation

Workpiece Length (mm)	Workpiece Diameter (mm)	Natural Frequency Hz(first mode)
900	67	140
900	63	140
900	55	140
900	44	139
900	38	136
900	31	133
1000	59	124
1000	52	121
1000	42	115
1070	71	115
1070	54	110

There are numerous methods for estimating the parameters in eq. (15). Taking into consideration the present requirement of continuously monitoring the changing dynamics of the machining process, an adaptive or recursive estimation procedure which updates its parameter estimates each time a new sample of the measured variable becomes available is more amenable. In light of this, the procedure of adaptive modeling for identification of chatter given in [16] is adopted.

Based on the simulation and real data testing results, an AR(6) model can be estimated as follows:

$$\phi_i(k) = \phi_i(k-1) + \mu X(k-i)e(k)$$

$$e(k) = X(k) - \hat{X}(k-1,1) \quad (16)$$

$$\hat{X}(k-1,1) = \phi_1(k-1)X(k-1) + \phi_2(k-1)X(k-2) + \phi_3(k-1)X(k-3)$$
$$\phi_4(k-1)X(k-4) + \phi_5(k-1)X(k-5) + \phi_6(k-1)X(k-6)$$

After obtaining ϕ_i, the modified PSD can be calculated by

$$S_m(f,k) = \frac{1}{\left|1 - \sum_{i=1}^{} \phi_i(k)B^i\right|^2} \qquad B = e^{-j2\pi f\Delta} \quad (17)$$

and the peak of PSD can be obtained using the golden section technique [17].

This new series of data for Sm, serving as a measure of stability of the machining process, can now be used to predict its future value if a suitable model can be found. Based on the preliminary off-line analysis [13], an AR(2) model was chosen for this purpose. Once again, the adaptive modeling technique was implemented to fit the new series of Sm; therefore,

$$S_{m,t} = \phi_1' S_{m,t-1} + \phi_2' S_{m,t-2} + a_t \tag{18}$$

and $\phi_j'(k') = \phi_j'(k'-1) + \mu' S_m(k'-j)\varepsilon'(k')$

where

$$\varepsilon'(k') = S_m(k') - \hat{S}_m(k'-1,1)$$

$$\hat{S}_m(k'-1,1) = \phi_1'(k'-1)S_m(k'-1) + \phi_2'(k'-2)S_m(k'-2)$$

The ℓ step ahead forecast of Sm is given by the general equations derived in eq. (12). For a second order model, the one and two-step ahead forecasts are derived separately by

$$\hat{S}_{m,t}(1) = \phi_1' S_{m,t} + \phi_2' S_{m,t-1} \tag{19}$$

$$\hat{S}_{m,t}(2) = \phi_1' \hat{S}_{m,t}(1) + \phi_2' S_{m,t} \tag{20}$$

To illustrate the above derivations, the vibration record measured at the tailstock while turning a slender SAE 1045 steel workpiece, 900 mm x 40 mm, with a depth of cut of 3 mm, feed 0.3 mm/rev, and spindle speed 680 rpm, as shown in Figure 5, is implemented.

Sampling the signal at 1 ms intervals and implementing the adaptive modeling strategy given by eq. (16), and the golden section technique, the change of these peak values is given in Figure 6. The dynamic stability limit value Smc which is calculated from the derivation in the previous section is also shown in the figure. It can be seen that peak of PSD clearly shows the development of chatter.

The predicted value of the PSD and the prediction error for $\ell=2$ are shown in Figure 6 and Figure 7, respectively. Although the prediction error increases as chatter charts to develop, it remains less than 10% of the PSD at the onset level of chatter.

DEVELOPMENT OF CHATTER SUPPRESSION STRATEGY

Predicting the generation of chatter has been discussed in previous sections, the other remaining important task is to study

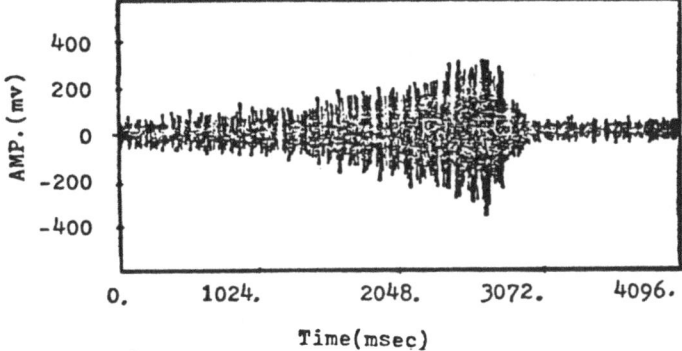

Figure 5. Vibration signal.

the cutting conditions which are free of chatter for a given machine tool.

The practical approach for identifying the machining process by relating the tailstock vibration signal to the dynamics of cutting process and machine tool structure has been developed in the previous section. Based on this signal, the peak of PSD corresponding to the dynamic mode of the workpiece fundamental natural frequency was used as the measure of the stability of the machining process. Experimental cutting tests can now be conducted to obtain the variation of this criterion with the change of speed and feed, and a stabilizing trend for the machining process can be determined.

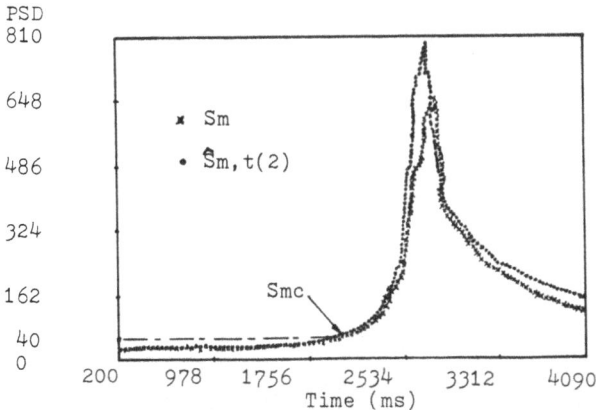

Figure 6. Peak of PSD, Smc, and $\hat{S}_{m,t}(2)$.

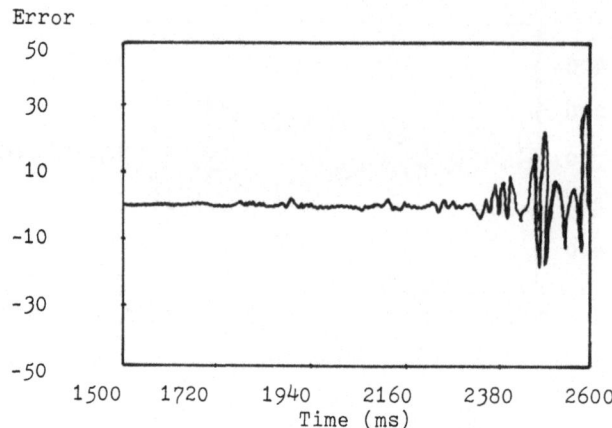

Figure 7. Two step ahead prediction erors in the stage between stable and unstable cutting.

Influence of Speed on Stability

Two workpieces of SAE 1045 steel with dimensions of 1000 mm x 65 mm and 900 mm x 65 mm were used to investigate the influence of speed on stability with the cutting conditions shown in Table 3. The vibration signal taken from the tailstock was passed through the 70~170 HZ band pass filter, sampled with a sampling interval of $\Delta=0.001$ seconds, and fitted to an ARMA (2,1) model. The variation of the damping ratio with a change of speed is plotted in Figure 8.

Table 3. Summary of the Cutting Conditions for Tests with Fixed Feed

Workpiece Dimensions	Case 1: 1000mm x 65mm	Case 2: 900mm x 65mm
Depth of cut (mm)	3	3
Feed (mm/r)	0.4	0.5
Cutting speed (M/min)	80,100,120,140	80,90,100,110,120,130
Coolant	none	none
Cutting tool	Holder-Kennamental KTGPR-164C INS TP-43 Insert-TPG433 KC810	

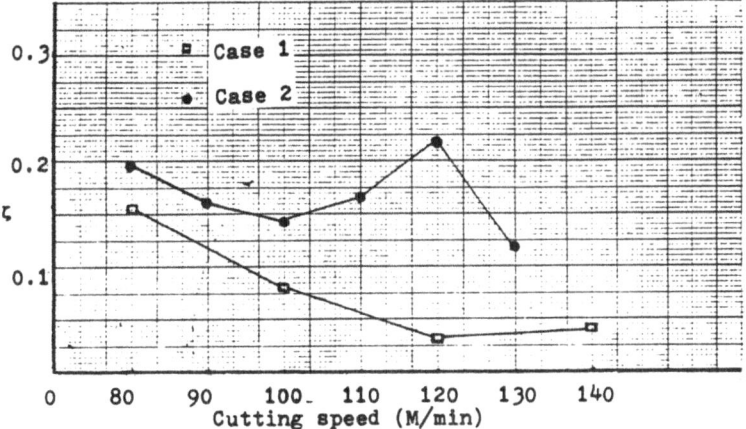

Figure 8. The variation of damping ratio with change in feed.

This experiment indicates a general tendency for increasing stability against chatter as cutting speed decreases and for cutting speeds lower than 120 M/min in case 1 and 100 M/min in case 2. This trend of changing in the damping ratio in low speed range is quite the same as shown in references [18,19].

Influence of Feed on Stability

Cutting conditions shown in Table 4 were used to investigate the influence of feed on stability.

Table 4. Summary of the Cutting Conditions for Tests with Fixed Cutting Speed

Workpiece Dimensions	Case 1: 1000mm x 65mm	Case 2: 900mm x 65mm
Depth of cut (mm)	3	3
Cutting speed (M/min)	100	100
Feed (mm/r)	0.2 0.4 0.6 0.8	0.1 0.3 0.5 0.7 0.9
Coolant	none	none
Cutting tool	Holder-Kennametal KTGPR-164C INS TP-43 Insert-TPG433 KC810	

The vibration signal taken from the tailstock was sampled and fitted to an ARMA (2,1) model. The variation of damping ratio with the change of feed is plotted in Figure 9. From this figure it can be seen that in the higher feed range, the damping ratio is comparatively larger than in the lower feed range.

Based on this study and the results as shown in references [3,19,20], it can be concluded that there is an increasing tendency towards stability against chatter as feed increases.

Chatter Suppression Strategy

Based on the results discussed in the above and the empirical experience obtained from experimental investigations, the strategy whose flowchart is shown in Figure 10 was proposed for the workpiece of SAE 1045 steel.

The first step to be taken in the strategy when chatter is imminent is to reduce the speed (IC1 on the flowchart). If this succeeds in preventing chatter, this reduced speed and the initial feed are maintained for a certain interval of time and taken as the first step of stable cutting conditions for this process (n=1 on the flowchart). After the certain interval of time, the cutting conditions will return to their initial values to satisfy the production criteria.

If the first step of reducing speed fails to suppress chatter, the second step, which consists of increasing the feedrate by 15% (IC2 on the flowchart), will be taken. If reducing speed and increasing the feedrate by 15% succeeds in suppressing chatter,

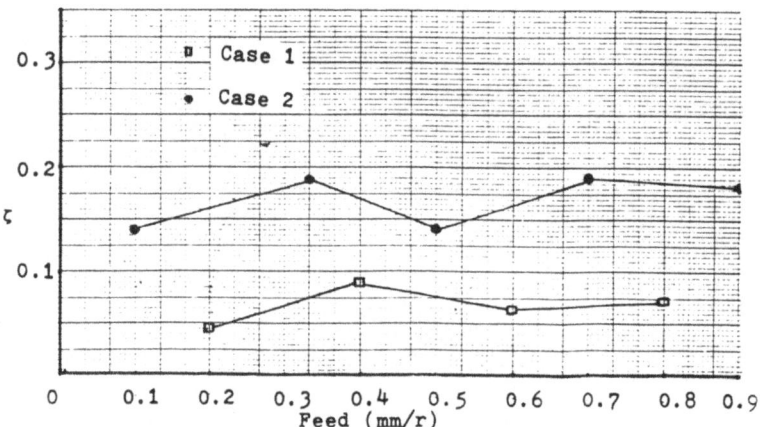

Figure 9. The variation of damping ratio with change in feed.

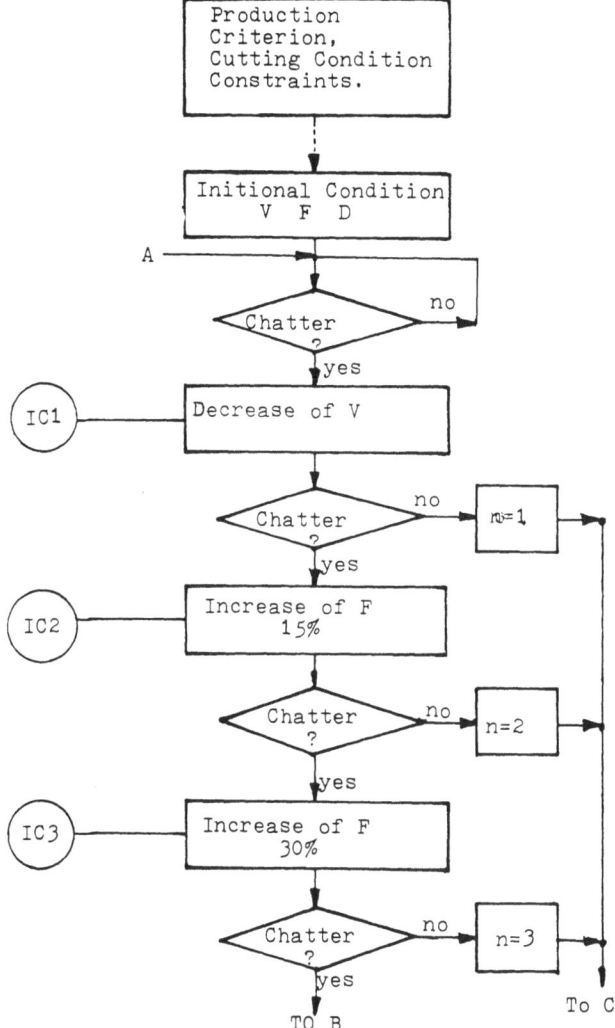

Figure 10. Flowchart of the chatter suppression strategy.

these values become the second step of stable cutting conditions (n=2 on the flowchart). Once again, after the certain interval of time, the cutting conditions will return to their initial values. If chatter occurs again, the cutting conditions will be changed to IC2 directly.

The procedure of changing speed and/or feedrate incrementally to find stable cutting conditions will continue until the last step, IC5, which contains the programmed increasing and decreasing of speed for a certain cycle.

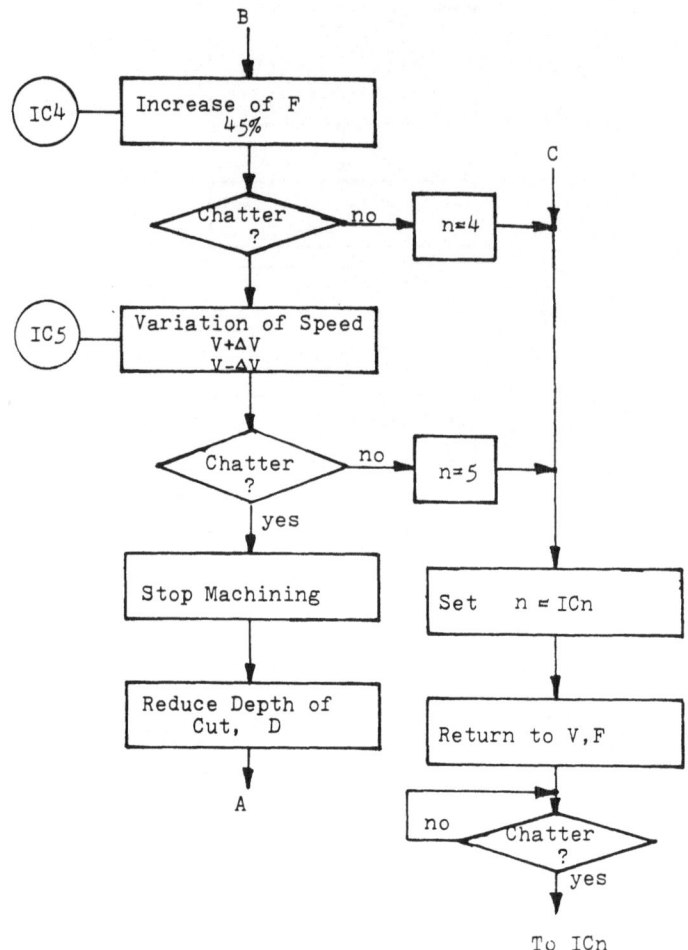

Figure 10. Continued.

If these proposed steps fail to suppress chatter, the machine will be stopped and the depth of cut will be reduced before cutting.

THE DESIGN OF A CHATTER SUPPRESSION CONTROLLER

A microcomputer based chatter suppression controller shown in Figure 11 is proposed. The vibration signal is passed through a charge amplifier and a 70~170 HZ band pass filter. The signal is then digitalized by an A/D converter. The digital data are implemented in the forecasting chatter system for on-line prediction and suppression of the generation of chatter. The monitoring system

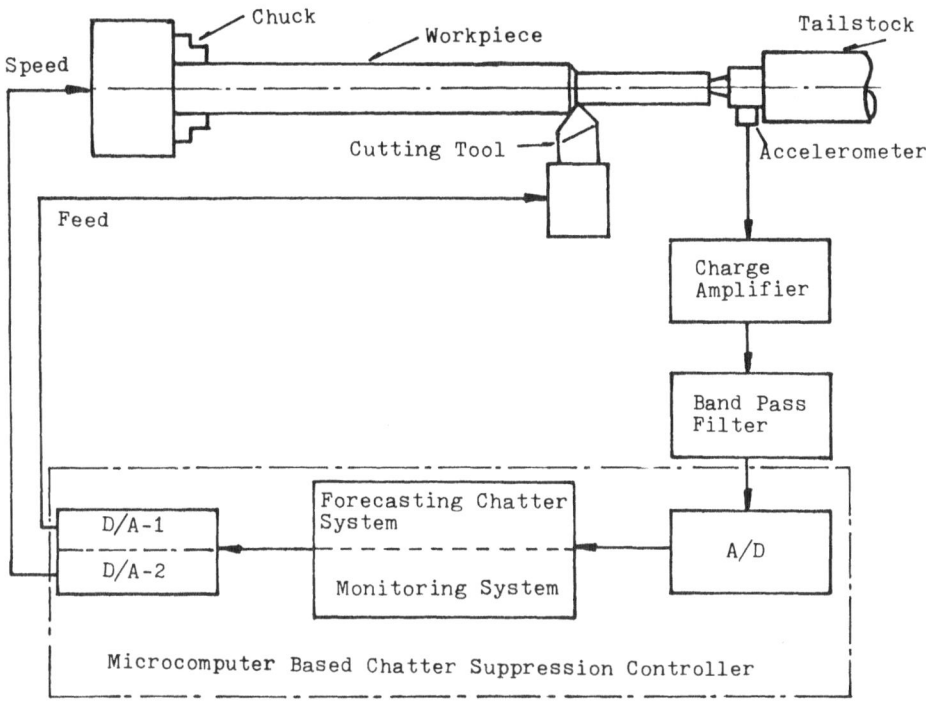

Figure 11. Schematic diagram of the chatter suppression controller.

uses the same digital data for detecting the abnormal cutting process, such as continuous chip entanglement and tool breakage.

Forecasting Chatter System

The detail of forecasting chatter system is shown in Figure 12.

At given time intervals, the microcomputer system samples the vibration signal representing the machining process. Based on these values and implementing adaptive modeling technique, the parameters of the AR(6) model are estimated. Once the AR(6) model parameters are determined, the peak of PSD, $S_m(f,k)$, is calculated by means of the golden section technique. The first set of data-say, 20-of $S_m(f,k)$ is then used to calculate the mean value, M, and the deviation, σ. The estimation value of the upper control limit, M+σ, is chosen as the dynamic stability limit, i.e., Smc, for each cutting process.

After the value of Smc is obtained, the continuous incoming Sm(f,k)'s are used to fit the forecasting model, AR(2). This

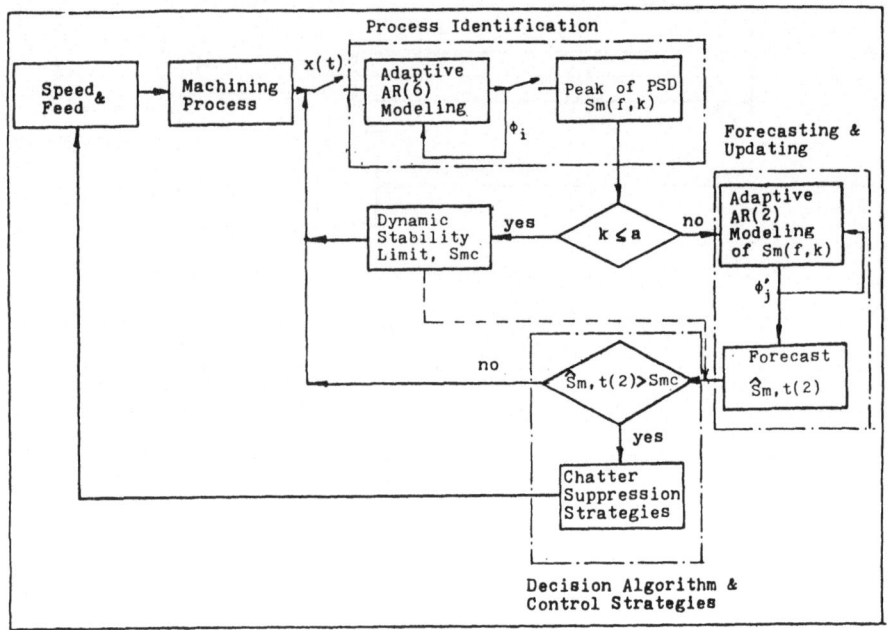

Figure 12. A detailed block diagram of the chatter forecasting scheme.

allows the possibility of predicting the future systems behavior and take the control actions accordingly.

Using the forecasted value $\hat{S}_{m,t}(2)$ and the dynamic stability limit, S_{mc}, which represents the stability limit of the current cutting process, the new speed and feed commands are determined and issued to the appropriate drives if corrective action is necessary, i.e., if the process is heading towards unstable conditions.

Monitoring System

The monitoring strategy through a microprocessor is given in Figure 13. The details of the procedures are as follows:

(1) Obtain the digitalized vibration data.

(2) In each cycle of the signal, the largest positive value of the amplitude is chosen as the amplitude of signal.

(3) Approximately the first 200 data points are used to estimate the mean value, M, of the vibration signal.

(4) The deviation, σ, is then calculated.

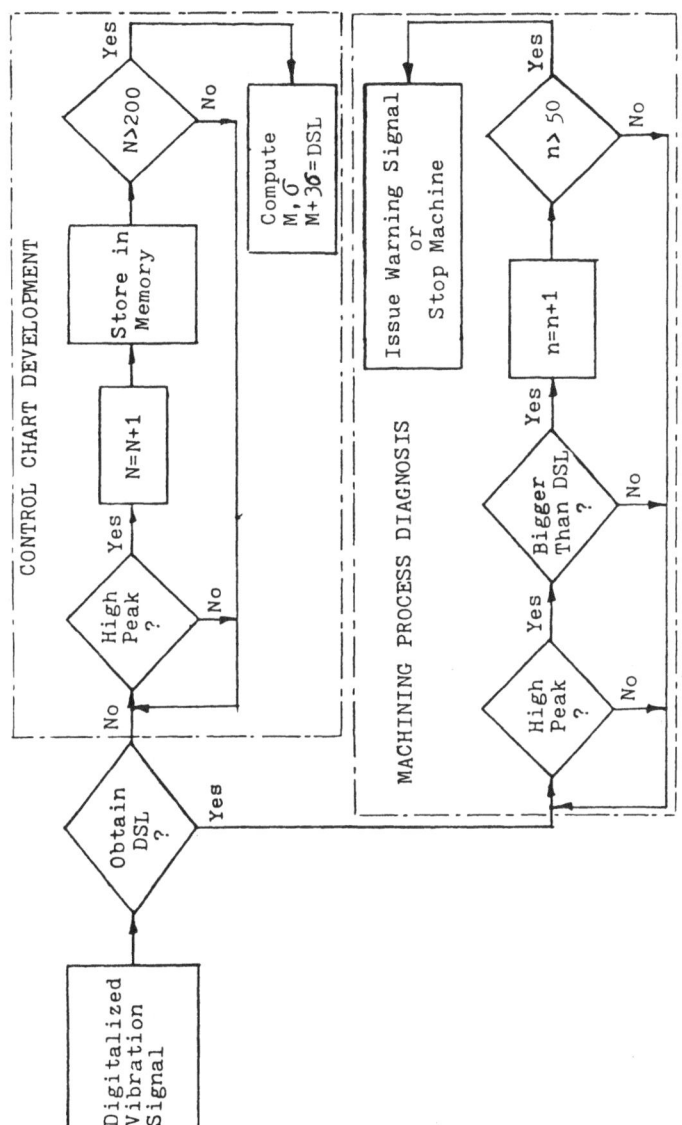

Figure 13. A scheme for monitoring the abnormal machining process.

(5) The DSL value for the current cutting process is defined as M+3σ.

(6) After obtaining the DSL, the largest incoming value of the data in each cycle is compared to the DSL.

(7) If successive data points, e.g., 50 points, exceed the DSL, an abnormal cutting state may exist.

(8) A warning signal is issued or the machine is stopped.

CUTTING TESTS AND RESULTS

Many experimental cutting tests were conducted to evaluate the entire control system. To demonstrate its effectiveness a series of cutting tests for two groups of the same length of workpiece are listed in Table 5. Out of eleven cases the suppression scheme was effective on ten occasions, except the very slender workpiece with 970 mm x 41 mm.

An example of the records which shows the actual response of speed, feedrate, and two step ahead forecasts of $S_{m,t}(2)$ is shown in Figure 14. It was taken from Experiment #8.

Table 5. Summary of Cutting Tests

No.	Workpiece L x D(mm)	Nominal Conditions			Chatter?	Controller Performance?	
		s(rpm)	f(mm/min)	d(mm)		Pred.	Suppr.
1	940 x 62	520	140	3	no	—	—
2	940 x 56	540	150	3	no	—	—
3	940 x 50	600	150	3	yes	yes	yes
4	940 x 44	620	150	2.5	yes	yes	yes
5	940 x 39	680	150	2.5	yes	yes	yes
6	970 x 71	460	100	3	yes	yes	yes
7	970 x 65	490	110	3	yes	yes	yes
8	970 x 59	530	120	3	yes	yes	yes
9	970 x 53	560	140	3	yes	yes	yes
10	970 x 47	620	150	3	yes	yes	yes
11	970 x 41	660	160	2	yes	yes	no

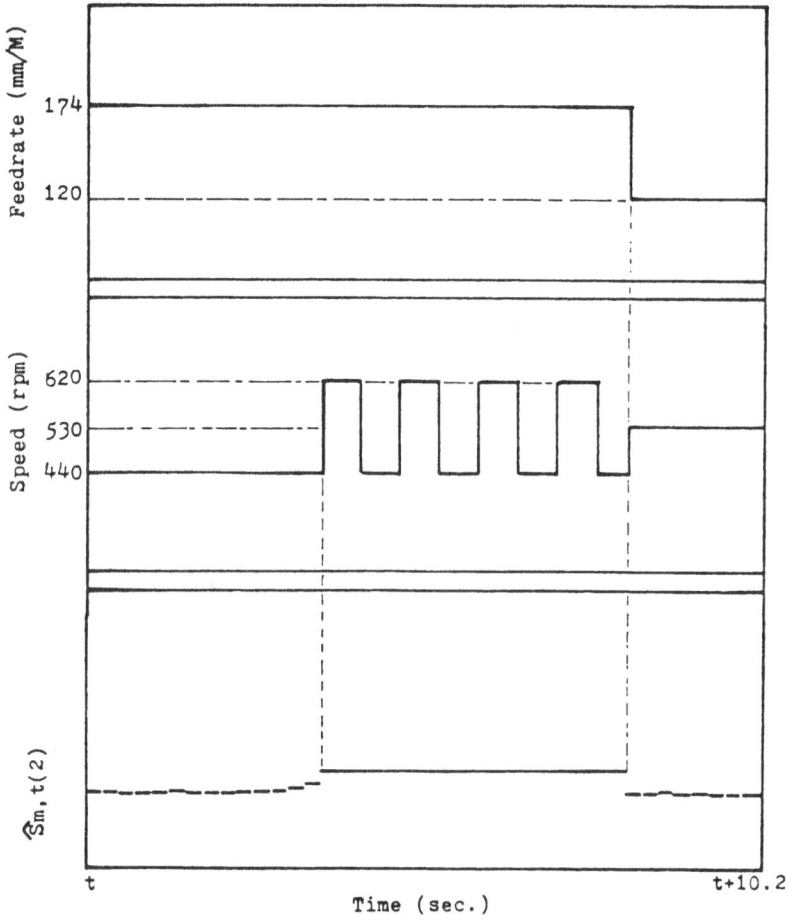

Figure 14. Records of the actual response of speed and feedrate - Experiment #8.

CONCLUSIONS

1. The peak of the power spectral density (PSD) function is a reliable indicator of the overall stability of the machining process.

2. The two step ahead prediction of the PSD yields a relatively small prediction error resulting in an accurate prediction of future trends.

3. The adaptive modeling technique is suitable for fast on-line implementation.

4. The cutting tests have proven the effectiveness of the proposed control scheme.

5. The self-learning scheme to determine the dynamic stability limit based on the quality control chart technique has been successfully implemented for on-line control of chatter.

6. The theoretical derivation and developed strategy presented in this study have provided a solid basis for developing a chatter free lathe, and it can also be applied in a special purposed machine to improve the safety and productivity of the machining process.

REFERENCES

[1] H. E. Merrit, "Theory of self-excited machine tool chatter," J. Eng. Industry, Trans. ASME, Vol. 87, No. 4, pp. 447-454, 1965.
[2] S. A. Tobias, Machine Tool Vibration, John Wiley, 1965.
[3] F. Koenigsberger and J. Tlusty, "Machine Tool Structures," Vol. 1, Pergamon Press, 1968.
[4] T. R. Comstock, F. S. Tse, and J. R. Lemon, "Application of controlled mechanical impedance for reducing machine tool vibration," J. Eng. Industry, Trans. ASME, Series B, Vol. 91, 1969.
[5] C. L. Nachitigal and N. H. Cook, "Active control of machine tool chatter," J. Basic Eng., Trans. ASME, Series D, Vol. 92, No. 2, pp. 238-244, June 1970.
[6] E. E. Metchell, "Design of a hardware observer for active machine tool control," J. Dynamic System, Measurement, and Control, Trans. ASME, Series G, Vol. 99, No. 4, 1977.
[7] F. A. Barney, S. M. Pandit, and S. M. Wu, "A new approach to the analysis of machine tool system stability under working conditions," J. Eng. Industry, Trans. ASME, Series B, Vol. 99, No. 3, pp. 585-590, August 1977.
[8] F. A. Barney, S. M. Pandit, and S. M. Wu, "A stochastic approach to characterization of machine tool system dynamics under actual working conditions," J. Eng. Industry, Trans. ASME, Series B, Vol. 98, No. 4, pp. 614-619, November 1976.
[9] T. L. Subramanian, M. F. DeVries, and S. M. Wu, "An investigation of computer control of machining chatter," J. Eng. Industry, Trans. ASME, Series B, Vol. 98, No. 4, pp. 1209-1214, 1976.
[10] K. F. Eman and S. M. Wu, "A feasibility study of on-line identification of chatter in turning operations," J. Eng. Industry, Trans. Vol. 102, 1980.

[11] K. F. Eman, "Machine tool system identification and forecasting control of chatter," Ph.D. Thesis, University of Wisconsin-Madison, 1979.
[12] S. Y. Tsai, K. F. Eman, and S. M. Wu, "Chatter suppression in turning," Eleventh NAMRC Proceedings, May 1983.
[13] S. Y. Tsai, "On-line identification and control of machining chatter in turning through dynamic data system methodology," Ph.D. Thesis, University of Wisconsin-Madison, August 1983.
[14] S. M. Wu, "Dynamic data system: A new modeling approach," J. Eng. Industry, Trans. ASME, Series B, Vol. 99, No. 3, pp. 708-714, August 1977.
[15] S. M. Pandit and S. M. Wu, "Time Series and System Analysis with Application," Wiley, 1983.
[16] W. Q. Yang, S. H. Hsieh, and S. M. Wu, "Adaptive modeling and characterization for control of chatter," 103rd Winter Annual Meeting, ASME, Nov. 14-19, 1982.
[17] A. A. Serig, "Optimum design of mechanical elements and system," Course Handouts, University of Wisconsin-Madison, 1984.
[18] J. Tlusty, "Measurement of the dynamic cutting force coefficient," NAMRC Conference, 1974.
[19] M. M. Nigm and M. M. Sadek, "Experimental investigation of the characteristic of dynamic cutting process," J. Eng. Industry, ASME, 1977.
[20] J. Peters and P. Vanherck, "Machine tool stability test and the incremental stiffness," Annals of CIRP, Vol. XVII, pp. 225-232, 1969.

MICROPROCESSOR AND PROGRAMMABLE

CONTROLLER BASED INDUSTRIAL AUTOMATION

M. F. Rahman

National University of Singapore
Singapore

ABSTRACT

The paper sets out to identify application areas in processing industries in which microprocessors and programmable controllers (PC) may be used selectively to automate the process lines. Comparative advantages of these two systems are discussed. In particular automation of a steel sheet classifier which uses two microprocessor based systems is described. Other examples of PC based automation are briefly described which bring out the salient points which are foremost in the system designer's mind when considering automation of any plant, particularly those requiring fast response from the controller.

INTRODUCTION

In many sections of industry, such as material tracking in metal processing industries, quality sorting in manufacturing, automatic testing, etc., microprocessors and more recently programmable controllers are increasingly being used to automate such processes. In the past microprocessor based systems dominated those areas where substantial data manipulation and speed of response were required. Microprocessor based systems, commonly termed as microcomputers (MCs), have the advantage of full computing power which may be utilized by using one of the high level languages and fast speed of response when interrupt inputs are used or when base level programs are kept short. A fast speed of response is often required when material at high speed has to be tracked accurately. Programmable controllers have a limited repertoire of instruction sets, and its scanning speed is not much faster than about 5 msec/scan. However, PCs have grown tremendously in recent years [2] to

the extent that these are also being introduced in automating many
processing lines. This paper describes a few application examples
where MCs and PCs have been used. Fuller description is given of a
sheet cut-up (shear) and classifier automation using two
microprocessor based systems.

MICROCOMPUTER VS PROGRAMMABLE CONTROLLERS - APPLICATION
CONSIDERATIONS

In the early seventies, PCs were an array of logic modules
which could be programmed in some sequence. Since then PCs have
grown tremendously due to incorporation of LSI chips, notably
microprocessors, to perform such tasks as input/output management,
communication links, programming aids and so on. From the similarity of architecture of PCs and MCs (see Figure 1) it could be argued
that PCs are actually special purpose computers such as process
control computers [1]. The fundamental differences are the degree
of prepackaging and operator interface - the main component of
which is programming and reprogramming ability on site without
extra hardware and skilled manpower.

Prepackaging of today's MCs is done to an extent which allows
the systems designer to use the full power of microprocessor hardware and software. This means that system designers have to have
specialized knowledge of microprocessor systems hardware and

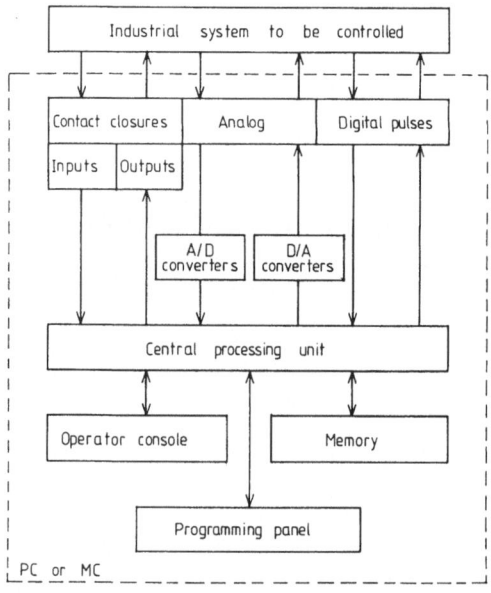

Figure 1. Architecture of an MC or PC.

software and systems development aids. Consequently, control and automation system design for a plant employing MCs usually takes place at the system designer's level.

Further prepackaging of MCs leads them to become special purpose computers and the price paid for this is in processing speed and power. PCs are MCs prepackaged to such an extent that virtually no knowledge of computers is required to develop or alter an application program. Application program is written usually in the relay ladder language. Operator interface with the plant, such as monitoring analogue and digital input-outputs of the plant, settings and current values of timers, counters, sequencers and other controlled variables, can be established through the ladder diagram. Figure 2 shows a flow diagram of the application systems design process for MC and PC. The fact that PCs require little systems development aids and specialized manpower has led to their phenomenal growth in the last few years.

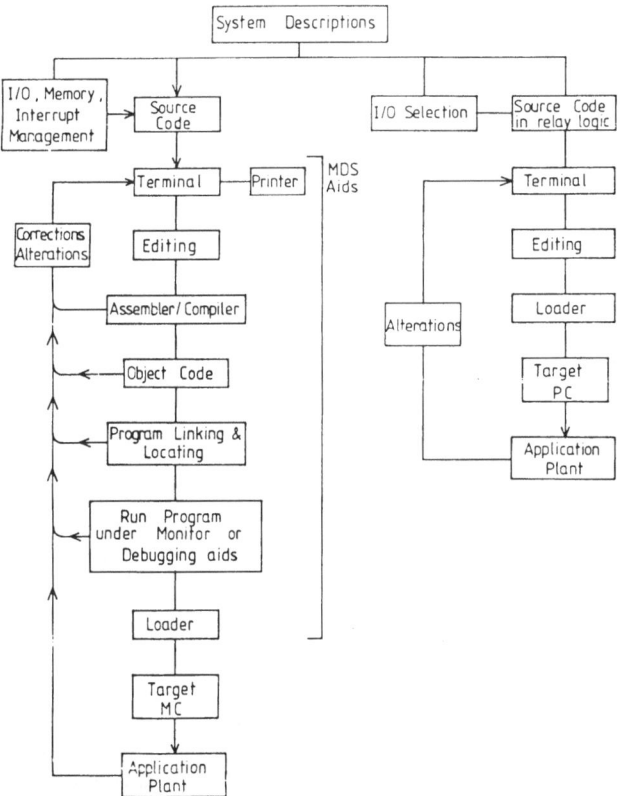

Figure 2. MC and PC system development.

Many applications are not critically dependent on extensive data manipulation and controller speed of response. Adaptability to changing conditions in the plant with least system development aids (manpower + hardware and software) is what is required in such cases. It is in such cases that PCs will become popular as controllers very next to the automated process line. In the next few sections a high speed sheet tracking application and other applications are discussed that highlight some of these aspects.

AUTOMATION OF A SIMULATED SHEET STEEL CLASSIFIER

A shear and classifier line was simulated in the laboratory (see Figure 3) and it was automated by two MCs (Intel 8080A band) and a PC (Modicon 484). The simulated classifier consists of a clock to represent advance of material through the line and a set of switches to introduce defects into the travelling strip of steel. Another set of switches are used to set up conditions on the sheet deflectors which are magnetic rolls (MR). A third set of switches are used to indicate counter reset and conveyor lock-out conditions to the controller. Such a setup effectively simulates actual classifiers found in steel mills and can be considered as a suitable application for comparing characteristics of MCs and PCs for industrial automation.

MC Based Automation of the Classifier

The classifier consists of two sections each controlled by its own MC (refer to Fig. 3). The defect tracking and cut-up section has an uncoiler and levellers which lead continuous strips of steel into the flying shear. The strip passes through the pinhole detectors which employ photo-arrays to detect pinhole defects in the strip. Only one detector is used at one time, the other being kept for standby or for checking purposes. An X-ray gauge detects any thickness defect and also a viewing window through which surface quality is monitored. A pinhole marker puts a mark as close as possible to the detected pinhole. Sometimes these are not visible easily to the naked eye. The shear cuts the strip into sheets of preset sizes. Control of the shear is not part of the classifier, but the instant of cut is detected by a proximity switch and indicated to the strip microprocessor. A new coil is also indicated to this microprocessor when new material enters the line. For the simulated line all these signals are either generated by software or manually.

The classifier section has four piles into which defect and prime sheets are channelled according to their quality. Deflection of these sheets is arranged by two magnetic rolls for each pile. Detection of arrival of sheet at each mag roll pair is in practice detected by photoelectric switches. In the simulator it was

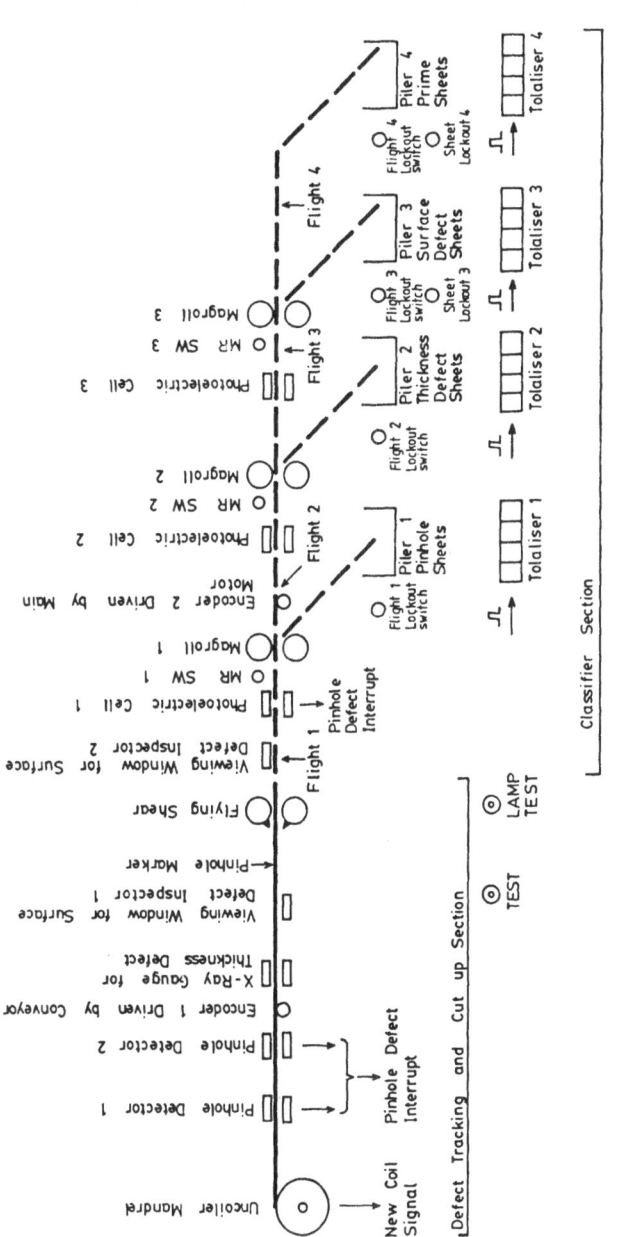

Figure 3. Steel sheet classifier system.

obtained from software by checking on distances of these switches from the shear. According to a defect type the top or bottom magnetic roll is energized. Sheets without any defect, which are called prime sheets, proceed to the prime pile. A viewing window is also available for a second check for surface quality by a second inspector.

At each pile flight (conveyor) and sheet lock-out switches are used as shown in Figure 3 to inhibit flight conveyors or the controller from sending sheets to the relevant pile. The sheet microprocessor sends pulses to the counters for each pile as sheets are directed to the piles. Operator interface with the line is through the RUN switch, magnetic roll (MR) drive switches which can be used to force sheets to certain piles, counter reset swiches and displays, sheet and flight lock-out switches, pinhole detector selector switches and surface defect inspector switches.

Requirements of Defect Location and Classification

Typically, the requirements that would necessitate a microprocessor based system are the following:

A. A high line speed - top speeds of such lines are often as high as 40 Km per hour.

B. Pinhole defect location with 2 mm of defect may be required. This requires a response time less than 200 µsec.

C. Thickness and surface gauge defect location within 20 mm of defect. This requires a response time of 2 milliseconds.

In the past automation of such classifiers has been carried out by hardwired controllers which basically consisted of shift registers which were clocked by the line encoder. These are inflexible and lacked the sophistication and facilities required of today's modern classifiers. Considering B above it was felt that only a controller with interrupt processing capability can be used. Even requirement C cannot be met by today's PCs as minimum scan times of less than 1 msec do not seem to be available.

Classifier Microprocessor Hardware Organization

It is obvious that in order to meet both requirements B and C microprocessors have to be used. An implementation uses two MCs for the purpose. Hardware configuration of each of these Mcs are shown in Figure 4. The strip MC tracks the strip for any defect up to the shear and indicates to the sheet MC the quality of the sheet just being led into the classifier section. The strip MC is interrupted by the pinhole detector and the shear. The shear signal has the higher priority. Other inputs to this MC are checked once

CONTROLLER BASED INDUSTRIAL AUTOMATION

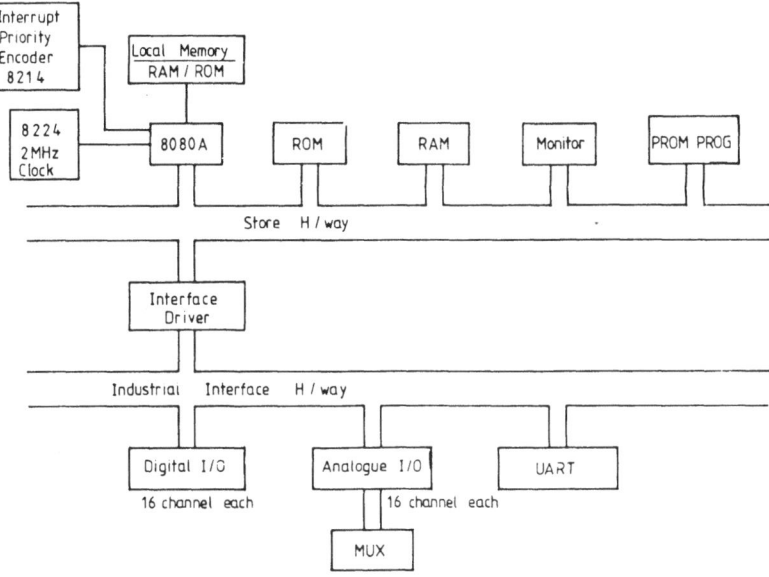

Figure 4. System hardware schematic.

every scan. The strip MC is also interrupted by the sheet MC when data about sheet quality is passed to this MC.

The underlying principle behind defect tracking depends on the knowledge of correct distances of the various defect detection switches from the shear. These are made available to the MC. The MC reads the encoder in every scan to register the progress of defects through the line. It also reads the encoder once every 5 msec to determine line speed. This is required in accounting for the energization and denergization times of the pinhole marker so that pinholes can be correctly marked.

The MC based automation discussed below met all requirements A, B and C discussed earlier adequately for simulated line speeds of 40 Km/hr and for each count representing 2 mm of movement. With the present generation of faster microcontrollers, such as the Intel 8051, it should be possible to meet these requirements by one MC and still have time available for such functions as fault diagnostics, line testing, data logging and so on.

<u>The Strip MC Software Organization</u>. This MC generates a table of defects as these are detected. Each entry for a defect consists of 4 bytes of which the first byte indicates whether the defect is valid or invalid and defect type. Second and third bytes hold the 16 bit data about position of the defect. On every shear interrupt the encoder pulse-counts are substracted from this data plus the

detector distance to shear. A negative result would indicate that the defect has passed the shear and the quality of the sheet is indicated to the sheet MC. Figure 5 shows the basic operation of this MC. Defects apart by not more than 500 mm are considered continuous. Appropriate error bands are allowed for each of these calculations so as to give allowances for such factors as slippage and displacement of material due to the cutting action of the shear. Arrival of pinhole defects to the pinhole marker is detected in a similar manner, but energization of the marker is dependent on line speed and marker response time.

Sheet MC Software Organization. As for the sheet classifier MC, the distances of the mag rolls and their associated photoelectric switches that detect arrival of sheets are made available to the MC. When the line is started or restarted all the mag rolls are kept in 'DOWN' position so that the first few sheets are rejected as pinhole. This also clears the classifier section of any sheets remaining from previous operation.

On DATA AVAILABLE interrupt from the strip MC a defect table is generated as in the case of the strip MC. Arrival of the sheets at a mag roll is also determined as before but now mag roll energization is dependent on line speed and mag roll energization time. However, now three different pointers are used for the three

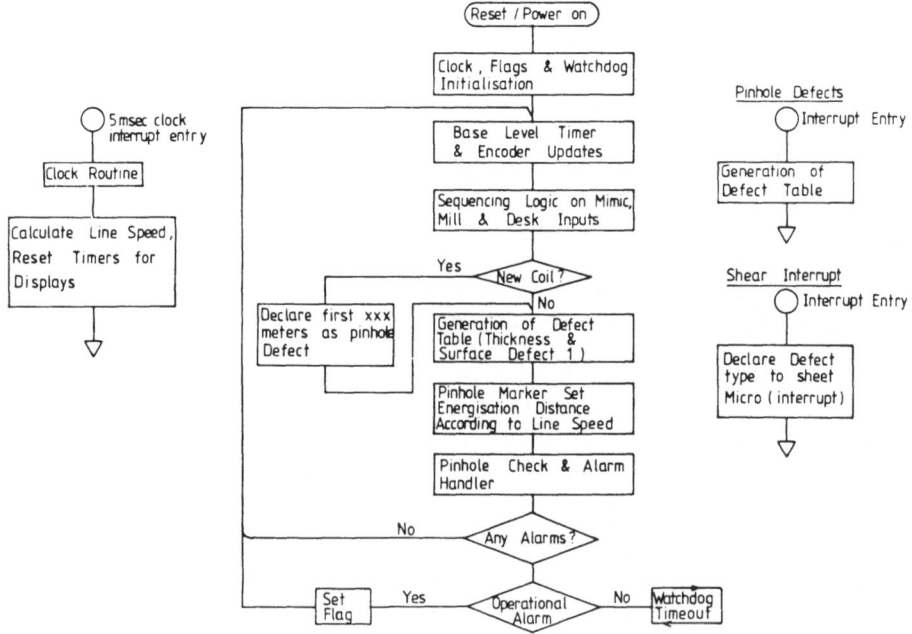

Figure 5. Strip micro flow chart.

simulated photoelectric switch interrupt routines as interrupts from the three mag roll deflectors are caused by different sheets. When a sheet is deflected by a mag roll the defect entry is invalidated as no further interrupt will take place due to these sheets. Figure 6 shows the basic operation of the sheet MC software.

PC BASED AUTOMATION OF THE CLASSIFIER

The above classifier simulator was automated by a Modicon 848 with arithmetic capability. After structuring the application program into many subroutines and avoiding unnecessary scanning of unused inputs and outputs, the fastest scan time was found to be about 19 msec. Simple arithmetic will show that if deflect location within 2 mm (typical) is desired, line speed would have to be no more than about 380 m/hr - which is too low for a practical classifier. However, there are other applications where a PC would be more suited. A few examples are discussed below.

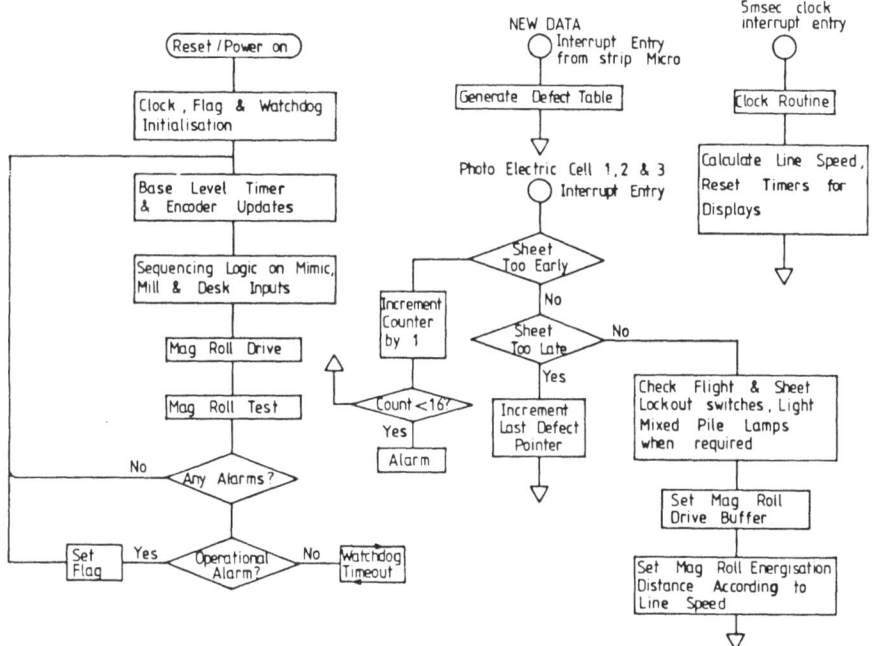

Figure 6. Sheet micro flow chart.

OTHER PROCESSES

One notable example of a PC based automation is for the rolling mill of Figure 7. The gauge and entry speed of metal coming into the mill, temperature of the hot metal, its yield stress, absorption coefficient, width and density are made known to the PC. From these the PC works out a rolling schedule and calculates the speed and tension references for subsequent rolling sections and also for the looper angle motor which also helps to create the required tension in the metal. In addition to these the PC does all the sequencing of the plant, prints out prodution log and checks data for alarm conditions. In this case an MC is used to provide a CRT based operator interface through which the operator can monitor mill conditions and set up data for metal from the following roll.

PCs are also finding applications in furnace charging where it determines the right mix of ingredients from a set of curves in real time. The PC then sends references to the valves which control the flow of these ingredients in their local closed loops. With present generation PCs it is also possible to implement such control algorithms as PID directly, thus dispensing with analogue PID controllers. The references to such PCs may be given by another PC or a more sophisticated control computer through communication links currently being available - thus forming a network.

Another application of a PC is in automating and monitoring a ladle car drive that transports molten metal between two stations. Here a PC sequences the drive and its thyristor converter. Required movement of the car is transmitted to the PC from another computer and the PC applies these to the drive and checks for operation within defined limits. In addition the PC checks out the thyristor converter and its control loops for proper functioning each time the drive contactor is closed. Furthermore, the PC monitors heating of the motor by calculating the motor temperature from current feedback and using a motor thermal model. An automated electroplating line

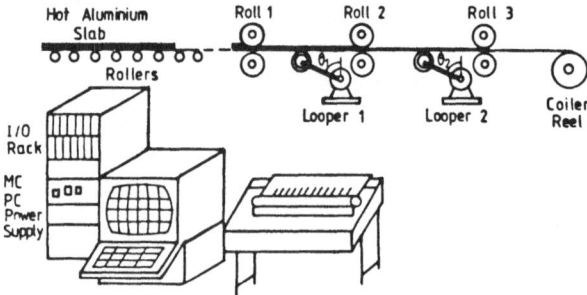

Figure 7. Using a PC for data setup and automation of a rolling mill.

shown in the schematic of Figure 8 is another application where workpieces are dipped into plating baths containing chemicals for predetermined amounts of time under control of a PC. The PC keeps track of workpieces as they progress through the line.

In each of these cases PCs were chosen for the following reasons:

1. The PC had ample time to respond to the process signals.

2. Application program generation and system integration efforts were minimal.

3. Application system monitoring by the operator is direct through the ladder diagram.

4. The ladder diagram programming is flexible without much hardware overhead requirement should modifications be required.

5. The PC is suitable for the harsh conditions prevalent in such industries.

CONCLUSION

Some of the application areas which can be automated by using a programmable controller or microcomputer or both have been discussed. Comparative advantage of a PC and an MC have been highlighted by some application examples. These controllers can also be considered to complement each other in that such functions as operator CRT interface and high speed calculations in real time can be left to the MC whereas sequential operation of the process, low speed calculations for setting up of data and limited closed loop control for the process can be left to the PC.

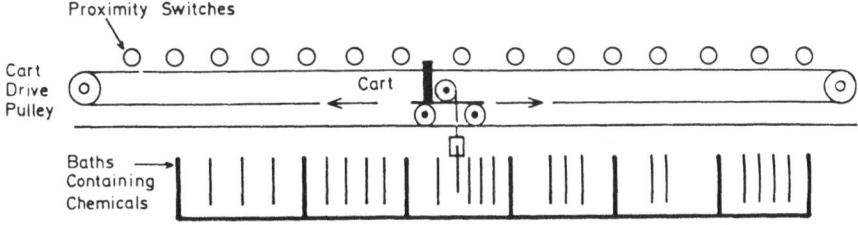

Figure 8. PC based automation of an electroplating line.

REFERENCES

[1] V. J. Maggioli, "The microcomputer/programmable controller ayndrome," IEEE Transactions on Industry Applications, Vol. IA-18, No. 3, May/June 1982.
[2] M. F. Rahman and C. C. Hang, "A review of programmable controllers in industrial automation and control," Instrument Asia 82 Conference, Instrument Society of America, Singapore Section, November 1982.

PRODUCTION SCHEDULING AUTOMATION IN AN ALUMINUM METAL PRODUCTS PLANT

Keh-Lon Thomas Lee and Arvind Jain

Kaiser Aluminum and Chemical Corporation
Oakland, California   U.S.A.

ABSTRACT

The authors developed a computerized capacity planning model for an aluminum metal products plant. This model is meant to automate the major segment of the production scheduling process and thus aid the chief planner in determining the feasibility of a proposed production plan, to identify purity bottlenecks and develop plant strategies. Given appropriate input, the model determines an optimum assignment of metal available from all sources to each finished product.

A linear programming formulation has been used to form the nucleus of the model. An interactive module is incorporated into the model to allow the user to enter the input data easily and to enhance his communication with the model.

This computerized model is an effective automated tool in not only devising alternative production plans, but also in making subsequent modifications.

DESCRIPTION OF THE PROBLEM

In an aluminum smelter, molten aluminum is produced through the electrolytic reduction of alumina. In an aluminum metal products plant, molten aluminum is alloyed with small amounts of other metals and cast into various shapes of finished metal products. Among these products, rolling ingots are shipped to fabricating plants to be rolled into sheet, plate, or aluminum foil. Extrusion billets and redraw rods are shipped to other plants to be processed into mechanical, structural, or electrical products. Other forms of products are remelt ingot and sow. Their primary use is as remelted material for casting. Remelt ingots weigh about 35 pounds each while sows are 1,200-pound blocks of aluminum.

Kaiser Aluminum's Chalmette, Louisiana, metal products plant is a casting facility located next to a Kaiser Aluminum smelter which supplies the metal products plant's hot metal, i.e., the molten aluminum fresh out of the reduction process. In addition, the plant can get cold metal from on-site inventory which is in the form of remelt ingot or sow. It can also request external cold metal in the form of ingots from certain farther away Kaiser facilities. The above mentioned hot metal, on-site inventory and external metal constitute the total raw metal supply of the plant.

The planning horizon of the metal products plant's metal planner is one month. In the middle part of every month, he receives instructions from the corporate metal planning department regarding the finished product requirements in terms of millions of pounds of the various final products. Usually about twenty different products are required. The planner's objective is to obtain a feasible production plan. He starts by listing the standard product specifications. Next he collects the forecast of hot metal supply, i.e., the reduction potline data. He checks the on-site inventory for scrap and remelt sow data and determines what external metal supplies are expected to be available. He also considers the scheduled down time for plant maintenance as well as other plant operating conditions. With all the above information at hand, the planner draws upon his experience to derive a feasible plan through hand calculation and trial and error. The manual planning process is depicted in Figure 1. If no feasible plan is obtained after several tries, the planner informs the corporate metal planning department of the situation and jointly with them arrives at a revised schedule of required finished products requirements.

FORMULATION OF THE MODEL

The authors' assignment was to automate the production scheduling process. After several discussions with the planner, we concluded that a linear programming formulation seemed appropriate and went ahead to formulate the problem in an L.P. framework. The various pieces of data are illustrated below in the form of input tables which are used to construct the coefficients of the linear program. A solution of an example problem described by these tables will be presented later in this paper.

The metal purity level is determined by its silicon and iron content. For example, a 0405 aluminum alloy is aluminum containing 0.04% silicon and 0.05% iron. The composition of the hot metal

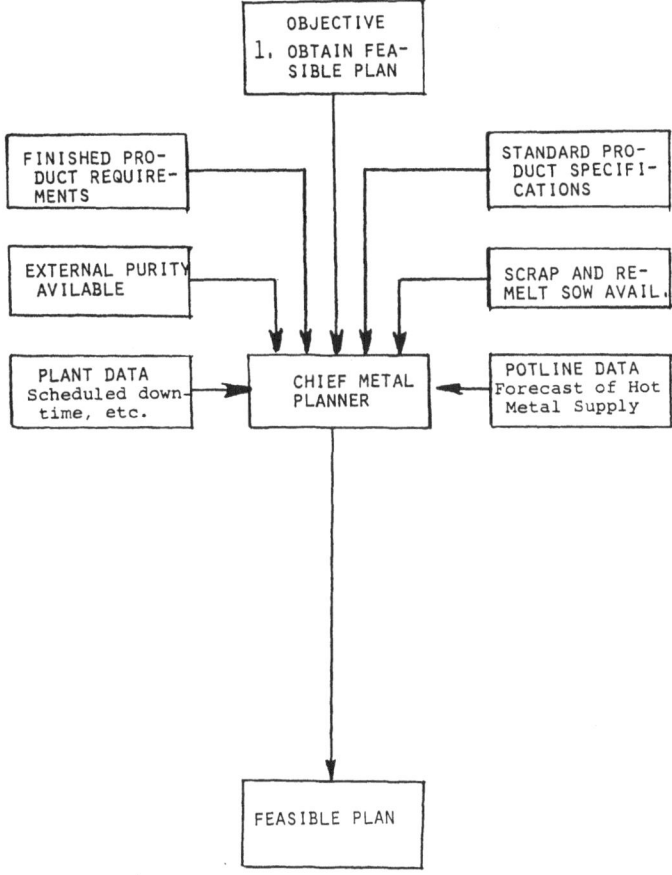

Figure 1. The manual planning process.

supply in terms of purity levels varies from month to month. The hot metal supply can be described by a 5-column table. Such a table in our example problem is shown in Table 1. The first four columns indicate the metal grade, quantity (in millions of pounds), silicon and iron levels (%). The last column indicates the desirability of a particular grade of metal if there is leftover. An entry of 1 indicates that it is a desired, or prime grade; a blank otherwise.

The remelt supply from on-site inventory and the external sources can be specified in two similar tables as illustrated in Table 2.

The fourth input table of the model is used to specify the quantity (in millions of pounds) of each finished product required during the planned month as well as five other product parameters:

Table 1. Input Data – Hot Metal Supply

TABLE	SUPPLY MMLBS	SI	FE	PM
0404	1.267	.040	.040	
0406	6.603	.040	.060	
0408	4.331	.040	.080	
0410	1.108	.040	.100	
0506	0.135	.050	.060	
0510	.596	.050	.100	
0615	1.190	.060	.150	
0620	.360	.060	.200	
0635	.160	.060	.350	
0699	.140	.060	.990	
06199	.115	.060	1.990	
06200+	.047	.060	2.000	
1010	.122	.100	.100	
1020	6.067	.100	.200	1
1520	2.852	.150	.200	
1535	3.882	.150	.350	
2035	.398	.200	.350	
3099	.854	.300	.990	
30199	.193	.300	1.990	
30200+	.150	.300	2.000	

- the maximum and minimum levels (%) of iron and silicon allowed, and

- a "recovery rate" (usually between 89% and 100%) which is the ratio of finished product weight to the weight of input metal.

Such a table in our example problem is displayed in Table 3.

The fifth input table indicates the composition of the iron and silicon alloys to be charged into the remelt furnace when required. The example shown in Table 4 shows that the iron alloy has 75% iron and 25% aluminum while the available silicon has 0.3% iron and 99% silicon.

The sixth and final input table specifies certain operating parameters. An example is shown in Table 5. The PURITY entry indicates the maximum purity assignment. That is the maximum percentage of the purer metal (with iron content less than a value specified by the user) can be assigned to high grade products (with iron content less than a particular value specified by the user). This is a very significant idea in making the model realistic, since in the real world there is a timing difficulty in having the purer hot metal out of the smelter continuously for the production of purer products when the quantity required is too large. The

Table 2. Input Data – Remelt and External Metal Availability

TABLE	REMELT MMLBS	SI	FE
220	.200	.100	.180
S-DRAINS	.300	.250	.300
OTHERS	.300	.440	.220
TABLE	EXTPUR MMLBS	SI	FE
S0506	0.000	.050	.050
S0510	.000	.07	.08

Table 3. Input Data - Finished Product Requirements and Specifications

TABLE	PRODUCTS MMLBS	LOSI	HISI	LOFE	HIFE	REC
6061	6.000	.600	.670	.130	.220	.898
6063	4.500	.380	.430	.160	.230	.898
H16M57	4.600	.030	.050	.100	.140	.974
H14M7	3.000	.030	.060	.140	.180	.974
I1020	.700	.050	.080	.110	.180	.990
11535	.600	.070	.120	.150	.320	.990
ALMG35	.600	.030	.060	.030	.060	.990
71A	.500	.030	.060	.030	.060	.990
356	.300	6.500	7.150	.120	.170	.990
220	.700	.050	.150	.100	.200	.990
A356	.400	6.500	7.150	.030	.080	.990
OTHERS	.300	.090	.150	.200	.430	.990

.900 in the SCRAP and SOW columns indicate that, again due to timing constraints, no more than 90% of the new scrap is available for remelt and no more than 90% of the sow generated internally during the planned month is available for remelt. These three parameters can be adjusted as practical operating conditions are tightened or loosened. The MELT entry indicates the efficiency of the remelt process. The CEIL entry is the remelt ceiling parameter (in millions of pounds), i.e., the limit on the total amount of cold metal that can be utilized.

The essential linear constraints are:

- The amount requirement of all the products are exactly met

- The amount of each grade of hot metal required is within the supply limit

- The amount of each grade of on-site inventory required is within the supply limit

- The amount of each grade of external metal required is within the supply limit

- The silicon level of each product is within the specified range

- The iron level of each product is within the specified range

- The total amount of remelt required is within the remelt capacity

Table 4. Input Data - Alloying Materials

TABLE	ADD FE	SI	AL
AFE	75.000	0.000	25.000
ASI	.300	99.000	0.000

Table 5. Input Data - Operating Conditions

TABLE UTIL	OPS PURITY	SCRAP	SOW	MELT	CEIL	DELTA
	.800	.900	.900	.990	9.500	-.00001

- No more than 80% of the purer hot metal is assigned to purer products

- No more than 90% of the new scrap generated is assigned for remelt

- No more than 90% of the (internal) sow generated is assigned for remelt

We also found out from the planner that implied in his manual trial and error process are certain additional considerations. A feasible production plan usually results in some metal being left over in the form of sow. Sow of high-iron content is of low value. On the other hand, residual metal of low-iron content is usually marketable. Indeed, on occasion, certain grades of low-iron sow are desired. Furthermore, when high purity external metal is necessary to make a feasible production plan, the planner tries to keep the required amount small. These considerations are made more explicit in our model as objectives:

Obtain feasible plan and/or
Maximize total residual metal quantity of specified grades or
Minimize high-iron residual metal quantity or
Minimize high purity external metal required.

MATHEMATICAL FORMULATION

Let

X_{ij} = the amount of hot metal of purity grade i to be assigned to product j,

$i = 1, 2, \ldots, I,$
$j = 1, 2, \ldots, J;$

Y_{kj} = the amount of on-site inventory grade k to be assigned to product j,

$k = 1, 2, \ldots, K,$
$j = 1, 2, \ldots, J;$

$Z_{\ell j}$ = the amount of external metal supply grade ℓ to be assigned to product j,

$\ell = 1, 2, \ldots, L,$
$j = 1, 2, \ldots, J;$

V_j = the amount of silicon added into product j,

$j = 1, 2, \ldots, J;$

W_j = the amount of iron added into product j,

$j = 1, 2, \ldots, J;$

α_i = silicon content of grade i hot metal,

$i = 1, 2, \ldots, I;$

β_i = iron content of grade i hot metal,

$i = 1, 2, \ldots, I;$

γ_k = silicon content of grade k on-site inventory,

$k = 1, 2, \ldots, K;$

δ_k = iron content of grade k on-site inventory,

$k = 1, 2, \ldots, K;$

η_ℓ = silicon content of grade ℓ external metal supply,

$\ell = 1, 2, \ldots, L;$

θ_ℓ = iron content of grade ℓ external metal supply,

$\ell = 1, 2, \ldots, L;$

μ_1 = silicon content of added silicon;

μ_2 = iron content of added silicon;

ν_1 = silicon content of added iron;

ν_2 = iron content of added iron;

(a_j, A_j) = required range of silicon content of product j,

$j = 1, 2, \ldots, J;$

(b_j, B_j) = required range of iron content of product j,

$j = 1, 2, \ldots, J;$

c_j = recovery rate of product j,

$j = 1, 2, \ldots, J$;

H_i = the amount of grade i hot metal available,

$i = 1, 2, \ldots, I$;

T_k = the amount of grade k on-site inventory available,

$k = 1, 2, \ldots, K$;

E_ℓ = the amount of grade ℓ external metal supply available,

$\ell = 1, 2, \ldots, L$;

P_j = the given amount of product j required,

$j = 1, 2, \ldots, J$;

C = the ceiling amount of remelting of cold metal;

κ = the maximum percentage of purer hot metal which can be assigned to purer products.

Then the basic problem can be formulated as:

Minimize $\sum_\ell \sum_{j=1}^{J} Z_{\ell j}$ for certain selected ℓ (usually of purer grade),

i.e., minimize the amount of external purity metal required;

or

Minimize $\sum_i (H_i - \sum_{j=1}^{J} X_{ij})$ for selected low grade i,

i.e., minimize high-iron residual metal;

or

Maximize $\sum_i (H_i - \sum_{j=1}^{J} X_{ij})$ for selected high grade i,

i.e., maximize selected prime or low-iron residual metal;

subject to the following constraints:

$$c_j \left(\sum_{i=1}^{I} X_{ij} + \sum_{k=1}^{K} Y_{kj} + \sum_{\ell=1}^{L} Z_{\ell} + V_j + W_j \right) = P_j ,$$

$$j = 1, 2, \ldots, J;$$

$$\sum_{j=1}^{J} X_{ij} \leq H_i , \qquad i = 1, 2, \ldots, I;$$

$$\sum_{j=1}^{J} Y_{kj} \leq T_k , \qquad k = 1, 2, \ldots, K;$$

$$\sum_{j=1}^{J} Z_{\ell j} \leq E_\ell , \qquad \ell = 1, 2, \ldots, L;$$

$$a_j \leq c_j \left(\sum_{i=1}^{I} \alpha_i X_{ij} + \sum_{k=1}^{K} \gamma_k Y_{kj} + \sum_{\ell=1}^{L} \eta_\ell Z_{\ell j} + \mu_1 V_j + \nu_1 W_j \right) \leq A_j ,$$

$$b_j \leq c_j \left(\sum_{i=1}^{I} \beta_i X_{ij} + \sum_{k=1}^{K} \delta_k Y_{kj} + \sum_{\ell=1}^{L} \theta_\ell Z_{\ell j} + \mu_2 V_j + \nu_2 W_j \right) \leq B_j ,$$

$$j = 1, 2, \ldots, J;$$

$$\sum_{j=1}^{J} \left(\sum_{k=1}^{K} Y_{kj} + \sum_{\ell=1}^{L} Z_{\ell j} \right) \leq C;$$

$$\sum_i \sum_j X_{ij} \leq \kappa \sum_i H_i$$

for selected pure grade i's and selected pure grade j's, i.e., not more than κ of the purer hot metal can be assigned to purer products.

The mathematical model described above is only the basic formulation. There are several other practical constraints which have been added in the implementation. For example, in some cases certain grades of hot metal are not allowed to be assigned to certain products. In others, use of cold metal may be completely prohibited. These additional constraints can easily be enforced in a linear programming framework. Also implemented is a treatment of scrap recycling. This is done by including scrap recycling variables and scrap balance constraints. The scrap of each type generated is modeled as a known fraction of the corresponding product and a stated percent of it is treated as available inventory for remelt.

INTERACTIVE MODULE

It is not easy for a non-technical user to communicate directly with the linear programming model. An interactive module written in FORTRAN was developed to make the linear programming formulation transparent to the user. Simple high-level commands allow the user to add, delete, change the input tables as well as to specify one of a number of objectives, specifically:

- maximize selected prime or low-iron residual metal quantity;
- minimize high-iron residual metal quanitity;
- minimize high purity external metal required.

The following example illustrates the uses of some of the commands in building an input table.

Example:
                    ADD TABLE SUPPLY

informs the system that the user intends to set up a table and name it SUPPLY.

                    ADD C MMLBS; ADD COLUMN SI; ADD C FE

tells the system that there should be the following columns for table SUPPLY: MMLBS, SI and FE. Notice that COLUMN can be abbreviated as C in the commands.

                    ADD ROW 0404; ADD R 1020

tells the system to add rows named 0404 and 1020 to table SUPPLY. Notice that the system accepts R as an abbreviation of the keyword ROW.

The current contents of table SUPPLY can be displayed by issuing the following ECHO command:

                    ECHO TABLE SUPPLY

In response, the system displays:

	TABLE	SUPPLY		
		MMLBS	SI	FE
0404		.000	.000	.000
1020		.000	.000	.000

The command

                    ROW 0404 5.607 .04 .05

will enter the values 5.607, .04 and .05. Now if we want to look at table SUPPLY, we can issue another ECHO command:

                    ECHO TABLE SUPPLY

The system responds with:

TABLE	SUPPLY MMLBS	SI	FE
0404	5.607	.040	.050
1020	.000	.000	.000

To change iron content of 0404 from 0.05% to 0.04%, we can enter:

        ROW 0404 FE .04

If we want to see if the change has been made, we can enter:

        ECHO TABLE SUPPLY

and the system responds with:

TABLE	SUPPLY MMLBS	SI	FE
0404	5.607	.040	.040
1020	.000	.000	.000

Certain other commands are as follows:

        DELETE parameter1 parameter2

which deletes a column, a column, a row or an entry. Parameter1 refers to the column. Parameter2 refers to the row. When both parameters are present, they are used conjunctively to denote the entry to be deleted.

        WRITE filename

saves a file with the indicated file name.

        INPUT filename

loads a file with the indicated file name.

        STOP

ceases the data entry process.

The input data are entered in the framework of various tables. Each table is set up by using the commands as illustrated above. As stated earlier, the input tables of an example problem are shown in Tables 1 through 5.

In addition to the commands described above, certain higher level commands are available to the user to run the associated CDC linear programming package [1,2]. For example, command

        -GENMAT

will trigger the system to generate the matrix for the linear programming program and the system will respond with

        **CALL PRIME, LOSOW, HISOW OR EXTERN?

which asks the user to specify whether he wants to maximize the
selected prime or low-iron residual metal quantity, or to minimize
the high-iron residual metal quantity or to minimize the high
purity residual metal required. If the user chooses to respond

-PRIME

the system will obtain a solution that maximizes the selected prime
residual metal. If there is no feasible solution, the system will
respond with:

**INFEASIBLE, CALL DATA?

to indicate the infeasibility and ask if the user wants to go back
to data entry mode to change the input data. Otherwise, the system
will display the optimal solution and then ask

**CALL HISOW OR LOSOW OR EXTERN, OR DATA?

to find out if the user wants to try another objective function or
change the input data.

The user can also select the output tables to be displayed.
The complete output consists of:

- a summary of the metal employed from all available sources
  (hot metal, on-site inventory, external metal supply);

- detail assignment from each grade of hot metal to products;

- a detail assignment description of how the new scrap is
  utilized;

- detail assignment from each grade of on-site inventory to
  products;

- external purity metal requirements and utilization;

- detail new residual metal generation.

A solution with PRIME as the objective function, i.e., maxi-
mizing the selected prime grade residual metal, for the input data
stated in Tables 1 through 5 is illustrated in Tables 6 through
9. Notice that in this example, there are no output tables for
external metal or on-site inventory utilization. This is because
no metal from these sources is required.

After entering the data and running the model, the user is
able to scrutinize the proposed solution. If he feels that certain
assignments and/or consequences are not desirable, he can communicate this to the model in a conversational manner and ask for a
revised solution. This interactive procedure is continued until
the user has developed a plan to his satisfaction.

Table 6. Output – Metal Utilization Summary

81/12/22.   16.04.49.   MAXIMUM RESIDUAL PRIME SOW RESULTS

EXTERNAL PURITY -.000 MM LBS
RESIDUAL LOW-IRON SOW GENERATED   .000 MM LBS
HIGH-IRON SOW GENERATED 7.977 MM LBS
RESIDUAL PRIME SOW GENERATED 6.067 MM LBS
TOTAL AMOUNT REMELTED 1.179 MM LBS

81/12/22.   16.04.49.   TABLE 1   METAL UTILIZATION SUMMARY ('000 LB)

PRODUCT	FROM HOT METAL	FROM REMELT INVTRY	FROM EXT. PURITY	FROM NEW SCRAP	FROM NEW SOW	FROM SILI-CON	FROM UCAR	TOTAL	TO MELT LOSS	TO NEW SCRAP REMELT	TO SCRAP INVTRY	FINISH-ED PRODUCT	% SI	% FE
6061	5745	0	0	0	970	0	0	6749	67	551	131	6000	.67	.20
6063	5044	0	0	0	0	33	0	5062	50	413	99	4500	.43	.23
H16M57	4592	0	0	0	178	17	0	4770	47	108	15	4600	.05	.14
H14M7	3111	0	0	0	0	0	0	3111	31	70	10	3000	.06	.18
11020	708	0	0	0	0	1	0	714	4	6	1	700	.08	.16
11535	606	0	0	0	6	0	0	612	6	5	1	600	.12	.28
ALMG35	613	0	0	0	5	0	0	612	6	5	1	600	.06	.06
71A	506	0	0	0	0	0	0	510	5	5	0	500	.04	.06
356	273	0	0	0	12	21	0	306	3	3	0	300	7.15	.17
220	714	0	0	0	0	0	0	714	7	6	1	700	.15	.20
A356	379	0	0	0	0	29	0	408	4	4	0	400	7.15	.08
OTHERS	302	0	0	0	3	0	1	306	3	3	0	300	.15	.43
TOTAL	22593	0	0	0	1179	101	1	23874	238	1179	257	22200	.53	.19

Table 7. Output – Hot Metal Utilization

81/12/22.   16.04.49.   TABLE 2   HOT METAL UTILIZATION ('000 LB)

PRODUCT	0404	0406	0408	0410	0506	0510	0515	0620	0635	0699	06199	06200	1010	1520	1535	2035	3099	TOTAL
6061	0	0	1674	0	0	0	0	0	0	0	0	0	0	2852	1219	0	0	5745
6063	920	2793	0	0	0	0	0	0	0	0	0	0	122	0	0	398	811	5044
M16M57	0	1398	0	1108	0	596	1190	0	160	140	0	0	0	0	0	0	0	4592
M14M7	0	645	2246	0	0	0	0	58	0	0	115	47	0	0	0	0	0	3111
11020	0	328	0	0	135	0	0	0	0	0	0	0	0	0	245	0	0	708
11535	0	0	165	0	0	0	0	0	0	0	0	0	0	0	441	0	0	606
ALMG35	2	611	0	0	0	0	0	0	0	0	0	0	0	0	0	0	0	613
71A	0	506	0	0	0	0	0	0	0	0	0	0	0	0	0	0	0	506
356	0	0	189	0	0	0	0	0	0	0	0	0	0	0	84	0	0	273
220	346	0	0	0	0	0	0	0	0	0	0	0	0	0	368	0	0	714
A356	0	323	56	0	0	0	0	0	0	0	0	0	0	0	0	0	0	379
OTHERS	0	0	0	0	0	0	0	302	0	0	0	0	0	0	0	0	0	302
TOTAL	1268	6604	4330	1108	135	596	1190	360	160	140	115	47	122	2852	2357	398	811	22593

PRODUCTION SCHEDULING AUTOMATION

Table 8. Output - New Scrap Utilization

```
81/12/22.     16.04.49.      TABLE 3     NEW SCRAP UTILIZATION ('000 LB)
```

PRODUCT	6061	6063	H16M57	H14M7	11020	11535	ALMG35	71A	356	220	A356	OTHERS	TOTAL
6061	551	413	0	0	0	0	0	0	0	6	0	0	970
6063	0	0	0	0	0	0	0	0	0	0	0	0	0
H16M57	0	0	108	70	0	0	0	0	0	0	0	0	178
H14M7	0	0	0	0	0	0	0	0	0	0	0	0	0
11020	0	0	0	0	6	0	0	0	0	0	0	0	6
11535	0	0	0	0	0	5	0	0	0	0	0	0	5
ALMG35	0	0	0	0	0	0	0	0	0	0	0	0	0
71A	0	0	0	0	0	0	0	5	0	0	0	0	5
356	0	0	0	0	0	0	5	0	3	0	4	0	12
220	0	0	0	0	0	0	0	0	0	0	0	0	0
A356	0	0	0	0	0	0	0	0	0	0	0	0	0
OTHERS	0	0	0	0	0	0	0	0	0	0	0	3	3
TOTAL	551	413	108	70	6	5	5	5	3	6	4	3	1179

DEVELOPMENT AND TESTING

The authors worked very closely with the corporate production planning department and the chief planner at the metal products plant. The planner was involved in every stage of the development. Many extensive interviews were conducted with the planner who had thirty years experience in that plant. Over thirty sets of actual data were used to test the model. Throughout the process, modifications to the model were made till finally it performed to the

Table 9. Output - New Sow Utilization

```
81/12/22.     16.04.49.      TABLE 6     NEW SOW UTILIZATION ('000 LB)
```

PRODUCT	1020	1535	3099	30199	30200+	TOTAL
6061	0	0	0	0	0	0
6063	0	0	0	0	0	0
H16M57	0	0	0	0	0	0
H14M7	0	0	0	0	0	0
11020	0	0	0	0	0	0
11535	0	0	0	0	0	0
ALMG35	0	0	0	0	0	0
71A	0	0	0	0	0	0
356	0	0	0	0	0	0
220	0	0	0	0	0	0
A356	0	0	0	0	0	0
OTHERS	0	0	0	0	0	0
TOTAL	0	0	0	0	0	0
RES SOW	6067	1524	43	193	150	7977
G TOTAL	6067	1524	43	193	150	7977

planner's satisfaction. The entire process could be viewed as an effort to crystallize the planner's experience into the model. In this case, the extracted knowledge turned out to be fairly structured and thus avoided the necessity of taking the rule-based approach [3].

CONCLUSION

The capacity of the metal products plant is a function of the parameters of the raw material supply. Therefore, as the parameters of the supply fluctuate or when the finished products requirements are altered, the planner must modify his production plan to accommodate these changes while still optimizing his objectives. The computerized model is an effective automated tool not only in devising alternative production scheduling plans, but also in making subsequent modifications. The linear programming model discussed above is of the classical product mix type. However, it incorporates several specialized features which are crucial to its successful implementation. With suitable modifications, such a model can be effectively used in other similar production scheduling automation systems.

ACKNOWLEDGMENTS

The authors wish to express their deep gratitude to Mr. William I. Stevens and Mr. Joseph C. White for their contributions in making this effort possible.

REFERENCES

[1] PDS-MAGEN - A General Purpose Problem Descriptor System, Reference Manual, Version 1.4, Control Data Corporation, Minneapolis, MN, 1983.
[2] APEX-IV Reference Manual, Control Data Corporation, Minneapolis, MN, 1983.
[3] R. Davis and D. B. Lenat, Knowledge-Based Systems in Artificial Intelligence, McGraw-Hill, New York, NY, p. 235, 1982.

MACHINABILITY DATA BASE SYSTEMS FOR AUTOMATED MANUFACTURING

P. Balakrishnan and M. F. DeVries

Department of Mechanical Engineering
University of Wisconsin-Madison
Madison, Wisconsin 53706   U.S.A.

ABSTRACT

A machinability data base system, which forms a part of the common manufacturing data base and is also capable of adapting and optimizing the machining data, is an important component of automated manufacturing systems. In this paper, the current status of machinability data base systems is analyzed. Several drawbacks of the present systems and the need for new developments are discussed. A generative type machinability data base system is proposed for automating the adaptation and optimization of the machining data. Various elements of these types of systems such as the machinability data base design, model builder, optimization algorithm, and adaptation algorithm are discussed. A typical machining problem is formulated and analyzed to illustrate the proposed adaptive optimization methodology.

INTRODUCTION

During NC programming, appropriate machining data such as spindle speed, feedrate, and cutting depth must be selected and specified in the part program for each pass of the cutting tool. The high cost of NC machines prompts the use of optimum machining data for improving productivity. However, finding the optimum machining data is not an easy task since they are influenced by a large number of process variables such as part material, quality requirements, part configuration, part rigidity, tool material, tool geometry, tool rigidity, fixture rigidity, machine tool specification, machine rigidity, desired tool replacement strategy, cost and time constraints, etc. The development of adaptive control was an effort to eliminate the need for specifying the optimum machining data in the NC program.

The practical implementation of Adaptive Control Optimization (ACO) systems has been difficult because of the problems in measuring the feedback parameters on-line. The systems currently used in production are of the Adaptive Control Constraint (ACC) type in which the feedrate is adaptively adjusted based on the on-line measurement of a process response (such as force, torque, or power). The performance of this type of adaptive control system depends on the algorithms and the mathematical models used to select the maximum feedrate limit and other machining conditions [1,2,3]. These algorithms and models are developed based on machinability data relating machining conditions to tool wear, cutting forces, cutter breakage forces, surface finish, part deflection, etc. Automated manufacturing systems are expected to use CNC machines and CNC machines with adaptive control systems. A suitable machinability data selection module is necessary in these systems for automating the machinability data selection.

In spite of the rapid developments in the area of CAD/CAM, machinability data selection in most CAM systems is still manually programmed. Even though many types of automated machining data selection systems were developed since the early 1960's, the developments in this area have not kept pace with the advancement in other areas of automated manufacturing. The major thrust in the current automated manufacturing systems is to use a common manufacturing data base which serves as the central repository for all the information needed [4]. Since the machinability data must form a part of this manufacturing data base, the machinability data selection system for use in automated manufacturing must be compatible with the present data base or data driven approaches and must have advanced capabilities.

This paper summarizes the present status of machinability data base systems. Some drawbacks of the existing types of systems and the need for new developments in order to enhance the suitability of these systems for use in automated manufacturing are discussed. A suitable machinability software selection scheme which generates the optimum machining data by algorithmic calculation based on the data base approach is presented. The possibility of automatically adapting this type of machinability data base to a particular machining environment using feedback data from the shop floor is discussed.

PRESENT STATUS OF MACHINABILITY DATA BASE SYSTEMS

Based on the method by which the required machining data are arrived at, the current systems [5,6] can be classified into the following types:

- Storage and retrieval systems, and
- Generalized empirical equation systems.

In the storage and retrieval type systems, a series of recommended cutting speeds, feeds, and other related information are stored in the computer storage media for various combinations of machining operations, workpiece materials, and tool materials. In order to have centralized control and data independence, some form of Data Base Management System (DBMS) software is used as illustrated in Figure 1. The numerical control programmer uses an application program to retrieve the appropriate data which have been stored previously by the Data Base Administrator (DBA). The data for use in these types of systems are generally gathered from handbooks, other books and literature, shop experience, machine tool and cutting tool manufacturers' catalogues, etc. Although this type of data can be used for starting recommendations, it normally tends to be conservative in order to cope with worst case machining situations.

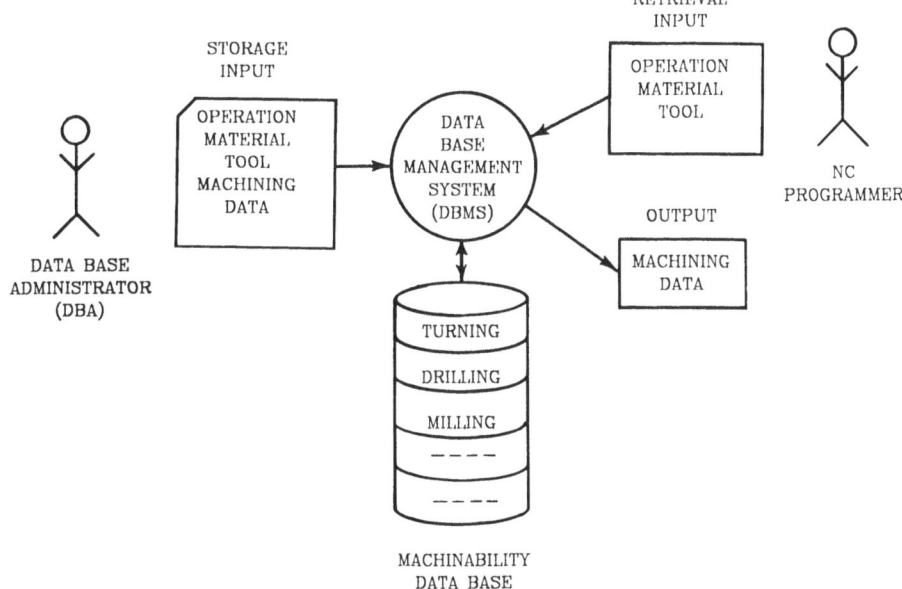

Figure 1. Storage and retrieval type system.

The generalized empirical equation system eliminates the need for the storage and retrieval of large amount of data. In this case empirical equations for speed and feed are used for computing the required data. These equations are developed based on experimental data. The data for a particular machining operation are reduced into empirical equations for speed and feed as a function of the important process variables as shown in Figure 2. These equations are directly used in the computer program. The various factors, constants, and exponents used in the equations are either given as input or they can be stored in a data base for retrieval when required. The reliability of the results from this type of system depends upon the reliability of the various factors, constants, and exponents used in the equations which, in some cases, might be too general for specific conditions. So far, this type of system has been developed for only simple machining operations such as single point turning.

DRAWBACKS OF THE PRESENT MACHINABILITY DATA BASE SYSTEMS AND THE NEED FOR NEW DEVELOPMENTS

By its very nature, the problem of machinability data selection involves the handling and manipulation of large amount of data. In the past, the principles of the systems have been kept simple because of the non-availability of suitable data handling tools. Presently, an advanced DBMS with powerful data handling capabilities is available for automated manufacturing systems. In view of this the

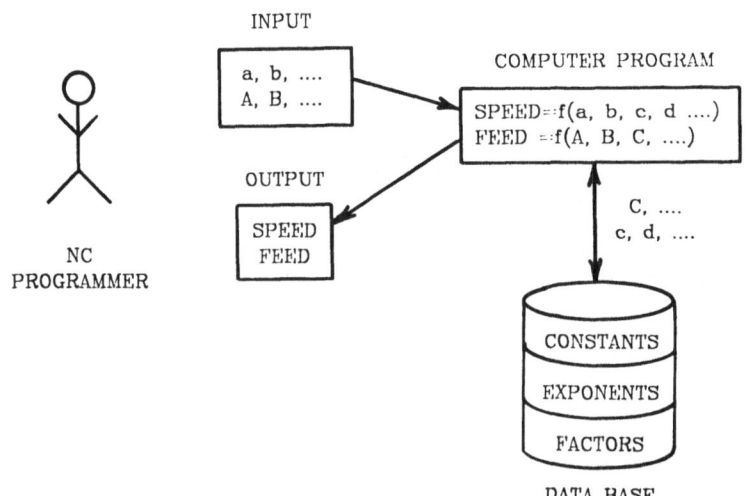

Figure 2. Empirical equation type system.

machinability data base structures and the data selection methodologies developed earlier should be reexamined in order to make them more compatible with the current developments.

The complexity of the machining process and the difficulties in describing it mathematically has made it difficult to deal with the influence of machining process variables on machinability data. The main problem is that the data gathered under one machining environment are not applicable to other environments where some of the conditions are different. Because of the enormous variety in machining operations, none of the available machining data are generally applicable. The only way out of this situation seems to be that automated manufacturing system users have to create their own machinability data base, based on the environment of their systems. Data from any other source must be viewed as a starting set only and must be suitably adapted to the environment in which it is to be used. The machining data selection system used in an automated manufacturing system has to satisfy this important requirement.

Most of the systems developed to date are of the data storage and retrieval type, since it is comparatively easy to conceive, design and implement such a system. However, since the data used in these systems are general in nature, the system must be continuously updated to make it optimum for the specific environment. Significant engineering analysis is required, not only to set up this type of data base, but also to update and maintain current recommendations. The lack of adequate maintenance is the main reason given for the failure of this type of system. Also, if the recommended data from the system are acceptable in the shop floor, no attempt is made to improve the conditions towards optimum, even though the recommended data may not be optimum.

Another major drawback in current machinability data base systems is the lack of feedback from the shop. Alterations in the shop to the data recommended by the system must be fed back into the data base for future use. Since manual updating is not compatible with the objective of automated manufacturing, methods must be developed for automating this process.

In order to reduce the burden of software in a complex automated manufacturing system, one of the methods suggested [7] is to use intelligent software which generates the data by algorithmic calculations instead of simply retrieving the data from files. Presently, efforts are underway to make the machining data selection more algorithmic and also to eliminate most of the drawbacks present in the current systems. Towards this goal, a generative type data selection method using mathematical models of the machining responses and optimization algorithms has been proposed [8,9]. Relevant details of such a system are discussed below.

GENERATIVE MACHINABILITY DATA BASE SYSTEMS

In the generative type system, instead of simply retrieving the required machining data from the data base, the data are generated each time they are needed using an algorithmic procedure. This algorithmic procedure is based on the general logic which is used to assess the suitability of a particular data set. For a given data set to be the optimum one for a specific machining situation, the data have to be within the feasible region of the machining variables and also it must satisfy the desired economic or performance objective. The feasible region is normally bounded by various constraints imposed by the machine tool, cutting tool, workpiece, and the machining process. If these constraints and the objectives are modeled as a function of the machining variables, the optimum machining data can be obtained by solving this optimization problem. The generative type system uses an algorithm based on this logic to generate the machining data. Since the engineering logic used to generate the data is programmed into the computer, this method is expected to minimize the updating and maintenance effort.

Mathematical modeling of the machining responses and an optimization algorithm which can make use of these models are the essential requirements for a generative type machining data selection system. The various machining response data or the mathematical model parameters data are used as the basic data in this case. The essential elements of a generative type machining data selection system are shown in Figure 3. The data base used in this system contains the data on parts, materials, machines, cutting tools, cost parameters, and various machining responses such as tool life, forces, power, surface finish, etc. When the request for machining data is issued from the NC programmer or processor to the machining data selection software, a search for the data corresponding to the input requirements is carried out by the DBMS. Further processing depends upon the kind of data available. If model parameters are available, the optimum cutting data are computed using the data. If models are not available the corresponding response data are modeled by the model building algorithm and these model parameters are stored in the data base. In case the response data are also not available, the default recommended machining data stored in the data base are given as output.

Elaborate machining response data as a function of machining variables are required for the generative type of systems. These data have to be gathered from expensive machinability experiments. Even then, the experimental data are not directly transferrable to the actual shop conditions. Because of this difficulty in gathering the required data, the usefulness of the generative type systems has been limited so far. With the rapid advances in sensors for machining processes, it is now possible to overcome some of the problems encountered in implementing a generative type machining data selection system. One possible approach is to gather the required machining data from the machine during the actual machining operations. This approach has been made possible because of the availability of CNC machines fitted with various machining parameter sensors in automated manufacturing systems.

Initially, the machinability data base is provided with a starter set of model parameter data similar to the handbook data presently used. When the machine is operating under the cutting conditions determined by the initial data set, response data from the various sensors in the machine are gathered by a data logger or a data acquisition system and these data are fed back into the data base as shown in Figure 3. Since these data represent the true state of the process, the future recommendations from the data base are to be suitably modified based on these feedback data. One of the advantages of using the generative type system is that this adaptation and updating of the data base can be automatically performed using the algorithmic procedure. When sufficient feedback data are available, the initial model parameters in the data base can be modified based on the feedback data using the model building module. The new model parameters are stored in the data base for

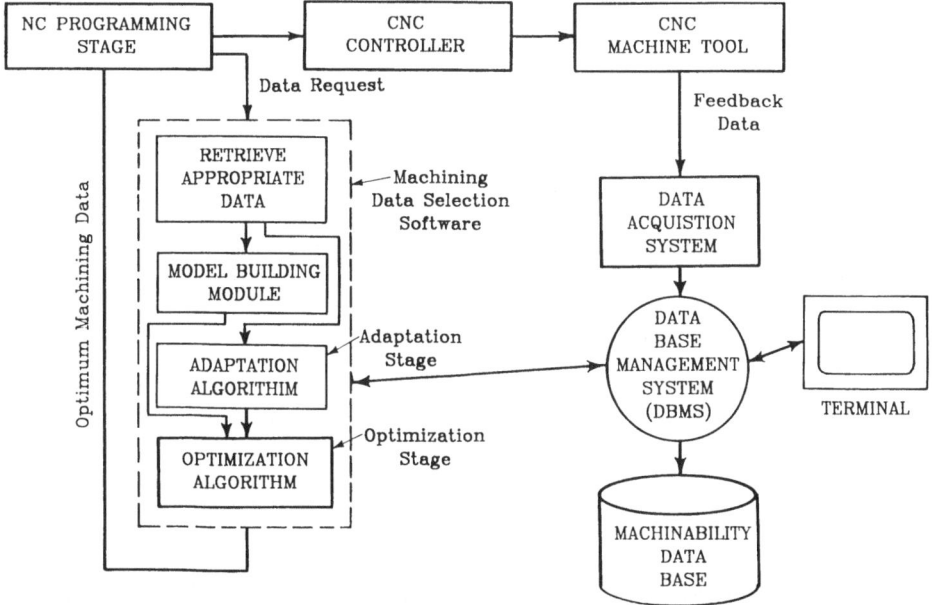

Figure 3. Adaptive optimization of machinability data base systems.

future use. With this arrangement starting from a rather limited data base, every user is able to obtain data adapted to the conditions in their own shop and at the same time optimize the cutting conditions. This concept of adaptation and optimization of machining operations is emerging as one of the important areas of research [10,11] needed for successful implementation of automated manufacturing systems.

Some of the details to be investigated and analyzed for developing a suitable generative type machinability data selection system are the following: 1) A machinability data base has to be designed to store, manipulate, and update the various machining response data. The DBMS available in the automated manufacturing environment can be used for this purpose. An appropriate logical design is to be arrived at by analyzing all the functional requirements; 2) Suitable mathematical modeling procedures are to be investigated for obtaining reliability machinability models from the response data; 3) Suitable optimization techniques which can make use of the machinability model parameters are to be selected for obtaining the optimum machining data; 4) An appropriate parameter adaptation algorithm is to be incorporated in the system for efficient adaptation of the model parameters based on the shop floor feedback data. Some of the details regarding these investigations are discussed below.

LOGICAL DESIGN OF MACHINABILITY DATA BASES

Logical data base design provides the essential information needed for the physical implementation of a data base system [12,13,14]. Since the machinability data base has to form a part of the common manufacturing data base, the various manufacturing functions which require data from this data base are identified. Once these functions are known, the various data items needed for these functions are collected. Related data items are grouped into suitable entity types and the relationships between the various entity types are established based on the type of DBMS used. A network model type DBMS is normally used in many automated manufacturing systems [15]. With the use of this powerful data driven approach, it is now possible to consider many more variables which influence machining data selection.

Figure 4 shows one possible logical design of the machinability data base confirming to the Conference on Data Systems Languages (CODASYL) requirements. The data structure diagram shown in Figure 4 illustrates the various record types needed and their relationships. Different types of machining data such as handbook data, laboratory data, response data, model parameter data, etc. are stored in the appropriate records. The handbook data are normally general in nature and are not related to any particular part, machine, and tool. The remaining machining data sets are connected

MACHINABILITY DATA BASE SYSTEMS

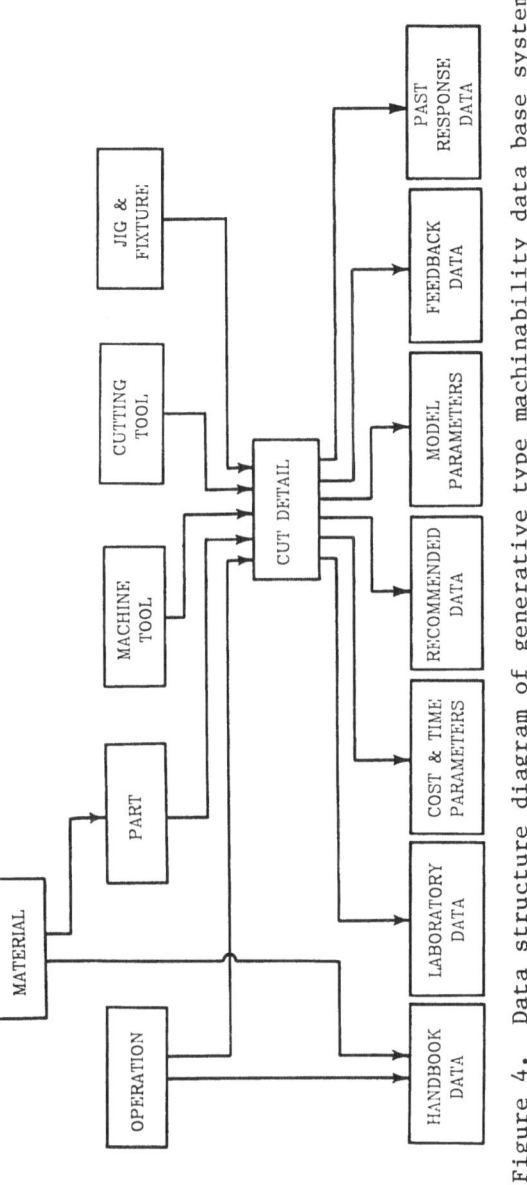

Figure 4. Data structure diagram of generative type machinability data base system.

to the appropriate part, machine tool, tool, fixture, and cut detail records such that they can uniquely represent the cutting data for a specific machining environment. Once the data base schema, subschemas, and areas are properly designed, the DBMS software can handle large number of data sets efficiently. The data storage and manipulation functions are performed using suitable application programs written in a high level language such as FORTRAN, PL/1, etc. The machining data selection software shown in Figure 3 forms a part of the application programs.

MODEL BUILDING ALGORITHMS

Mathematical models of tool life and various constraints are needed for use in the optimization algorithm. The model form frequently used for representing the various machining responses is a multiplicative nonlinear model which can be written in a general form as:

$$R = C \left(\prod_{j=1}^{p} \xi_j^{\beta_j} \right) \varepsilon' \tag{1}$$

where R is the measured response, ξ_j are the machining variables, C and β_j are the model parameters, p is the number of independent variables, and ε' is a multiplicative random error. Taking natural logarithms converts the above intrinsically linear type nonlinear model into the standard linear form:

$$\ln R = \ln C + \sum_{j=1}^{p} \beta_j \ln \xi_j + \ln \varepsilon' \tag{2}$$

which can be written as:

$$Y = \beta_0 + \sum_{j=1}^{p} \beta_j X_j + \varepsilon \tag{3}$$

where $Y = \ln R$, $\beta_0 = \ln C$, $X_j = \ln \xi_j$, and $\varepsilon = \ln \varepsilon'$. This equation is a linear polynomial model of first order, and multiple regression analysis can be applied to obtain the model parameters β_0 and β_j's using the experimental machining response data. Based on the estimated parameters, the model for the responses becomes,

$$\overline{\ln R} = b_0 + \sum_{j=1}^{p} b_j \ln \xi_j \quad \text{or} \quad \hat{R} = e^{b_0} \prod_{j=1}^{p} \xi_j^{b_j} \tag{4}$$

MACHINABILITY DATA BASE SYSTEMS 545

where b_o and b_j are the estimated model parameters. These model parameters are to be stored in the data base for use in the optimization algorithm.

In case the first order model is not adequate for some response data, the model can be extended to second order form given by:

$$\overline{\ln R_i} = b_o + \sum_{j=1}^{p} b_j \ln \xi_{ij} + \sum_{j=1}^{p} \sum_{k \geq j}^{p} b_{jk} \ln \xi_{ij} \ln \xi_{ik}; \quad i = 1, 2, \ldots, n. \tag{5}$$

All the variables may not be necessary to adequately model the data. Including unnecessary variables increases the standard error of all the model parameters. The main objective in machinability model building is to obtain the best model with a minimum of irrelevant variables. Many techniques are available for selecting the best subsets of variables. They are: 1) all possible subsets regression, 2) forward selection, 3) backward elimination, and 4) stepwise regression. A comparative analysis of these four model building techniques was performed [16] with the objective of identifying the most appropriate technique for use in machinability data base systems. The results indicate that it is possible to automate the model building using stepwise and backward elimination methods. Computer programs based on these techniques are commercially available.

OPTIMIZATION ALGORITHMS

The machining optimization procedure generally consists of the following sequence of operations:

- Formulate the desired objective function. Commonly used objective functions are production cost per piece, production rate, metal removal rate, and number of parts produced between tool changes.

- Define all the constraints applicable to the machining system.

- Maximize or minimize the objective function, subject to the constraints, using a standard optimization technique.

The various objective functions mentioned above can be mathematically expressed as:

$$C_u = C_o t_m + \frac{t_m}{T} (C_o t_{tct} + C_t) \tag{6}$$

$$T_u = t_m + \frac{t_m}{T} t_{tct} \qquad (7)$$

$$Q = f(V,f,d) \qquad (8)$$

$$N_p = \frac{T}{t_m} \qquad (9)$$

where

C_u = production cost per piece, $/piece
T_u = production time per piece, min/piece
Q = metal removal rate, m^3/min
N_p = number of parts produced between tool changes, parts
C_o = operating cost, $/min
C_t = tool cost, $/edge
t_{tct} = tool change time, min
t_m = machining time, min
T = tool life, min
V = cutting speed, m/min
f = feed, mm/rev
d = depth of cut, mm

In the above models, the tool life, T, and the machining time, t_m, can be expressed as a function of the machining variables. In practice the desired objective function is to be optimized based on various prevailing constraints such as the available range of speeds and feeds, allowable forces, power, and torque, desired surface finish, dimensional accuracy, tool life, etc.

The machining optimization problem can be formulated in the following general mathematical form:

Maximize or minimize:

$$\text{Objective function } f(\xi_1,\xi_2,\ldots,\xi_p) \qquad (10)$$

Subject to constraints:

$$g_i(\xi_1,\xi_2,\ldots,\xi_p) \leq C_i; \quad i = 1,2,\ldots,m \qquad (11)$$

$$LB(\xi_j) \leq \xi_j \leq UB(\xi_j); \quad j = 1,2,\ldots,p \qquad (12)$$

where

g_i = nonlinear or linear constraint functions
C_i = allowable constraint values

$LB(\xi_j)$ = lower bound on ξ_j variables
$UB(\xi_j)$ = upper bound on ξ_j variables
m = number of constraint functions.

The objective function is nonlinear and the various constraints could be linear or nonlinear. Several methods of solution for this problem have been proposed based on various mathematical programming techniques such as the gradient method, penalty function method, geometric programming, goal programming, and dynamic programming [17,18]. A standard nonlinear programming software package [19] is incorporated in the system for solving the optimization problem.

ADAPTATION OF MACHINABILITY DATA BASE SYSTEMS

Automatic adaptation and optimization of machining conditions is an important task to be performed in automated manufacturing. For designing an Adaptive Control Optimization (ACO) system, the exact machining process models must be known during the design stage. However, the process model parameters of the actual machining situation are often not known a priori to optimize the system. Under these conditions, a self-organizing control system [20] is needed. This type of system is capable of modifying the process model parameters on-line using the data generated during the actual machining operation. The use of adaptive control techniques for on-line optimization of the machining process have been limited to laboratory applications so far because of the problems encountered in measuring feedback parameters on-line during the actual production. To overcome this problem the use of off-line or postprocess method instead of the on-line method has been proposed [21,22].

Some of the response parameters such as tool life and surface finish are easier to measure off-line whereas parameters such as forces, power, and torque, etc. can be conveniently measured on-line. In off-line adaptation the cutting conditions are not adapted during the current machining operation. However, based on the feedback data, the future optimum data obtained from the machinability data base system are modified towards the true optimum for the next batch of parts. Over a period of time the data base is adapted to the particular machining environment and the machining data obtained will be the true optimum for that situation. Since the system learns during the actual operations, this class of systems is also called learning systems [20]. The purpose of learning in the case of machinability data base systems are to adapt the various response model parameters to be the actual machining situation. Once these parameters are adapted the optimization algorithm provides the true optimum machining data.

The major hardware requirement for the implementation of this adaptive optimization method is the availability of suitable sensors and data collection systems for acquiring the process performance data from the machine tool. Since the machine tools used in automated manufacturing systems can be provided with advanced level sensors for constraint type adaptive control (ACC) and machine diagnostics purposes [23], they can also be used for collecting the feedback data needed for off-line adaptation. Also, shop floor data collection devices are expected to be used extensively in automated manufacturing for shop floor feedback planning and control [24]. Until now these devices have been designed primarily for collecting data needed for management, inventory, and accounting requirements. With suitable expansions of the data collection systems, the process performance data could also be fed back to the common manufacturing data base. The response data gathered during actual production are to be properly identified and stored in the data base based on the part, machine tool, tool, and other cutting details. Machining response data required for off-line adaptation include forces, power, surface roughness, workpiece accuracy, temperature, vibration, chip form, tool life or tool wear information, etc.

The building of adequate response models and the adaptation of the model parameters based on the feedback data are the basic functions to be performed by the model building module in the generative type system. Multiple regression analysis techniques suitable for model building are well established, and a wide range of techniques are available for obtaining consistent and reliable models for the response data. However, all the multiple regression techniques are of one shot or batch processing types in that model parameter estimates are calculated based on the entire data set. Even though the regression methods are useful for building an adequate model for a given data set, they exhibit considerable disadvantages for parameter adaptation. If new feedback data subsequently become available and new parameter estimates are desired, all the previous response data have to be retrieved and added to the new data. This mode of solution is not computationally desirable. Also, regression model building techniques are not suitable if only subjective information about the model parameters is available as the start-up data.

To overcome the above mentioned problems various types of on-line recursive or sequential parameter estimation procedures well known in control engineering can be used. In sequential estimation, the model parameters are estimated using one observation at a time. The model parameters are sequentially updated based on each new feedback data. The advantage in using this type of model building technique is that during parameter adaptation only the previous model parameters are retrieved from the data base, thus eliminating the need for the previous response data set. Also, the sequential estimation procedure is capable of using subjective prior information regarding the model parameters. A model building software module based on this sequential estimation procedure

MACHINABILITY DATA BASE SYSTEMS

provides the model parameters needed for optimization, and it is also capable of adapting the model parameters using the feedback data.

The use of a Sequential Maximum A Posterior (MAP) estimation procedure based on the Bayesian statistical approach for parameter adaptation was analyzed [25]. The recursive equations for this procedure are given as [26]:

$$A_{u,i+1} = \sum_{k=1}^{p} X_{i+1,k} P_{uk,i} \tag{13}$$

$$\Delta_{i+1} = \sigma_{i+1}^2 + \sum_{k=1}^{p} X_{i+1,k} A_{k,i+1} \tag{14}$$

$$K_{u,i+1} = \frac{A_{u,i+1}}{\Delta_{i+1}} \tag{15}$$

$$e_{i+1} = Y_{i+1} - \sum_{k=1}^{p} X_{i+1,k} b_{k,i} \tag{16}$$

$$b_{u,i+1} = b_{u,i} + K_{u,i+1} e_{i+1} \tag{17}$$

$$P_{uv,i+1} = P_{uv,i} - K_{u,i+1} A_{v,i+1} \tag{18}$$

where $u = 1,2,\ldots,p$, $v = 1,2,\ldots,p$, p is the number of parameter estimates, X is the independent variable, Y is the dependent variable, σ^2 is the variance of Y, and P is the covariance of estimators.

In order to investigate the effectiveness of the adaptive optimization procedure, a typical axial turning problem from reference [27] was used. This problem is formulated as:

Maximize:
$$C_u = 0.2 + 1.26 X_1^{-1} X_2^{-1} + 0.1713 \times 10^{-6} X_1^3 X_2^{0.16} X_3^{1.4} \tag{19}$$

Subject to:
$$g_1 = e^{0.873} X_1^{0.91} X_2^{0.78} X_3^{0.75} \leq 4, \text{ hp} \tag{20}$$

$$g_2 = e^{19.137} X_1^{-1.52} X_2^{1.004} X_3^{0.25} \leq 100, \mu\text{in} \tag{21}$$

$$50 \leq X_1 \leq 500 \tag{22}$$

$$0.001 \leq X_2 \leq 0.01 \tag{23}$$

$$X_3 = 0.2. \tag{24}$$

The various response models used in this problem are given as:

Tool life, $\quad T = e^{17.528} X_1^{-4} X_2^{-1.16} X_3^{-1.4} \tag{25}$

Tool wear, $\quad V_B = e^{-17.711} X_1^{3.2} X_2^{0.928} X_3^{1.12} X_4^{0.8} \tag{26}$

Power, $\quad HP = e^{0.873} X_1^{0.91} X_2^{0.78} X_3^{0.75}$, hp $\tag{27}$

Surface Finish, $\quad SF = e^{19.137} X_1^{-1.52} X_2^{1.004} X_3^{0.25}$, μin. $\tag{28}$

The optimum solution for this machining optimization problem was found to be [27]: $X_1^* = 377$ fpm, $X_2^* = 0.006$ ipr, and $C_T^* = \$1.16$ per part.

For this adaptive optimization analysis, the start-up models available in the data base for tool wear, power, and surface finish were assumed to be different from the ones given by Equations 26 to 28. If the actual feedback data corresponds to the models given by Equations 26 to 28, then the adaptive enhancement methodology must converge the machining conditions towards the optimum solution mentioned above for this example problem.

The models for power and surface finish constraints were sequentially adapted during the adaptive optimization analysis. The feedback data used for adaptation was simulated using the following models:

$$V_{Bf} = e^{-17.711} X_1^{3.2} X_2^{0.928} X_3^{1.12} X_4^{0.8} \left[1 + (RN \times K) \right] \tag{29}$$

$$HP_f = e^{0.873} X_1^{0.91} X_2^{0.78} X_3^{0.75} \left[1 + (RN \times K) \right] \tag{30}$$

$$SF_f = e^{19.137} X_1^{-1.52} X_2^{1.004} X_3^{0.25} \left[1 + (RN \times K) \right] \tag{31}$$

where RN is the random number sequence [28] and K is the coefficient of variation.

The start-up models available in the data base were assumed to be different from the above and are of the form:

$$V_{Bs} = e^{-16.0} X_1^{4.5} X_2^{2.5} X_3^{1.12} X_4^{1.5} \tag{32}$$

$$HP_s = e^{1.0} X_1^{0.75} X_2^{0.60} X_3^{0.75} \tag{33}$$

$$SF_s = e^{19.0} X_1^{-1.0} X_2^{.15} X_3^{0.25} \tag{34}$$

For obtaining the optimum machining conditions for this constrained problem, the nonlinear programming subroutine GPMNLC [29] was used. The three response models were adapted after each tool change using the recursive Equations 13 to 18. Adaptive optimization analysis of this constrained problem was performed for various K values using the following initial parameters:

Model	$\underline{P}_{uv,o}$	s_i	K
Flank Wear	diag[$10^4, 10^3, 10^3, 10^3$]	0/0.33/0.66	0/0.1/0.2
Power	diag[$10^4, 10^3, 10^3$]	0/0.16/0.32	0/0.1/0.2
Surface Finish	diag[$10^4, 10^3, 10^3$]	0/0.5/1.0	0/0.1/0.2

The changes in the cutting speed, feed, and the cost values during the adaptive enhancement analysis are shown in Figure 5 for ten tool changes.

It is evident from the results that the optimum values obtained based on the start-up response models are different from the true optimum values corresponding to the feedback data. The adaptive enhancement procedure gradually adapts and optimizes the machining conditions to the actual machining environment. For this particular problem, the machining cost is reduced by half at the end of the adaptation. Since the response models have a maximum of four parameters to be updated, these models and the cutting conditions are suitably adapted after the fourth tool change. The use of a proper initial covariance matrix of the model parameters is important for this adaptation since the diagonal elements of this matrix contribute mostly to the parameter modification. Specification of small values implies that changes in the parameters are not expected to be large. However, if the model parameter changes considerably, as in this example problem, the variance values are to be perturbated in order to follow these changes closely. It is

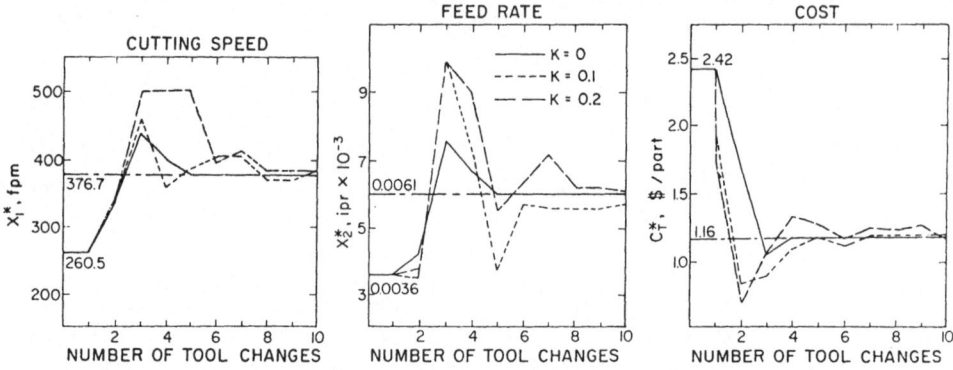

Figure 5. Changes in the optimum values during adaptive enhancement.

evident from the results that this variance perturbation method is useful for gradual enhancement of the machinability data base.

Comparison of the optimum values for two different coefficients of variation with that of the ideal case indicate that the adaptive optimization of the constrained problem is not affected by the presence of process variations. When the process variation is present, the optimum values fluctuate. However, the machining conditions are always improved. The optimum values at the end of the adaptive enhancement are very close to the expected values of: $X_1^* = 377$ fpm, $X_2^* = 0.006$ ipr, and $C_T^* = \$1.16$ per part. The results of this simulation analysis indicate the suitability of this off-line adaptive enhancement methodology for use in Automated Manufacturing Systems.

CONCLUSION

In automated manufacturing there is a need for a machinability data base which forms an integral part of the common manufacturing data base and an automated machining data selection system which requires minimum human effort for proper maintenance and updating. This paper presented the current status of machinability data base systems and evaluated their suitability for use in automated manufacturing systems. Several drawbacks of the existing types of systems were identified and the desired features which are to be incorporated in a system for use in automated manufacturing were discussed. The recent concept of generative data selection and the various elements of this type of system such as data base, model builder, and optimization algorithm have been described and their functional requirements analyzed. The availability of CNC machines fitted with advanced sensors will overcome some of the problems encountered in implementing a generative type system. If the process performance data are gathered and fed back to the machinability data base, it is possible to adapt the machinability data base to the particular environment of the user using suitable adaptive optimization algorithms. Relevant details of this adaptive optimization concept have been presented. It is expected that the generative type system based on the data base approach and utilizing the adaptive optimization concept will satisfy the machinability data requirements of the present and future automated manufacturing systems.

REFERENCES

[1] V. A. Tipnis, "Development of mathematical models of adaptive control systems," Proc. 13th NC Society Conference, pp. 149-156, 1976.

[2] V. A. Tipnis, S. C. Buescher, and R. C. Garrison, "Mathematically modeled machining data for adaptive control of end milling operations," Proc. NAMRC-IV, pp. 279-286, 1976.

[3] R. A. Mathias and R. B. Ludwig, "A machinability routine that optimizes CL file feeds and speeds prior to postprocessing," SME Technical Paper, MS79-403.

[4] D. S. Appleton, "Measure twice; cut once," Datamation, Vol. 28, No. 2, pp. 126-136, February 1982.

[5] M. F. DeVries, P. Balakrishnan, and J. Agapiou, "A study of two tasks applicable to an automated manufacturing research facility-Volume I: Task A; A study of machinability data banks," Report for the National Bureau of Standards, University of Wisconsin-Madison, 1982.

[6] P. Balakrishnan and M. F. DeVries, "A review of computerized machinability data base systems," Proc. NAMRC-X, pp. 348-356, 1982.

[7] H. Yoshikawa, "Study on the structure of software for fully automated factory," Proc. CIRP Intl. Sem. on Manufacturing Systems, Vol. 7, No. 2, pp. 101-109, 1978.

[8] M. Y. Friedman, M. Field, and K. J. Kahles, "Machinability data bank design," Ann. CIRP, Vol. 23, No. 1, pp. 171-172, 1974.

[9] W. J. Zdeblick, "Real time manufacturing data selection system," Proc. CIRP Intl. Sem. on Manufacturing Systems," Vol. 9, No. 4, pp. 243-263, 1980.

[10] J. Stanic and V. Solaja, "On an adaptive optimization model of manufacturing processes," Ann. CIRP, Vol. 27, No. 1, pp. 419-423, 1978.

[11] W. J. Zdeblick, "An adaptive planning methodology for machining operations," SME Technical Paper, MR 82-243, 1982.

[12] R. A. Ross, "Logical data base design," Data Base Journal, Vol. 11, No. 4, pp. 2-8, 1981.

[13] S. Arun, "Logical data base design: A management oriented approach," Information & Management, Vol. 5, No. 2, pp. 77-85, June 1982.

[14] D. C. Tsichritzis and F. H. Lochovsky, Data Models, Prentice-Hall, Inc., 1982.
[15] R. J. Miner, M. E. Grant, and R. J. Mayer, "Decision support for manufacturing," Proc. IEEE Winter Simulation Conf., pp. 543-549, 1981.
[16] P. Balakrishnan and M. F. DeVries, "Analysis of mathematical model building techniques adaptable to machinability data base systems," Proc. NAMRC-XI, pp. 466-475, 1983.
[17] R. H. Philipson and A. Ravindran, "Application of mathematical programming to metal cutting," Mathematical Program Study, Vol. 11, Enginering Optimization, M. Avriel (Ed.), North Holland Publishing Co., pp. 116-134, 1979.
[18] D. L. Kimbler, R. A. Wysk, and R. P. Davis, "Alternative approaches to the machining parameter optimization problem," Compt. & Indus. Eng., Vol. 2, No. 4, pp. 195-202, 1978.
[19] A. D. Waren and L. S. Ladson, "The status of nonlinear programming software," Operations Research, Vol. 27, No. 3, pp. 431-456, May-June 1979.
[20] G. N. Saridis, Self-Organizing Control of Stochastic Systems, Marcel Dekker, Inc., 1977.
[21] T. Sata and K. Matsushima, "A proposal of the multilayered control of machine tools for fully automated machining operations," Proc. IFAC Symposium on Information Control Problems in Manufacturing Technology, Y. Oshima (Ed.), Pergamon Press, pp. 173-181, 1978.
[22] S. Yonetsu and I. Inasaki, "Optimization of turning operation," Proc. NAMRC-VI, pp. 17-23, 1978.
[23] S. K. Birla, "Sensors for adaptive control and machine diagnostics," Technology of Machine Tools-Machine Tool Task Force, Vol. 4, Chapter 7-12, October 1980.
[24] P. Link, "Shop floor control," Commline, Vol. XI, No. 6, pp. 16-18, November 1982.
[25] P. Balakrishnan and M. F. DeVries, "Sequential estimation of machinability parameters for adaptive optimization of machinability data base systems," paper submitted for presentation at the ASME Winter Annual Meeting, November 1983.
[26] J. V. Beck and K. J. Arnold, Parameter Estimation in Engineering and Science, John Wiley & Sons, 1977.
[27] D. S. Ermer and R. K. Pradhan, "Economic selection of cutting conditions for constrained single or multipass operation," Proc. NAMRC-VII, pp. 355-361, 1979.
[28] IMSL Library Reference Manual, IMSL, Inc., 1982.
[29] GPM/GPMNLC: Extended Gradient Projection Method Nonlinear Programming Subroutine for ASCII Fortran, User Manual for the UNIVAC 1100, MACC, University of Wisconsin-Madison, 1983.

AN INTEGRATED APPROACH TO DESIGN, IMPLEMENTATION,

AND TESTING OF DIGITAL SYSTEMS

>Manzer Masud
>
>Department of Computer Science & Engineering
>University of Petroleum and Minerals
>UPM Box 1996
>Dhahran, Saudi Arabia

ABSTRACT

Recent advancements in the micro-electronic technology have made it possible to manufacture digital systems of considerable complexity at a very low per-system cost. Full potential of this manufacturing capability, however, can be utilized only if the associated digital hardware design automation systems can be upgraded to reduce the cost associated with the design process. This is especially true of special purpose low-volume digital systems.

From initial requirement specification, the design of a digital system progresses through several levels of refinements until it reaches the final fabrication phase. A modern digital hardware design automation system must be able to support all phases of design activities, including testing, from a single description of the digital system which is to be designed. Since the complexity of digital systems is ever-increasing it is necessary that the design automation system provides an abstraction mechanism for clear, concise and unambiguous description of the digital system. Such abstraction is provided by the computer hardware description languages (CHDLs).

This paper discusses the design and implementation of a CHDL based automation system that provides an integrated environment to support all phases of design activities from initial specification to final fabrication and testing of digital system. Current industrial applications of the automation system and future research plans will also be discussed.

INTRODUCTION

Recent advancements in the field of integrated circuit technology have made it possible to manufacture digital systems of considerable complexity at a low per-system cost. Full potential of this manufacturing capability, however, can be utilized only if digital hardware design automation systems can be upgraded to reduce the cost associated with the design of digital systems.

From initial system specification, the design of a digital system progresses through several levels of abstractions until it reaches the final fabrication phase. Requirement specifications of a system to be designed is a conceptual model detailing what the system is required to do. It is the most abstract model of the system and provides very little, if any, detail of how the system will be implemented. The next step is to develop an overall system layout in terms of storage elements, data paths, and control unit which manipulates information in proper order to accomplish the specified task. This level of system description is generally referred to as the register transfer level. The abstract register transfer description must then be expanded into logical design of the system in terms of flipflops, logic gates and their interconnection. Finally, the logical design is converted into physical implementation. This last step requires partitioning of the digital system into subassemblies, placement of components on printed circuit boards, and routing of wires and connectors. If the digital system is to be implemented as an LSI or VLSI package then a detailed circuit design, in terms of transistors, etc., is derived from the logical design followed by the actual chip layout, masking, and fabrication.

Several other activities are associated with design. Proper checkup points and tests must be developed to verify that the design at each level meets the original specification. Test sequences must be developed to detect manufacturing defects in the final product. Diagnostic programs may be provided to help in the maintenance. And finally, proper documents, for example, maintenance manuals and user's guides must be prepared.

The usual approach is to consider each of the above mentioned activities individually. Computer aided design (CAD) systems are available to help the designers at each stage of the design activity. However, CAD systems for different activities were developed independent of each other, and each one has its own front end. That means that each CAD system will require the designer to re-express his design in a different notation (language). The loss of time and effort is obvious; a more serious problem is that one is never sure that descriptions at various stages really represent the same design.

AN INTEGRATED APPROACH TO DESIGN

A large digital system may use a very large number of components. These components are tested individually, but yet another CAD system is needed to test the whole digital system. Models of individual components must be prepared again and their interconnections specified. This again, translates into loss of time and possibility of errors.

Many of the CAD systems in use today require the design engineer to provide a very detailed description of his circuit. These systems do not provide much help to the designer in organizing his design and do not aid in his thinking process. Because the descriptions are lengthy and tedious, they are prone to error and their debugging is difficult. All this results into a high design cost and low productivity.

From the above discussion, it is evident that a unified and integrated approach to the design problem must be taken. A complete automation system is needed in which the circuit to be designed will have to be described only once and at an adequate level of abstraction. Such abstraction is provided by Computer Hardware Design/Description Languages (CHDLs). It may be true to say that if the realization of special purpose low-production digital devices is to be feasible it will be through the use of CHDLs.

Several CHDLs have been developed in the recent past. A few major ones are: AHPL [9], CDL [6], DDL [7], ISP [1], and RTS [19].

The digital hardware design automation system discussed here is based on the universal AHPL. An extension of a very well-known computer hardware description language - AHPL.

AN OVERVIEW OF AHPL

Like APL [12], AHPL operates on vectors. Basic AHPL operators are shown in Table 1. Some of the commonly used operations are shown in Table 2.

Several examples showing how AHPL can be used to describe complex digital systems are given in reference [10].

AHPL programs (circuit descriptions) have a one-to-one hardware correspondence. That is, each statement or step can unambiguously be translated into hardware.

AHPL statements are executed sequentially. The sequence may be altered by a branch statement or subprogram invocation.

AHPL allows one to describe clock mode sequential circuits with ease and brevity, the two major parts of an AHPL description

Table 1. AHPL Operators

Operator	Description
&	Logical And
+	Logical Or
@	Logical Exclusive Or
&/	All bits And
+/	All bits Or
@/	All bits Exclusive Or
↑	Complement
←	Transfer
=	Connection
→	Branch
A	Simple register, memory, or bus
A,B	Column concatenation
A!B	Row concatenation
A[J]	J^{th} bit (column) of A
A[M:N]	Bits M thru N of A
A<J>	J^{th} row of A
A<M:N>	Rows M thru N of A
*	Conditional

of a sequential circuit, called module, are the declaration part and the control sequence part. In the declaration section, various buses, registers, inputs, outputs, and combinational logic units are declared. The control sequence is a list of numbered statements, each consisting of an action followed optionally by a branch. All register transfers and bus connections are specified in the action part. The branch part specifies which actions are to be performed next. The branch may be conditional on the value of any line or register.

The primary description segment in AHPL is the module. To further organize the design, two additional description segments, the combinational logic unit (CLU) and functional register (FNREG), are also allowed.

A module has the highest level of intelligence. It consists of memory elements, I/O lines, etc. It cannot be invoked by other modules. A functional register has combinational logic and memory elements, but has no control sequence. It is active only when invoked and retains the results of an invocation until invoked again. CLUs are pure combinational logic. They are active when invoked but, being memory-less, do not retain the results of an invocation.

The universal AHPL is a much enhanced version of the original language. It allows one to specify asynchronous data transfer,

Table 2. Commonly Used AHPL Operations

Transfer & Connection

A ← B	Load A with contents of B. B stays unchanged.
X,Y ← A,B	Multiple source and destination registers
X ← (A!B) * (f,↑f)	Conditional selection of source
X* K ← A	Conditional transfer of data

Program Control

→ n	Unconditional branch
→ (f,↑f)/(n_1,n_2)	Conditional branch
→ (n_1,n_2,n_3)	Multiple branch, diverge for parallel execution

Sub Program Invocation

X ← INC(X)	New value of X is the result of performing the 'INC' operation on X.

global branches (resets), multiple clocks and application dependent details in the register transfer description of a circuit. Readers interested in a detailed presentation of the language may refer to [15].

APPROACH

The design automation system discussed in this paper employs the universal AHPL as the main user interface. The functional behaviour of the digital circuit to be designed is expressed in the language. The automation system processes this abstract description and produces the desired output. The output of the automation system depends on the task it has to perform. It may be a simple interconnection list of primitive logic elements - AND, OR, INVERTERS, etc.; or it may be a sequence of commands to drive a numeric controller for wire-wrapping or drafting; or it may be a complete layout for VLSI implementation of the device.

A viable approach would be to design a separate processor for each new application. Each such processor (compiler) will do the necessary input/output transformation. However, in order to minimize the cost of software development, it is necessary to identify and separate the common aspects of the compilation process from those which are application dependent. To this end, the compilation process has been divided into three stages as depicted in Figure 1. Several intermediate representations (outputs of intermediate stages)

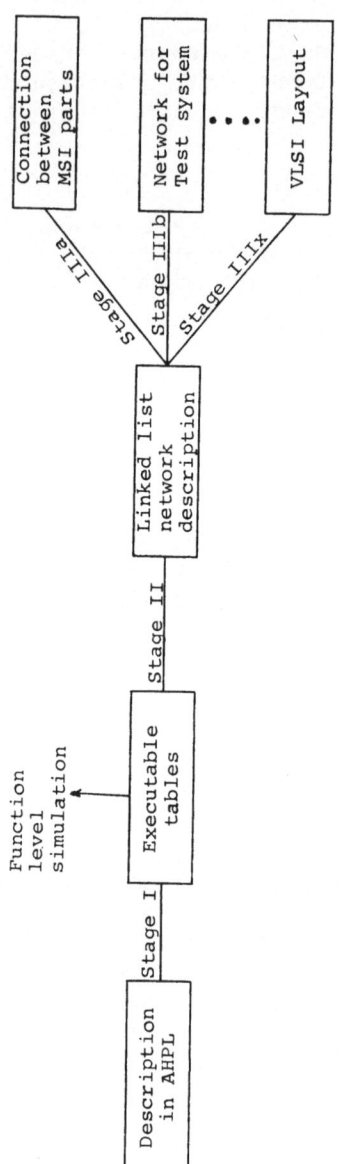

Figure 1. Three-stage hardware compiler.

are possible. Most useful are the ones which minimize the programming effort and are easily understandable.

Stage I of the compiler decomposes the source text into a quadruple table and other tables which help in keeping track of variables and determining the referencing environment. Quadruples generated by the first stage are not like those generated by compilers of software languages. These quadruples are just a tabular representation of the original AHPL source program.

The second stage processes the output of the first stage. It assigns control states and translates the description into a transfer/connection list for each data element (register, bus, etc.). The description is stored in a linked list.

The two databases produced by the first two stages of the compiler are independent of the application. The third stage of the compiler will depend on the application - it will use one or both databases to generate the output appropriate to a particular application. It may be a wire list inter-connecting integrated circuit parts or an abstract interconnection list suitable for test sequence generation or a complete design and layout for VLSI implementation.

Clearly, the requirements of a wide range of applications including those mentioned earlier cannot be accomplished by a single stage III processor. Instead, a separate stage III is written for each application, as illustrated in Figure 1. New versions of stage III may be prepared at any time a need for a significantly different output becomes evident. Often a stage III processor acts as an interface between stage II and existing CAD software available at the point of application.

Decomposing the compilation into stages has significant advantages. The task of syntax recognition and decomposition of source program into an internal form is performed by the first two stages. Thus, it becomes easier to write a custom tailored stage III. Also, since stages I and II are free from application dependent details they can be made more modular and easily modifiable. If new features are added in the language at a later stage, most stage III processors for applications not requiring the added features will stay unmodified. On the other hand, stages I and II will not have to be modified for each new application of the design automation system.

Application Specific Details

Since the design automation system will be used for a large variety of applications it is necessary to provide a method by which a user may enter details which are applicable to one application but not to the other. Application dependent details may be entered by two methods. One, as a part of the declaration statements in the original text. And two, as a separate input (may be a language in itself) to a stage III processor.

Separation of application specific details from the main language has greatly reduced the burden of the language designer and at the same time has enhanced the flexibility of the design automation system. The mechanism, to some extent, makes it unnecessary to anticipate in advance all features of the various applications for which the design automation system will be used.

A Unified Database

Outputs of the stage II compiler will be used by several stage III processors. For this reason it is necessary to store the output in a database which is flexible and efficient. Information regarding various physical, logical and other attributes of commonly used digital components is also stored. Mechanism to update and modify these attributes and insert and delete information regarding digital components is to be provided. A carefully designed database enhances the efficiency, reliability and flexibility of the hardware design automation system.

CURRENT STATUS AND APPLICATIONS

The multistage compiler for the Universal AHPL has been implemented on a DEC-10 machine. Effort is underway to implement it on other machines, particularly IBM/370, CDC cyber, and VAX-11/780. Research is being made to make further improvements in the mechanism to access and update the unified database.

Several stage III processors are under consideration and some are at various stages of completion. As new applications are sought more stage III processors will be developed. A parallel effort is directed towards incorporating more and more available CAD application software into the design automation system. Incorporation of available CAD systems into the design automation system is a quick, relatively easier, and direct method to enhance its capabilities and to increase its sphere of application.

At present the automation system is being successfully used for research and industrial applications. Some of these are discussed below.

Device Modeling System for Test Engineers

Testing of a digital device is accomplished by introducing test signals, pattern of ones and zeroes, at its inputs and measuring the corresponding output waveform. The input/output behavior of the unit under test is then compared with the expected behavior of a good device.

Testing of a digital circuit board having several devices (IC packages) is a more difficult task and requires sophisticated computer-based automatic test equipment. Several types of test equipment with various degrees of sophistication are available. One such system requires the user to prepare models of individual IC packages on the circuit board which is to be tested. The models are processed by the software provided with the test equipment. The software includes a gate-level simulator, a test sequence generator, and a diagnosis and analysis package. The test sequence generator automatically generates input patterns which will detect failures in the circuit. The simulator gives the response of the circuit for the applied test sequence. The same sequence is also applied to the actual device, and the two responses are compared. The diagnosis and analysis program determines if it is necessary to continue the testing in order to achieve the required reliability. The system may be used for go/no-go test as well as for isolating faulty IC packages on the circuit board.

A major problem with the automatic test system was the input language. To prepare a device model it was necessary to describe the whole device in terms of basic components such as gates and flipflops. Each element must be assigned a unique number and the whole model must be prepared using the format [20]: "(element number) (element type)/(input list)//". To model an LSI device of the complexity of a microprocessor using the above format is not very practical. It is tedious, time consuming, and prone to error. Modeling a device like Motorola's M6800 would need several man-months of effort. More complex VLSI devices will take much more time. Thus, the need of a more sophisticated method to enter the device model became evident. A method which would not only simplify the coding, but will also help the test engineer to develop well-organized models.

CDHLs provide a formal mechanism to express the internal organization and working of digital devices. However, to develop an entirely new test system based on a CHDL would have have been an expensive and time-consuming endeavor. Instead, a method which allowed the use of the AHPL-based automation system as a front end to the existing test system software was developed. The model of the device to be tested is described in AHPL and is processed by the stage I and stage II compilers of the automation system. The linked list output of stage II is then converted into the input format of the existing test system software. The interface between stage II output and the available test system software is straightforward and needed only a few weeks of programming effort. This clearly indicates that existing CAD software can be integrated with the design automation system to enhance its capabilities.

Test Sequence Generation

A digital device is said to be functioning properly if it produces the desired output sequence for a given input sequence for all acceptable input sequences. A sequence of 2^n patterns is sufficient to test a purely combinational n-input device. Considerably fewer patterns will be needed if special techniques [14] are employed. Testing of a sequential device, however, is much more difficult because its response depends on all previous inputs.

Several methods have traditionally been used to test sequential devices. These include: converting the sequential circuit into an iterative network [3]; comparing the state table of the unit under test with that of a good circuit [8,13]; and simultaneous simulation of all single fault versions of the circuit for user-supplied input patterns. The AHPL-based automation system may be used to support any of the above mentioned methods. However, the above methods are not adequate for highly sequential LSI and VLSI devices with very few output pins. This observation led to the search for a new method to generate test sequences. The method is based on the design language AHPL.

AHPL partitions a circuit into control and data sections. State table of the control section may be searched exhaustively while only a small portion of the much larger data section needs to be searched. Furthermore, the compact design language description would aid in guiding the heuristics test sequence generation operation. This is the approach taken by SCIRTSS (sequential circuit test search system) [2,4,11]. Reference [5] gives an overview of this approach and an analysis of SCIRTSS' performance.

The two databases generated by the stage I and the stage II processors are used by the SCIRTSS software. Stage I tables are used to sensitize and propagate errors while the linked list output of stage II is used for simulation and diagnosis purposes.

Implementation of VLSI

Very large scale integration technology of 1980's would make it possible to manufacture devices with more than a million elements on a single chip [16]. This means that highly complex special purpose devices and super micro-computers will be fabricated on a single VLSI package. Current ad-hoc design methods employed by several manufacturers will not be adequate to handle devices of such complex structure. Also, testing of highly sequential devices designed by ad-hoc methods will not be an easy task. In order to make a VLSI realization of special purpose low-volume devices feasible, it is necessary to automate the design process. An AHPL-based VLSI design automation system is under development.

The linked list output of stage II contains all the information that is necessary to fabricate the complete circuit. Stage III is being written to convert the stage II output into physical hardware to be implemented as a VLSI chip. Such a program must perform the task of automatic routing and placement to make the most efficient use of the available chip area. It is difficult to write a good placement algorithm for a complex VLSI device. However, if some constraints are put on the circuit layout then the placement algorithm will be relatively simple. The storage logic array (SLA) [18] constraints the layout by distributing logic between rows and columns of memory elements. The constraints imposed by SLA simplify the task of automatic layout with a relatively small penalty in chip area utilization [17].

CONCLUSIONS

Applications of the design automation system discussed above indicate that the system has the potential to meet the challenge of ever-increasing complexity of digital devices. The three systems developed separately can be unified to provide an integrated design and manufacture environment. Additional sub-systems may be developed and attached later. Thus, the combined system is able to support all phases of design activities, including testing, from a single AHPL description of the device to be fabricated.

ACKNOWLEDGMENTS

The author would like to acknowledge the contributions of Professor Frederick J. Hill, Dr. Z. Navabi and Mr. D. Chen towards the project.

REFERENCES

[1] M. R. Barbacci, "Instruction set processor specifications (ISPS): the notation and its application," IEEE Trans. on Computers, vol. C-30, pp. 24-40, Jan. 1981.
[2] J. E. Belt, "An heuristic approach to test sequence generation for AHPL described sequential circuits," Ph.D. dissertation, University of Arizona, 1973.
[3] M. A. Breuer, "A random and algorithmic technique in fault detection test generation for sequential circuits," IEEE Trans. on Electronic Computers, Vol. C-20, pp. 1364-1370, Nov. 1971.
[4] E. A. Carter, "Fault test generation for sequential circuits described in AHPL," Ph.D. dissertation, University of Arizona, 1973.

[5] C. H. Chiang, F. Hill, A. Mohseni, and D. Chen, "Fault detection test generation at the register transfer level," in Proc. IEEE First Annual Phoenix Conf. on Computers and Communications, pp. 58-63, May 1982.

[6] Y. Chu, "An Algol-like computer design language," Communications ACM, pp. 607-615, October 1965.

[7] J. R. Duley and D. Dietmeyer, "A digital system design language (DDL)," IEEE Trans. on Computers, Vol. C-17, pp. 850-861, Sept. 1969.

[8] D. E. Farmer, "Algorithms for designing fault detection experiments for sequential machines," IEEE Trans. on Computers, Vol. C-22, pp. 159-167, Feb. 1973.

[9] F. J. Hill, "Updating AHPL," in Proc. 1975 Int. Symp. on Hardware Description Languages and Their Applications," pp. 22-29, Sept. 1975.

[10] F. J. Hill and G. R. Peterson, Digital Systems: Hardware Organization and Design, 2nd ed., John Wiley & Sons, New York, 1978.

[11] B. M. Huey, "Search directing heuristics for sequential circuit test system (SCIRTSS)," Ph.D. dissertation, University of Arizona, 1973.

[12] K. Iverson, A Programming Language, John Wiley & Sons, New York, 1962.

[13] C. R. Kine, "An organization for checking experiments on sequential circuits," IEEE Trans. on Elect. Computers, Vol. EC-15, pp. 113-115, Feb. 1966.

[14] Z. Kohavi, Switching and Finite Automata Theory, 2nd ed., McGraw-Hill, New York, 1978.

[15] M. Masud, "A modular implementation of a digital hardware design automation system," Ph.D. dissertation, University of Arizona, 1981.

[16] C. Mead and L. Conway, Introduction to VLSI System, Addison Wesley, Massachusetts, 1981.

[17] Z. Navabi, M. Masud, W. Knapp, and F. Hill, "Impact of VLSI technology on the hardware description language AHPL," in Proc. 1980 Conf. on Circuits and Computers, 1980.

[18] S. S. Patil and T. Welch, "A programmable logic approach for VLSI," IEEE Trans. on Computers, Vol. C-28, pp. 594-601, Sept. 1979.

[19] R. Piloty, "Segmentation constructs for RTS III," in Proc. 1975 Int. Symp. on Hardware Description Languages and Their Applications," pp. 115-124, Sept. 1975.

[20] K. Wacks, F. Hill, M. Masud, and P. deBruyn Kops, "An integrated system for LSI device modeling," in Proc. Automatic Testing 80, Paris, France, Sept. 1980.

Local Area Networks, Data-Bases, and Graphics

IMPLEMENTING PRIORITY FUNCTIONS IN LOCAL AREA NETWORKS

Lionel M. Ni

Department of Computer Science
Michigan State University
East Lansing, MI 48824   U.S.A.

ABSTRACT

Prioritizing the handling of various data and control messages is important for network performance. Among the three most popular media access protocols that are likely to be standardized, CSMA/CD protocol is the only one that does not provide priority mechanism in handling various priority levels of packets. In CSMA/CD protocols, all messages are treated equally in competing for the communication channel. Thus, important or time-critical messages may be severely delayed. A new CSMA/CD protocol implemented with message-based priority functions is proposed. In this scheme, a distributed priority-code comparison algorithm is developed to determine the highest priority class that can compete for the communication channel. Both nonpreemptive and preemptive disciplines are discussed. In each discipline, both the single mode and the batch mode are considered and compared. The overhead in implementing the proposed protocol has shown, through simulation, to be much less than previously proposed methods. Among four different operating alternatives, the preemptive single mode scheme, in general, provides a better performance.

INTRODUCTION

Local area networks have captured the interest and inspired the imagination of a host of people in the last five years. Some proprietary local networks are known by such names as Ethernet (Xerox), ARCNet (Datapoint), LocalNet (Sytek), WangNet (Wang), and IBM Ring. These local networks are different mainly in their media access methods in the data link layer. Three most popular media access methods that are likely to be standardized are CSMA/CD [4], token-passing bus [5], and token ring [1] protocols. These protocols try to give equal priority to all attached stations and to avoid having one station monopolize the network. However, prioritizing the handling of various data and control messages is important for network performance.

Defining more than one distinct priority level for packets is useful in decreasing the queueing delay for important packets and to ensure that the most critical network functions are not delayed. For example, to balance the load of various computers in a local network, control messages which carry the current load information of host computers must be transmitted with the highest priority; otherwise, the current load information could be outdated when received [7,13]. Many other applications, such as sensor data, process control, interactive applications, etc., are all time-critical. Both token-passing bus and token ring protocols do provide optional priority mechanism in handling various priority levels of packets. However, the CSMA/CD protocol does not incorporate priority functions. Furthermore, the undeterminism nature may make the network performance even worse.

Two kinds of priority functions have been adopted. A station-based priority function was proposed by Franta and Bilodeau [2]. In this scheme, each station has a fixed priority. Message-based priority functions provide a more dynamic approach to be assigned to messages rather than stations [8,11]. Station-based priority functions can be considered as special cases of message-based priority functions which are studied in this paper. Rom and Tobagi [9] have indicated four requirements in designing prioritized protocols: 1) hierarchical independence of performance; 2) fairness; 3) robustness; and 4) low overhead. However, some of these requirements are often contradictory goals, for example, fairness and low overhead. Thus, we should try to find trade-offs between them. Here, fairness is defined for packets within each priority class.

This paper is organized by first introducing the concepts and terminologies of CSMA/CD and prioritized CSMA/CD (PCSMA/CD) protocols. A previously proposed PCSMA/CD protocol is then briefly discussed in Section 2. A new PCSMA/CD protocol based on priority-code comparison scheme and its operating alternatives are detailed in Section 3. System performance measurements and comparisons are shown in Section 4.

PRIORITIZED CSMA/CD PROTOCOLS

A carrier sense multiple access with collision detection (CSMA/CD) protocol is characterized by a single channel with a

IMPLEMENTING PRIORITY FUNCTIONS

distributed contention resolution scheme. A station is ready if it has messages ready for transmission; otherwise, it is an idle station. All stations are able to listen to the communication channel and to detect the existence of carrier over the channel. If the channel is busy (a carrier is sensed), the ready station defers transmission. If no station's carrier is sensed, the channel is free and the ready stations may initiate transmission. Stations that are eligible to compete for the channel are called contending stations. In CSMA/CD, all ready stations are contending stations.

A slot time is defined as the channel round-trip propagation time plus some carrier sensing time. Once a transmission begins, it has to propagate over the entire channel so that all other stations can sense its carrier and defer their own transmissions. Once past the slot time, the transmission continues to completion. On the other hand, if another transmission begins during the slot time - it not yet having sensed the first transmission's carrier - the two transmissions are said to collide and the protocol enters contention state. A busy channel thus can be further classified into either success or collision states. While transmitting, the station listens for a collision. If one is detected, all transmitting stations cease transmitting and each waits for a random period of time, based on the truncated binary exponential backoff algorithm. The backoff periods are integral multiples of the slot time [4].

As their backoff periods expire, the stations will try to compete for the channel again. In the contention period, the channel is slotted and each slot is either idle or collision depending on the lengths of different backoff periods. The contention period ends when a success slot is detected. A success slot indicates that a station has successfully seized the channel and marks the beginning of a successful transmission. In summary, the CSMA/CD protocol has three states: idle, contention, and transmission [10].

In CSMA/CD protocols all messages are treated equally in competing for the single channel. As the channel throughput approaches the channel capacity, the average packet delay will be rapidly increased. Thus, some important or time-critical messages may be severely delayed. A prioritized CSMA/CD (PCSMA/CD) protocol incorporates priority functions in the CSMA/CD protocol. The central scene in implementing a message-based priority function is the strategy used in the assessment of the current highest priority class among all ready stations. An extra state, priority assessment state, is added to determine the current highest priority class that can be transmitted. In this study, we assume that there are N distinct priority classes and M stations, where class-N is the highest priority class and class-1 is the lowest priority class.

Each station in the network maintains a waiting queue in which all arriving packets are ordered according to their priority classes. If more than two packets have the same class, they are ordered according to their arrival times. During the priority assessment period (PAP), the station priority of the m-th station (m=1,2,...,M) is defined to be the priority class of the packet in the front of the waiting queue and is denoted by p(m), where p(m) is in the range of 0 and N. p(m)=0 indicates that the m-th station is an idle station.

The channel priority, k, is defined as the current highest priority class determined in the PAP. Thus, k=max{p(i) for i=1,2,...,M}. The channel priority may be changed in each PAP. Once the channel priority is determined in the PAP, all packets residing in the waiting queue of each station with the same priority as the channel priority are registered and are referred to as registered packets. The m-th station is a contending station if p(m)=k. In other words, a contending station has at least one registered packet in its waiting queue.

The PCSMA/CD protocols can be characterized by the state transition diagram depicted in Figure 1. The communication channel has four states (periods): idle, contention, priority assessment, and transmission. When all stations are quiet, the channel is in the idle state. Any packet arrival will cause the channel to change state from idle to contention as that of the CSMA/CD. Collision and idle may occur alternatively and repeatedly in the contention

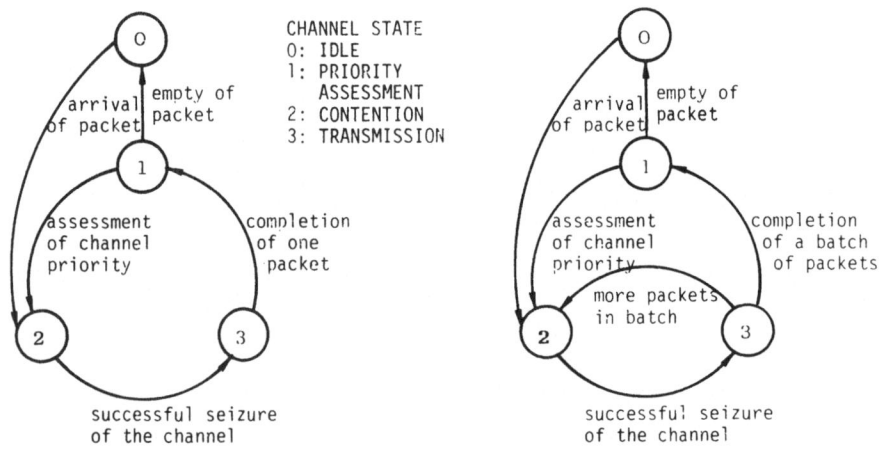

Figure 1. The state transition diagram of prioritized CSMA/CD (PCSMA/CD) protocols.

state. Once a station acquires the channel, the channel enters the transmission state and that station will transmit one packet.

Two operation modes can be considered in the transmission state. In the single mode, the channel will enter the assessment state after successfully having transmitted one packet. In the batch mode, all registered packets must be transmitted before entering the assessment state. Thus, the channel will enter the contention state again if there are more registered packets.

In the PAP, the channel priority is determined and all registered packets and contending stations are identified. Only contending stations are eligible to compete for the channel in the contention state. In the PAP, if all stations are idle, that is, the channel priority k is zero, the system will enter the idle state. Note that the assessment period is triggered at the end of a transmission. If all stations are perfectly synchronized, the channel may repeat the assessment period until k>0. However, this is impractical. Thus, once entering the idle state, a pure CSMA/CD, regardless of the priority, is used to make the channel eventually entering a transmission state as shown in Figure 1.

Each station in the network is able to monitor the channel and to detect three kinds of events over the channel [12]: (1) "idle" - no transmission; (2) "success" - successful transmission; and (3) "collision" - two or more transmissions. Three procedures are defined in this paper. Procedure Start-Transmit(j) indicates the starting of transmission at the beginning of the j-th slot in the PAP. Procedure Make-Reservation(j) sends a "1" (a short burst) in the j-th slot of the PAP. Procedure Detect(Event(j)) detects the channel event at the end of the j-th slot.

Two PCSMA/CD protocols were proposed in the past and have shown that priority function can reduce packet delay significantly for those urgent packets. The major difference between these schemes and our proposed scheme lies in the strategy used during the PAP. Iida et al. use various lengths of preamble signals to indicate different priority levels of messages [3]. The preamble length of the i-th priority class is linearly proportional to the priority level i. Thus, the overhead caused by implementing such a preamble priority scheme is increased linearly as the number of priority classes increases. However, in case of a collision between two equal priority messages, these are rescheduled into the future, resulting in an operation which violates the requirement of hierarchical independence of performance [11].

The second PCSMA/CD protocol based on the priority reservation scheme was proposed by Rom and Tobagi [9,11]. In this scheme, each station has four states: idle, ready, contending, and abort. Furthermore, the assessment period is slotted. Let Station(m) denote the current state of the m-th station. During the PAP, a ready station, say the m-th station, will perform the priority reservation algorithm shown below.

Priority Reservation Algorithm:

```
Station(m) := Ready; (* p(m)>0 *)
j := 1; (* the first slot *)
while j < (N-p(m)+1) and Station(m)=Ready do
  {Detect(Event(j));
   if Event(j)=idle then j := j+1 else Station(m) := Abort}
if station(m)=Ready then {Station(m) := Contending;
                          Start-Transmit(N-p(m)+1)}
```

In this scheme, each priority class is assigned with a fixed reservation slot. The lower the priority class is, the later the slot will be gained. A station is aborted if its priority is lower than the channel priority. The aborted station will become ready in the next PAP. The length of the PAP is in the range of [0,N-1] slots. If only one station starts to transmit at the (N-p(m)+1)-th slot, a successful transmission is guaranteed; otherwise, collision will occur and these contending stations must be rescheduled. Note that to make reservation at the (N-p(m)+1)-th slot instead of starting to transmit may waste one slot if there is only one contending station. An example based on the priority reservation scheme is shown in Figure 2, where K_i indicates the number of ready stations with the station priority being i. In this scheme, the overhead caused by reservation becomes significant as the number of distinct priority classes increases, particularly if low priority messages are major components of the traffic in the network.

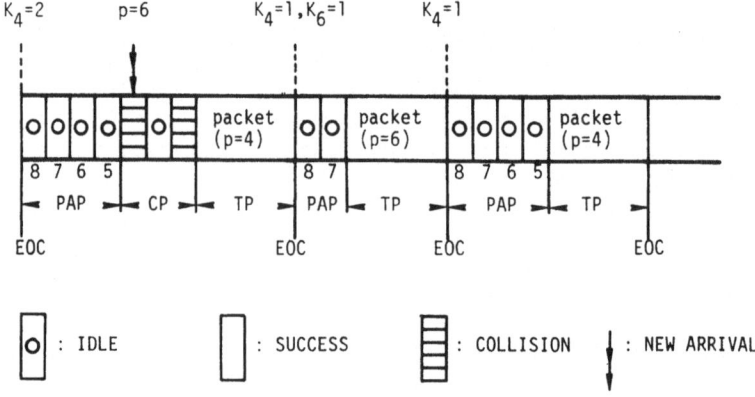

Figure 2. An example of the priority reservation scheme.

IMPLEMENTING PRIORITY FUNCTIONS

THE PRIORITY-CODE COMPARISON SCHEME

A priority-code (p-code) comparison scheme is proposed. In this scheme, the station priority is represented as a binary code $p(m)=p_{n-1}\cdots p_0$, where $n = \lceil \log_2(N+1) \rceil$. Note that $p(m)=0$ is reserved for indicating an idle station. During the assessment period, a ready station, say the m-th station, will perform the following algorithm.

Priority-Code Comparison Algorithm:

```
Station(m) := Ready;
if p(m)=N=2^(n-1) then {Station(m) := Contending;
                        Start-Transmit(1)};
if p(m)<>N then {Detect(Event(1));
                 if Event(1) <> Idle
                    then Station(m) := Abort}
j := 2;
while j <= n-1 and station(m)=Ready do
  if p_(n-j)=0 then
      {Detect(Event(j));
       if Event(j)<>Idle then Station(m) := Abort
       else j := j+1}
  else (* p_(n-j)=1 *)
     {Make-Reservation(j);
      Detect(Event(j));
      if Event(j)=Success then
          {Station(m) := Contending; Start-Transmit(j+1)}
      else j := j+1}
if Station(m)=Ready then
  {if p_0=0 then {Detect(Event(n));
                  if Event(n)=Idle then
                     {Station(m) := Contending;
                      Start-Transmit(n+1)}
                  else Station(m) := Abort}
   else {Station(m) := Contending; Start-Transmit(n)}}
```

During the PAP, each slot will complete the comparison of one binary bit starting from the most significant bit of the p-codes of all ready stations. The basic idea behind this scheme is that during the PAP, only higher priority classes will survive and lower priority classes will be aborted. For a given n, N can be in the range of $[2^{n-1}, 2^n-1]$. If $N=2^{(n-1)}$, the highest priority is 100...000. Thus, the most significant bit, $p_{n-1}=1$, is sufficient to identify the highest priority class. In this case, the algorithm further favors the urgent packets by starting their transmission at the first slot of the PAP.

In general, the number of slots required in the PAP depends on the priority class and the event occurring in each slot. In the worst case, n slots are needed. In the best case, no slot is needed (a success in the first slot of PAP wastes no channel bandwidth). In the following discussions and simulation experiments, 4 bits of priority-code (n=4) and 8 (N=8) different priority classes are assumed unless otherwise specified.

After the PAP, contending stations will enter the contention period. Two scheduling disciplines are considered: nonpreemptive discipline and preemptive discipline. In each discipline, both single mode and batch mode are studied. First, the nonpreemptive discipline is described in which the current transmitting packet cannot be preempted by newly generated higher priority packets.

Nonpreemptive Discipline

Nonpreemptive Single Mode. In this mode when the transmission is over, the end of carrier (EOC) marks the time epoch of the next PAP. Figure 3 gives an example of the channel activity in this mode. Let K_i represent the number of stations with priority class i. For ease of explanation, each station is assumed to have only one packet with the same priority as the station priority. At the beginning, two stations are ready to transmit and the priority for both of them is 4. Four slots are needed in the PAP in which three of them are idle and the second slot has a collision. In the contention period, one of them will seize the channel. The channel then enters the transmission period.

Suppose that in the contention period, a station has a newly generated packet with priority level 6 as marked in Figure 3. Based on the nonpreemptive discipline, this station with a newly generated higher priority packet will keep silent. At the end of the successful transmission, the EOC signals another PAP. Now one station has priority 4 and another station has priority 6. In the PAP, only three slots are necessary. The state of the first slot is idle because both $p_3=0$. The state of the second slot is collision because both $p_2=1$. The state of the third slot is success

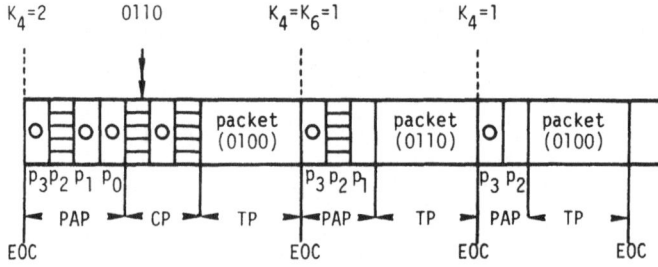

Figure 3. An example of the nonpreemptive single mode PCSMA/CD.

because only the station with priority 6 puts a "1" in that slot. Thus, the station with priority 4 is aborted and the PAP is truncated. In this case, the contention period can be skipped and the packet with priority 6 can be successfully transmitted. At the third EOC, only one station is ready to transmit with priority 4. In this case, two slots are needed and a packet of class 4 is finally transmitted.

Nonpreemptive Batch Mode. In the batch mode, all registered packets must be transmitted before the channel enters the PAP. For this purpose, one continuation slot (c-slot) must immediately follow every successfully transmitted packet. Every station has to put a "1", Make-Reservation(c-slot), in the c-slot if it has more registered packets to transmit. If Event(c-slot) is not idle, the channel state returns to the contention period from the transmission period; otherwise, an empty c-slot marks a time epoch, end of batch (EOB), which indicates the beginning of a new PAP. Figures 4 and 5 demonstrate the flowchart description and an example of the nonpreemptive batch mode.

The concept of registered packets can partially improve the fairness because registered packets have arrived at a past time interval. New arrivals will not be registered until the next PAP. In this sense, partial fairness can be achieved according to packet arrival time intervals.

Preemptive Discipline

The major disadvantage of nonpreemptive discipline is that it cannot satisfy the first design requirement: hierarchical independence of performance. In other words, under nonpreemptive discipline, increasing loads from lower priority classes may degrade the performance of higher priority classes. This is true especially in the nonpreemptive batch mode, in which a higher priority packet arriving after the PAP has to wait until the whole batch of lower

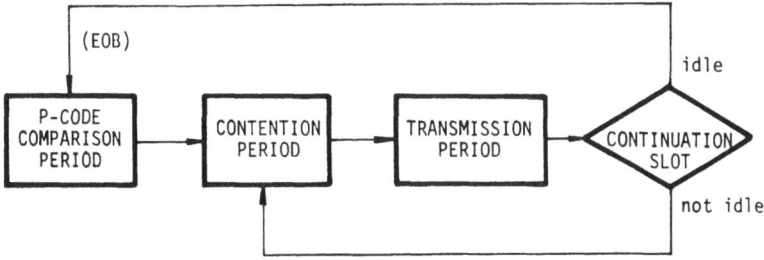

Figure 4. Flowchart description of the nonpreemptive batch mode PCSMA/CD.

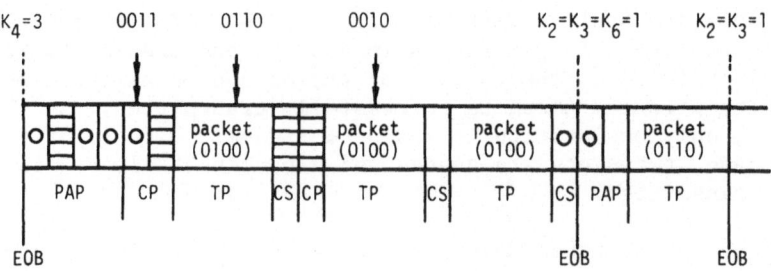

Figure 5. An example of the nonpreemptive batch mode PCSMA/CD.

priority packets have completed their transmissions. A station is referred to as a preempting station if it has a newly generated packet with a higher priority than the channel priority.

In Tobagi's scheme [11], a newly generated packet may preempt an ongoing transmission, i.e., in a transmission period of a lower priority level. In this case, the channel utilization is further decreased due to wasted transmission. Although this scheme may be useful in the military environment, preemption of ongoing transmissions will not be considered in this study.

<u>Preemptive Single Mode</u>. Since preemption is not allowed in the transmission period, preemption can only occur at idle times of the contention period. The preempting station will keep on sensing the channel and will start to transmit if the channel is sensed idle. If a collision is detected, the preempting station involved in the collision will wait until the next PAP (one try). The collision may be caused by preempting stations or by contending stations or a mix of them. However, there is no way to distinguish these types of collisions. The preempting stations can try only once; otherwise, different higher levels of preempting stations may contend for the channel without knowing the priority of each other. In the latter case, the channel will become a CSMA/CD protocol among those contending and preempting stations. Allowing preemption in the contention period can partially improve the channel utilization by decreasing the idle time due to rescheduling delay. An example of the preemptive single mode is illustrated in Figure 6.

<u>Preemptive Batch Mode</u>. In the batch mode, preemption is allowed in the contention period as it is in the preemptive single mode. Furthermore, a preemption slot (p-slot) is added between the transmission period and the continuation slot. A preempting station will put a "1" in the next p-slot, Make-Reservation(p-slot). Three events may occur in the p-slot: (a) If Event(p-slot) is "success," it means that only a single preempting station made a reservation.

IMPLEMENTING PRIORITY FUNCTIONS

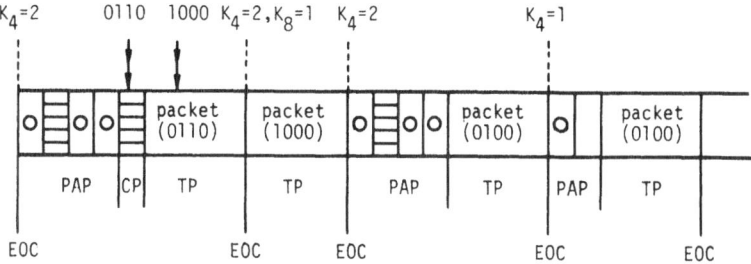

Figure 6. An example of the preemptive single mode.

The preempting station may transmit the packet immediately and the channel will be in the transmission period. At the end of the transmission, another p-slot is reserved as usual; (b) If Event(p-slot) is "collision," it means that more than one preempting station attempts to transmit. The channel changes its state to PAP immediately; (c) If Event(p-slot) is "idle," the next slot will be the continuation slot as in the normal case.

If a preempting station detects a collision in the contention period during its first try, that station will wait for the next preemption slot to make a reservation. A flowchart description and an example of the preemptive batch mode are illustrated in Figures 7 and 8, respectively.

SIMULATION RESULTS AND PERFORMANCE ANALYSIS

A simulator has been developed to verify the network performance. Details of the simulation experiments can be found in [6]. The simulator is able to measure the performance of the PCSMA/CD protocol for different operating modes and different workload distributions. The priority reservation scheme is also included for comparison purpose. Various performance measures are gathered in the course of simulation. Main performance data include: 1) the average packet delay and variance of different priority classes, $D[i]$ and $V[i]$; 2) the average packet delay and the variance of all packets, D and V; 3) the throughput of each priority class, $S[i]$; and 4) the total channel throughput, S.

A series of simulation experiments have been conducted. The experiments are based on the following workload distributions and system parameters: 1) the packet generation is a Poisson process for all stations with the same arrival rate; 2) the packets are uniformly distributed among different priority classes; 3) the number of stations is 5; 4) the channel bandwidth is 1 Mbps; 5) the end-to-end propagation delay, T, is 5 µsec; and 6) the packet length is fixed with 500 bits or $L=100T$.

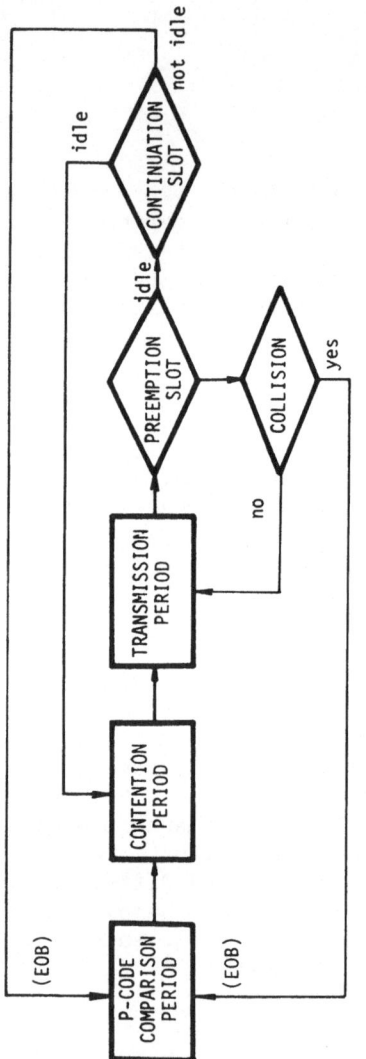

Figure 7. Flowchart description of the preemptive batch mode PCSMA/CD.

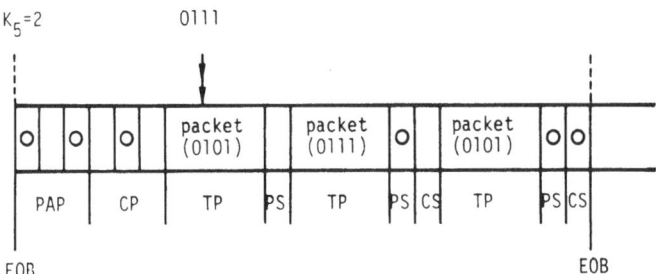

Figure 8. An example of the preemptive batch mode.

Throughput-Delay Characteristics

The throughput-delay characteristics, D[i] vs. S, of the nonpreemptive single mode protocol is shown in Figure 9 for 8 different priority classes. The higher the priority is, the smaller the delay will be as was expected. The delay versus the total system throughput, D vs. S, is also indicated by the dash line. As the system throughput approaches the channel capacity, the packet with urgent information has to suffer an intolerable delay in a nonprioritized CSMA/CD protocol as revealed by the dash line. However, with the prioritized CSMA/CD protocol, those important packets with a higher priority can still be transmitted with much less delay in such a heavy-load situation. Obviously, the gain is paid by increasing the delay of lower priority packets.

Effect of Different Priority Assessment Schemes

The priority reservation scheme is compared with the proposed p-code comparison scheme by observing their throughput-delay characteristics. Figure 10 shows the throughput-delay curves with respect to two different priority assessment schemes for N=8, where both of them assume a nonpreemptive single mode operation. The following two conclusions are made: 1) the p-code comparison scheme provides a lower overhead than that of the priority reservation scheme and 2) the larger the number of priority classes is, the better the advantage will be.

The difference between these two assessment schemes is mainly due to the different number of slots required to assess the highest priority class. Let S_{pc} and S_{pr} represent, respectively, the number of slots required for p-code comparison scheme and priority reservation scheme. In the worst case, we have

$$S_{pc} = n = \lceil \log_2(N+1) \rceil \quad \text{for all classes}$$

$$S_{pr} = N-i \quad \text{for } i=1,2,\ldots,N$$

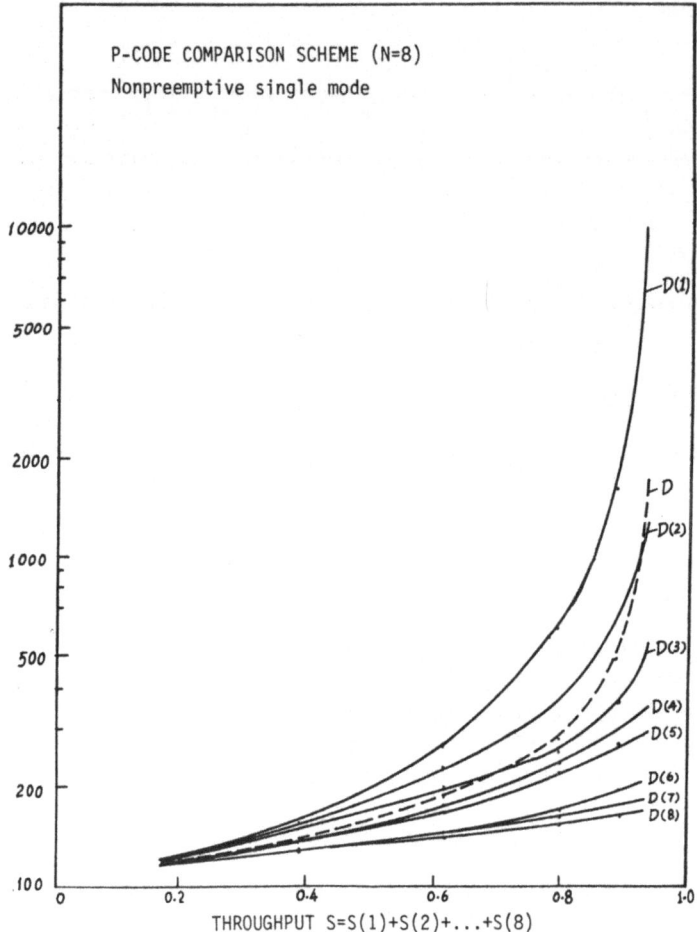

Figure 9. Throughput-delay characteristics for the nonpreemptive single mode PCSMA/CD.

For lower priority classes in a system with a large number of priority classes, i.e., $i \ll N$, we have

$$S_{pc} / S_{pr} = \lceil \log_2 N \rceil / N \qquad \text{for } i \ll N$$

In this case, the p-code comparison scheme takes a lower overhead. More precisely, the p-code comparison scheme is better if the priority class i is within the following range:

$$1 \leq i \leq N - \lceil \log_2(N+1) \rceil$$

If i is not within the above range, the priority reservation scheme may perform better. However, this drawback can be alleviated

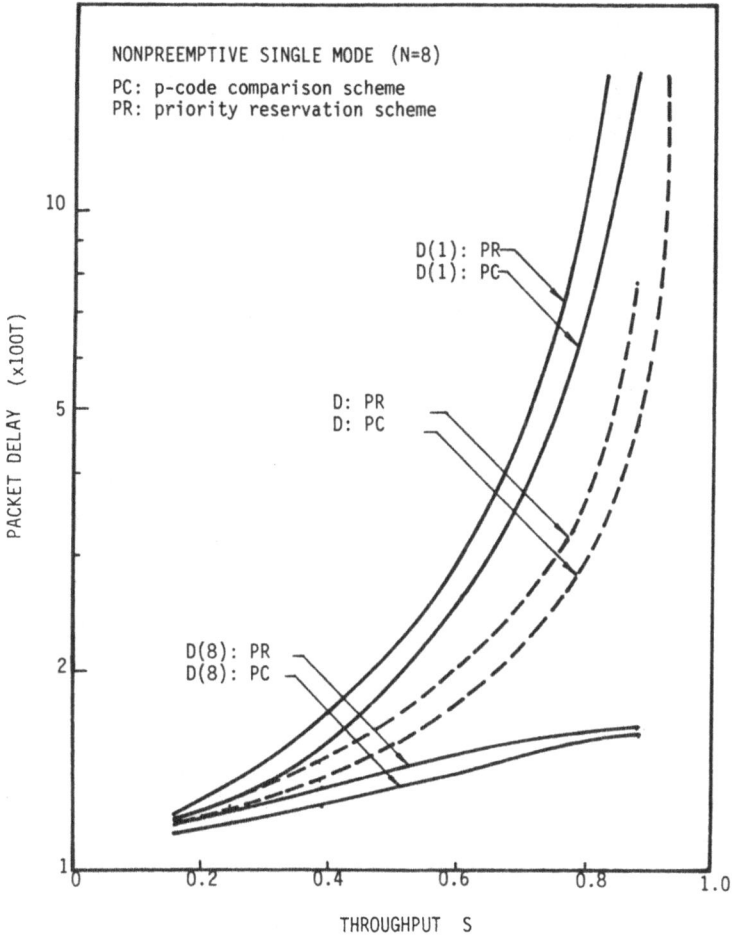

Figure 10. Sensitivity of packet delay to two priority assessment methods in nonpreemptive single mode PCSMA/CD with N=8.

by the truncation capability provided in the p-code comparison scheme. In the priority reservation scheme, the overhead for each priority class is fixed. But in the p-code comparison scheme, the overhead may vary depending on the event in the comparison slot. The number of comparison slots may be truncated if a success event is detected. The truncation capability favors, particularly, higher priority classes.

Effect of Preemption on the Performance

The drawback of nonpreemptive discipline is that higher priority packets may be blocked by current transmissions of lower priority packets. In the batch mode, the interval between two consecutive p-code comparison periods is even larger. Thus, the probability of the number of newly generated higher priority packets is greater, especially at the heavy-load condition. This implies that the higher priority packets have to wait a longer time because of the transmission of lower priority packets, which violates the design requirements of hierarchical independence.

Figure 11 shows the effect of preemption on the performance for the cases of single mode and batch mode, respectively. The throughput-delay curve of two extreme priority classes, D[1] and D[8], is displayed. The objective of preemption is to prevent

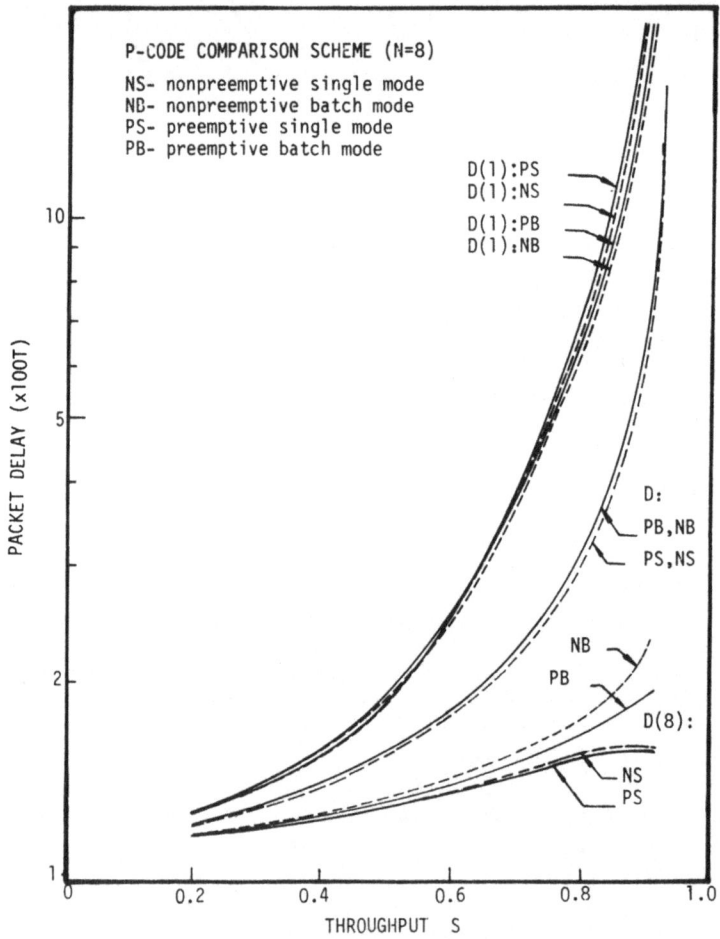

Figure 11. Throughput-delay characteristics for nonpreemptive single mode, nonpreemptive batch mode, preemptive single mode, and preemptive batch mode PCSMA/CD with N=8.

lower priority packets from blocking higher priority packets. In both operation modes, the delay of the higher priority classes is further decreased at the expense of increasing the delay of lower priority classes. In the batch mode, performance improvement is more significant than in single mode. This is because an extra preemption slot reserved at the end of each successful transmission allows another preemption opportunity. The average channel utilization is also improved by adding the capability of preemption because the channel idle time due to rescheduling delay is partially eliminated by higher priority classes. Note that the extra preemption slot may be eliminated if the channel priority equals the highest priority class, N.

Figure 12 displays the relationship between the variance of the delay to the system throughput in the nonpreemptive batch mode (NB), nonpreemptive single mode (NS), preemptive batch mode (PB), and preemptive single mode (PS) PCSMA/CD for the highest priority level. As revealed in Figure 12, the delay of the highest priority packets becomes increasingly worse for the batch mode when the offered load increases. The improvement of performance and fairness is shown by the decreasing of the variance of the delay. In general, the preemptive discipline provides a smaller variance. But the difference is insignificant in the single mode for the highest priority packets and will be explained in the next section. The variances of the delay for lower priority classes are very close in all four modes and are much greater than higher priority packets.

Comparison of Single Mode and Batch Mode

In general, the batch mode reduces the occurrence of comparison periods at the expense of providing one more continuation slot at the end of each transmission period. Let $c[i]$ represent the average number of comparison slots required for the priority class i, where $0 \leq c \leq n$ and $1 \leq i \leq N$. In the preemptive single mode, the number of overhead slots is $b[i]c[i]$ if $b[i]$ packets are transmitted. In the preemptive batch mode, the number of overhead slots will be $2b[i]+c[i]$ because one continuation slot and one preemption slot are added. The success of the preemptive batch mode relies on $b[i]$ and $c[i]$ being large enough such that

$$b[i]c[i] > (2b[i]+c[i]) \qquad i=1,2,\ldots,N$$

For lower priority classes, c is usually close to n and the batch size b is usually greater than 1 at heavy load. The reason for the latter case is the blocking of lower priority classes; thus, more lower priority packets are waiting in the queue. For a uniform distribution of priority classes, we have $b[i] \geq b[i+1]$. Thus, the batch mode provides a better performance for the lower priority classes as shown in Figure 11 at heavy-load condition. In the light-load condition, the batch size b is very likely to be

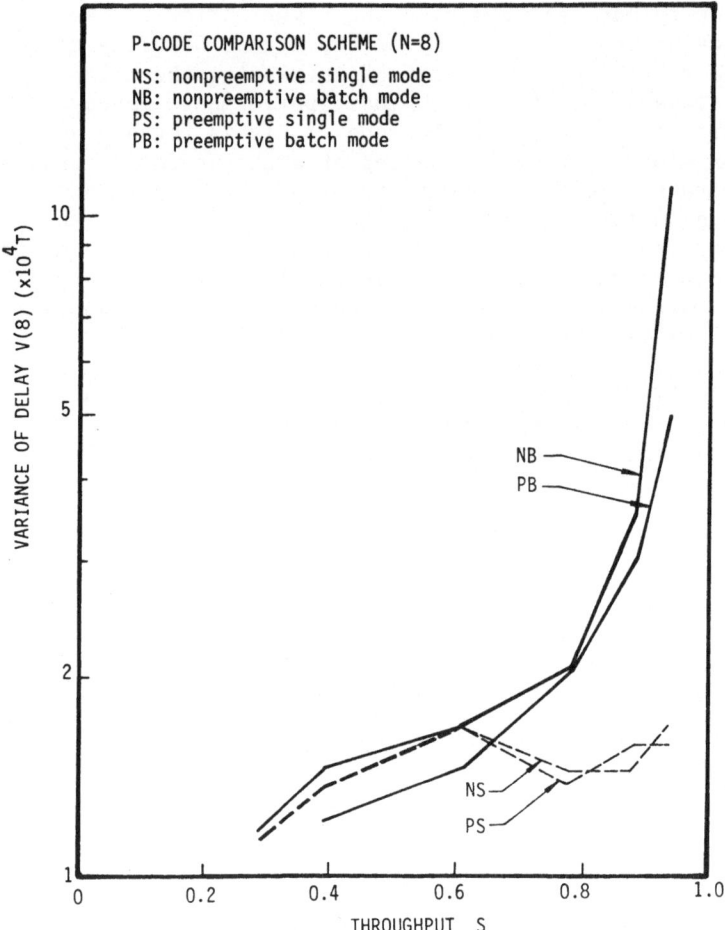

Figure 12. Variance of packet delay, V[8], for nonpreemptive single mode, nonpreemptive batch mode, preemptive single mode, and preemptive batch mode PCSMA/CD with N=8.

1. In this case, the single mode is always better than the batch mode. However, in the light-load condition, the channel is idle for most of the time. The performance difference, thus, is insignificant as shown in Figure 11, which is based on the uniform distribution of priority classes. If most of the packets are lower priority classes, such as the geometric distribution, the batch mode will show a better improvement.

For the highest priority class, c[N]=0, the above criterion is never satisfied. For other higher priority classes, c is somewhere between 1 and n depending on the offered load. The simulation results show that the delay of priority classes greater than 3 in batch mode is slightly longer than in single mode. This is caused by a smaller value of b for these higher priority classes.

Another interesting property is that the delay of the highest priority class will decrease a little bit when the channel approaches the maximum throughput as shown in Figure 11 under the single mode operation. At such very heavy-load condition, the channel is totally dominated by the highest priority packets. The situation in which the highest priority packet is waiting for the completion of a lower priority packet will never occur because the lower priority packets have no chance to get the channel. However, this fact does not appear in the batch mode. In preemptive batch mode, a higher priority packet may be transmitted immediately if it is the only one made reservation in the preemption slot. When the channel throughput further increases, the probability that more than one station wants to make reservation in the preemption slot is high. In this case, the channel has to enter the comparison period and, of course, the delay will be increased for those higher priority packets.

The advantage of the batch mode is its higher degree of fairness. In CSMA/CD protocols, a new arrival may be transmitted earlier than an old packet. Furthermore, it is possible to have an indefinite wait during a heavy load if a packet fails to get the channel in every try, although the probability of this situation is extremely slight. The same situation may also occur in the single mode. However, in the batch mode only registered packets can be transmitted. Even the new arrival which has the same priority as the channel priority cannot be registered until the next comparison period. In this sense, an indefinite wait within the same priority class can be eliminated. An indefinite wait by lower priority classes of packets is still possible if higher priority classes of packets keep on holding the channel.

CONCLUSION

This paper is concerned with the problem of prioritizing packet transmission in carrier sense multiaccess local communication systems. A message-based priority function is implemented by adding a p-code comparison period to the original CSMA/CD protocol. Different design alternatives including single mode, batch mode, preemptive discipline, and nonpreemptive discipline have been carefully investigated and compared. A preemptive single mode PCSMA/CD protocol displays its superiority above other protocols.

The additional overhead caused by the introduction of the priority scheme was discussed. The overhead is also shown to be much less compared to other schemes. In addition to low overhead, the proposed prioritized CSMA/CD protocol also satisfies the requirements of fairness, hierarchical independence of performance, and robustness.

ACKNOWLEDGMENT

This research was supported in part by the U.S. National Science Foundation under grant ECS-83-04967.

REFERENCES

[1] D. W. Andrews and G. D. Schultz, "A Token-Ring Architecture for Local Area Networks: An Update," COMPCON Fall 82, pp. 615-624, September 1982.
[2] W. R. Franta and M. B. Bilodeau, "Analysis of a prioritized CSMA protocol based on staggered delays," ACTA Information 13, pp. 299-329, 1980.
[3] I. Iida, M. Ishizuka, Y. Yasuda, and M. Onoe, "Random access packet switched local computer network with priority function," Proc. Nat. Telecommun. Conf., pp. 37.4.1-37.4.6, December 1980.
[4] IEEE Project 802 Local Area Network Standards, Draft IEEE Standard 802.3: CSMA/CD Access Method and Physical Layer Specifications, Draft D, IEEE Computer Society, December 1982.
[5] IEEE Project 802 Local Area Network Standards, Draft IEEE Standard 802.4: Token-Passing Bus Access Method and Physical Layer Specifications, Draft E, IEEE Computer Society, July 1983.
[6] X. Li and L. M. Ni, "A simulation model of prioritized CSMA/CD protocols," Proc. of the IASTED Int'l Symp., Applied Simulation and Modelling (ASM'83), pp. 55-59, May 1983.
[7] L. M. Ni, "A distributed load balancing algorithm for point-to-point local computer networks," Proc. of the COMPCON 82 Fall, pp. 116-123, September 1982.
[8] L. M. Ni and X. Li, "Prioritizing packet transmission in local multiaccess networks," Proc. of the 8th Data Communications Symposium, pp. 234-244, October 1983.
[9] R. Rom and F. A. Tobagi, "Message-based priority functions in local multiaccess communication systems," Computer Networks, pp. 273-286, 1981.
[10] A. S. Tanenbaum, Computer Networks, Prentice-Hall, Inc., 1981.
[11] F. A. Tobagi, "Carrier sense multiple access with message-based priority functions," IEEE Trans. on Communications, Vol. COM-30, pp. 185-200, January 1982.
[12] J. F. Shoch, Y. K. Dalal, D. D. Redal, and R. C. Crane, "Evolution of the Ethernet local computer network," Computer, pp. 10-27, August 1982.

[13] B. W. Wah and J. Y. Juang, "An efficient protocol for load balancing on CSMA/CD networks," Proc. of the 8th Conf. on Local Computer Networks, October 1983.

THE DESIGN AND IMPLEMENTATION OF A DISTRIBUTED DATABASE SYSTEM

Jyh-Sheng Ke, Ching-Liang Lin, Hsing-Lung Chen, Chiou-Feng Wang, Chia-Hsiang Chang, Yow-An Pan, Chien-Chun Lu and Shyi-Ting Kang

Institute of Information Science
Academia Sinica, R.O.C.

ABSTRACT

This paper describes what has been learnt from the experience of implementing a local area network and a distributed data base system. Solutions for those typical issues arising from the distribution of data and processing power, including the connection of different operating systems, distributed query processing, and concurrent control, are presented.

INTRODUCTION

Institute of Information Science, Academia Sinica, is a newly founded research organization. At this institute, we have three minicomputers, one PDP-11/70, two PDP-11/23's, and a number of microlevel computers. In September 1980, we initiated the project DDBS to implement a distributed database on top of these mini- and microcomputers. Figure 1 illustrates the network architecture of the installed system. The operating system installed in the PDP-11/70 is RSX-11/M, and the operating system installed in the two PDP-11/23's is UNIX. The project has been scheduled into three progressive stages. The first stage is to implement a relational data base management system using C language under UNIX. The second stage is to implement a local area network by connecting the three minicomputers and a Z80-based front end processor with relevant layered protocol structures. The third stage is to implement a distributed database system basing on the results of the earlier stages by tackling all issues arising from the distribution of data, including the problems of concurrency control and query processing.

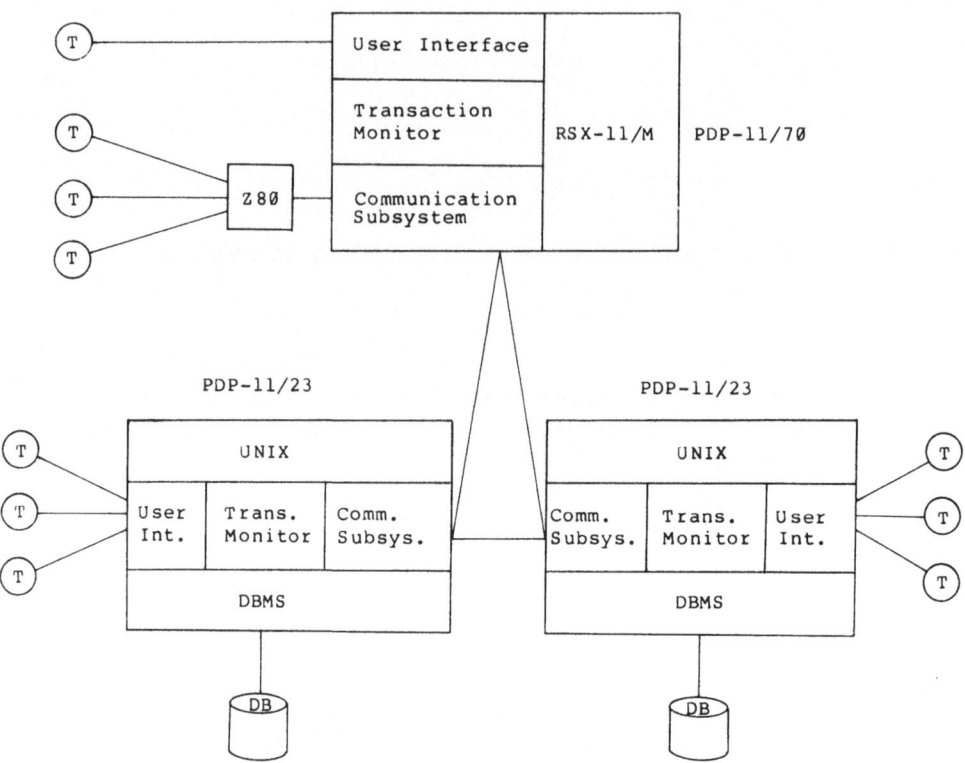

Figure 1. Local area network architecture.

This paper describes what we have learnt from the experience of converging the two diversified technologies, namely the computer network and the data base. The implementation is definitely not new research but a piece of bread and butter work designed to provide a distributed data base system for a variety of users in which the network connections and data distribution are made to be transparent to the end user.

CONNECTING UNIX AND RSX-11/M

As we mentioned before, the PDP-11/70 has been installed with RSX-11/M operating system, and the two PDP-11/23's have been installed with UNIX operating system. The physical connection among these three minicomputers are RS232C serial lines and will be replaced by a high-speed cable and its associated interface in the near future. The protocol hierarchy in the network is shown in Figure 2.

DISTRIBUTED DATABASE SYSTEM

DDCMP is a simulation program of DEC-standard link level protocol handler which ensures correct frame transmissions. NSP is responsible for the flow control message exhcange between two processes in different computers. FTP is responsible for the remote file transfer. RJE allows terminal users to use the resources of remote computer environments.

The unidirection 'pipeline' interprocess communication mechanism of the UNIX system is not suitable for implementing the protocol hierarchy of the computer network. We have enhanced the interprocess communication capability of the UNIX system with the message-based 'port' mechanism and some relevant system calls for synchronizing asynchronous events [9].

The main implementation problem when connecting RSX-11/M to a network is where to site the protocol handler. It can either be installed as part of the terminal I/O driver; or it can be run as a system task communicating with the network through some interface to the terminal I/O driver. Due to the complexity of the original RSX-11/M terminal I/O driver, it was decided that the protocol handler must be run as a background task and communicating with the network through an add-in virtual terminal driver as an interface to the terminal I/O driver. Figure 3 shows the terminal I/O driver function of the standard RSX-11/M system. Figure 4 shows the modified RSX-11/M in which three additional software modules have been incorporated to function as a virtual terminal protocol handler via which multiple terminals can be connected to the PDP-11/70 through a Z80-based concentrator. Essentially the integration of VTMON and VTDRV is the protocol handler at NSP level.

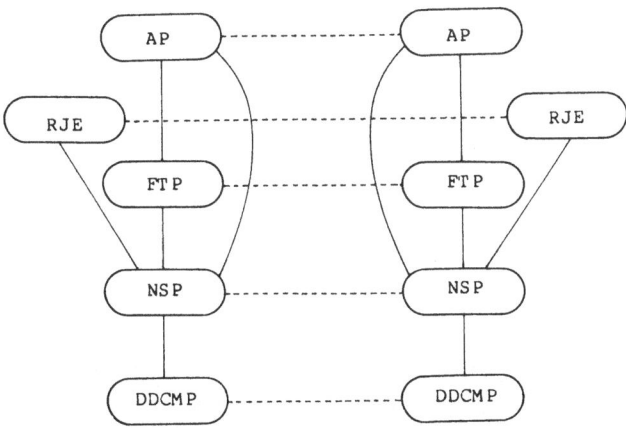

Figure 2. Network protocol hierarchy.

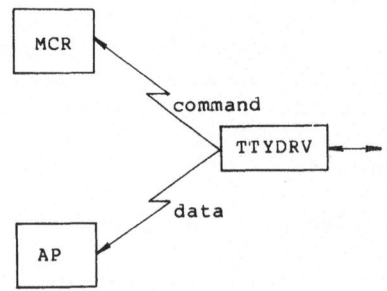

Figure 3. Standard RSX-11/M terminal I/O.

All I/O requests from the application tasks will be sent to the Virtual Terminal Driver, VTDRV, which in turn will queue an AST (Asynchronous System Trap) routine in the TCB (Task Control Block) of the Virtual Terminal Monitor task, VTMON, so that VTMON will reversely issue an O/I request to VTDRV and send a message to the DDCMP task upon the execution of the AST routine. The DDCMP task will then issue relevant physical I/O requests to TTYDRV accordingly. Figure 5 shows the relationships between the application program, high-level protocol (i.e. FTP and RJE), and the low-level protocol handlers. Detail of these protocol programs can be found in [7].

DISTRIBUTED RELATIONAL DATA BASES

There has been a lot of debate over the inefficiency of the relational model data bases. However, since our major concerns in selecting data model were in choosing a data model with high-degree data independency and strong mathematical foundation for query processing, we were convinced that the relational model is the only model which can meet our goal.

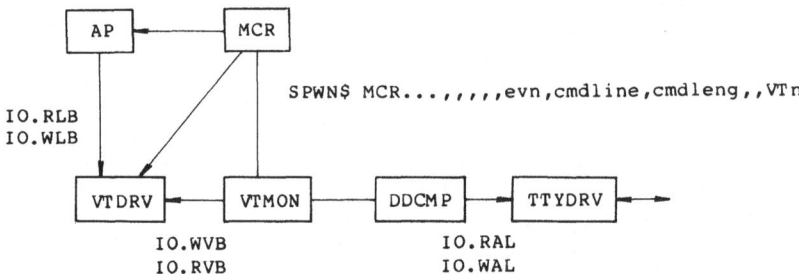

Figure 4. Modified RSX-11/M terminal I/O.

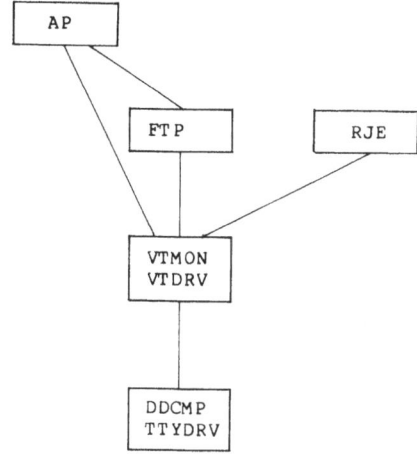

Figure 5. Protocol handlers under RSX-11/M.

Implementation of the DBMS has been strongly influenced by INGRES [3] and System R [1]. The DBMS includes two subsystems, namely Relational Interface System (RIS) and Physical Storage System (PSS) (see Figure 6). The RIS provides high-level, data-independent facilities for data definition, manipulation, and retrieval. The data definition facilities of the RIS allow the DBA to define the data at conceptual level as well as internal level. They also allow a variety of external views to be defined upon common underlying data to provide different users with different

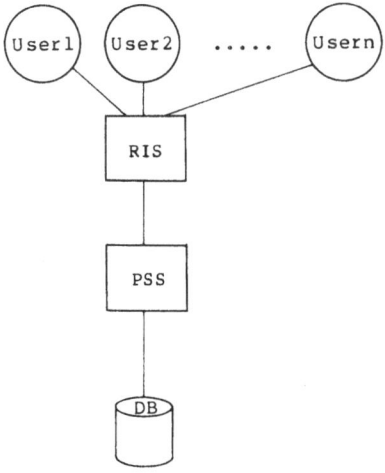

Figure 6. Architecture of the DBMS.

views of the data base. The external views of the common data also
facilitates some authorization checks and privacy control. One of
the important features of the DDL facility is that each attribute
in the relational database may be optionally associated with an
integrity control routine which will be triggered to execute upon
the corresponding update operation. The data manipulation facilities of the RIS allow data to be inserted, deleted, modified,
manipulated, and retrieved by using a set of high-level relational
operators which exempt the users from knowing the internal storage
structure of the stored data. Moreover, a high-level, nonprocedural query language has also been implemented to provide a
friendly interface for the novice users. Each query statement is
translated into a transaction constructed by a sequence of relational algebra operators with some heuristic optimization
techniques.

The PSS manages disk storage allocation, indexed file access.
B-tree structure has been used for indexing secondary keys of each
relation. The disk space contains a set of files each of which is
constructed by a set of fixed-size pages. For the purpose of saving
space and speeding up access, we employ the technique of mapping
logical space (what the relational operator sees) into physical
space (what the data is really stored). A page map table has been
maintained to map logical page number into physical page number.
This mapping technique can reduce the problems of hashing dispersed
records and is particularly useful in overflow control. The PSS
also supports concurrency control by using shared-read-lock and
exclusive-write-lock with lock-all-or-none strategy.

For the purpose of providing end users with a consistent data
base and making the user have an illusion that they are using a
large unified data base, a transaction monitor has been implemented
on top of the data base management system. The transaction monitor
receives transactions from the end users and serializes the DML
statements of multiple transactions to achieve high-degree
parallelism of executing data base operations.

Due to the address space limitation of PDP-11/23 (64K bytes),
the DBMS under UNIX has been structured into several related processes. The process structure of the DBMS is shown in Figure 7.
The function of each process is self-explanatory and will not be
recounted here. Routines for indexed access, integrity control,
and error recovery will be called by these processes at proper
time. More detail of the DBMS can be found in [6].

QUERY PROCESSING

As we pointed out before, one of the objects of the DDBS is to
make the distribution of data and processing power be transparent

DISTRIBUTED DATABASE SYSTEM

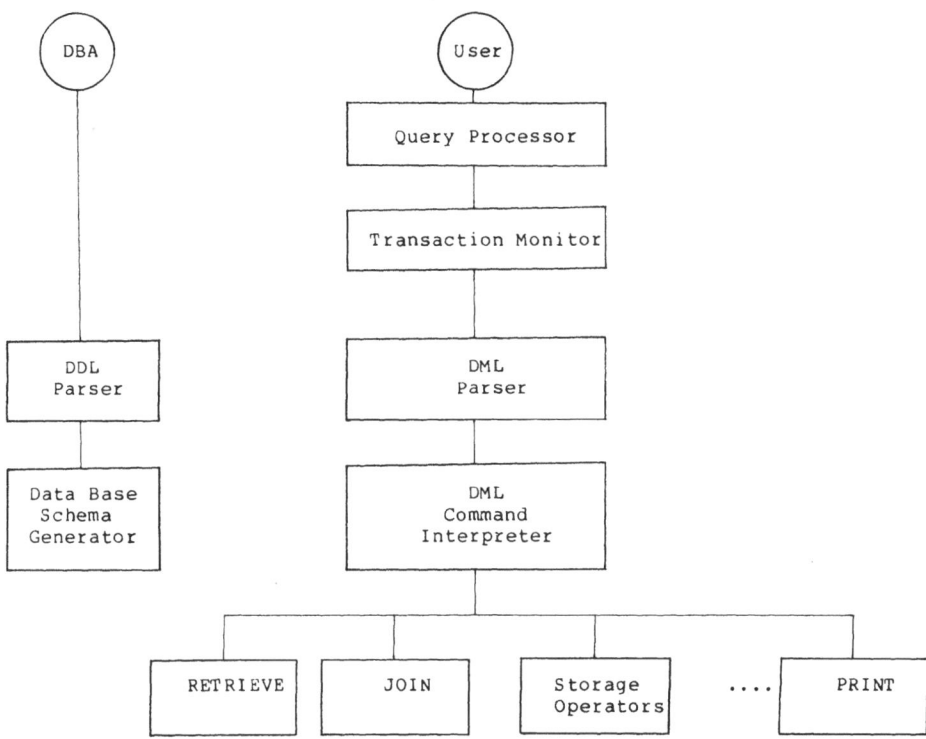

Figure 7. Process structure of DBMS.

to the user. In DDBS, we achieve this goal by providing the user with a high-level query language which in turn is supported by the 'VIEW' definition facility.

A VIEW is defined by the DBA and is a subschema of a set of related data items in the problem domain. To the user, each VIEW represents the relationships among those data items that the user is allowed to access. The set of data items defined in a VIEW may be distributed over several sites, however, the locations of stored sites and the data access paths are transparent to the user. In the following we will use an example to illustrate the technique of query processing in the DDBS.

Assume that three sites, A, B, C, are stored with data as described below:

Site A:
  BOOK(BID,BNAME,BAUTHOR,PUBLISHER,PUB DATE,CATEGORY,
       STATUS,L#,S#)

  JOURNAL(JID,JNAME,VOL,NO,DATE,JPUBLISHER,JCATEGORY,L#,S#)

```
              PROCEEDING(PID,PNAME,CONFERENCE,DATE,LOCATION,L#,S#)

         Site B:
           PAPER(PNO,TITLE,PAUTHOR,JPID,DATE,PAGE)

         Site C:
           SUPPLIER(S#,SNAME,SCITY,SPHONE)
           LIBRARY(L#,LCITY,ADDRESS,DIRECTOR,LPHONE)
```

Basing on this knowledge of data distribution, the DBA can define a VIEW, named VIEW_LIB, as follows:

```
         DEFINE
         FROM      BOOK.A,JOURNAL.A,PROCEEDING.A,PAPER.B,SUPPLIER.C,
                   LIBRARY.C
         INTO      VIEW_LIB(*)
         WHERE     BOOK.S#==SUPPLIER.S#
                   AND JOURNAL.S#==SUPPLIER.S#
                   AND PROCEEDING.S#==SUPPLIER.S#
                   AND BOOK.L#==LIBRARY.L#
                   AND JOURNAL.L#==LIBRARY.L#
                   AND PROCEEDING.L#==LIBRARY.L#
                   AND PAPER.JPID==JOURNAL.JID
                   AND PAPER.JPID==PROCEEDING.PID
```

In response to this VIEW definition a semantic graph (see Figure 8) will be generated and can be used as the foundation of query translation.

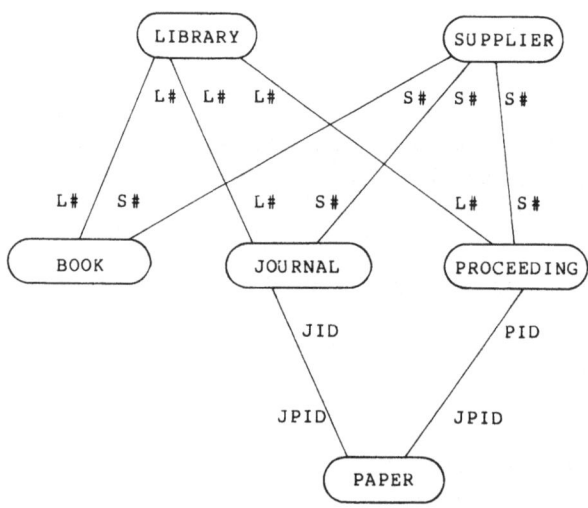

Figure 8.   Semantic graph.

DISTRIBUTED DATABASE SYSTEM

In this example, we have included all data items in VIEW_LIB. However, this is not always the case. The DBA may define a VIEW which contains only a subset of data items depending on the security degree of the user.

Now, suppose that a user at site C issues a query, to find the director of the library which owns a proceeding that contains the paper "On Modeling Relational Databases", by specifying the query as follows:

```
RETRIEVE
FROM VIEW_LIB INTO R(L#,DIRECTOR)
WHERE TITLE == 'On Modeling Relational Databases'
    AND PID==JPID
```

Basing on the semantic graph in Figure 8, the access path for this query can be found as shown in Figure 9.

From the access path shown in Figure 9, a transaction for retrieving data to answer the user's query can be generated as shown below:

```
BEGIN_TRAN
B: RETRIEVE FROM PAPER INTO R1(JPID) WHERE TITLE ==
   'On Modeling Relational Databases'
```

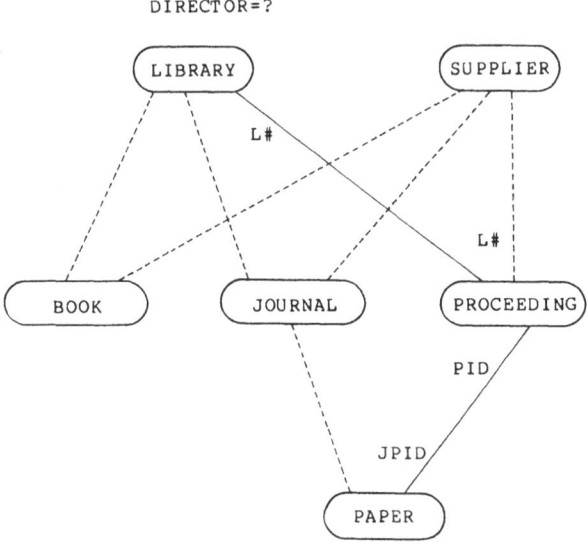

Figure 9. Access path.

```
A:  RETRIEVE
    FROM PROCEEDING INTO R2(PID,L#)
B:  MOVE B.R1 TO C.R1A:MOVE A.R2 TO C.R2
C:  JOIN (R1,R2) INTO R3 WHERE JPID=PID
C:  JOIN (R3,LIBRARY)
    INTO R(L#,DIRECTOR)
    WHERE R3.L#==LIBRARY.L#
END_TRAN
```

The ordering of DML statements has been optimized by first reducing the relations with restriction and projection operations. The reduced data will be moved to the query site, and join operations will then be performed if necessary. The technique of using semi-join [4] has not been employed in this implementation. However, we have developed a technique, called graph projection [5], which hopefully can minimize the overhead of semi-join operations. We plan to implement this technique in DDBS query processing in the near future.

CONCURRENCY CONTROL

In DDBS, a transaction is essentially constructed by a sequence of relational commands and is the basic unit of computation from the user's viewpoint. In this section we will propose a technique for handling concurrency control problem, with granularity at relation level, in our DDBS. This technique is similar to SDD-1's [2] approach of handling concurrency control problem. Figure 10 shows a model for processing transactions in the distributed database environment.

At each site, there is a Data Base Management System (DBMS), a Transaction Scheduler (TS), a Transaction Monitor (TM), a Communication Subsystem (COMM). Each user is assigned with a Query Processor (QP) to translate the user's query into a Transaction Command (TC) which in turn is constructed by a set of Relational Commands (RC). All transaction commands are sent to TM for being analyzed and then distributed to proper execution sites via COMM. The TM first puts a timestamp on each transaction command. This timestamp is generated by concatenating the system clock with the identification number of TM. TM also classifies every relational command into READ, WRITE, and EXECUTE operations. This classification can be shown below:

```
.join R1 and R2 into R3

      READ   R1
      READ   R2
      EXECUTE
    * WRITE  R3
```

DISTRIBUTED DATABASE SYSTEM

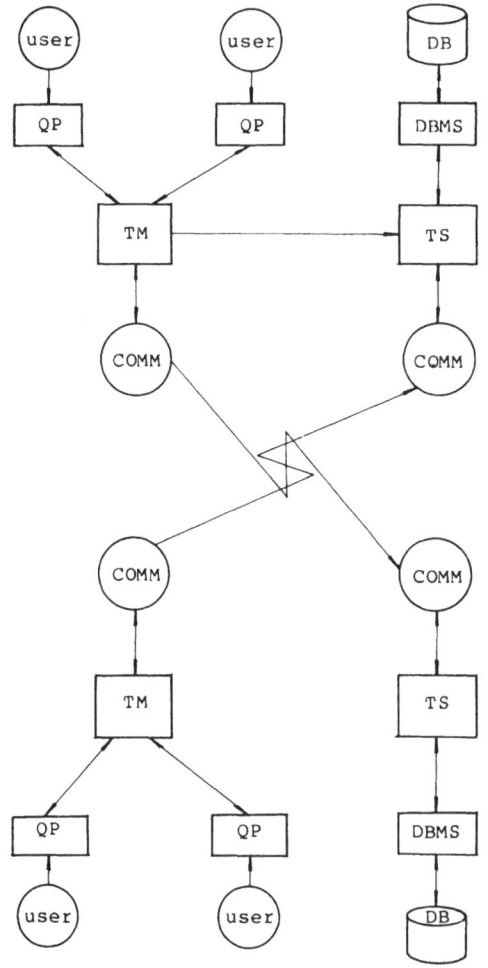

Figure 10. Transaction processing.

```
.retrieve from R1 into R2

      READ    R1
      EXECUTE
    * WRITE   R2

.move A.R1 to B.R2

      at A: READ    R1
    * at B: WRITE   R2
```

.modify R1

>   READ R1
>   EXECUTE
>   WRITE R1

Those statements preceded with a '*' are concerned with intermediate relational files and can be ignored when considering concurrency control problem. Other relational commands can be classified similarly.

TM collects all READ(WRITE) commands in a transaction into a READ-ALL(WRITE-ALL) command. It then submits these READ-ALL/WRITE-ALL commands to TS's by following some synchronization protocols. Figure 11 shows the synchronization protocol between TM and TS.

TS processes the received READ-ALL command by reading all the desired data into a workspace (one for each transaction per site). After finishing the READ-ALL command, an acknowledgement is sent to the TM which submitted the READ-ALL command. The EXECUTE commands are then processed on the workspace independently. When all EXECUTE commands of a transaction have been completed, TM submits the WRITE-ALL command to TS for updating with new data.

Every relational file has an associated timestamp. This timestamp is the timestamp of the last WRITE command that updated it. All relational files can only be updated by the WRITE command with a larger timestamp. All READ commands can only read the relational files with a smaller timestamp.

When TS receives commands from either the remote TM (via COMM) or the local TM, it will serialize the READ-ALL commands and WRITE-ALL commands and guarantees the effective total serial ordering of executing transactions. In the following we will use the example transaction of Section 3 to illustrate the process of serialization and synchronization.

For illustration, we add a DELETE command in the example transaction as below:

```
BEGIN_TRAN
B: RETRIEVE FROM PAPER INTO R1(JPID)
   WHERE TITLE==
     'On Modeling Relational Databases'
A: RETRIEVE FROM PROCEEDING
   INTO R2(PID,L#)
B: MOVE B.R1 TO C.R1
A: MOVE A.R2 TO C.R2
C: JOIN (R1,R2) INTO R3 WHERE JPID==PID
C: JOIN (R3,LIBRARY)
```

DISTRIBUTED DATABASE SYSTEM

TM:

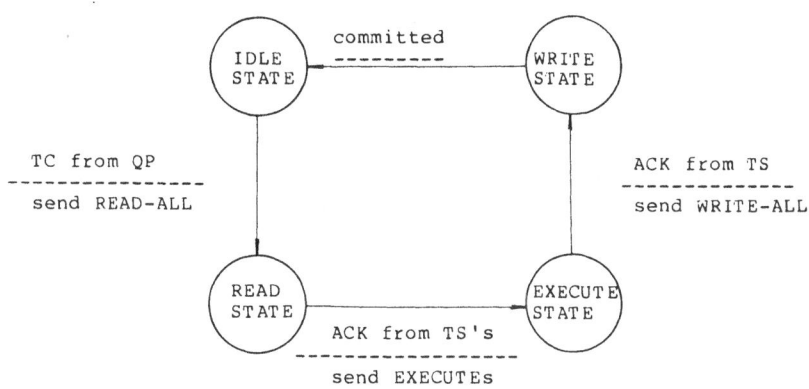

TS:

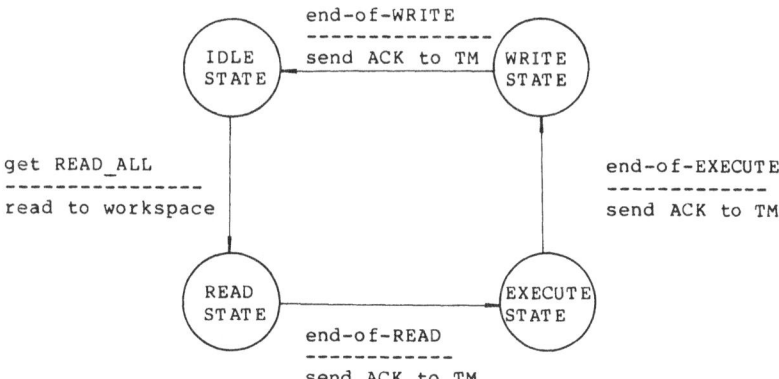

Figure 11. Synchronization protocol between TM and TS.

```
    INTO R(L#,DIRECTOR)
    WHERE R3.L#==LIBRARY.L#
C: DELETE FROM LIBRARY
    WHERE LIBRARY .L#==R.L#
END_TRAN
```

This transaction includes seven relational commands. The READ-ALL commands and WRITE-ALL commands that will be submitted to the execution sites are described below:

    To TS of Site A: READ-ALL(PROCEEDING)
                    WRITE-ALL(none)

```
    To TS of Site B: READ-ALL(PAPER)
                     WRITE-ALL(none)
    To TS of Site C: READ-ALL(LIBRARY)
                     WRITE-ALL(LIBRARY)
```

At READ-STATE, the three READ-ALL commands will be sent to A, B, and C, respectively. These READ-ALL commands will be serialized with other transactions by the respective TS receiving them. After all READ-ALL commands and EXECUTE commands have been processed, the TM will enter WRITE-STATE and send the WRITE-ALL commands to each TS. These WRITE-ALL commands will also be serialized by the respective TS receiving them.

CONCLUSION

This paper describes the design and implementation of a heterogeneous local area network and a relational data base management system. On top of the computer network and the DBMS, a distributed data base system has been built up.

A high-level nonprocedural query language facility has been implemented to facilitate the novice users. Using this query facility, the concept of access path is transparent to the user.

A concurrency control model has also been proposed to be implemented for our DDBS system. We understand that the proposed model may not be flexible enough for real-time applications. However, we believe that this model is very reliable and customizable. More flexible and efficient techniques are under study.

The current computer network consists of one PDP-11/70 with RSX-11/M and two PDP-11/23's with UNIX. Two VAX's are planned to join the network in the near future. The highly portable communication software will make the required effort of this extension become trivial. To construct a computer network under the existing operating systems (i.e. UNIX and RSX-11/M), several modifications of the operating systems have been made to enhance the capability of interprocess communication.

A library application system is now run to test the reliability and performance of the DDBS. The statistical results of this test will be presented in a further report.

REFERENCES

[1] M. M. Astrahan, M. W. Blasgen, D. D. Chamberlin, K. P. Eswaran, J. N. Gray, P. P. Griffiths, W. F. King, R. A. Lorie, P. R. McJones, J. W. Mehl, G. R. Putzolu, I. L. Traiger, B. W. Wade, and V. Watson, "System R: A relational approach to database management," TODS, Vol. 1, No. 2, June 1976.

[2] P. A. Bernstein, D. W. Shipman, and J. B. Rothnie, "Concurrency control in a system for distributed databases (SDD-1)," ACM TODS, Vol. 5, No. 1, March 1980.

[3] M. Stonebraker, E. Wong, P. Kreps, and G. Held, "The design and implementation of INGRES," TODS, Vol. 1, No. 3, September 1976.

[4] P. A. Bernstein and D. M. Chiu, "Using semi-joins to solve relational queries," JACM, Vol. 28, No. 1, January 1981.

[5] P. T. Chang and J. S. Ke, "Optimization of query processing in distributed database systems," Proc. of ICS 82, Taichung, December 1982.

[6] C. H. Chang, C. F. Wang, and J. S. Ke, "Query processing in a distributed database system," Proc. of NCS 83, Hsinchu, Taiwan, R.O.C., 1983.

[7] J. S. Ke, L. C. Liu, H. L. Chen, and Y. A. Pan, "Design and implementation of a front end processor in the RSX-11/M environment," Proc. of NCS 83, Hsinchu, Taiwan, R.O.C., 1983.

[8] C. L. Lin, C. C. Lu, and J. S. Ke, "IISNET: A unix-based computer network operating system," Proc. of NCS 83, Hsinchu, Taiwan, R.O.C., 1983.

[9] C. C. Lu, C. L. Lin, and J. S. Ke, "Enhanced interprocess communication mechanisms for UNIX," Proc. of ICS 82, Taichung, December 1982.

[10] J. B. Rothnie, Jr., P. A. Bernstein, S. Fox, N. Goodman, M. Hammer, T. A. Landers, C. Reeve, D. W. Shipman, and E. Wong, "Introduction to a system for distributed databases (SDD-1)," ACM TODS, Vol. 5, No. 1, March 1980.

A HIGH LEVEL GRAPHIC DATABASE FOR CAD

San-Cheng Chang

Department of Civil Engineering
National Taiwan University
Taipei, Taiwan, R.O.C.

ABSTRACT

Efficient CAD (Computer-Aided Design) software development and maintenance relies heavily on a convenient and versatile graphic database manager. Such a management system does not replace the CAD program, but is driven by the host CAD program to perform a number of database management and editing functions common to all applications. A protocol for such a data management system is proposed in this paper. The main characteristics of this system lie in the object organization, the ability to edit graphic data conveniently, the overall object-tree data structure, and the implementation of a number of high-level interactive functions. Overall, this new generation of software represents a step toward the direction of distributed processing while allowing the host CAD program the maximum flexibility possible in data management.

INTRODUCTION

A well-structured CAD program usually has a hierarchy easily distinguishable into the following three levels:

(1) The part that performs application-specific functions and data management.

(2) The part that performs graphic data management and editing.

(3) The image-builder that constructs the display from the graphic database.

The designer of an interactive CAD program often finds himself faced with all three tasks above. Rarely is there a general data management system for the task of the first level without significantly downgrading the overall efficiency of the CAD program. However, a proper system may be designed to take over the responsibility of the latter two categories, and for the sake of ease of software development and maintenance as well as hardware independency of the program, such a system should be devised. Furthermore, through the proper design of this system, the load of application-specific data management in the host can be partially alleviated. The data management system described in this paper is designed on these premises.

DATA STRUCTURE

In this system, the graphic content of an image is organized as an "object-tree" as shown in Figure 1. The host CAD program constructs the tree through the database manager, and the manager builds the image by scanning the tree when the tree definition is complete.

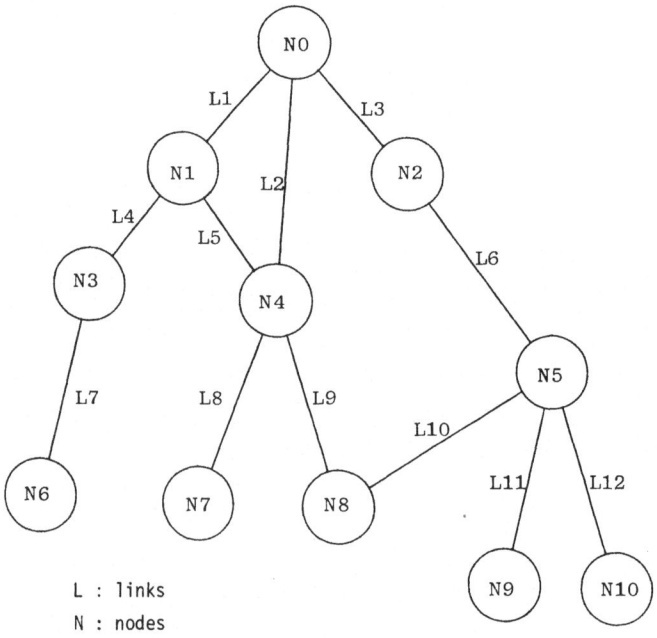

L : links
N : nodes

Figure 1. A typical object-tree.

Graphic commands and associated data are categorized into three types: First are those that appear in the final image, such as lines, dots, and tests. Then are those that set attributes of the display, such as the blinking of image, sizes of character strings, and viewport used. Finally are the transformation specifications that affect orientation of the displayed image. These graphic commands and data constitute the node or link (branch) of the tree depending on their types. Specifically, coordinate and text definitions of the first type are confined to reside in nodes of the tree, while the other two types must locate on links and are implictly referred as "link attributes." Both nodes and links may be empty or may contain multiple graphic commands. They are collectively called "objects" and the corresponding tree so formed is an "object-tree."

Scanning Object Tree

All graphic data must be properly transformed and mapped into the screen coordinate system before display. The content of a node is affected by attributes of all links between the node and the top-node, i.e., the root. The root itself cannot contain any graphic data, however, since the transformation to its content is not definable by links.

While building up the graphic display, the tree is scanned from the root toward the branches. Several status blocks are kept and modified when corresponding commands in links are encountered. Each time a node is scanned, the current transformation accumulated from links above is applied to its content. Specifically, coordinate, viewport, and text transformation data are concatenated to the status block except for explicit reset commands. Attribute setting commands are always loaded, thus only those closest to a node will take effect on the nodal content. The connection between two nodes is not limited to a single link. Multiple links are permitted, thus allowing the content of a node to be scanned and hence displayed more than once. To enable multiple links between nodes, the tree is stored in the form of a link-table, i.e., a table showing the nodes above and below each link.

Object Manipulations

Facilities are provided for the host CAD program to manipulate the tree at run-time so as to achieve the effect of varying the image interactively. Each object is identified by its object ID assigned dynamically at run-time upon object creation. In addition to the preliminary functions of creating objects and specifying object locations in the tree, the following have been implemented:

(1) object deletion: when the content of an object is no longer needed for display, it can be deleted so that its object storage and object ID can be reclaimed.

(2) object replacement: an object may be fully replaced by its updated version. During a replacement, the new object may be assigned a different location in the object-tree from the old one, and further memory release or allocation is performed automatically depending on the size of the new object.

(3) object activation/deactivation: it often arises in various applications that an image is not needed temporarily, so it would be desirable to "turn-off" the object from display. When a node is turned-off, its content is not scanned for display, but it still serves its structural function as a connection to branches of lower levels. On the other hand, if a link is turned-off, not only will its content no longer affect the final display, but also the tree structure below it is not scanned anymore.

(4) link relocation: this feature is intended mainly for the case of menu page switching for a multi-page host program. A link can be relocated from its original position in the tree while retaining its complete attribute definition. The connections between nodes of the old link are destroyed when the link is relocated.

(5) link generation: as described previously, the current system imposes no limit on the number of links between two nodes. For cases where the content in a node is to be displayed a large number of times at a high regularity, automatic generation of links would be appropriate. In the current system, it is possible to state the differences between links and the number of links desired to generate multiple links at the specified location.

OBJECT STRUCTURE

An object is a collection of graphic commands and data. Internally, these commands and data are arranged as a linked-list within each object, be the object a link or a node, as shown in Figure 2. Each cell in the list consists of a command ID, command length, and relevant data. The command length is used to guide the scanning module to jump to the next command-cell in the list until a zero command ID indicating the end of object is reached. The graphic data varies in length depending on the command type.

Storage Allocation

Storage to objects are allocated in terms of pages. A page is a 512-byte memory or disk segment. A page may belong only to

HIGH LEVEL GRAPHIC DATABASE

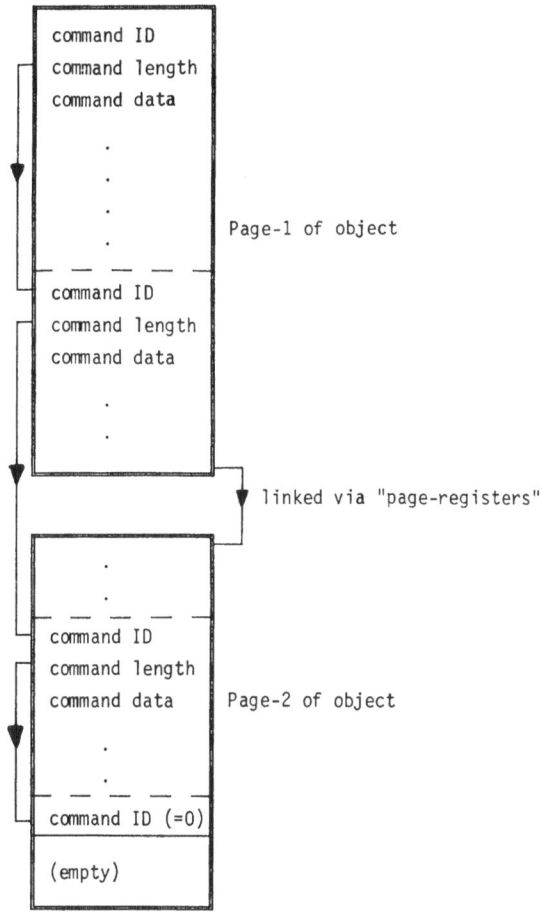

Figure 2. Linked-list structure of an object.

one object, even though the object may not use up the entire page. The pages of an object need not be contiguous to one another, but is linked through an array of page-registers storing the owner of each page. When an object is deleted, the corresponding registers are cleared. The search for free pages when a new object is formed requires only a single scan through this array. Since allocation is always performed sequentially, the overhead introduced by this scheme is expected to be trivial.

Label

To enable the host CAD program to edit object content, a set of artificial reference points may be placed in an object. These reference points are called "labels." A label must be assigned an ID unique within the object. A label of ID 0 is always implicitly placed in the beginning of each object, whereas placement of other labels is entirely determined by the host to facilitate convenient location reference in the future. Internally, a label is treated as a special type of command that possesses no graphical meaning. The label is a helpful tool to identify a command in an object, and speeds up the search for commands significantly. Command in the object-tree is identified by a command address that consists of the object ID, the label ID, and the sequence count of the target command after the label. The command address can be thought of as a common language between the host CAD program and the current system for designating commands, and plays an important role in structuring an object.

Object Editing

A major distinction between the object in this system and the "segment" widely used elsewhere is that an object may have its content edited, while in most other systems segments must be fully replaced. The content of an object may be partially removed, modified, or enhanced. This feature allows the size of an object to increase significantly and thus the total number of objects required for an image may be reduced. Accordingly, bookkeeping effort for objects in the host is appreciably less than that required in using segments. Operations to an object currently include append and insert new data, remove or truncate existing data. Replacement of existing data with new ones is also possible, and the replacing and replaced parts do not need to be of equal or comparable length. All these editing activities are performed with the help of command addresses that enable the location of editing to be specified.

CAD WORK-STATION FUNCTIONS

A mid-range CAD graphic work-station usually includes, in addition to the graphic display monitor, a data tablet, function keys, and dials. These devices serve as the basic channels for interactive input. In addition, they can be used to manipulate the displayed image through local capabilities. The manipulation is device dependent to a very high extent. To shield the CAD program from exposing to such undesirable situation, software that performs "canned" interactive functions should be introduced.

Tablet Input Facility

The most basic input device of a CAD station is the data tablet. However, despite its wide acceptance, tablets differ in their resolutions and sizes, so conversion on the tablet input is always required. The current system adopts a scheme of Crane [1] which uses a universal tablet coordinate system of range 0 to 1. A tablet may be divided into a number of rectangular regions called "tablet windows." Each tablet window is specified by the universal tablet coordinates and is assigned a linear mapping function that maps the input to a desirable range. Tablet definition may be made to vary from time to time in different stages of the program to achieve improved conveniency. During the sampling of tablet stylus, the tablet window ID in which the stylus is located and the transformed stylus coordinates are returned to the host. The host can then use the information directly without further processing. This scheme therefore renders the CAD program hardware independent with respect to tablets, and only the proper tablet driver needs to be coded when new tablets are introduced into the CAD work-station.

Hit-Detection

It is a common practice in CAD programs to allow the user to point to a certain item that is to be modified. To identify the item, be it a point or a line, a technique called "hit-test" or "hit-detection" must be performed. The current system allows the user to identify such item through a single call to the hit-test utility. Basically, a user issues a "pen-down" request (using the tablet facilities) to obtain the viewport location of the item hit. With this location given, the system automatically forms a small window around it, and specified portion of the object-tree is scanned to determine if items of any object is displayed in this part of the viewport. The command address of the item hit, if any, is then returned to the host. Usually, this address is closely related to the application-specific database, and modifications can be made easily.

Interactive Functions

There is also a level of software in current system that allows the host program to perform highly interactive local functions without resorting to elaborated procedures or device dependent codes. The host designates the interactive input device as well as to which portion of the object-tree this input is to enter, the latter being given in the form of a command address. The control is then completely transferred to the system until a specified interrupt is sensed, after which the graphic database is automatically updated to reflect the change of image, and the control is returned to the host. The host is therefore relieved from the need of intervention during the entire interactive process. Presently, high level functions such as dragging, inking, rubber-banding, and most importantly, real-time viewing, have been implemented. This list continues to grow as experience from CAD program development accumulates.

IMPLEMENTATION

To prevent the system from being too device dependent, the modules in the current system are grouped into three types: (1) information module, (2) data management module, and (3) device driver module. As shown in Figure 3, information module interfaces directly with the host program to move graphic data from host to a temporary buffer. When the host issues a command to update either the graphic database or display, the program control is transferred to the data management module to move and update object data, and then the device driver to update the display. All system dependent codes are collected in the driver module. Presently, this system has been implemented on a PDP-11 and a VAX computer. In the former, object storage are allocated on disk, while virtual storage capability is used on VAX. Both implementations use "task-spawning" notion for the device driver. This not only reduces host program size, but also new driver may be activated without relinking the host. The host program is therefore relatively device independent.

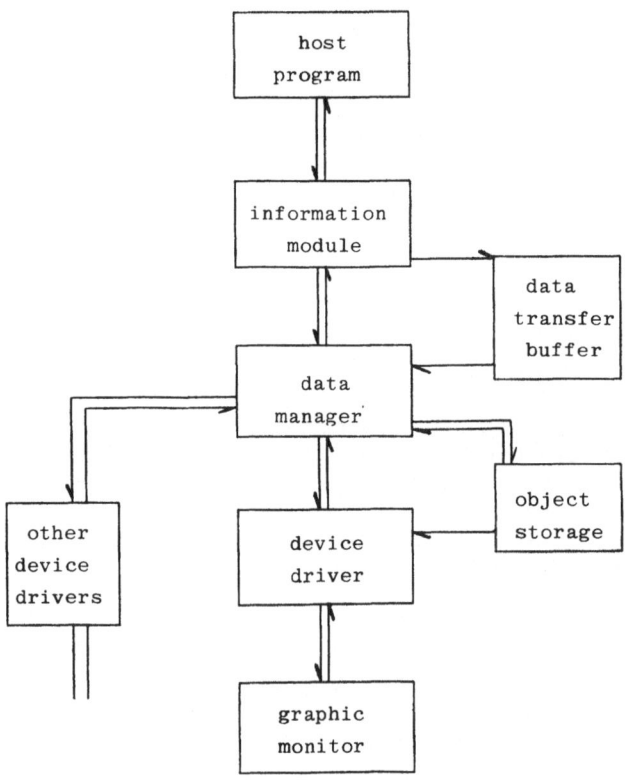

Figure 3. Software structure block diagram.

Device drivers for TEKTRONIX 4014 and Evans & Sutherland PS300 graphic stations have been developed. CAD programs can run without modification on both devices. For the PS300 model essentially a database manager and a communication system that translates commands between the host and work-station are needed. For TEKTRONIX 4014, in addition to the basic database manager, an object-tree decoder, image builder and local function emulator are required.

CONCLUSION

The presented paper is a concise description of the principles underlying the database system. The system represents the recognition of the distributed processing architecture gaining increasing popularity in CAD work-station configurations. Nevertheless, application of the system to diverse graphic stations such as PS300 and TEKTRONIX 4014 is an example of how work-station of little local intelligence is not in any way prevented from the scope of application as long as proper device driver is coded. Through the use of this system, the host CAD program can deal with application-specific problems more effectively, and development as well as maintenance of software can be done in a much better environment than available before.

REFERENCES

[1] T. Crane, "PICSYS Reference Guide, Version 0.01," Program of Computer Graphics, Cornell University, Feb. 1981.

A GRAPH-THEORETIC APPROACH TO THE HIDDEN LINE ELIMINATION PROBLEM

W. H. Chu and D. T. Lee

Solo Systems, San Jose, CA 95134, U.S.A.
Department of Electrical Engineering and Computer Science, Northwestern University, Evanston, IL 60201, U.S.A.

ABSTRACT

We consider the hidden line elimination problem for displaying 3D solid objects represented as polyhedra on 2D display devices. The objects are assumed to be composed of polygonal faces. The algorithm presented is based on a graph-theoretic approach in that a certain graph representing the projections of the edges of the polyhedron is constructed and visible edges are found using depth-first search technique in the graph. The way in which visible edges are found is different from the previous ones in that it first assumes all the edges of the polygonal faces are invisible from the viewer and based on some criteria seeks out visible ones, whence the efficiency of the algorithm is, to some extent, independent of the number of hidden edges. In this sense it is more appropriate to refer to the algorithm as visible line determination algorithm.

The worst case time complexity of the algorithm is shown to be $O(n_b n \log n) + F$, where n is the total number of edges, n_b the number of boundary edges and F the number of visible edges displayed. However, if the number n_b of boundary edges is much smaller than n, which is true in most cases, the algorithm is more efficient than previous approaches whose worst case complexity may be $O(n^2)$.

INTRODUCTION

The hidden line elimination (HLE) problem arises in computer graphics where 3D objects are to be displayed on a 2D display device. To enhance comprehensibility of the display, all the hidden lines or surfaces of the objects are eliminated from display. The problem was first studied by Roberts [15], who used a mathematical model that utilized volume inequality matrices to determine whether a point is inside or outside a volume. The test was extended by linear inequality solutions to tell which segment of a line is behind a volume. Galimberti and Montanari [11] and Loutrel [14] separately proposed similar algorithms to tackle this problem. Their approach involves tracing each edge of the object and keeping track of the number of polygons covering the edge. When the count is zero, the edge is visible. This approach requires extensive computation for each edge.

Since then, research in object representations has proceeded in several directions. More realism was introduced by representing the object by curved surfaces rather than polygons [13], or by adding shadows or textures [2]. Attention was also directed towards CRT display of images, in which the main concern in representing an object was by pixels [7] and HLE algorithms were developed based on the consideration of the pixels [19]. The interest in representing objects by polyhedra continues because of its simplicity and potential applications in many areas [1,4,5,6,9,17]. Fuchs et al. [10] proposed an a priori "binary space partitioning tree" to speed up computations of scenes consisting of polygons, many of whose relative geometric relations are static. Weiler and Atherton [20] used the idea of polygon area sorting to trim or clip obstructed polygons. Franklin [8] developed an algorithm that overlays a grid on the scene. The fineness of the grid is commensurable with the complexity of the overlapping surfaces. Edges and faces are sorted in the grid cells, and the visible edges are determined within each cell by an algorithm similar to that of Weiler and Atherton [20].

There is a class of scan line algorithms that follow a somewhat different approach in treating the HLE problem. This class of algorithms divide the picture plane into strips and trace across each strip to determine the visibility of edges within the strip. Hamlin and Gear [12] developed two algorithms that operate in the image space. One algorithm traces along each scan line and determines the visibility of edges cut by the scan line. The other algorithm traces each scan line and determines the visibility of each vertex encountered and the visibility of edges joining the vertex. Sechrest and Greenberg [16] developed an object space scan line algorithm that uses variable-width strips to divide the picture plane. A scan line is passed across the plane at each vertex or intersection point. The visibility of the edges within each strip is determined from low y-value vertices to high y-value vertices. For a summary of work on this topic the reader is referred to Sutherland et al. [18].

The approach proposed in this paper, however, is different from the previous ones in the process of distinguishing hidden lines from visible lines. Previous algorithms basically examine each edge to determine if any part of it is visible. Any edge which is blocked from the viewer is eliminated. Hence, these algorithms are known as Hidden Line Elimination (HLE) algorithms. The algorithm to be presented assumes that all edges are invisible and attempts to seek out the visible edges. In this sense it would be more apt to refer to the algorithm as Visible Line Determination algorithm. However, the traditional name will be followed. The edges that are determined to be visible will be included in a list and therefore at the end of the process those edges not included in the list are ignored. So long as an edge, or part of it, is obstructed, no effort is made in determining how many surfaces shield it from the observer, nor is effort made in studying the geometric detail behind a visible surface. Thus, a large amount of computational work is saved as compared to the other algorithms.

The problem is formally defined as: given a solid body composed of polygonal surfaces, determine the 2D projection of the object on a plane between the object and an observer. No edge or segment of an edge that is obstructed from the viewer should be included in the 2D projection. The polyhedron is described by its surfaces and each surface is a polygon described by its edges. Each edge, in turn, is described by its two endpoints. The left-handed Cartesian coordinate system is chosen to represent the object, and hence each endpoint is represented by its coordinates (x,y,z) in the 3D space. The viewing direction is assumed to be along the direction of the positive z-axis. The technique used here is by propagation. First, a visible edge is located. Based on certain criteria, the search for visible edges propagates from this point on. The propagation spreads from edge to edge until the entire polyhedron is covered. Before we give detailed description of the algorithm, some terminology and observations are in order.

An inward normal is a vector that is perpendicular to a surface and points towards the interior of the solid body. A face is either a front face, if the inward normal points away from the viewer, or a back face, otherwise. A front face is potentially visible to the viewer, and a back face is definitely invisible. An edge is a boundary edge if it borders a front face and a back face. If these two faces have an angle (containing the exterior) less than 180°, then the boundary edge is invisible; otherwise, it is potentially visible. The contour edges are the boundary edges that border the projection of the 3D object and the "background", i.e., they define the boundary of the projection, which may be a simply- or multiply-connected region in the view plane. For example, in Figure 1, edges (e,f), (e,d), (f,g) are boundary edges and (c,d), (d,e), (e,h) are contour edges.

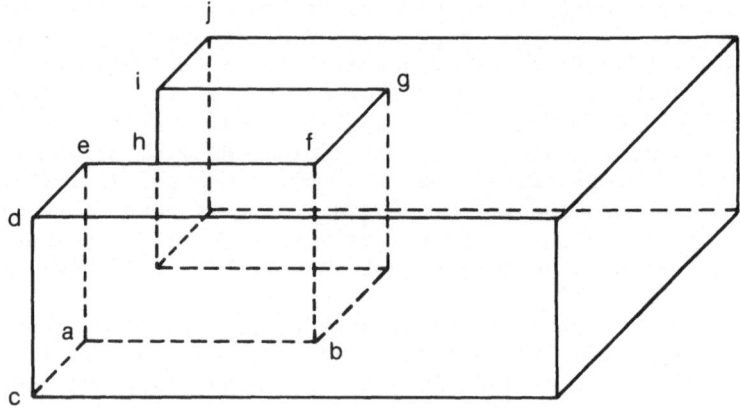

Figure 1. Illustration of boundary edge and contour edge.

Observation 1: If a face is totally obstructed by another face, at least one vertex of the obstructing face is closer to the viewer than any vertex of the obstructed face.

Observation 2: If a vertex is obstructed by a face, at least one vertex of the obstructing face is closer to the viewer than the vertex.

Observation 3: An edge of a front face can only be partially blocked by a boundary edge.

Observation 4: If an edge is partially visible, the boundary edge that blocks it must itself be visible at the point of blocking. To be precise, if the segment (p,q) of an edge (s,t) is visible, then the boundary edges that intersect (s,t) at p and at q must be visible at p and at q.

DATA STRUCTURE

The polyhedron is assumed to be no self-penetrating, without holes, and composed of polygonal faces that are stored in a Polygon Table, each entry of which points to a description of its associated edges. The edges are represented by entries in the Edge Table (Figure 2). Edges belonging to a polygon are grouped together in clockwise order such that the interior of the polygon lies to the right when the edges are traversed. Each edge description contains the coordinates of its endpoints and an indicator of its type. Note that each edge is represented twice, once for each of the two polygons that share the edge. A pointer in each edge description links the two copies of the same edge. For convenience, for each vertex v we also keep an adjacency list ADJ(v) of all the vertices

HIDDEN LINE ELIMINATION PROBLEM

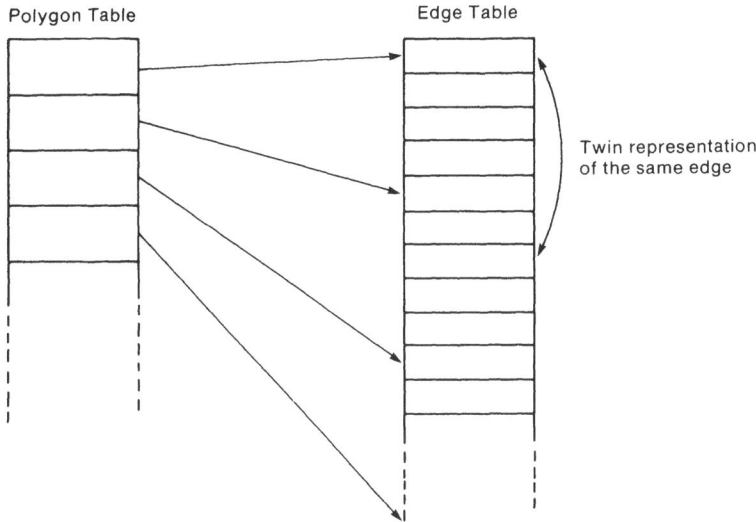

Figure 2. Representation of the object.

that are adjacent to v and the indexes of the polygons sharing edge (v,t) are stored in the entry for vertex t in ADJ(v).

The output of the algorithm to be given below consists of a list, called Visible Edge List, of visible edges and a graph, called Contour Graph, containing the contour edges. Besides, associated with each visible edge there are two pointers to the polygon table identifying the two faces sharing the edge. This information will be useful for subsequent shading algorithm if desired.

The 2D projection of the polyhedron is initially represented by a graph, called the intersection graph, each node of which either represents a vertex or an intersection point, referred to, respectively, as the end node and the intersection node. Each intersection node has four links, two for each of the intersecting edges. The links connecting the dominated edge (i.e., the edge that is blocked or farther away from the viewer), is referred to as in or out links depending on whether it leads into the dominating polygon. Figure 3 shows two intersecting polygons where polygon A is the dominating polygon and the graph representation of the two polygons. The intersection graph is built after the 2D projection of the polyhedron is made and the intersection points computed. As the visible edge list is gradually built, links will be removed from the intersection graph and the contour graph constructed.

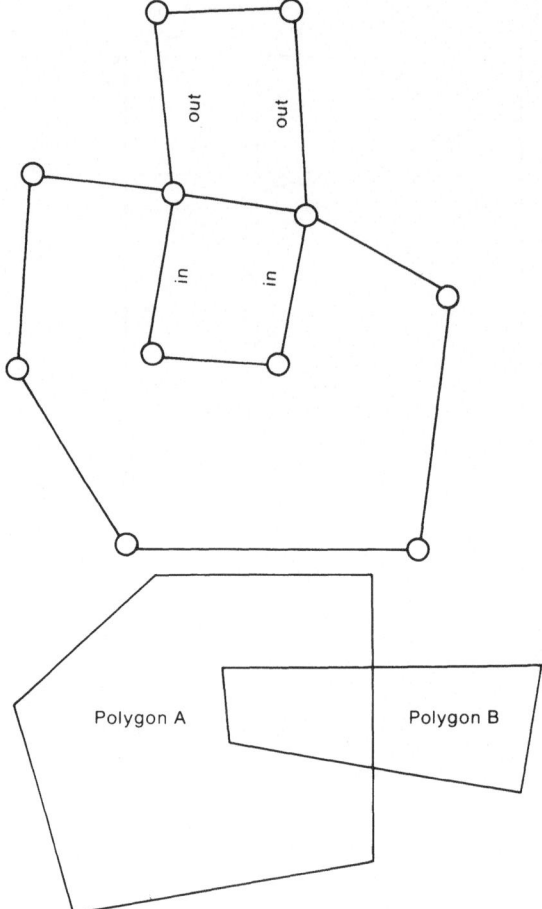

Figure 3. Graph representation of two intersecting polygons.

THE ALGORITHM

As mentioned earlier, we assume all the edges are invisible and seek out the visible ones as the algorithm proceeds by adding them into the visible edge list one at a time. We now give a step-by-step description of the algorithm.

Step 1: Compute the inward normal for all faces and record in the polygon table.

Step 2: Remove the edges of back faces and if the edge removed is a boundary edge, then record that in the entry of its counterpart, the copy that belongs to the adjacent front face. Delete also those boundary edges that are invisible.

HIDDEN LINE ELIMINATION PROBLEM

Step 3: For each edge not eliminated, project its two endpoints onto the view plane and compute their coordinates. Note that the z-coordinate of the endpoint is retained for depth computation to be performed later.

Step 4: Compute the intersections of the line segments on the view plane. Note that only intersections that involve boundary edges will affect the result of the algorithm, so we need not compute all possible intersections. Each intersection appears as a point $\langle x,y \rangle$ on the view plane. Let line segments L_1 and L_2 intersect at point $\langle x,y \rangle$, where L_i is the 2D projection of the edge E_i, i=1,2, in 3D space. The point $\langle x,y \rangle$ corresponds to $\langle x,y,z_1 \rangle$ on E_1 and to $\langle x,y,z_2 \rangle$ on E_2. One of the edges E_1 and E_2 is closer to the view plane and thus blocks or dominates the other. Therefore, we need to compute z_1 and z_2 in order to determine which edge is the dominating edge. For reasons that will become clear later we only record the intersection point when the boundary edge is dominating. The intersection node is now regarded as one of the vertices and the intersecting edges are each cut into two new edges. These new edges are added to the edge table and the adjacency list of the new vertex added to the data structure as well.

Step 5: Build the intersection graph that consists of all the vertices including intersection points and the (new) edges of front faces. Figure 5 shows the initial intersection graph of the polyhedron shown in Figure 4.

Step 6: Initialize the Next Visit Stack by placing the edges that are incident upon the vertex which is the closest to the

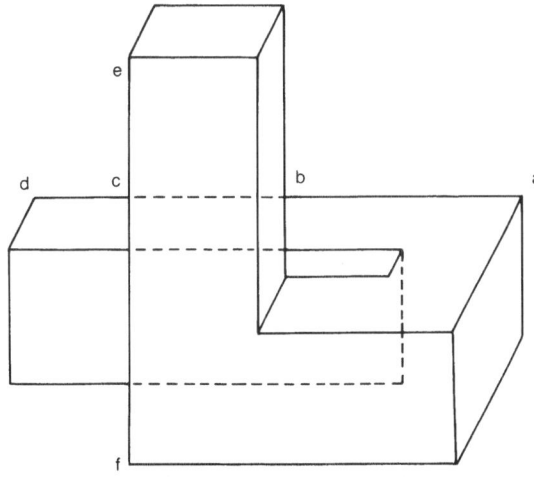

Figure 4. A polyhedron whose visible edges are shown in solid lines.

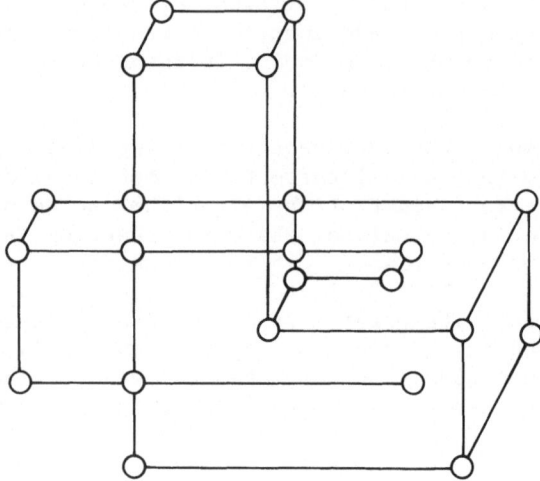

Figure 5. The intersection graph of the polyhedron in Fig. 4.

viewer. We shall then start with one of the edges on the stack and gradually expand outward, searching for all visible edges by placing them in the visible edge list.

Step 7: Select the next edge on the stack and trace it. When an edge is traced, it terminates at a node, which may be either an intersection node or an end node. Suppose it is an intersection node. At least one of the two intersecting edges is a dominating boundary edge. There are two possible cases depending on the directions in which the edge is traced. Consider the case shown in Figure 6a, where edges ab and cd intersect at x and edge ab is the dominating boundary edge. Let the direction of trace be along ax and the node x is encountered. The in link of the dominated edge at the node x, i.e., edge xd, is deleted from the intersection graph. Nothing is said about edge cx at this time. The edge xb is then placed on the stack and edge cx placed on the Contour List. The contour list stores the node x and the associated information such as the edge type and the faces sharing the edge. The other case (Figure 6b) where the tracing is along edge cd is handled similarly, except that both xa and xb are placed on the stack. (cf. Step 9)

If the node is an end node (Figure 6c), then all the links meeting here must be visible. Therefore, all untraced links not yet on the stack are added to the stack.

Assume that we are tracing along edge ad in Figure 4. The edge is segmented at b and ab is included as visible, and the rest of it is ignored at this point. The untraced line segment cd,

HIDDEN LINE ELIMINATION PROBLEM

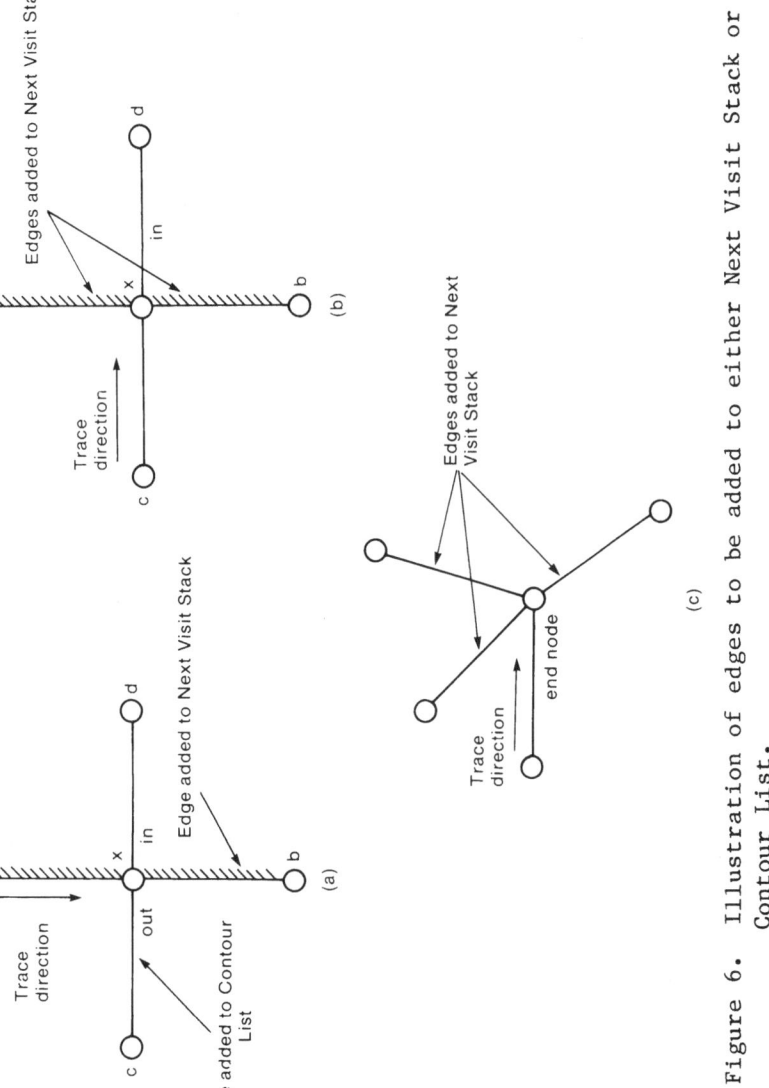

Figure 6. Illustration of edges to be added to either Next Visit Stack or Contour List.

which is visible, will reappear and emerge only at a node intersected by a visible (dominating) boundary edge. This ensures that the segment cd will be picked up at a later time when the dominating boundary edge is traced.

Step 8: As each edge is traced, it will be deleted from the intersection graph. For the edge whose two associated polygons in the projection are not known we include it in a temporary contour graph and consider it as a temporary contour edge. Note that only boundary edges can possibly be included in the graph. For other edges traced, they are placed in the visible edge list.

Steps 7 and 8 are repeated until the next visit stack is exhausted.

Step 9: At this point we may have a number of links in the contour list whose visibility is yet to be determined (due to the propagation process that never goes from a dominating edge to a dominated edge). Note that these edges are all "attached" to visible boundary edges already included in the temporary contour graph. Consider the case shown in Figure 7, where we have four groups of polygons representing the four possible situations. Outside the boundary edge a_0a_{12} polygons A, A', A" are completely visible; B, B', B" are visible but enclosed within the boundary of polygon C; C, C' and C" are visible except for the edges obstructed by group A, and D, D' and D" are completely obstructed. A systematic approach is necessary to differentiate each of these cases. Up to this point the edge a_0a_{12} of polygon E is a temporary contour edge. In fact, at this point all the temporary contour edges can be viewed as edges forming a polygon. Edge a_0a_{12} is shown as a straight line in Figure 7 for ease of presentation only. It can be a portion of the polygon boundary formed by temporary contour edges.

From the 12 intersection points along the contour boundary we need to select a visible edge to start the propagation process. This is where we need to make use of the contour list. Recall that this list is managed by adding any intersection node on a newly formed contour edge where the visibility of the dominated edge is yet to be determined. A node is deleted from the list when the visibility of the dominated edge at the corresponding intersection node has been determined. We can select the node closest to the viewer (the point with the smallest z-coordinate) by examining all the nodes sequentially in the list. If a large number of these searches is expected, a height balanced binary tree may be used to implement the list. The polygon associated with this point must be visible at this node. Place this link on the next visit stack and start the tracing as in step 7.

HIDDEN LINE ELIMINATION PROBLEM 627

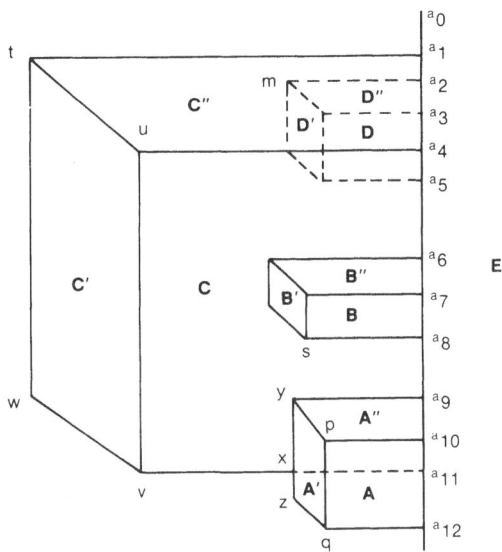

Figure 7. Four types of edges attached to contour edges.

Referring to Figure 7, say we pick edge a_4u to start the trace. When node u is reached, tracing proceeds along links ut and uv. When node t is reached, tracing is along ta_1 and tw. When nove v is reached, tracing is along vw and vx. All these links traced are visible edges. Node x is an intersection node and edge va_{11} of polygon C is blocked at x. Based on observation 4 the blocking boundary edge yz must be visible at x and the tracing proceeds along xy and xz. Note that when tracing along xy is performed, the edge is not considered as a temporary contour edge since both faces (polygons C and A') sharing it in the projection are known. (In fact, this is always true when tracing goes from a dominated edge to a dominating edge). Hence, edge xy, although it is a boundary edge, will not be included in the contour graph. Note also that the following situation will happen when tracing proceeds from dominated edges to dominating edges. The dominating edge, which is necessarily a boundary edge, may have been traced before, i.e., they belong to the contour graph. For example, after link ya_9 is traced, we would proceed to trace a_9a_8 and a_9a_{10}, which have been traced before. At this point we "trace" them (i) to remove them from the contour graph since we have determined the identities of the faces sharing the edges, e.g., polygons C and E sharing edge a_9a_8 and polygons A" and E sharing a_9a_{10}, and the information recorded, and (ii) to pick up nodes on the contour list by tracing in the contour graph and decide whether the associated edges are visible as described below. Take, for instance, the node a_8 and its associated edge a_8s. Since we know that a_8 is shared in the projection by polygons C and E and edge a_8s is an out link, whether

a_8s of polygon B is visible depends on whether polygon C blocks it. The point $a_8 = \langle x,y \rangle$ corresponds to the point $e_C = \langle x,y,z_C \rangle$ on polygon C and $e_B = \langle x,y,z_B \rangle$ on edge a_8s of polygon B. From these two projections we can determine if a_8s is blocked by polygon C. In our example a_8s is visible and will be placed on the next visit stack and this picking-up operation continues until no more nodes can be processed in this manner. Consider the situation when tracing proceeds from edge ta_1 to the dominating links a_1a_0 and a_1a_2. a_1a_0 remains to be contour edge but a_1a_2 does not. We carry out the operations described earlier and find that edge a_2m is blocked by polygon C". Hence, node a_2 is deleted from the tree and link a_2m deleted from the intersection graph. The process continues until node a_4 is reached. At that point edge a_1a_4 is recorded in the visibl edge list as shared by polygons C" and E. Figure 8 shows the final contour graph after the above operations are performed.

This step is repeated until the contour list is empty. At that time the visible edge list will have included all visible projections of the edges of the polyhedron and the faces bordering each of them. The temporary contour graph will contain only the contour edges of the given polyhedron. The algorithm is summarized as follows.

ALGORITHM VISIBLE

Begin

1. For all polygons
      Compute inward normal
      Record the type, front or back, of the polygon
   end for

2. For all front face edges
      If edge E borders a front face and a back face and if it is
      not invisible then mark E as boundary edges.

3. Project edge E to view plane and retain the depth information
   associated with its two endpoints.
   end for

4. For all front face edges
      If edge E is a boundary edge
         then do
               Compute intersection with all other front face edges
               If edge E is dominating then record intersection
               endo
   end for

HIDDEN LINE ELIMINATION PROBLEM

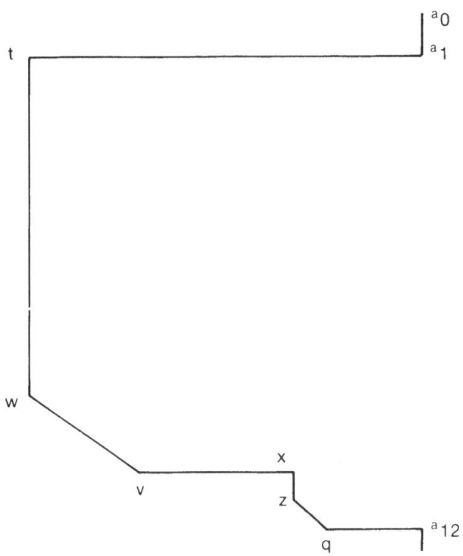

Figure 8. Final contour graph of Fig. 7.

5. Build initial intersection graph IntG with end nodes and intersection nodes. Initialize contour list L to be an empty height balanced binary tree and initialize a contour graph ConG.

6. Find vertex v closest to the viewer.

6a Place edges joining at node v on next visit stack S.

7. Do while S is nonempty
    Pop link E off of stack S
    Let E be denoted by (M,N) with N being the terminating node. Assume polygons Q and R share the link E which is part of edge E'.
    If N is an intersection node of edges E' and E"
    then if E' is dominating
        then do
            Delete the in link of E" from IntG
            Add out link of E" to contour list L
            Place the other link of E' at N on S
            endo
        else do
            Let (N,A) and (N,B) be the two links of the dominating edge E" associated with which is the information of its front face P. Link (N,A) is shared by polygons P and R, and link (N,B) shared by polygons P and Q.

```
                    If (N,A) in ConG then call TRACE (N,A)
                                    else Add (N,A) to S
                    If (N,B) in ConG then call TRACE (N,B)
                                    else Add (N,B) to S
                endo
        else add all links not included in S to S.

8.      If the two polygons that share E are not known then add E to
        ConG.

8a      Delete E from IntG.
    end while

9.      Do while L is nonempty
        Pick from L the node X closest to viewer and remove it from L.
        Place corresponding link on S.

9a      Repeat steps 7 and 8.
    end while
end

PROCEDURE TRACE (N,X)

    Begin
        Let P and Q denote the two polygons sharing the link (N,X)
        and P is the polygon of which link (N,X) is part of its
        boundary.
        Add link (N,X) to visible edge list along the information P
        and Q.
        Delete (N,X) from ConG.
        If there is no more link (X,Y) in ConG then return
        else do
                Get the next link (X,Y) in ConG
                If X is an end node then call TRACE (X,Y) else
                do
                    Let (X,Z) be the link associated with the node X in L
                    If (X,Z) is visible with respect to polygon Q
                    then do
                            Place link (X,Z) on S /* (X,Z) shared by polygons
                            Q and R */
                            Call TRACE (X,Y) /* (X,Y) shared by polygons P and
                            R /*
                            endo
                    else call TRACE (X,Y) /* (X,Y) shared by P and Q */
                    endo
                endo
        end
```

Remarks

In case holes are allowed to be part of a given object as shown in Figure 9, we need to modify our input data structure. Normally, polygons with holes are represented by separate lists with the outermost boundary edges represented by edges in, say, clockwise direction and the boundary edges of holes by edges in counterclockwise direction. With this representation alone we would not be able to trace the boundary edges of the polygon by the method given above. Consider the example shown in Figure 9. The hole abcd is detached from the rest of the outer boundary edges of face A and hence cannot be reached by propagation. An easy fix of it is that we add a pseudo-edge between the outer boundary and the boundary of each hole. The sole purpose in doing so is to allow the tracing of visible edges to propagate from one set of detached edges to the other. Once the pseudo-edges are introduced the tracing mechanism can proceed. However, there is a catch in the algorithm that needs to be fixed. That is, through a hole we may "see" some other faces of the object. If we can see more than one face, then we have no problem, since at least one visible edge will be picked up and tracing can continue. The trouble is when only one face is visible. In this case we need to distinguish that face from the "background", which is actually the exterior "polygonal" face whose boundary is shared by all the outermost visible faces. The algorithm will produce in this case a contour graph which is not connected. The outermost graph represents the contour of the object when displayed on the view plane; while the inner ones represent holes and may be subject to modifications. Observe that if through the hole represented by an "inner" subgraph of the contour graph we can see a polygonal face, then all the vertices of

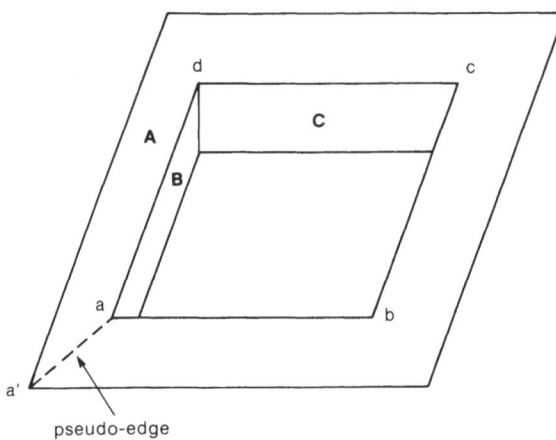

Figure 9. Pseudo-edges are added to connect detached polygons.

the subgraph must all lie in the interior of the polygon when projected onto the view plane. To determine which front face can be seen through a hole we simply take any vertex of the subgraph and check if any projection of the front face contains the projection of the vertex on the view plane. If more than one such projection is found, we select the one which is the closest to the viewer. This process repeats until all inner subgraphs are taken care of.

TIMING ANALYSIS

Let us now analyze the running time of the algorithm. Steps 1 to 3 take $O(n)$ time, where n is the number of edges (or faces). Step 4 takes $O(n_b n)$ time, where n_b ($< n$) is the number of boundary edges if it is implemented in a straightforward manner. If the algorithm of Bentley and Ottmann [3] is used, the step takes $O((n+s)\log n)$ time, where s is the total number of intersection points resulted from the projection of all the uneliminated edges. Step 5 also runs in time $O(n_b n)$, which is the complexity of the intersection graph. In step 6 to find the vertex closest to the viewer takes $O(n)$ time. Manipulation of the contour list in steps 7 to 9 can be implemented to run in $O(n_b n + v \log n) + F$ time, where v is the number of vertices in the contour graph and is at most $O(n_b n)$ and F is the number of edges placed in the visible edge list. If a linear list, rather than balanced binary tree, is used, the searches take time $O(vt)$, where t is the number of times step 9 is executed. A linear list structure is much preferred to a binary tree for reasons that the overhead involved in the latter for performing insertions and deletions is far greater than in the former case. If the number t is small, compared to $\log_2 n$, then binary tree structure should definite be avoided. If holes are allowed as part of the input, then an additional $O(h\, n)$ time is needed, where h is the number of holes.

As can be easily seen, the most time-consuming part is step 7 through step 9. Therefore, the overall worse-case time complexity is $O(n_b n \log n)$.

CONCLUSIONS

We have presented an algorithm that determines the visible edges (faces) of a 3D object when displayed on a 2D display device. It departs from previous approaches by assuming that all edges of the polygonal faces of the polydron are invisible to the viewer and by some criteria identifying all visible ones. It produces as output the visible edges along with the information of their associated polygons so that it can be used together with a shading algorithm when display of colors or halftoning of the object is desired. The worst case time complexity is shown to be $O(n_b n$

log n) + F, where n is the number of edges, n_b the number of boundary edges and F the output size.

ACKNOWLEDGMENT

This research is supported in part by the National Science Foundation under Grants MCS 8202359 and MCS 8342682.

REFERENCES

[1] J. K. Aggarwal and R. O. Duda, "Computer analysis of moving polygonal images," IEEE Trans. Comput., C-24,10, Oct. 1975.
[2] P. Atherton, K. Weiler and D. Greenberg, "Polygon shadow generation," Computer Graphics, 12,3, pp. 275-281, August 1978.
[3] J. L. Bentley and T. Ottmann, "Algorithms for reporting and counting geometric intersections," IEEE Trans. Comput., C-28,9, pp. 643-647, Sept. 1979.
[4] D. K. Brotz, "Intersecting polyhedra with successive planes," Computer & Graphics, 2, 1976.
[5] J. H. Clark, "Hierarchical geometric models for visible surface algorithm," Comm. ACM, 19,10, pp. 547-554, Oct. 1976.
[6] M. Cyrus and J. Beck, "Generalized two- and three-dimensional clipping," Computer & Graphics, 3, 1978.
[7] C. Eastman, J. Livdini and D. Stoken, "A database for designing large physical systems," AFIPS44, pp. 603-611, 1975.
[8] W. R. Franklin, "A linear time exact hidden surface algorithm techniques," Computer Graphics, 14,3, pp. 117-123, July 1980.
[9] H. Freeman and P. P. Loutrel, "An algorithm for the solution of the two-dimensional hidden-line problem," IEEE Trans. Electronic Comput., EC-16,6, pp. 784-790, Dec. 1967.
[10] H. Fuchs, Z. Kedem and B. Naylor, "On visible surface generation by a priori tree structures," Computer Graphics, 14,3, pp. 124-133, July 1980.
[11] R. Galimberti and U. Montanari, "An algorithm for hidden line elimination," Comm. ACM, 12,4, pp. 206-211, April 1969.
[12] G. Hamlin, Jr. and C. W. Gear, "Raster-scan hidden surface algorithm techniques, Computer Graphics, 11,2, pp. 206-213, Summer 1977.
[13] M. Lee and R. V. DiMarco, "Computer graphics with hidden surfaces," Computer & Graphics, 3, 1978.
[14] P. P. Loutrel, "A solution to the hidden line problem for computer-drawn polyhedra," IEEE Trans. Comput., C-19,3, pp. 205-213, March 1970.
[15] L. G. Roberts, "Machine perception of three-dimensional solids," Optical and Electro-Optical Information Processing, MIT Press, Cambridge, Mass., pp. 159, 1964.

[16] S. Sechrest and D. P. Greenberg, "A visible polygon reconstruction algorithm," Computer Graphics, 15,3, pp. 17-27, Aug. 1981.
[17] I. E. Sutherland and G. W. Hogman, "Reentrant polygon clipping," Comm. ACM, 17,1, pp. 32-42, Jan. 1974.
[18] I. E. Sutherland, R. F. Sproull and R. A. Shumacker, "A characterization of ten hidden surface algorithms," Computing Surveys, 6, pp. 1-55, Mar. 1974.
[19] G. S. Watkins, "A real time visible surface algorithm," Univ. of Utah, Comput. Sci. Dept., UTEC-CSc-70-101, NTIS AD 762-004, June 1970.
[20] K. Weiler and P. Atherton, "Hidden surface removal using polygon area sorting," Computer Graphics, pp. 214-222, Summer 1977.

List of Contributors

Numbers in parentheses indicate the pages on which the authors' contributions begin.

RUZENA BAJSCY (387), Integrating Vision and Touch for Grasping of an Object, Computer & Information Science Department, University of Pennsylvania, Philadelphia, Pennsylvania 19104, U.S.A.

P. BALAKRISHNAN (535), Machinability Data Base Systems for Automated Manufacturing, Manufacturing Systems Engineering, University of Wisconsin, Madison, Wisconsin 53706, U.S.A.

Y. K. CHAN (153), Low Cost CAD/CAM System for Machine Parts, Department of Computer Science, The Chinese University of Hong Kong, New Territories, Hong Kong

CHIA-HSIANG CHANG (591), The Design and Implementation of a Distributed Database System, Institute of Information Science, Academia Sinica, Taipei, Taiwan, R.O.C.

CHUNG-JYE CHANG (37), CAD Application on PC Board Design, Telecommunication Laboratories, Ministry of Communications, Chung-Li, Taiwan, R.O.C.

SAN-CHENG CHANG (607), A High Level Graphic Database for CAD, Department of Civil Engineering, National Taiwan University, Taipei, Taiwan, R.O.C.

CHENG CHEN (119), PLAMG: An Automatic PLA Minimizer and Generator for VLSI System Design, Department of Computer Engineering, National Chiao Tung University, Hsin-Chu, Taiwan, R.O.C.

HSING-LUNG CHEN (591), The Design and Implementation of a Distributed Database System, Institute of Information Science, Academia Sinica, Taipei, Taiwan, R.O.C.

Z. CHEN (169), Computer-Assisted Methods for Defining 3D Geometric Structure of Mechanical Parts, Institute of Computer Engineering, National Chiao Tung University, Hsin-Chu, Taiwan, R.O.C.

JACK M. CHENG (3), Automatic Generation of Knowledge Base from Electronic Diagrams for Computer-Aided Design, Center for Information Research, University of Florida, Gainesville, Florida 32611, U.S.A.

LINFU CHENG (267), Design of a Computer Aided Robot Design System, Department of Electrical & Computer Engineering, University of Miami, Coral Gables, Florida 33124, U.S.A.

BRYAN A. CHIN (411), Automatic Welding: Infrared Sensors for Process Control, Department of Mechanical Engineering, Auburn University, Auburn, Alabama 36849, U.S.A.

W. H. CHU (617), A Graph-Theoretic Approach to the Hidden Line Elimination Problem, Department of Electrical Engineering, Northwestern University, Evanston, Illinois 60201, U.S.A.

D. D. COWAN (213), A Theoretical Proposal for a CASD System Extending the Jackson's Method for Program Construction, Department of Computer Science, University of Waterloo, Waterloo, Ontario, Canada

M. F. DEVRIES (535), Machinability Data Base Systems for Automated Manufacturing, Manufacturing Systems Engineering, University of Wisconsin, Madison, Wisconsin 53706, U.S.A.

W. K. EPPLE (433), Present State and Future Trends in the Development of Programming Languages for Manufacturing, Department of Computer Science, University of Karlsruhe, Karlsruhe, Germany

K. S. FU (305), A Graphic-Theoretic Approach to 3D Object Recognition and Estimation of Position and Orientation, School of Electrical Engineering, Purdue University, West Lafayette, Indiana 47907, U.S.A.

R. C. GONZALEZ (345), Industrial Computer Vision, Electrical Engineering Department, University of Tennessee, Knoxville, Tennessee 37996-2100, U.S.A.

K. A. HWANG (95), BRUTUS: An Interactive Graphic Editor for IC Layout, Computer Center, National Chiao Tung University, Hsin-Chu, Taiwan, R.O.C.

ARVIND JAIN (519), Production Scheduling Automation in an Aluminum Metal Products Plant, Kaiser Aluminum & Chemical Corporation, Oakland, California 94611, U.S.A.

CHEIN WEI JEN (59), SPICE-2P: A SPICE-2 Incorporated with a Partition Scheme, Institute of Electronics, National Chiao Tung University, Hsin-Chu, Taiwan, R.O.C.

J. M. JOU (73), A CAD Program for VLSI Placement and Routing, Department of Electrical Engineering, National Cheng Kung University, Tainan, Taiwan, R.O.C.

LIST OF CONTRIBUTORS

SHYI-TING KANG (591), The Design and Implementation of a Distributed Database System, Institute of Information Science, Academia Sinica, Taipei, Taiwan, R.O.C.

JYH-SHENG KE (591), The Design and Implementation of a Distributed Database System, Institute of Information Science, Academia Sinica, Taipei, Taiwan, R.O.C.

CHUNG LEN LEE (59), SPICE-2P: A SPICE-2 Incorporated with a Partition Scheme, Institute of Electronics, National Chiao Tung University, Hsin-Chu, Taiwan, R.O.C.

D. T. LEE (617), A Graph-Theoretic Approach to the Hidden Line Elimination Problem, Department of Electrical Engineering, Northwestern University, Evanston, Illinois 60201, U.S.A.

JAU-YIEN LEE (73), A CAD Program for VLSI Placement and Routing, Department of Electrical Engineering, National Cheng Kung University, Tainan, Taiwan, R.O.C.

KEH-LON THOMAS LEE (519), Production Scheduling Automation in an Aluminum Metal Products Plant, Kaiser Aluminum & Chemical Corporation, Oakland, California 94611, U.S.A.

YOUNG-LI LEE (119), PLAMG: An Automatic PLA Minimizer and Generator for VLSI System Design, Department of Computer Engineering, National Chiao Tung University, Hsin-Chu, Taiwan, R.O.C.

YEH JIAN LIANG (119), PLAMG: An Automatic PLA Minimizer and Generator for VLSI System Design, Department of Computer Engineering, National Chiao Tung University, Hsin-Chu, Taiwan, R.O.C.

CHING AN LIAW (59), SPICE-2P: A SPICE-2 Incorporated with a Partition Scheme, Institute of Electronics, National Chiao Tung University, Hsin-Chu, Taiwan, R.O.C.

C. C. LIN (95), BRUTUS: An Interactive Graphic Editor for IC Layout, Computer Center, National Chiao Tung University, Hsin-Chu, Taiwan, R.O.C.

CHING-LIANG LIN (591), The Design and Implementation of a Distributed Database System, Institute of Information Science, Academia Sinica, Taipei, Taiwan, R.O.C.

D. C. LIU (95), BRUTUS: An Interactive Graphic Editor for IC Layout, Computer Center, National Chiao Tung University, Hsin-Chu, Taiwan, R.O.C.

CHIEN-CHUN LU (591), The Design and Implementation of a Distributed Database System, Institute of Information Science, Academia Sinica, Taipei, Taiwan, R.O.C.

C. J. LUCENA (213), A Theoretical Proposal for a CASD System Extending the Jackson's Method for Program Construction, Departamento de Informatica, Pontificia Universidade Catolica, Rio de Janeiro, Brazil

NELS H. MADSEN (411), Automatic Welding: Infrared Sensors for Process Control, Department of Mechanical Engineering, Auburn University, Auburn, Alabama 36849, U.S.A.

R. C. B. MARTINS (213), A Theoretical Proposal for a CASD System Extending the Jackson's Method for Program Construction, Departamento de Informatica, Pontificia Universidade Catolica, Rio de Janeiro, Brazil

MANZER MASUD (555), An Integrated Approach to Design, Implementation, and Testing of Digital System, Department of Computer Science & Engineering, University of Petroleum & Minerals, Dhahran, Saudi Arabia

P. MONTAGUE (245), Coordinating Multiple Robot Arms to Increase Productivity, Department of Computer Science, Virginia Polytechnic Institute & State University, Blacksburg, Virginia 24601, U.S.A.

M. N. NAGRAJ (139), CAD of Truss Structures by the Interactive Simplex Method, Division of Computer Applications, Asian Institute of Technology, Bangkok, Thailand

W. B. NGAI (153), Low Cost CAD/CAM System for Machine Parts, Department of Computer Science, The Chinese University of Hong Kong, New Territories, Hong Kong

LIONEL M. NI (569), Implementing Priority Functions in Local Area Networks, Computer Science Department, Michigan State University, East Lansing, Michigan 48824, U.S.A.

T. T. NIAN (73), A CAD Program for VLSI Placement and Routing, Department of Electrical Engineering, National Cheng Kung University, Tainan, Taiwan, R.O.C.

KOZO OKAZAKI (399), Spherical Shading Correction of Eye Fundus Image by Parabola Function, Department of Information & Computer Sciences, Osaka University, Toyonaka, Osaka, Japan

LIST OF CONTRIBUTORS

YOW-AN PAN (591), The Design and Implementation of a Distributed Database System, Institute of Information Science, Academia Sinica, Taipei, Taiwan, R.O.C.

D. B. PERNG (169), Computer-Assisted Methods for Defining 3D Geometric Structure of Mechanical Parts, Institute of Computer Engineering, National Chiao Tung University, Hsin-Chu, Taiwan, R.O.C.

M. F. RAHMAN (507), Microprocessor and Programmable Controller Based Industrial Automation, Department of Electrical Engineering, National University of Singapore, Kent Ridge, Singapore

U. REMBOLD (433), Present State and Future Trends in the Development of Programming Languages for Manufacturing, Department of Computer Science, University of Karlsruhe, Karlsruhe, Germany

JOHN ROACH (245), Coordinating Multiple Robot Arms to Increase Productivity, Department of Computer Science, Virginia Polytechnic Institute & State University, Blacksburg, Virginia 24601, U.S.A.

WEN ZEN SHEN (119), SPICE-2P: A SPICE-2 Incorporated with a Partition Scheme, Institute of Electronics, National Chiao Tung University, Hsin-Chu, Taiwan, R.O.C.; PLAMG: An Automatic PLA Minimizer and Generator for VLSI System Design, Department of Computer Engineering, National Chiao Tung University, Hsin-Chu, Taiwan, R.O.C.

LI-WEN SHIH (119), PLAMG: An Automatic PLA Minimizer and Generator for VLSI System Design, Department of Computer Engineering, National Chiao Tung University, Hsin-Chu, Taiwan, R.O.C.

SHINICHI TAMURA (399), Spherical Shading Correction of Eye Fundus Image by Parabola Function, Department of Electrical Engineering, Tottori University, Tottori, Tottori, Japan

JULIUS T. TOU (3), Automatic Generation of Knowledge Base from Electronic Diagrams for Computer-Aided Design, Center for Information Research, University of Florida, Gainesville, Florida 32611, U.S.A.

SHING-YUAN TSAI (481), On-Line Identification and Suppression of the Time-Varying Machining Chatter in Turning Via Dynamic Data System (DDS) Methodology, Department of Mechanical Engineering, University of Wisconsin, Madison, Wisconsin 53706, U.S.A.

P. A. S. VELOSO (213), A Theoretical Proposal for a CASD System Extending the Jackson's Method for Program Construction, Departamento de Informatica, Pontificia Universidade Catolica, Rio de Janeiro, Brazil

CHIOU-FENG WANG (591), The Design and Implementation of a Distributed Database System, Institute of Information Science, Academia Sinica, Taipei, Taiwan, R.O.C.

PAUL S. WANG (203), Computer-Aided Finite Element Analysis: Interfacing Symbolic and Numerical Computational Techniques, Department of Mathematical Sciences, Kent State University, Kent, Ohio 44242, U.S.A.

E. K. WONG (305), A Graphic-Theoretic Approach to 3D Object Recognition and Estimation of Position and Orientation, School of Electrical Engineering, Purdue University, West Lafayette, Indiana 47907, U.S.A.

CHI-HAUR WU (287), A CAD Tool for the Kinematic Design of Robot Manipulator, Department of Electrical Engineering, Northwestern University, Evanston, Illinois 60201, U.S.A.

H. C. WU (73), A CAD Program for VLSI Placement and Routing, Department of Electrical Engineering, National Cheng Kung University, Tainan, Taiwan, R.O.C.

SHIEN-MING WU (481), On-Line Identification and Suppression of the Time-Varying Machining Chatter in Turning Via Dynamic Data System (DDS) Methodology, Department of Mechanical Engineering, University of Wisconsin, Madison, Wisconsin 53706, U.S.A.

V. WUWONGSE (139), CAD of Truss Structures by the Interactive Simplex Method, Division of Computer Applications, Asian Institute of Technology, Bangkok, Thailand

INDEX

ACO (Adaptive Control Optimization)
  system, 536, 547
ADA language, 465-468, 475
Adaptive Control Optimization (ACO)
  system, 536, 547
AHPL language
  current status, 562
  overview of, 557-559
AL language 453, 456
All vicinity pattern search, 401-402
Aluminum metal products plant, 519-534
APT language, 435-437
ARMA (Autoregressive Moving Average Model), 482-483, 490-491, 494-495
Automatic generation of knowledge base, 3-36
  data files for, 29, 32-33
  illustrative example, 29, 30-31
  introduction, 3-5
  literature review in, 5-29
  SPICE program, 33-34
Automatic welding, 411-429
  depth of penetration, 418-422
  experimental procedure, 413-414
  experimental results, 414-416
  introduction, 411-412
  moving arc measurements, 416-418
  results discussed, 422-425
AUTOPASS language, 450
AUTORED system, 9-12
Autoregressive model (AR), 490-491

Autoregressive Moving Average Model (ARMA), 482-483, 490-491, 494-495
Autorouting, 50-51

Backlighting methods, 348-349, 379
Backtracking problems, 214
Backward-chaining system, 249
Backward differential changes, 291-292
Bajcsy, Ruzena, 387-398
Balakrishnan, P., 535-554
BELCAD system, 281
Binary Merging algorithm, 142
Blending, 81-82
Boolean equations, PLAMG inputs, 119, 120
Boolean function minimization, 122-126
Boolean preprocessor (BPRER), 121-122
Boundary Data Schemes, 171
BRUTUS, 95-117
  box manipulation in, 104-105
  cell manipulation in, 108-109
  CIF file, 109-110
  cursor control in, 102-104
  introduction, 95-96
  label manipulation in, 108
  layout plot construction, 97-99
  layouts of, 112-116
  outlines of, 96-97, 98
  PLT float, 110
  remarks on, 111-112

         status table setting for,
            99-102
         zigzag manipulations in,
            105-108
      BUILD(4), 153

      CAD. See Computer-aided design
      CAEE system, 281
      CAESAR system, 95, 111
      Calibration methods
         error minimization, 300-301
         manipulators design, 288
      Camera model, 328, 329
         industrial computer vision,
            349-352
      Cameras
         automatic welding, 413
         imaging methods and, 347-348
         infrared sensors, 411
         lighting methods and, 348 349
         moving arc measurements, 416
         See also TV cameras
      CARD (computer-aided robot design)
            system.
            See Robot design system (CAD)
      Carrier sense multiple access with
            collision detection. See
            CSMA/CD
            protocols
      Cartesian coordinates, 287, 288,
            289
      Cell Decomposition Schemes, 171
      Chan, Y. K., 153-168
      Chang, Chia-Hsiang, 591-605
      Chang, Chung-Jye, 37-58
      Chang, C. Y., 73-94
      Chang, San-Cheng, 607-615
      Channel routing, 74, 85-89
      Chatter identification and
            control, 481-505
         See also Time varying machining
            chatter
      Chatter suppression controller
            design, 498-502
      CHDL. See Computer hardware
            description language
      Chen, Cheng, 119-138
      Chen, Hsing-Lung, 591-605
      Chen, Z., 169-202

Cheng, Jack M., 3-36
Chin, Bryan A., 411-429
Chu, W. H., 617-634
CIF input/output files, 95
Circuit design. See VLSI circuits
Circuit diagrams, machine recognition, 6
Circuit simulation
   CPU time for, 59-60
   partitioning model for, 60-61
   results of, 64-67
Circular elements, 10
Cobol programs, 473-474
Commercial data processing
   languages, 435
   programming languages development, 473-474
Computer-aided design, 3-241
   automatic generation of knowledge
      base, 3-36
   BRUTUS, 95-117
   data transform programming
      method, 213 241
   digital manufacturing systems,
      556-557
   finite element analysis, 203-212
   high level graphic database for,
      607-615
   machine parts design, 153-168
   PC board design application, 37-57
   PLAMG, 119-138
   robot manipulators, 287-302
   robot system, 267-286
   SPICE-2 modification, 59-71
   three-dimensional structure of
      mechanical parts, 169-202
   truss structures, 139-151
   VLSI circuit development, 73-94
Computer-aided manufacturing, 433-566
   controller based industrial automation, 507-518
   digital systems, 555-566
   machinability data base systems,
      535-554
   production scheduling automation,
      519-534
   programming languages development,
      433 479
   time varying machining chatter,
      481-505

INDEX

Computer hardware description languages (CHDLs), 555, 557, 563
Computer vision. See Industrial computer vision; Vision and image processing
Concurrent computing, 246, 258-259
Conference on Data Systems Languages, 542
Connecting elements (circuits)
  machine recognition of, 9
  representation of, 7, 9
Constructive Solid Geometry Schemes, 171
Controller based industrial automation, 507-518
  applications for, 516-517
  introduction, 507-508
  microcomputers and programmable controllers compared, 508-510
  programmable controller based automation, 515
  simulated sheet steel classifier, 510-515
Coordination of robot arms. See Robot arms coordination
Cowan, D. D., 213-241
CSMA/CD protocols, 569, 570-574
Cursor, 102-104
Cutting tests, 502-503

Data acquisition system, vision and touch integration, 393
Data base
  digital systems manufacturing, 562
  high level graphic database, 607-615 (See also High level graphic database)
  robot design systems, 276, 283
Data Base Management System (DBMS), 537, 538
Data base systems
  distributed, 591-605 (See also Distributed database system)
  machinability database system, 535-554

Data entry, CAD/CAM systems, 4
Data flow design, 217-219
Data storage and retrieval systems, 537, 539
Data transform programming method, 213-241
  data flow design, 217-219
  data transform method, 219-223
  file processing programming, 223-227
  glossary of functions in, 237-238
  introduction 213-215
  Jackson method description, 215-217
  matrix transposition problem, 235-236
  system log problem, 233-235
  telegram analysis problem, 227-232
Decision-theoretic methods, 367-369
Decision tree, 6
Denotation file, 32
Denotations
  multiple-pass extraction, 21-23
  representations of, 7
Depth of penetration (welding), 418-422
Design Automation (DA), 37
Designer's Workbench system, 281
DeVries, M. F., 535-554
Diffuse-lighting methods, 348
Digital systems manufacturing, 555-566
  AHPL overview, 557-559
  approach to, 559-562
  current status, 562-565
  introduction, 556-557
Distributed database system, 591-605
  concurrency control, 600-604
  distributed relational databases, 594-596
  introduction, 591-592
  query processing, 596-600
  UNIX and RSX-11/M connection, 592-594
Distributed relational databases, 594-596
Dynamic Data System (DDS) methodology, 481-505
Dynamic Stability Limit (DSL), 489-490

Edge detection techniques, 357
Editing net list. See Net list editing
Effective capacitance, 61-63
Electronic circuit diagrams
  anatomy of, 6-7
  automated interpretation design of, 7-12
  See also Multiple-pass extraction
Electronic diagrams, automatic generation of knowledge base from, 3-36
Electronic scheme analysis, 6
Element matrices, 205-208
EMULA, 269
End-effectors
  robot manipulators design, 287
  vision and touch integration, 392, 293
  See also Robotics
Epple, W. K., 433-479
Error envelopes, 297-299
EXAPT system, 438-441
Execmon, 260-262
Eye fondus image, 399-409

Feature vectors, 10, 12
Figures of merit, 75-77
File processing, 214, 223-227
File systems, 281-283
Finite element analysis, 203-212
  FORTRAN code expression growth and efficiency, 210-212
  FORTRAN code generator, 208-210
  generation of element matrices, 205-208
  introduction, 203-205
Finite element generator, 204-205
Finite-state automata, 6
Folding technique, 126-128
Forecasting chatter system, 499-500
Forging operations, 155
FORTRAN
  finite element analysis, 207-212
  production scheduling automation, 528

Fourier transform, 355-356
Frequency-domain preprocessing methods, 355-357
Fu, K. S., 305-343
Functional element file, 32
Functional elements extraction, 18, 20-21

Generalized cones, 364-365
Generalized empirical equation systems, 537, 538
Generative machinability database systems, 540-542
GENFORTRAN, 209
Global routing, 84-85
Gonzalez, R. C., 345-385
Gradient use, 360-362
Graphics
  hidden line elimination problem, 617-634
  high level graphic database, 607-615
  robot design systems, 275-276, 281-282
Graphs, 314-318
Grasping abilities
  vision and touch integration for, 387-398
  See also Robotics

Hardware
  controller based industrial automation, 512-513
  machinability database systems, 547-548
  microprocessors, 508-509
  region growing approach, 358
  robot languages, 456
  SIR machine, 274
  thresholding approach, 357
Hidden line elimination problem, 617-634
  algorithm, 622-632
  data structure, 620-622
  introduction, 617-620
  timing analysis, 632
Higher-dimensional grammars, 376-378
High level graphic database, 607-615
  CAD work-station function, 612-613

data structure, 608-610
implementation, 614-615
introduction, 607-608
object structure, 610-612
High-level vision, 345, 346
Hill climbing method, 400-401
Hit-detection, 613
Homogeneous transformation, 289-291
Horizontal connecting lines extraction, 15, 17-18, 19
Hwang K. A., 95-117

IC layout, 95-117
Illumination. See Lighting
Image processing techniques, 174-175
Image segmentation, 357-362
Imaging devices, 346-347
Imaging methods, 347-348
Incomplete (imperfect) line drawings, 325-326
Industrial computer vision, 345-385
  camera model, 349-352
  decision-theoretic methods, 367-369
  description problem in, 362-365
  frequency-domain preprocessing methods, 355-357
  generalized cones, 364-365
  higher-dimensional grammers, 376-378
  illumination, 348-349
  imaging devices, 346-347
  imaging methods, 347-348
  interpretation, 378-381
  introduction, 345-346
  line and junction labeling, 363-364
  recognition process, 365-378
  segmentation, 357-362
  semantics use, 374-375
  spatial-domain preprocessing methods, 352-355
  string grammars, 372-374
  structural methods, 369-371
  See also entries under Vision
Industrial robots
  CAD of, 258
  manipulators design, 288
  See also Robotics
Information processing, 393-394
Infrared sensors 411-429
Inspections, 348-349
Interaction, 153-154
Interactive graphic editors, 95-117
Interactive processing, 82-83
Interactive Simplex method, 139-151
Interactive symbolic processing, 441-444
Interpretation, 345, 378-381

Jackson's method
  data transform programming method and, 213-241
  description of, 215-217
Jain, Arvind, 519-534
Jen, Chein Wei, 59-71
Joint coordinate frames, 292
Joint differential changes, 292-294
Jou, J. M., 73-94
Junction and line labeling, 363-364
Junction dot extraction, 12, 14-15, 16
Junction dot file, 32
Junction types constraints, 312-314

Kang, Shyi-Ting, 591-605
Ke, Jyh-Sheng, 591-605
Kinematic design, 287-302
Kinematic errors
  calibration for, 299-300
  joint differential changes due to, 292-294
Kinematics, 289-291
Knowledge base. See Automatic generation of knowledge base

Language(s)
  BRUTUS, 95
  digital systems manufacturing, 555, 557
  machinability database systems, 542
  microprocessor/programmable controllers, 509
  picture manipulation languages, 23-28
  PLAMG, 131

production scheduling automation, 528
robot design systems, 275-276, 278, 279, 283
robotics, 270
SIR machine, 274
See also entries under names of languages
LARS language, 472-473
Lee, Chung Len, 59-71
Lee, D. T., 617-634
Lee, J. Y., 73-94
Lee, Keh-Lon Thomas, 519-534
Lee, Young-Li, 119-138
Liaw, Ching An, 59-71
Lighting
  industrial computer vision, 348-349, 379
  region growing approach, 358
  thresholding approach, 358
Light pen
  BRUTUS, 96, 102
  robot design systems, 276
Lin, C. C., 95-117
Lin, Ching-Liang, 591-605
Line and junction labeling
Line drawings
  machine recognition of, 5-6
  problems of, 325 326
Line segment file, 32
Liu, D. C., 95-117
Loading effect, 61-63
Local area networks, 569-589
  distributed database system and, 591-605
  introduction, 569-570
  prioritized CSMA/CD protocols, 570-574
  priority-code comparison scheme 575-579
  simulation results and analysis 579-587
Logical database design, 542-544
Logic symbols, 6
Low-level vision, 345-346
Lu, Chien-Chun, 591-605
Lucena, C. J., 213-241

Machinability database system, 535-554
  adaptation of, 547-552
  generative type system, 540-542
  introduction, 535-536
  logical design of, 542-544
  model building algorithms, 544-545
  optimization algorithms, 545-547
  present drawbacks of, 538-539
  present status of, 537-538
Machine parts design (CAD), 153-168
  components examples in, 160-162
  cutter-path for N.C. milling, 158-160
  further enhancements for, 162-163
  introduction, 153-154
  physical characteristics, 157-158
  solid primitives in system, 154-157
Machine tools programming, 435-447
  APT language, 435-437
  EXAPT system, 438-441
  generative programming by machine tool control, 446-447
  interactive symbolic programming, 441-444
  special purpose languages, 444-446
Machine vision. See Industrial computer vision; entries under Vision
Machining, 481-505
MACSYMA system, 204
Madsen, Nels H., 411-429
MADSIM system. See Robot design system (CAD)
Manipulator design. See Robotics; Robot manipulators design
Manufacturing, 433
  See also Computer-aided manufacturing
MARS method, 470 472
Martins, R. C. B., 213-241
Masud, Manzer, 555-566
Matrix transposition problem, 235-236
Mechanical parts. See entries under Three-dimensional
Medium-level vision, 345 346
Memory storage, 59, 60
Merge Insertion algorithm, 142
Microcomputers. See Microprocessors
Microprocessors

INDEX

controller based industrial automation, 507-518
  programmable controllers compared, 508-510
MINI-algorithm, 122-123
Minimum Steiner tree, 85
MODCON, 153-154
Modelling, 275, 277
Model matching, 318-320
Montague, Paul, 245-265
Mouse, 276
Moving arc measurements, 416-418
Multiple-pass extraction, 9
  denotations recognition, 21-23
  functional elements extraction 18, 20-21
  horizontal connection lines extraction, 15, 17-18, 19
  junction dot extraction, 12, 14-15, 16
  picture manipulation languages 23-28
Multiple robot arms. See Robot arms coordination
Multiple-view representations, 306
Multi-threading problem, 233

Nagraj, Mahendra Narayan, 139-151
Natural language programming, 458
N.C. milling operations, 154, 156 158-160
Net list editing, 46-48
Net partitioning, 86-87
Networks. See Local area networks
Ngai, W. B., 153-168
Ni, Lionel M., 569-589
Nian, H. T., 73-94
NMOS technology, 99
Nonpreemptive discipline, 576-577
Numerical computational techniques, 203-212

Objective function, 77
Object structure, 610-612
Object tree, 608-609
Occlusion problems, 379-380
Ocular fundus image. See Eye fundus image
Ikazaki, Kozo, 399-409

On-line chatter identification and control, 481-505
  See also Time varying machine chatter
Open-loop robots
  error minimization, 300-301
  kinematic errors calibration, 299-300
  kinematic error model of, 294-297
  See also Robotics
Optimization-based CAD systems, 140

PADL(5), 153
Pan, Yow-An, 591-605
Parabola function, 399-409
Parallel search algorithm, 324-325
Pattern recognition principles, 5-6
Pattern search method, 401-403
PC board design (CAD applications), 37-57
  autorouting, 50-51
  clean-up 51
  factors affecting, 40-44
  initial placement, 39-40, 48
  introduction, 37-44
  manufacturing aids, 51-53
  net list editing, 46-48
  physical creation, 44-46
  ratsnet checking, 48-50
  signal prerouting, 50
PCSMA/CD, 579-587
PEARL language, 464-465
Perng, D. B., 169-202
PERT charts, 246
Perturbations, 412
Photoelectric switches, 510, 512
Physics, 248
Pick and place tasks, 245
Picture manipulation languages, 23-28
"Pin-hole" camera, 328, 329
PLA. See Programmable Logic Array
PLACE (McAuto) system, 269
PLA folding technique 126-128
  See also Programmable Logic Array Minimizer and Generator (PLAMG)
Planar patches, 358-359
Planner, 249-258
Polyhedra, 170
Precondition Restraint Graph (PCG), 249-250

Preemptive discipline, 577-579
Preprocessing methods
  computer vision, 345
  frequency-domain, 355-357
  spatial-domain, 352-355
Primitives
  feature vectors, 10, 12
  representation of, 13
Printed circuit board design. See PC board design (CAD applications)
Priority-code comparison scheme, 575-579
Priority functions, 569-589
Process control languages, 435, 460-473
  ADA, 465-468
  characteristics of, 460-464
  development tools for, 468-473
  extension of existing languages 64
  PEARL, 464-465
Production scheduling automation, 519-534
  development and testing of, 533-534
  interactive module, 528-533
  mathematical formulation, 524-527
  model formulation for, 520-524
  problem description, 519-520
Productivity
  computer-integration automation v
  manipulators design, 288
  manufacturing and, 433
  PC board design, 41
  robot design systems, 277, 280
  robotics, 268
  VLSI circuit development, 73-74
Programmable controller
  classifier automation, 515
  controller based industrial automation, 507-518
  microprocessors compared, 508-510
Programmable Logic Array (PLA), 120
Programmable Logic Array (PLA) folding technique, 126-128

Programmable Logic Array Minimizer and Generator (PLAMG), 119-138
  Boolean prepreocessors, 121-122
  discussion and conclusions, 135-136
  introduction, 119-120
  PLA folding techniques, 126-128
  PLAGEN, 128-131
  PLA product term minimization, 122-126
  test, evaluation, verification, 131-135
Programming languages development, 433-479
  commercial data processing, 473-474
  future trends in, 474-477
  introduction, 433-435
  machine tools, 435-447
  process control, 460-473
  robots, 448-460
Projection-graphs, 314-318
Pure Primitive Instancing Schemes, 171

Quality control
  controller based industrial automation, 510, 512
  PC board design, 41-44, 51
  production scheduling automation, 519-534
  robot design systems, 277
  robotics, 268
Query processing, 596-600

Rahman, M. F., 507-518
Ramtek 6211 color display device, 185
Range data, 358-359
Range sensing, 358
Razor search, 403-405
Recognition, 345, 365-378
Region growing approach, 357-358
Relay ladder language, 509
Rembold, U., 433-479
Repartition, 81-82
Roach, John, 245-265
Robot, defined, 270
Robot arms coordination, 245-265

INDEX

introduction, 245-246
need for, 247-249
operating system for, 258-263
overview of, 246-247
planning system design, 249-258
Robot design system (CAD), 267-286
  architecture and design objectives, 278-280
  considerations and plans, 280-283
  introduction, 268-269
  methodology and requirements, 270-271
  relevant work on, 269-270
  SIR machine, 271 275
  system requirements, 275-277
Robotics, 245-302
  computer-aided design system, 267-286
  kinematic design of robot manipulators, 287-302
  languages and 434-435
  programming languages, 448-460
  robot arms coordination, 245-265
  three-dimensional object recognition, 328
  vision and touch integration, 287-298
  vision systems and, 348-349
Robot languages, 448-460
  design considerations, 448-450
  existing languages surveyed, 450-451
  implicit programming languages, 458
  natural language programming, 458
  new concepts for, 451-458
  programming aids, 459-460
Robot manipulators design (kinematic), 287-302
  backward differential changes, 291-292
  calibration for kinematic errors, 299-300
  error envelopes, 297-299
  error minimization, 300-301
  error model of an open-loop robot, 294-297
  introduction, 288 289
  joint differential changes, 292-294
  kinematics and, 289-291
ROMULUS(6), 153
Routing. See Signal routing
Row cluster development, 80-81
RSX-11/M system, 592-294

SARS Information System, 469-470, 473
Search time speed up, 320-323
Seed blocks selection, 80
Segmentation
  computer vision, 345
  industrial computer vision, 357-362
Seiko 9500 system, 154
Selection rule, 75-77
Semantics, 374-375
Semiconductors
  visual sensors, 347
  VLSI, 119
Sensing and sensors
  computer vision, 345
  infrared, 411-429
  machinability database systems, 547-548
  machine tool programming, 446-447
  robotics, 448
Serial order perturbation, 402-403
Shaded display, 171
Shading correction. See Spherical shading correction
Shadowing effects, 379
Shen, Wen Zen, 59-71, 119-138
Shih, Li-Wen, 119-138
Signal routing, 50
Silhouette object recognition, 306
Simple Instructional Robot. See SIR machine
Simplex method, 140, 142
  See also Interactive simplex method
Simulation
  robotics, 269
  See also Circuit simulation
Simulation-based CAD systems, 140
SIR machine, 271-275
  described, 272
  implementation, 274-275

instructions, 273-274
registers, 273
system requirements, 274
Slicing method, 405-406
Software
  controller based industrial
    automation, 513-515
  high level graphic database,
    607, 613
  machinability database systems,
    539, 548-549
  microprocessors, 508-509
  robot design systems, 277, 280
  robot languages, 459
  SIR machine, 274, 275
Software design
  robotics, 268
  robot design systems, 277,
    279-280
Solid state diode arrays, 346-347
Spatial-domain preprocessing
    methods, 352-355
Spatial Occupancy Enumeration
    Schemes, 171
Special purpose languages, 444-446
Spherical shading correction,
    399-309
  hill climbing method, 400-401
  introduction, 400
  pattern search method, 401-403
  razor search, 403-405
  slicing method, 405-406
SPICE program, automatic
    generation of knowledge
    base for 33-34
SPICE-2, SPICE-2P compared, 71
SPICE-2P, 59-71
  circuits simulation results in,
    64-67
  implementation of, 63-64
  introduction to, 59-60
  loading effect considerations
    in, 61-63
  partitioning model, 60-61
  partitioning rules, 63
  SPICE-2 compared, 71
SRL language, 453, 456
Stability
  feed and, 495-496

speed and, 494-495
Statistical pattern recognition,
    305
Stereo vision
  imaging methods, 348
  industrial computer vision, 358
  vision and touch integration,
    389
Storage and retrieval systems,
    537,539
String grammars, 372-374
String matching, 369-371
STRIPS systems, 246, 247, 249
Structure clash, ee7, 233, 235
Structured lighting, 349, 379
  <u>See also</u> Lighting
Subgraph searching, 318-320
Superchannel, 75
Sweeping Schemes 171
Symbolic computational techniques,
    203-212
Symbols
  electronic circuit diagrams, 7, 8
  PC board design, 44-45, 48
Symbol tree, 11
Syntactic methods, 371
Syntactic pattern recognition, 305
System log problem, 233-235

Tactile module, 392
Tactile sensory data, 393-394
Tamura, Shinichi, 399-409
Teach-in programming, 448
Telegram analysis problem, 227-232
Testing
  digital devices, 562-563
  sequence generation, 564
Textual programming, 448-449
Thermography. See Infrared sensors
Three-dimensional geometric structure
    of mechanical parts, 169-202
  computer simulation results,
    185-186
  data structure definition, 171-172
  experimental results, 193-196
  introduction, 169-171
  parts design, 186-193
  reconstruction of parts from
    two-dimensions and, 179-186
  two-dimensional geometric
    structures and, 172-179

Three-dimensional images, 345
  camera model and, 349
  hidden line elimination problem, 617-634
  industrial computer vision, 358
  junction and line labeling, 363-364
  planar patches, 359-360
  vision and touch integration, 387-398
Three-dimensional mechanical systems, 269, 277, 281-282
Three-dimensional model graphs, 308-312
Three-dimensional object modeling and recognition, 305-343
  experimental results, 334-337
  graph-theoretic matching, 308-325
  incomplete line drawing, 325-326
  introduction, 305-308
  position and orientation estimation, 328-334
  verification, 326-328
Thresholding approach, 357
Throughput-delay characteristics, 581
Time varying machining chatter, 481-505
  chatter suppression controller design, 498-502
  chatter suppression strategy, 492-498
  cutting tests and results, 502-503
  dynamic analysis in, 489
  dynamic stability limit, 489-490
  introduction, 482
  machining process, 490-492
  machining process identification 485-489
  mathematical background, 482-485
Token-passing bus protocols, 570
Token ring protocols, 570
Tou, Julius T., 3-36
Touch and vision integration. See Vision and touch integration
Track assignment procedures. See Channel routing

Truss Structures (CAD), 139-151
  actual system, 143-145
  design procedure in, 145-146
  example of, 146-148, 149, 150, 151
  Interactive Simplex method and, 142-143
  introduction, 139-141
  problem description, 141-142
Truth table
  minimization, 122-126
  PLAMG outputs, 119, 120, 121
Tsai, Shing-Yuan, 481-505
TV cameras
  imaging devices, 346-347
  three-dimensional geometric structure, 170
  three-dimensional object recognition, 335
  See also Cameras
Two and one-half dimensional information, 358
Two-dimensional structures, 169-171, 172-186, 305, 306, 358

UNIX, 592-594
User interface
  computers and, 434
  digital systems manufacturing, 559
  high level graphic database, 612-613
  microprocessor/programmable controllers, 509
  production scheduling automation, 528
  robot design systems, 276-277, 279, 280, 282
  robot languages, 451, 458

VAL language, 453
VAXIMA system, 204
Veloso, P. A. S., 213-241
Vertical clusters placement, 84
Vertical constraint graph, 86
Vidicon cameras, 346-347
Virtual robot machines, 270-271
  SIR machine, 271-275
Vision and image processing, 305-429
  automatic welding, 411-429
  industrial computer vision, 345-385

spherical shading correction, 399-409
three-dimensional object modeling and recognition, 305-343
vision and touch integration, 387-398
Vision and touch integration, 387-398
  control strategies, 395-397
  data acquisition system, 393
  information processing of tactile sensory data, 393-394
  introduction, 387-388
  representation needed for, 395
  tactile module, 392
  three-dimensional object representation and feature extraction, 389-392
  vision module for, 388-389
  visual and tactile information integration, 394-395
VLSI circuits, 73-94
  BTUTUS, 95, 96, 97-99
  channel routing, 85-89
  digital systems manufacturing, 564-565
  figures of merit and, 75-77
  global routing, 85-86
  implementation/results, 89-91
  introduction, 73-75
  placement algorithm for, 78-84
  problem difinition, 77-78
  PLAMG for, 119-138

Wuwongse, Vilas, 139-151
Zoom camera, 332-333

Wang, Chiou-Feng, 591-605
Wang, Paul S., 203-212
Welding. See Automatic welding
Wire frame display, 171
Wire length, 48
Wong, E. K., 305-343
World coordinate system
  camera system, 349-352
  error minimization, 300-301
  kinematic error model, 294-297
  manipulators design, 287, 289, 291
Wu, Chi-haur, 287-302
Wu, H. C., 73-94
Wu, Shien-Ming, 481-505